Quaternary Soils

Edited by

W. C. Mahaney

York University

Geo Abstracts
Norwich
England

Information concerning the Quaternary Symposia can be obtained from:
Professor W. C. Mahaney / Department of Geography /
Atkinson College / York University / Toronto / Ontario M3J 2R7

Cover illustration:
Moraines at the Pinedale Type Locality of Blackwelder,
Fremont Lake, and the crest of the Wind River Range, Wyoming

Published by Geo Abstracts Ltd.,
University of East Anglia,
Norwich, NR4 7TJ,
England

ISBN 0 86094 012 8

IN MEMORIAM

This study is dedicated to the memory of our colleague Karen Brunson, who passed away October 11, 1976 at the beginning of a promising career.

Carolyn G. Olson & Robert V. Ruhe

Printed in Great Britain by Headley Brothers Ltd 109 Kingsway London WC2B 6PX and Ashford Kent from artwork supplied

Preface

The third conference on Quaternary Research was held May 21-23, 1976, under the auspices of Atkinson College of York University. Over a hundred Quaternary scientists from various institutions in North America attended, and 18 papers were presented in three sessions: soil dynamics, soil stratigraphy, and soil morphogenesis.

This volume begins with a discussion of a multiple-process model of soil genesis by R.W. Simonson. He stresses the evaluation of processes which promote horizonation in soils, such as, addition of variable quantities of organic matter, leaching of carbonates, weathering of silicate minerals, development of humic acids; and processes which retard horizonation, such as, animal burrowing, influx of eolian sediment, plant recycling of nutrients, and surface erosion. The soil is considered a natural entity developing from the balance over time of several processes operating together in soil genesis. J.A. McKeague, *et al.*, discuss the genesis of Podzolic soils stressing the soil-forming factors of parent material and climate. They place particular emphasis on organic complexes of Al and Fe; weathering of clay minerals, including the origin of smectite interstratified with mica in the Ae (A2) horizons; and the decomposition of clay-size primary minerals in the Ae horizons. S. Pawluk discusses the problems associated with interpretation of paleoenvironments as deduced from buried soils, as well as the physical and chemical changes in paleosols following burial, and the influence of climatic change on soil morphology.

The second session begins with a discussion of the concepts and methods of Quaternary soil stratigraphy by R.B. Morrison. He views the soil profile as forming on stable sites with minimal erosion and deposition and providing important environmental information for specific periods of geologic time. C.G. Olson, *et al.*, discuss the late-Quaternary soil record in Southern Indiana including the Sangamon Soil formed on Illinoian Till, Loveland Loess, and upper Paleozoic bedrock. They also discuss soils forming on Wisconsinan loess, and compare clay-mineral assemblages in ground soils with similar assemblages in the Sangamon Soil. L.R. Follmer reviews the evolution of the concept of the Sangamon Soil and discusses the *type area* in Central Illinois, the age of the Sangamonian Stage, and the need for a *type* Sangamon Soil transect. The sequence of buried soils of Yarmouthian, Sangamonian, and late-Wisconsinan age in Missouri, as discussed by W.H. Allen and R.A. Ward, involves reference to variations in physical and mineralogical properties of soil-stratigraphic units, recognition of buried soils, and problems with nomenclature and age-determination.

Preface

A soil-geomorphic reconnaissance of alluvial deposits in Southern California, by R.J. Shlemon, includes a discussion of Late-Tertiary fanglomerates and younger Quaternary piedmont fans and pediment surfaces, all differentiated on the basis of soil development. Research with paleosols in the Prairie Provinces of Canada, by J.F. Dormaar, begins with a survey of paleopedology, and ends with a discussion of current research methods employed in analysis of paleosols of post-glacial age. W.C. Mahaney examines the morphological, mineralogical and chemical variations found in five soil-stratigraphic units of Holocene and Late-Pleistocene age in Western Wyoming. This is followed by W.C. Wigley, *et al.*, who discuss the glacial history of the Copper Basin, Idaho, using deposit morphology, boulder provenance, surface weathering features, and soil-profile development to differentiate deposits in the glacial succession. They point out some of the difficulties of using soil morphology in differentiating deposits of Bull Lake and Pinedale age. A discussion of relict and buried paleosols in Central Yukon is presented by N.W. Rutter, *et al.* They deal with the morphology of relict soils forming in loess and eolian sand, and buried soils developed in till or outwash of Pre-Reid (early Pleistocene), Reid (Early or pre-Wisconsin), and McConnell (Late-Wisconsin) age, as well as thermal contraction cracks in the Reid and pre-Reid Surfaces, and paleoclimatology as deduced from relative degrees of soil development.

The last session on soil morphogenesis begins with L.J. Evans, and involves a discussion of the interrelationships between individual pedons and the soil-parent material, especially the problem of identifying lithologic discontinuities as they are related to soil property variations, and analysis of pedologic processes using ratios of stable/unstable minerals. T.R. Moore examines soil development along a transect through the boreal forest of Quebec to the subarctic and arctic zones of the Eastern Canadian Arctic. Podzolization is indicated by soil morphology (specifically the Ae horizon development and the degree of translocation of Fe, Al, and organic carbon), which varies as a result of organic matter production and decomposition as well as the length of time available for pedogenesis. A discussion of the Polar Desert Soil, by J.C.F. Tedrow, includes analysis of soil genesis, and the variation in individual pedons as related to vegetation, permafrost, and lithology. Comparative morphology and genesis of Arctic Brown and Alpine Brown soils in North America, by J.G. Bockheim, includes analyses of clay-mineral variations and relative differences in soil-chemical relationships. Organic soils, in Manitoba and the Northwest Territories, are discussed by C. Tarnocai, who emphasizes the dynamics of soil genesis as related to soil-forming factors of climate, biota and time. Soil development processes in the Western Cordilleran region of Canada, are discussed by L.M. Lavkulich and J.I. Sneddon, who center on the distribution of alpine soils as related to climate.

On Sunday, May 23, D.W. Hoffman led a field trip across the Peel Plain, Interlobate Moraine, and Schomberg Plain of Southern Ontario. He placed particular emphasis on the relationships between surficial geology and soil morphology. Soil profile descriptions are contained in *Quaternary Soils Symposium, Abstracts-with-Program,* Third York University Symposium on Quaternary Research, 87 p.

I am indebted to several individuals for assistance in the planning of the Symposium. M. Knittl, Dean of Atkinson College and J.M. Cameron, Chairman of the Department of Geography, Atkinson College, assisted with financial support; Professors W.E. Northover and J.M. Cameron helped in various ways with the organizational and logistical difficulties. Technical planning was completed with the assistance of A. Cote and his Facilities staff. E. Cassalman typed the entire manuscript, and along with L.M. Mahaney proofread the papers. Several students assisted with the organization of the symposium and are too numerous to name in full. In particular, I thank G. Augustus, A.B. Bagby, P. Bowden, G. Carr, L. Dominici, V. Elchuk, L.J. Gowland, D.G. McWilliams, J.D. O'Shea, and D. Wallace. Sessions were taped by J. Briggs and illustrations were drafted by G. Berssenbrugge, and C. Grounds. All illustrations were photographically reproduced by J. Dawson and B. Kanarens.

Contents

Contents

Contributing Authors

W.H. Allen	Water Rights Division Arizona State Land Department Phoenix, Arizona
J.G. Bockheim	Department of Soil Science University of Wisconsin Madison, Wisconsin
K.L. Brunson	Water Resources Research Center and Department of Geology Indiana University Bloomington, Indiana
J.F. Dormaar	Research Station Agriculture Canada Lethbridge, Alberta
L.J. Evans	Department of Land Resource Science University of Guelph Guelph, Ontario
E.B. Evenson	Department of Geological Sciences Lehigh University Bethlehem, Pennsylvania
L.R. Follmer	Illinois State Geological Survey Urbana, Illinois
A.E. Foscolos	Geological Survey of Canada Calgary, Alberta
D.S. Gamble	Soil Research Institute Central Experimental Farm Agriculture Canada Ottawa, Ontario
D.W. Hoffman	Centre for Resources Development University of Guelph Guelph, Ontario
O.L. Hughes	Geological Survey of Canada Calgary, Alberta
L.M. Lavkulich	Department of Soil Science University of British Columbia Vancouver, British Columbia
W.C. Mahaney	Department of Geography Atkinson College York University Toronto, Ontario

Contributing Authors

J.A. McKeague	Soil Research Institute Central Experimental Farm Agriculture Canada Ottawa, Ontario
T.R. Moore	Department of Geography McGill University Montreal, Quebec
R.B. Morrison	Department of Geoscience University of Arizona Tucson, Arizona
C.G. Olson	Water Resources Research Center and Department of Geology Indiana University Bloomington, Indiana
T.A. Pasquini	Department of Geological Sciences Lehigh University Bethlehem, Pennsylvania
S. Pawluk	Department of Soil Science University of Alberta Edmonton, Alberta
G.J. Ross	Soil Research Institute Central Experimental Farm Agriculture Canada Ottawa, Ontario
R.V. Ruhe	Water Resources Research Center and Department of Geology, Indiana University, Bloomington, Indiana
N.W. Rutter	Department of Geology University of Alberta Edmonton, Alberta
R.J. Shlemon	P.O. Box 3066 Newport Beach, California
R.W. Simonson, retired	Soil Classification and Correlation U.S. Department of Agriculture Washington, D.C.
J.I. Sneddon	Department of Soil Science University of British Columbia Vancouver, British Columbia
C. Tarnocai	Canada-Manitoba Soil Survey University of Manitoba Winnipeg, Manitoba

J.C.F. Tedrow
Department of Soils and Crops
Rutgers University
New Brunswick, New Jersey

R.A. Ward
Division of Geology and Land Survey
Missouri Department of Natural Resources
Rolla, Missouri

W.C. Wigley
Department of Geological Sciences
Lehigh University
Bethlehem, Pennsylvania

Sessions Chairmen

J.B. Bird	Department of Geography McGill University Montreal, Quebec.
B.D. Fahey	Department of Geography University of Guelph Guelph, Ontario
D.W. Hoffman	Centre for Resources Development University of Guelph Guelph, Ontario
J.A. McKeague	Agriculture Canada Central Experimental Farm Ottawa, Ontario
R.V. Ruhe	Water Resources Research Center and Department of Geology Indiana University Bloomington, Indiana
N.W. Rutter	Department of Geology University of Alberta Edmonton, Alberta
A. MacS. Stalker	Geological Survey of Canada Ottawa, Ontario
J.C.F. Tedrow	Department of Soils and Crops Rutgers University New Brunswick, N.J.

Soil Dynamics

A Multiple-Process Model of Soil Genesis

Roy W. Simonson

ABSTRACT

Soil genesis consists of a tangled complex of many processes. For convenience in discussion, soil formation can be considered two overlapping steps, namely, accumulation of parent materials and horizon differentiation. Some regolith must be present before horizon differentiation can begin, but it need be only centimeters thick. Moreover, immense quantities of rock have been weathered in the past. Weathering does not stop when horizon differentiation begins or as it continues, but the latter is the heart of soil genesis and the primary subject of this paper.

The individual processes operating in horizon differentiation can be grouped into four main kinds, viz., additions, losses, transfers, and transformations. As an open rather than closed system at the land surface, the soil mantle both receives and loses substances. Some are transferred within the soil profile or body and others are modified in composition or form or both.

Among the processes that operate over time in soil genesis, some promote and others offset or retard horizon differentiation. Examples of processes that promote differentiation of horizons are additions of organic matter, losses of carbonates, transfers of silicate clay minerals, and decomposition of organic residues with formation of humus. Processes offsetting or retarding horizon differentiation are the mixing of soil materials by burrowing animals, additions of airborne sediments, circulation of nutrient elements through flora, and the churning of soils by frost or by shrinking and swelling.

The balance over time among the many processes in soil genesis determines the nature of every soil profile and soil body. If the balance differed from one place to another, so do the kinds of soils. If differences were small, the kinds of soils are much alike, as are those of a pair of soil series in the same family. If the differences in balance were large, the kinds of soils are much different, as are a Spodosol (Podzol) in southern Quebec and a Mollisol (Chernozem) in North Dakota, although both were formed from glacial till. Yet the same basic processes operated in differentiation of horizons, *i.e.*, in genesis, of the two kinds of soils.

The balance among processes may remain the same for short or long periods. A given balance may persist long enough to result in distinct horizons of a certain character within soil profiles. The balance may then change or horizon differentiation may be interrupted. The pathway of horizon differentiation can thus be shifted, with superposition of new characteristics on the old in soil profiles. Clear evidence that there have been shifts in balance exists in a number of soils, which bespeak less well expressed effects in other soils. For Quaternary soils generally, there is always the distinct possibility that their present character reflects more than one interval of horizon differentiation.

INTRODUCTION

Soil genesis consists of a complex of many processes. Every soil profile and soil body is a reflection of one or more combinations of processes, each of which has persisted for some time. Examples of differing kinds of processes are given in this paper, and the balance among them is discussed. First, however, a concept of soil and the steps in soil formation are described briefly. Last are some remarks about multiple intervals of soil genesis and their significance to what we now find on the land surface.

CONCEPT OF SOIL

Soils form a continuum over the land surface except for steep and rugged mountains, rocky and sandy desert wastes, and fields of perpetual ice and snow. As a rule, soil constitutes the uppermost part of the regolith but may coincide with the latter if it is thin. Thickness of the soil mantle over the land surface ranges from a few centimeters to about 2 meters (Simonson, 1968).

The soil mantle can be considered a mosaic of three-dimensional bodies, small segments of the rind of the earth, as illustrated in Figure 1. The upper boundary of each body is distinct, but the lower one is not except where hard rock occurs at shallow depth. Boundaries between soil bodies are usually gradational and indistinct. The scarcity of distinct boundaries between soil bodies and between the soil and the remainder of the regolith is an important feature of the continuum for any model of soil genesis.

All soil bodies share a number of characteristics. All are three-phase systems, consisting of solid, liquid, and gas. Proportions of the three phases differ from one

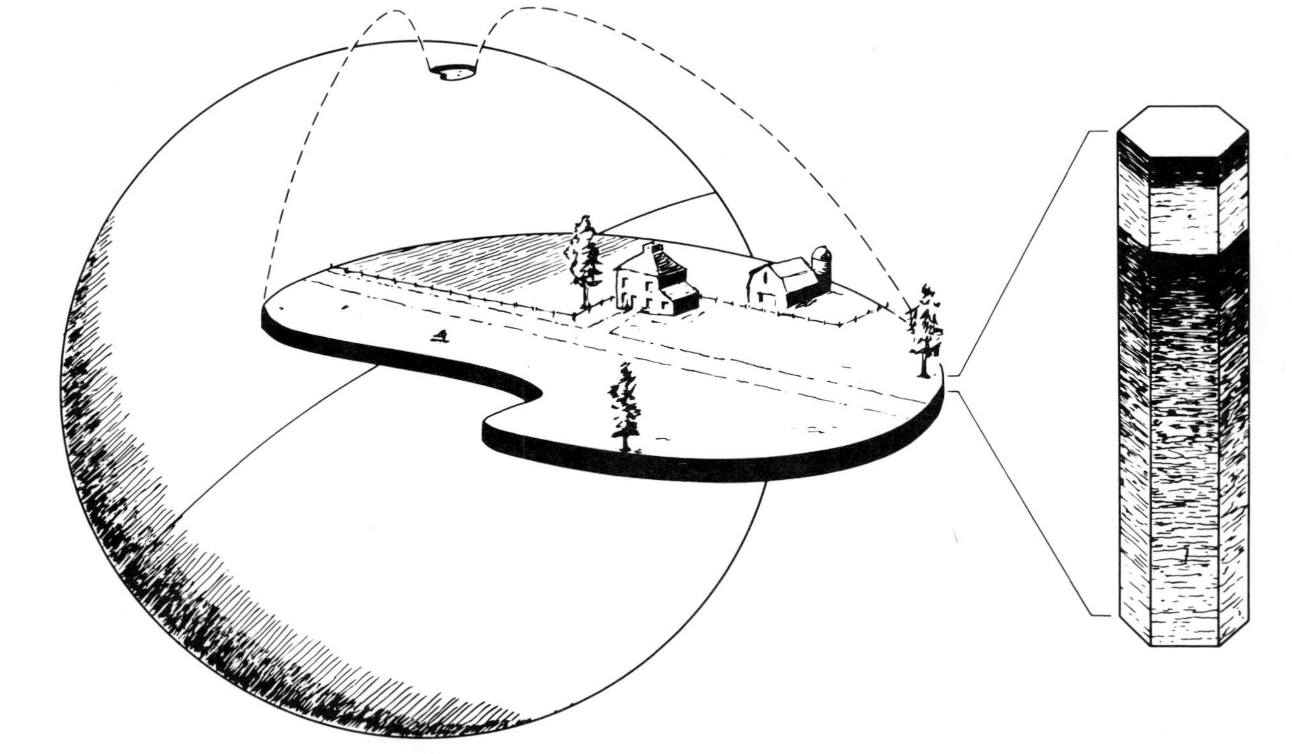

Figure 1 Diagrammatic illustration of the relationships of a single body of soil to the rind of the earth and to a single pedon. Such a soil body would be shown as one delineation on a detailed soil map. The body would consist of a large number of pedons such as the one that is illustrated plus inclusions of others. Sketch by Walter M. Simonson

soil body to another. All bodies consist of the same major components and the same elements, also with differing proportions among soil bodies. These shared characteristics have not been given the attention they deserve in theories of soil genesis held in the past. The focus has been on differences among soils with consequent neglect of similarities.

Despite the sharing of properties by all soils, there are local and regional differences. Some of these are substantial. The Spodosols of Quebec are much different from the Vertisols of the Deccan plateau of India.[1] Such differences must also be accommodated in any model of genetic processes, even as must shared characteristics.

Soil bodies are open systems (Simonson, 1970; Huggett, 1975). Substances and energy are added to and lost from every soil body as horizon differentiation proceeds. Examples of additions and losses are given later in this paper. A model of soil genesis must allow for the occurrence of soil bodies as open systems subject to change with time.

Individual soil bodies are identified and described chiefly on the basis of profiles or pedons, the second of which is illustrated in Figure 1. The profile of a Haplustalf (Alfisol) is shown in Figure 2. Information in addition to that on profiles or pedons is normally included as for example, configuration of the land surface, the setting in which a soil body occurs, and the underlying rock or regolith. Most of the evidence from which inferences about soil genesis can be drawn, however, comes from the nature of and relationships among horizons in soil profiles.

STEPS IN SOIL FORMATION

Soil formation consists of two overlapping steps. One is the accumulation of parent materials and the other is horizon differentiation (Simonson, 1957).

Parent materials are accumulated through the weathering of rocks in place or the deposition of weathered rock materials. Such accumulation is not considered in this paper because information on weathering is widely available (Ollier, 1969; Birkeland, 1974). The nature of the parent materials does affect pathways of horizon differentiation and hence is important. Furthermore, some regolith must be present before horizon differentiation can begin, although it need be only a few centimeters thick.

---

[1] The current soil classification system of the U.S. Department of Agriculture is followed throughout this paper (Soil Survey Staff, 1976).

Figure 2 Profile of a Haplustalf (Alfisol) in central Texas. This has a pale A2 horizon and a distinct reddish argillic (Bt) horizon. Major horizons are labeled at the left side. The scale along right side is numbered in feet and marked with ticks at 2- and 6-inch intervals. Each small interval is also 5 cm.

After a little regolith has accumulated, the differentiation of horizons can begin. That step is the essential one in soil genesis. Weathering can and does proceed without the formation of soil, as witness the disintegration of rock to depths of 25 m (80 ft) in places along the Rio-Sao Paulo highway in Brazil. As horizon differentiation proceeds, however, soils are formed whether the regolith be thick or thin.

Decomposition of minerals continues within a soil even after a profile has distinct horizons. Conversely, processes operating chiefly in differentiation of horizons, such as transfer of silicate clay minerals, can extend below the soil profile into weathering rock, becoming progressively attenuated with increasing depth. The two steps in soil formation are therefore not discrete and distinct; the breakdown simply provides manageable topics for discussion.

COMBINATIONS OF PROCESSES

The myriad of processes in horizon differentiation can be placed into four main groups, namely, additions, losses, transfers, and transformations (Simonson, 1959). That the number is large has long been recognized. Some 50 years ago, Marbut (1951) wrote that "a good many things" happened as soils were formed from weathered parent materials. His examples can all be placed into one or the other of the four groups listed.

Processes in horizon differentiation affect both substances and energy. For example, there may be additions, losses, transfers, and transformations of organic matter, silica, silicates, sesquioxides, carbonates, and water. Energy is involved in all of these processes in addition to the substances themselves. Organic matter is added as fresh plant residues and partial decomposition products are lost through leaching. Silicate minerals are added as airborne dust, are altered into clay and oxides, and are transferred within profiles. Water and dissolved elements are added in rain, some react to form new substances within soils, some move downward short distances, and some go entirely out of the system.

Some processes promote and others retard or offset horizon differentiation. For example, many species of soil fauna mix materials both vertically and horizontally and thus partly offset horizon differentiation. So does mixing of soil materials by frost action in very cold regions and by the churning associated with shrinking and swelling of "self-swallowing" soils such as the Regur of India. Rapid accretions of sediments or removal of soil materials can offset horizon differentiation. The recycling of nutrient elements by growing plants retards the development of horizons by slowing down losses due to leaching.

Because the processes operating to differentiate

horizons in all soils are in part conflicting, they can be considered as pushing and pulling in many directions. The present nature of every soil profile and every soil body is a resultant of the functioning of many and partly conflicting processes over time (Yaalon, 1971).

The relative importance of any one process or group of processes in a combination differs from place to place over the face of the earth. If the difference in balance among processes in two places is small, the resulting soils will be much alike, such as those of a pair of soil series in the same family. Conversely, if the difference in balance from one place to another is large, differences between soils will also be large, as is true of the Aridisols of Arizona and the Ultisols of Georgia.

ADDITIONS

Energy is added as heat from the sun and in combined forms in organic residues. Water is added mainly as precipitation. Further additions are silicates in airborne dust and salts dissolved in rain. These are simply examples rather than an exhaustive list. The examples in the remainder of this section are water and oxygen, organic matter, and silicates and silica.

Water and Oxygen

Added mainly in the form of rain and snow, water participates in many reactions in soils. Oxygen is present generally in the soil atmosphere and also in dissolved form. Recently, it has been reported that the weak acidity of rain near Pittsburgh, Pennsylvania, is not due to absorbed carbon dioxide nor to sulfuric and other acids emitted by factory smokestacks (Frohliger and Kane, 1975). This finding promises to be of interest in soil genesis if the observations can be confirmed elsewhere and the nature of the acids themselves can be established. Their characteristics could make the acids a distinctive factor in the differentiation of horizons in soil profiles.

In a study 30 years ago Haseman and Marshall (1945), using index minerals, concluded that a Grundy profile (Aquic Argiudoll from loess, a Mollisol) in north-central Missouri had gained 6.4% in mass and 7.7% in volume as horizons were differentiated. The gains are totals for a vertical section 1.6 m (63 in) deep. Gains in mass were ascribed chiefly to hydration of minerals and to oxidation of elements such as iron, in short, to additions of water and oxygen. The increased volume was ascribed to activities of living organisms and to microplastic movements associated with accumulation of clay in the B2 horizon.

Organic Matter

Bodies of dead and wastes of living organisms constitute the additions. The added organic matter also provides energy in combined forms for the soil population (Runge, 1973)

Generally speaking, the largest additions of organic residues consist of plant remains that fall on the surface. The wide range in amounts added is indicated in Table 1.

Table 1. Annual additions of litter, exclusive of roots, from several types of vegetation.

Vegetation and region	kg/ha/year--dry matter
Rainforest, western Africa (Nye, 1961a)	9,300-14,600
Savanna, western Africa (Greenland and Nye, 1959)	2,300
Tall grass prairie, Iowa (Jenny, 1941)	3,000
Shortgrass prairie, North Dakota (Redmann, 1975)	1,200
Hardwood forest, New York (Chandler, 1941)	3,000
Wormwood deserts, Asia (Rodin and Bazilevich, 1967)	30-150

Total quantities of organic matter added beneath the soil surface have a narrower range among types of vegetation than do additions of litter. Even so, the amounts are large enough to be important. In a rainforest of the humid tropics, the underground additions have been estimated to be half as large as those of litter (Nye, 1961b). Estimates were that the underground additions equalled the amounts of surface litter in shortgrass prairie (Redmann, 1975) and deserts (Rodin and Basivelich, 1967). The underground additions do not alter the relative rankings for differing types of vegetation, but they do expand the full range. Additions under rainforest are more than two orders of magnitude greater than those of deserts.

For the world as a whole, only a small share of the additions of organic matter in horizon differentiation can be attributed to animals. Kovda (1972) estimated that animals form no more than 1% of the total mass of living tissue on the land surface. Despite their small relative mass, animals add disproportionately larger shares of organic matter in some localities. One example is that of the Attines or fungus-culturing ants of Latin America (Weber, 1972). In Brazil, sauva ants will completely strip a tree of its leaves in one day. The ants use the fragments as substrates for their fungus gardens in ground chambers and tunnels.

Silicates and Silica

Additions are made in several ways. Within the U.S., for example, dust is transported both by large storms and as a nearly steady flux. Similar types of additions occur elsewhere in the world. Quantities added can be substantial.

From data available at the time, Free (1911) estimated that 770 million metric tons of dust were moved an average distance of 2400 km (1,500 mi) each year in the Great Plains, USA. This quantity is more than twice the average annual discharge of 310 million metric tons of sediments at the mouth of the Mississippi River. Approximately 1% of the transport capacity of the atmosphere is required to move the dust. Free and his colleagues had thoroughly screened existing literature; the bulletin has a 70-page bibliography on loess.

Several superimposed sets of ruins of prehistoric villages separated by loess deposits have been described in western Nebraska and Kansas (Wedel, 1961). These demonstrate movement of dust in quantity in the Great Plains for thousands of years.

Dust was reported in the air from Dubuque, Iowa, to Chelsea, Vermont, during a storm on March 9, 1918. Measured at Madison, Wisconsin, the dustfall was 4.8 g/sq m, or 4.8 metric tons/sq km (13.5 tons/sq mi). Allowing for differences in deposition from place to place, the figure suggests that somewhere between 1 and 10 million tons of dust were deposited over an area of 160 to 480 thousand/sq km (100 to 300 thousand sq mi). Particle size of the dust was mainly between 0.008 and 0.025 mm in diameter. The particles were chiefly quartz and feldspar grains coated with limonite and hematite, suggesting the arid Southwest as a source. Feldspar grains were clear and without signs of alteration. (Winchell and Miller, 1918)

Small, continuing additions of dust are being made to soils in many parts of the United States at the present time. My example covers only dust originating within the country and omits the aerosolic dust that circles the world (Clayton *et al.*, 1972). Quantities deposited at 14 stations during 27 months preceding July 1, 1966 were reported by Smith *et al.* (1970). Data from five stations are repeated in Table 2.

A continuing small flux of dust can add substantial amounts to soils over long periods of time. If a figure of 600 kg/ha/yr is taken as an average, a total of 600 metric tons/ha would be added in 1,000 years. That amount is more than half of the value of roughly 1,000 metric tons/ha commonly assigned to the surface 15 cm (6+-inch) layer of a

Table 2. Additions of dust and its proportions of sand plus silt at 5 points in the United States. (March 1964 to July 1966)

Location	kg/ha/mo (average)	kg/ha/yr (average)	% sand and silt
Sidney, Montana	69.7	838	65
Manhattan, Kansas	50.3	604	63
Oxford, Mississippi	61.6	739	62
Marcellus, New York	49.2	590	62
Marlboro, New Jersey	28.4	341	57

soil under cultivation. Additions of fine dust in this slow but continuing fashion could be responsible for the siltiness of A horizons in many soils of the eastern United States.

Past additions of silica and silicates are also indicated for some soils that have distinct horizons, seemingly formed long ago. An important component of the material in the solum of a Maury profile (Typic Paleudult, an Ultisol) in the Bluegrass region of Kentucky has been identified as loess on the basis of oxygen isotopes in the quartz (Syers *et al.*, 1969). The isotope ratios in quartz within the solum match those of loess in western Kentucky more closely than they do those of the silica in the underlying limestone. Yet Maury soils were long believed to have been derived entirely from underlying phosphatic limestone. No aeolian component within the solum was recognized. Present evidence for past additions of silica, and presumably silicates, seems good.

LOSSES

Substances and energy originally present in the regolith may be lost during horizon differentiation. So may any additions as soil genesis proceeds. Losses may be due to one or more of leaching, removal by erosion, and volatilization, to name some of the possibilities. Examples chosen for this section of the paper are organic matter and silica and silicates.

Organic Matter

Losses of added organic matter are largely through decomposition, which also constitutes one of the major kinds of transformations during horizon differentiation. Decomposition will thus be discussed in the section on transformations rather than here. Destruction of organic matter by fauna and losses by leaching are used as the examples in this subsection.

Estimates by Maldague (1964) were that termites in a rainforest in the Congo basin of Africa numbered 1,000 per sq m and had a mass of 100 kg/ha. Further, he concluded that the termites consumed and immobilized about 10 metric tons/ha/yr of fresh litter. Studies elsewhere (Edwards *et al.*, 1970) indicate that invertebrate soil fauna play an important role by mechanically breaking down plant litter so that it becomes more readily susceptible to microbial decomposition.

Some small lakes in Connecticut contain organic acids in large enough quantities to give waters a yellowish tinge (Shapiro, 1958). The amounts are as large as 5 mg/liter in water that is but faintly colored. Average molecular weights of the organic acids center around 450. The acids are believed to have been formed by partial decomposition of organic matter in soils, with subsequent leaching to transfer the acids to the lakes. Common soils in Connecticut are Dystrochrepts (Inceptisols) formed in glacial drift.

Dark humic substances are carried down and out of sandy Aquods (Spodosols in Coastal Plain materials) in eastern North Carolina. Amounts of organic matter moving down from the top 10 cm (4 in) of the A horizon were determined at monthly intervals for 2 years (Holzhey *et al.*, 1975). The greatest amount of organic matter recovered was 47 gm/sq m/yr, and it consisted chiefly of fulvic acid.

Seemingly, the fulvic acid is moving down and completely out of the soil. The profiles have dark layers up to several meters thick with upper boundaries at depths of 1.7 to 2.0 m (about 6 ft). The organic matter in one of these dark layers has a radiocarbon age of 31,000 years BP (Holzhey *et al.*, 1975). Such a large figure suggests that the organic matter now present in the dark layer accumulated long ago and that little accumulation is taking place now. Organic matter moving down from the A horizon seems to be leaving the soil completely. Further evidence that dark organic matter does move out in drainage waters is the occurrence of the "blackwater" swamps and streams in the lower Coastal Plain of the southeastern United States.

Silica and Silicates

Direct measurements of losses from soils are few, and indirect evidence must therefore be used, for the most part. One set of data available from lysimeters indicates the order of magnitude of silica losses. Those data are consequently given as one example in this subsection. Other examples are to illustrate losses of silicate clays.

Measurements of runoff, percolate, and substances lost were made over a period of years in Illinois with a group of eight lysimeters consisting of cylindrical cores of soil

90 cm (36 in) in diameter and 100 cm (4 in 40 in) long, cores collected with a minimum of disturbance (Stauffer and Rust, 1954). The average annual losses of silica and sesquioxides from four lysimeters over a 10-year period are given in Table 3.

Table 3. Average losses of SiO_2 and R_2O_3 (sesquioxides) in kg/ha/yr for 10 years from large soil cores in Illinois

Soil type	SiO_2	R_2O_3	Percolate (mm/yr)
Muscatine sil (Aquic Argiudoll)	16.6	4.8	299
Tama sil (Typic Argiudoll)	14.8	3.4	224
Cowden sil (Mollic Albaqualf)	3.8	1.4	64
Cisne sil (Mollic Albaqualf)	2.3	0.4	30

Losses of silica in kg/ha/yr ranged from 2.3 for the Cisne profile to 16.6 for the Muscatine profile. Both soils are derived from loess but the former has a prominent argillic horizon or claypan that restricts water movement. The amounts of silica lost seem directly related to total quantities of percolate, *i.e.*, to the amounts of water passing through the soil columns. If the average rates of loss were to persist for 1,000 years, the totals would be no more than a few metric tons per hectare. The prevailing climate does not promote strong weathering, even of parent materials which themselves are not yet greatly weathered.

Silicate clay minerals can be lost from soil profiles without first being broken down into their constituents. Near the type location for the Miramar series (Pachic Argixeroll from quartz diorite, a Mollisol) and along State Route 1 about 2 miles north of the village of Montara and 9 miles south of the city limits of San Francisco, California, a soil profile and the underlying saprolite are exposed in a cut about 7.5 m (25 ft) deep. Thickness of the soil profile is about 1.5 m (5 ft), below which the saprolite clearly shows rock structure. Extending from the lower part of the soil profile to the bottom of the cut are platy quartz veins oriented about 45 degrees from the vertical. The quartz plates range from about 1 to 2 cm in thickness and from 5 to 10 cm in length and have very small gaps between them. On the upper sides of the plates to the bottom of the exposure are distinct, dark and sticky coats up to 1 mm thick.

I interpreted these to be primarily silicate clays with subordinate organic matter. The coats are continuous with clay films on peds in the overlying argillic (Bt) horizon.

A second example lies along State Route 224 one mile southwest of the village of Marbury, Maryland. The site is about 18 miles south of the southeast corner of the District of Columbia. Exposed at the land surface in the roadcut is a Bourne profile (Typic Fragiudult, an Ultisol) formed partly in a loamy mantle and partly in the former A horizon (30 cm or 1 ft thick) of a Typic Paleudult (Ultisol) in Coastal Plain materials. The fragipan is partly in the loamy mantle and has completely taken over the former A horizon but stops at the top of the buried yellowish red argillic (B2t) horizon, that being too high in clay to permit formation of a fragipan. Extending down about 3 m (10 ft) from the bottom of the buried argillic horizon are distinct yellowish red clay films as linings in vertical partings in the Coastal Plain materials. These clay films are conspicuous against a mottled yellowish brown, gray, and red background. The films are continuous with those on ped surfaces in the buried argillic horizon. A few of the linings end in small cylindrical clay plugs with vertical axes about 25 mm (1 in) long and 6 mm (1/4 in) in diameter atop a horizontal iron pan 2-3 mm thick. The cylindrical plugs are very high in clay, comparable to the clay bands found in some prominent fragipans. I concluded from the positions of the linings and cylindrical plugs that the clay and iron oxides would have moved downward and completely out of the regolith except for the thin iron pan.

TRANSFERS

Transfers consist of both upward and downward movements of substances. Upward movements are more important than has been generally realized. Their role further suggests that horizon differentiation proceeds more rapidly than generally realized, given the numbers of soil profiles with distinct or prominent horizons. Substances transferred include soil materials *in toto*, water, organic matter, silica, silicates, and sesquioxides, to list part of the full set. Examples used here are soil materials, organic matter, and silicate clay minerals.

Soil Materials

Transfers of soil materials both upward and downward are accomplished by soil fauna. Earthworms transfer materials upward in goodly quantities in soils of the humid tropics. Nye (1955) measured amounts of worm casts placed on the surfaces of some soils in Nigeria during one 6-month rainy season. The quantities were 45 metric tons/ha during the half-year. Little or no transfer occurred during the dry season. Mostly, the earthworms operated in the top 30 cm (1 ft) of the soils, although termites transferred soil materials from greater depths. From his observations, Nye (1955) concluded that the uppermost 25 to 60 cm (10 to

24 in) of the profiles studied had all been handled by termites. No particles were present in the uppermost layers that could not be moved by a termite; anything too big was left behind as a "stone line."

We found pottery shards of various sizes, all low-temperature fired and without decoration, at and below a depth of 90 cm (3 ft) in a majority of our pits in Eutropepts and Dystropepts (Inceptisols considered Latosols at the time of the survey in 1948) in the Palau Islands, which are 800 miles southeast of Mindanao, Philippines. All told, we dug about 40 pits. Three-fourths of them had pottery shards at depth. Present on the soil surface at the time was a thin layer with medium-sized, strong granular structure, which puzzled me at first. Eventually, I concluded that the layer consisted of worm casts and that the pottery fragments had been buried by the upward transfer of soil materials over time (Vessel and Simonson, 1958).

Continuing upward transfers of soil materials from the B and C horizons are normal to Spodosols and Inceptisols of New England (soils formerly classified in the Brown Podzolic group). Immediately beneath a layer of litter and partly decayed organic matter are miniature ant mounds. These are conical, about 1 cm tall, and 7 cm in diameter, on the average. The mounds number one per sq ft or about 11/sq m. Amounts of soil transferred upward would equal a continuous layer 2 cm (slightly less than 1 in) thick in 250 years (Lyford, 1963). The upward transfers of materials from the deeper profiles could be an important factor in the lack of bleached or albic horizons in the soils.

Krotovinas, usually dark, roughly cylindrical fillings of former animal burrows, are common in the Mollisols of the world. Their presence in soils of the USSR is accepted as evidence that a given site was once grassland, irrespective of present vegetation.

Krotovinas are common in the Vermustolls (Mollisols) of eastern Romania. During excursions for the 8th International Congress of Soil Science in 1964, we were shown examples of these soils, classified as Chernozems of various kinds in that country. The pits were 10 m long, 2 m wide, and 2 m deep. I counted the krotovinas in one wall of each of two pits on the Dobrogea plateau not far from the Black Sea. In 2 sq m of wall below a depth of 1 m, I found 22 in one pit and 19 in the other. About half as many per unit area were present in the pit floors. Average diameter of krotovinas was near 8 cm (3+ in). Numbering about 10 per sq m, on the average, they formed approximately 5% of the lateral cross section of the pit walls below a depth of 1 m. Presumably, they would also form about 5% of the volume of the soils between depths of 1 and 2 m. That proportion identifiable at any one moment implies considerable mixing of soil materials over periods of centuries.

Further evidence of substantial transfers of soil

materials in the Vermustolls of eastern Romania was provided by the nature of their A1 horizons. These horizons were more than half earthworm and insect casts by volume. Moreover, many such horizons had been thickened appreciably by faunal mixing, leaving only remnants of cambic horizons (B horizons without accumulation) in the pit walls. The combined activities of burrowing animals and smaller fauna in the Vermustolls of Romania have effected much transfer of soil material, including both mineral and organic matter. Such mixing also proceeds in the Mollisols of North America, although not to the same extent.

Organic Matter

Transfers of organic matter within soil profiles as part of the materials moved about by fauna have already been described. Organic matter is also transferred in solution and suspension independently of soil mixing.

The distribution of organic carbon in profiles of Spodosols in Michigan and of certain Mollisols in Iowa indicates downward transfers from A to B horizons. For example, a profile of Blue Lake sand (Haplorthod from glacial drift) had the following contents of organic carbon by horizons: A1--5.82%, A2--0.40%, Bh--1.28%, and C--0.05% (Franzmeier and Whiteside, 1963). For a profile of Edina silt loam (Argialboll from loess), comparable figures are: A1--2.25%, A2--0.52%, B2t--0.85%, and C--0.18% (Simonson, 1959). In both profiles, the first and larger maximum marks the A1 horizon, with a second and smaller one in the B horizon. The two minima are in the A2 and C horizons. The contents are presumptive evidence of transfers of organic matter, an interpretation supported by other lines of evidence.

Direct evidence for transfers of organic matter within profiles has been obtained by measuring amounts moving down from A horizons of Spodosols. Observations in North Carolina were mentioned in the earlier section about losses. Similar measurements had been made earlier in Canada.

About 20 liters of pale yellow leachate were collected from the bottom of the A2 horizon of a Cryohumod (Humic Podzol) in Newfoundland over a period of 6 months (Schnitzer and Desjardins, 1969). The dry weight of organic matter dissolved in those 20 liters was slightly more than 1 gm. Approximately 90% of the organic matter was fulvic acid, which was the same as specimens extracted previously from Bh horizons of other Humods. The pale yellow color is of incidental interest because it was observed earlier in waters of small lakes in Connecticut (Shapiro, 1958).

Silicate Clay Minerals

Downward transfers of silicate clay minerals are considered responsible for argillic horizons (B horizons of clay accumulation), diagnostic for Alfisols and Ultisols and present in some Aridisols and Mollisols. Quantities transferred need not be large, but some transfer is required. Numerous observations over the years support inferences that clay is transferred.

Thick clay coats cover fired rocks in a buried hearth and clay films are present on ped faces in the overlying B horizon of a soil profile formed in alluvium along a tributary of the Willamette River in western Oregon (Reckendorf and Parsons, 1966). Radiocarbon dating of charcoal from the hearth give it an age of 5,250 ± 270 years BP. During that span of time, approximately 1.5 m (5 ft) of alluvium was deposited over the hearth and horizons differentiated in that alluvium to form a distinct Argixeroll (Mollisol) profile.

On the basis of particle size distribution, amounts of quartz, fractionation of the clay separate, and microscopic examination of thin sections, horizon by horizon, Kundler (1961) concluded that appreciable amounts of fine clay (less than 0.2 micron) had been transferred from the A to the B horizons of Hapludalfs and Glossudalfs (Alfisols) formed in glacial drift in northern Germany. Soils were identified there as Parabraunerde and Fahlerde.

Albic materials (pale sand and silt grains) finger into the upper parts of argillic (Bt) horizons of Udalfs (Alfisols) in New York (Bullock *et al.*, 1974). The thickest clay films are at the Bt-C horizon boundary rather than within the argillic horizon. These phenomena were attributed to stripping of silicate clays from upper parts of argillic horizons accompanied by downward transfers.

Related but not identical features exist in Elioak soils (Typic Hapludults from mica schist) in Maryland 10 to 20 miles northwest of Washington, D.C. The blocky peds in the argillic horizon have complete red coats, but only small patches on occasional ped faces can be identified as clay films. These are in contrast to distinct clay films that line vertical partings in the upper C horizon. The clay films in the partings can be traced upward to join the red coatings on peds in the argillic horizon. The distribution pattern and distinctness of the clay films in Elioak profiles suggest that clay is being transferred downward, some out of the argillic horizon. An example of this pattern of clay films can be found in an Elioak profile about .8 km (1/2 mi) north of the Rockville city limits. The site is just off State Route No. 355 along the road east to Derwood.

TRANSFORMATIONS

These entail changes in both composition and form. Examples of changes have been mentioned earlier in the paper, as for instance the hydration of minerals and oxidation of iron in a soil profile in Missouri, cited as illustrations of additions. Transformations may affect any or all substances constituting soils. Transformations can alter the appearance of portions of profiles to give rise to horizons. Among the many transformations that do occur, a few of those of organic matter and silicate minerals were selected as examples.

Organic Matter

Transformations of added organic matter consist chiefly of the decomposition of complex compounds into simpler ones. The decomposition passes through a number of intermediate stages, being a highly complex set of processes in itself. End products of this complex set of processes are water, carbon dioxide, and salts. A small share of the compounds formed during intermediate stages are condensed into new compounds of high molecular weight, reversing the general pattern. Such high molecular weight compounds then persist for long periods as humus in soils (Paul, 1970).

Most of the organic residues added during a growing season disappear within the year, but small fractions persist longer, some for centuries (Bartholomew and Kirkham, 1960). One example of the rate of turnover comes from a 32-year old Scots pine stand in England (Ovington, 1965). During the year, the litter fall was 9,630 kg/ha of which 8,330 kg disappeared.

Additions of organic matter and the consequent formation of humus provide the pigment for the dark, thick A horizons of Mollisols throughout the world. These horizons are of special importance to pedology because they captured the attention and stirred the imagination of Dokuchaeiv, thereby leading to establishment of a new school of soil science a century ago (Simonson, 1968).

Silicate Minerals

Transformations of silicates include the breakdown of primary minerals, the formation of secondary minerals, and the decomposition of those in turn. An example of each of these three is given in this subsection.

Decomposition of muscovite and biotite in the silt and sand fractions of some soils (probably Dystrochrepts) with formation of vermiculite as the principal end product was reported by Stephen (1952) in England. Muscovite lost potassium and gained water. Biotite lost both potassium and iron, the latter being largely oxidized and converted to goethite. The decomposition proceeded through several stages from the original muscovite and biotite to the end products of vermiculite and goethite.

Synthesis of small amounts of kaolinite at room temperature from silicon and aluminum in aqueous solution was reported not long ago (Hem and Lind, 1974). Quercetin, an organic flavone, was added to some solutions and the pH was adjusted to range between 6.5 and 8.5. After aging for 6 to 16 months, the precipitate included as much as 5% of well-formed flakes of kaolinite. In the solutions to which no flavone was added, however, the precipitate remained amorphous with no indication of kaolinite after 2 years. Synthesis of kaolinite in the laboratory suggests a possible mode for the formation of the mineral in soils with their wide assortment of organic compounds.

Thin sections from the A2 and Bt horizons of a sandy Aqualf (a suborder of the Alfisol order) formed in cover sand in southern Holland were examined with a petrographic microscope, X-ray diffraction, and an electron microprobe (Brinkman *et al.*, 1973). From their data, the authors concluded that smectite and illite present in clay films were decomposing to liberate silica. It was then being reprecipitated as microcrystalline quartz.

BALANCE AMONG PROCESSES OF HORIZON DIFFERENTIATION

In preceding sections, examples have been given of processes in each of the groups of additions, losses, transfers, and transformations. These four groups function in the genesis of all soils. Moreover, the components in each group are at least potential processes for every soil. The examples simply illustrate what happens and do not cover the full number or range of processes. Even so, the partial list does give some idea of the complexity of soil genesis -- that many processes are involved in horizon differentiation. Furthermore, the processes within combinations operate at cross purposes to some extent. Specimen processes promoting and offsetting or retarding horizon differentiation are illustrated in Figure 3. The diagram also suggests the complexity of soil genesis.

The balance over time among a host of individual processes is the key to the nature of every soil; that balance determines the characteristics of the soil profile and soil body. In other words, the ultimate character of a soil depends on the relative importance of all individual processes in each combination. For example, additions of organic matter are small and of little importance in the combinations of processes giving rise to Aridisols. Transfers of sesquioxides are also of little importance. These same processes are of the first importance in the genesis of Oxisols. Yet the processes operate in formation of both groups of soils.

The critical role of balance among many processes can be illustrated more fully by comparison of the soils of a pair of great groups, the Haplargids (Aridisols) and Hapludults (Ultisols). Soils of both classes have argillic (Bt) horizons. Haplargids have relatively thin sola, neutral or alkaline reactions, high base status, low contents of organic matter, little iron associated with silicate clays in argillic horizons, marked accumulations of carbonates below their sola, and low contents of soluble salts. Hapludults have relatively thick sola, strongly acid reactions, low base status, low contents of organic matter,

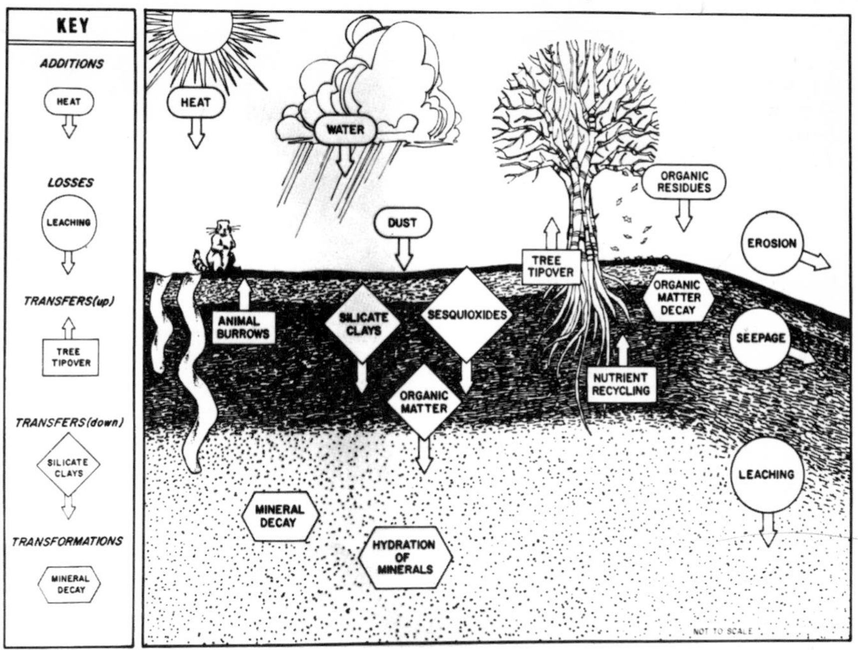

Figure 3 Diagram to illustrate a small share of the many and conflicting processes that differentiate horizons within soils. A key to the kinds of processes is given along the left side of the sketch. The complexity of horizon differentiation is suggested rather than recorded in the diagram. Sketch by Walter M. Simonson.

appreciable amounts of iron associated with silicate clays in argillic horizons, no accumulations of carbonates, and no soluble salts. Additions of silicates and bases are much higher and those of organic matter much lower for Haplargids than for Hapludults. Losses of bases and organic matter are much higher for Hapludults. Transfers of iron are appreciable in Hapludults and negligible in Haplargids. Transfers of carbonates are large in Haplargids, whereas such substances are lost in early stages by Hapludults. Transformations include precipitation of carbonates to form dense and hard layers below the sola of Haplargids as compared to breakdown of primary minerals and formation of iron oxides in Hapludults. The two kinds of soils are markedly different, and the distinctions hark back directly to differences in balance among the process of horizon differentiation. Basic processes are the same, but their relative importance in the combinations is not. The balance over time has been critical, as it is for all soils.

MULTIPLE INTERVALS OF HORIZON DIFFERENTIATION

On a geologic time scale, soils are temporary phenomena.

Roy W. Simonson

If we could go back and follow changes at the land surface, we would find that soils had been formed and destroyed a number of times. Soil profiles with the top several centimeters baked red occur locally immediately beneath basalt flows of the Columbia plateau of northwestern United States and the Deccan plateau of India. Buried profiles in all stages of truncation in the Mississippi valley demonstrate possibilities for destruction of soils (Simonson, 1954).

Soils are not necessarily destroyed by changes in the environment. Instead, the pathway of horizon differentiation may be altered so that new characteristics are superimposed on the old. Ollier (1959) concluded that soils in parts of eastern Africa had gone through a pair of genetic cycles in order to acquire their present character. That situation is not restricted to eastern Africa.

Polymorphic Soils

Examples of soils that bear obvious marks of at least two intervals of horizon differentiation are easy to find. In this paper, such profiles are called polymorphic to distinguish them from those that do not have characteristics due to two or more intervals. The term, polygenetic, might have been used here, but it originally carried the idea of a change in climate, whereas no change can be established for examples given here.

One set of polymorphic soils occurs in the Coastal Plain of eastern Virginia. Under some stands of shortleaf pine, the uppermost part of the soil profile consists of a sequence of miniature horizons, like those of a Spodosol except for size. The trees in such stands have ages of 40 to 50 years, according to increment borings. Moreover, the trees occupy former fields, cultivated within the memory of older local residents. The sequence of miniature horizons is 10 to 15 cm (4 to 6 in) thick within the A horizon of a Typic Paleudult (Ultisol). In adjacent fields still in cultivation, the Paleudult has an A2 horizon of light yellowish brown (10YR 6/4) loamy sand, 40 to 50 cm (16 to 20 in) thick. Below that is an argillic (Bt) horizon of yellowish brown (10YR 5/4) light sandy clay loam.

Within a pine stand is a thin surface covering of needles and twigs. Beneath that is an A2 horizon of pinkish gray (7.5YR 7/2) sand, 2 or 3 cm (an inch or so) thick over a B2 horizon of yellowish red (5YR 4/6) loamy sand, 7 to 10 cm (3 or 4 in) thick. Beneath that in turn is the normal A2 horizon of the Paleudult. The sequence of miniature horizons indicates a change in the pathway of genesis after the abandoned fields were occupied by pines.

A second example of polymorphic soils is taken from northwestern Kansas. This is a Mollisol profile (Pachic Haplustoll) about 2.5 km (1 1/2 mi) southeast of Colby, Thomas County. The profile has the following horizon sequence: A1 (0-15 cm), B2 (15-42 cm), B2b (42-74 cm), Cbca

(74-134 cm), and Cb (134-164 cm). The full profile is considered to have been formed in two loess deposits, a thin second mantle having been laid down on top of a Typic Haplustoll in a thicker deposit.

Several lines of evidence can be drawn from the nature of the present soil profile to support the inference that two intervals of horizon differentiation have occurred. For one thing, thicknesses of the mollic epipedon (a dark surface layer high in organic matter and bases) and the profile are both greater than normal for Typic Haplustolls. The A1 and B2 horizons are very dark brown (10YR 2/2) and the B2b very dark grayish brown (10YR 3/2) in color. At least the uppermost 42 cm and more likely 74 cm would thus qualify as a mollic epipedon. Thickness of the solum plus the Cbca horizon is 134 cm. Those dimensions would characteristically be about half as large for the mollic epipedons and profiles of Typic Haplustolls.

Color and structure of the B2 and B2b horizons and the accumulation of carbonates in the latter are further indications of a polymorphic nature. Color of the B2 horizon is darker than normal, whereas that of the B2b horizon is not for Typic Haplustolls. Grade of structure is also weaker than characteristics in the B2 horizon. It has weak, coarse prismatic structure in contrast to the moderate, medium subangular blocky structure with blocks grouped into prisms in the B2b horizon. Carbonates are beginning to accumulate as small flecks on ped faces in the lower part of the B2b horizon, suggesting that it rather than the Cbca horizon is now receiving the transfers.

From the character and sequence of horizons in the profile, I concluded that the A1 horizon and perhaps a little of the B2 horizon had been formed in the loess laid down over the former Typic Haplustoll. Most of the B2 horizon is believed to have been formed in the previous A1 horizon, which helps to explain its darker color and weaker-than-usual grade of structure. The B2b and Cbca horizons were the B2 and Cca horizons of a soil profile prior to the deposition of the latest loess.

After the second loess was deposited, horizon differentiation resumed, following the same pathway as before. Because the soil is now forming in part in a former profile, however, its horizons depart from those normal to Typic Haplustolls. The current profile is therefore considered polymorphic in nature, reflecting more than one interval of horizon differentiation.

Significance to Quaternary Soils

In all probability, horizon differentiation has followed several different pathways in the genesis of many Quaternary soils. The existence of soils which clearly show the effects of more than one interval of horizon differentiation bespeaks the occurrence of soils with less obvious, perhaps even obscure, effects. For that reason, it is important in the examination of soil profiles to keep in mind the possibility that their characteristics may reflect genesis along several pathways. As the knowledge of soil genesis increases, I expect that the number of recognized polymorphic soils will increase as well.

REFERENCES CITED

Bartholomew, W.V., and Kirkham, D., 1960, Mathematical descriptions and interpretations of culture induced soil nitrogen changes: Transac. 7th Internatl. Cong. Soil Science (Madison, Wis.), v. 2, p. 471-477.

Birkeland, P.W., 1974, Pedology, weathering, and geomorphological research: N.Y., Oxford Univ. Press, 285 p.

Brinkman, R., Jongmans, A.G., Miedema, R., and Maaskant, P., 1973, Clay decomposition in seasonally wet, acid soils: Micromorphological, chemical and mineralogical evidence from individual argillans: Geoderma, v. 10, p. 259-270.

Bullock, P., Milford, M.H., and Cline, M.G., 1974, Degradation of argillic horizons in Udalf soils in New York State: Soil Sci. Soc. America Proc., v. 38, p. 621-628.

Chandler, Jr., R.F., 1941, The amount and mineral nutrient content of freshly fallen leaf litter in the hardwood forests of central New York: Jour. Amer. Soc. Agron., v. 33, p. 859-871.

Clayton, R.N., Rex, R.W., Syers, J.K., and Jackson, M.L., 1972, Oxygen isotope abundance in quartz from Pacific pelagic sediments: Jour. Geophysical Research, v. 77, p. 3907-3915.

Edwards, C.A., Reichle, D.E., and Crossley, Jr., D.A., 1970, The role of invertebrates in turnover of organic matter and nutrients: *in* Reichle, D.E., ed., Analysis of temperate forest ecosystems: N.Y., Springer-Verlag, p. 147-172.

Franzmeier, D.P., and Whiteside, E.P., 1963, A chronosequence of Podzols in northern Michigan: II. Physical and chemical properties: Michigan Agr. Exp. Sta. Quaterly Bull., v. 46, no. 1, p. 21-36.

Free, E.E., 1911, The movement of soil material by the wind: USDA Bureau Soils Bull. 68, 173 p.

Frohliger, J.O., and Kane, R., 1975, Precipitation: Its acid nature: Science, v. 189, p. 455-457.

Greenland, D.J., and Nye, P.H., 1959, Increases in the carbon and nitrogen contents of tropical soils under natural fallows: Jour. Soil Science, v. 10, p. 284-299.

Hem, J.D., and Lind, C.J., 1974, Kaolinite synthesis at 25° C.: Science, v. 184, p. 1171-1173.

Haseman, J.F., and Marshall, C.E., 1945, The use of heavy minerals in studies of the origin and development of soils: Missouri Agric. Exp. Sta. Research Bull. 387, 75 p.

Holzhey, C.S., Daniels, R.B., and Gamble, E.E., 1975, Thick Bh horizons in the North Carolina Coastal Plain: II. Physical and chemical properties and rates of organic additions from surface sources: Soil Sci. Soc. America Proc., v. 39, p. 1182-1187.

Huggett, R.J., 1975, Soil landscape systems: A model of soil genesis: Geoderma, v. 13, p. 1-22.

Jenny, H., 1941, Factors of soil formation: N.Y., McGraw-Hill, p. 207.

Kovda, V.A., 1972, The world's soils and human activity, *in* Polunin, N., ed., The environmental future, Proc. 1st Conference on Environmental Future, Finland--1971: London, Macmillan, p. 359-393.

Kundler, P., 1961, Lessivés (Parabraunerden, Fahlerden) aus Geschiebemergel der Würm-Eiszeit in norddeutschen Tiefland: Zeitsch. Pflanzenernahrung, Düngung, Bodenkunde, v. 95, p. 97-110.

Lyford, W.H., 1963, Importance of ants to Brown Podzolic soil genesis in New England: Harvard Forest Paper 7, 18 p.

Maldague, M.E., 1964, Importance des populations des termites dans les sols equatoriaux: Transac. 8th Internatl. Cong. Soil Science (Bucharest), v. 3, p. 743-752.

Marbut, C.F., 1951, Soils: Their genesis and classification (Lectures to U.S. Dept. Agr. Grad. School, 1927): Madison, Wis., Soil Sci. Soc. America, p. 25.

Nye, P.H., 1955, Some soil-forming processes in the humid tropics: IV. The action of the soil fauna: Jour. Soil Science, v. 6, p. 73-83.

_________, 1961a, Organic matter and nutrient cycles under moist tropical forest: Plant and Soil, v. 13, p. 333-346.

_________, 1961b, Some effects of natural vegetation on the soils of West Africa and on their development under cultivation, Proc. Abidjan Symposium on Tropical Soils and Vegetation--1959: Paris, UNESCO, p. 59-63.

Ollier, C.D., 1959, A two-cycle theory of tropical pedology: Jour. Soil Science, v. 10, p. 137-148.

____________, 1969, Weathering: London, Oliver & Boyd, Ltd., 304 p.

Ovington, J.D., 1965, Woodlands: London, The English Universities Press, p. 93.

Paul, Eldor, 1970, Plant components and soil organic matter: Recent Advances in Phytochemistry, v. 3, p. 59-104.

Reckendorf, F.F., and Parsons, R.B., 1966, Soil development over a hearth in Willamette Valley, Oregon: Northwest Science, v. 40, p. 46-55.

Redmann, R.E., 1975, Production ecology of grassland communities in western North Dakota: Ecological Monographs, v. 45, p. 83-106.

Rodin, L.E., and Bazilevich, N.E., 1967, Production and mineral cycling in terrestrial vegetation: London, Oliver & Boyd, p. 185-191.

Runge, E.C.A., 1973, Soil development sequences and energy models: Soil Science, v. 115, p. 183-193.

Schnitzer, M., and Desjardins, J.G., 1969, Chemical characteristics of a natural soil leachate from a Humic Podzol: Canadian Jour. Soil Science, v. 49, p. 151-158.

Shapiro, J., 1958, Yellow acid-cation complexes in lake water: Science, v. 127, p. 702-704.

Simonson, Roy W., 1954, Identification and interpretation of buried soils: American Jour. Science, v. 252, p. 705-732.

________________, 1957, What soils are, *in* Soil: USDA Yearbook of Agriculture: Washington, U.S. Govt. Printing Office, p. 17-31.

________________, 1959, Outline of a generalized theory of soil genesis: Soil Sci. Soc. America Proc., v. 23, p. 152-156.

________________, 1968, Concept of soil: Advances in Agronomy, v. 20, p. 1-45.

________________, 1970, Loss of nutrient elements during soil formation, *in* Engelstad, O.P., ed., Nutrient mobility in soils: Accumulation and losses: Soil Sci. Soc. America Special Publ. 4, p. 21-45.

Smith, R.M., Twiss, P.C., Krauss, R.K., and Brown, M.J., 1970, Dust composition in relation to site, season, and climatic variables: Soil Sci. Soc. America Proc.,

v. 34, p. 112-117.

Soil Survey Staff, 1976, Soil taxonomy: A basic system of soil classification for making and interpreting soil surveys, U.S. Dept. Agric. Handbook 436: Washington, U.S. Govt. Printing Office, 754 p.

Stauffer, R.S., and Rust, R.H., 1954, Leaching losses, run-off, and percolate from eight Illinois soils: Agronomy Jour., v. 46, p. 207-211.

Stephen, I., 1952, A study of rock weathering with reference to the soils of the Malvern Hills: Part I. Weathering of biotite and granite: Jour. Soil Science, v. 3, p. 20-33.

Syers, J.K., Jackson, M.L., Berkheiser, V.E., Clayton, R. N., and Rex, R.W., 1969, Eolian sediment influence on pedogenesis during the Quaternary: Soil Science, v. 107, p. 421-427.

Vessel, A.J., and Simonson, Roy W., 1958, Soils and agri-culture of the Palau Islands: Pacific Science, v. 12, p. 281-298.

Weber, N.A., 1972, The Attines: The fungus-culturing ants: American Scientist, v. 60, p. 448-456.

Wedel, W.R., 1961, Prehistoric man on the Great Plains: Norman, Univ. Okla. Press, 355 p.

Winchell, A.N., and Miller, E.R., 1918, The dustfall of March 9, 1918: American Jour. Science, V. 274, p. 599-609.

Yaalon, D.H., 1971, Soil-forming processes in time and space, *in* Yaalon, D.H., ed., Paleopedology -- Origin, nature, and dating of paleosols: Jerusalem, Internatl. Soc. Soil Science and Israel Universities Press, p. 29-39.

Properties, Criteria of Classification, and Concepts of Genesis of Podzolic Soils in Canada

J. A. McKeague, G. J. Ross, D. S. Gamble

Soil Research Institute Contribution 605

ABSTRACT

The Podzolic order is one of nine major classes in the Canadian system of soil taxonomy. It includes those soils having a B horizon in which amorphous products of organic matter decomposition and mineral weathering, mainly organic complexes of Al and Fe, have accumulated beyond specified minimum amounts. Such B horizons are usually recognizable in the field by their color and consistence and by the associated horizons in the profile. However, laboratory criteria are necessary to decide the classification of some soils having properties close to the arbitrary boundary line, through the continuum of soil properties, that separates Podzolic soils from other conceptual classes at the order level.

The distribution of Podzolic soils in Canada is clearly related to the factors of soil formation, especially parent material and climate. Generally speaking, Podzolic soils occur on sandy parent materials in areas of cryoboreal to cool, humid to perhumid soil climate. Vegetation associated with Podzolic soils includes coniferous, mixed, and deciduous forest, as well as shrubs and mosses. A vast area of Podzolic soils is associated with coniferous vegetation but this may be due more to the relationship between climate and vegetation than to a direct relationship between soils and vegetation.

The genesis of Podzolic soils involves leaching, depletion of bases, the interaction of soluble organic matter with soil minerals and the translocation of materials. Weathering of clay-size minerals in Podzolic soils follows a characteristic pattern with some exceptions. Typically, the dominant clay-size layer silicates change from mica and chlorite in the C horizon to chloritized vermiculite in the B to beidellitic smectite, often interstratified with some mica layers, in the Ae horizons. Clay-size primary minerals, other than some quartz and feldspar usually are not found in the Ae horizons of Podzolic soils. Estimates of the capacity of organic solutes to complex and to transport Al and Fe in Podzolic soils are crude because of lack of knowledge of the chemical composition of soil solutions and fragmentary information on reactions of dilute organic solutions with soil minerals.

EVOLUTION OF CONCEPTS AND CLASSIFICATION OF PODZOLIC SOILS

The term "podzol" which originated in Russia before 1880, once meant the ashy gray, surface or near surface layer, Ae horizon, that occurs in several kinds of soil (Muir, 1961). It was not thought to be related to the underlying darker-colored layer, B horizon, until near the end of the 19th century. Since 1900 the terms podzol and podzolic soils, as used in Russia, have included several kinds of leached soils: those with B horizons of humus or humus-sesquioxide accumulation, those with B horizons of clay accumulation, and acid soils with weakly developed B horizons.

In Canada, the term "podzol" has been known and used since the initiation of soil surveys before 1920. The first system of "field classification for Canadian soils" named several soil zones that included the words Podzol or Podzolic (NSSC, 1945). The dominant soils of these zones were not defined clearly but broadly speaking: Podzol meant a soil with an Ae horizon and a B horizon of humus-Fe accumulation, Gray-Brown Podzolic soils had a B horizon of clay accumulation, Brown Podzolic soils were acid soils with brown B horizons and no Ae horizon.

In the first outline of a "taxonomic" system for Canadian soils (NSSC, 1955), a Podzolic class was named as one of 7 classes at the highest hierarchical level and defined briefly as follows: "Soils with (A_2) bleached eluvial horizons and (B_2) illuvial horizons having accumulations of R_2O_3 and/or organic matter or clay (not solonetzic B)." Podzolic soils were subdivided into 5 classes, 2 of which (Gray Brown Podzolic and Gray Wooded) had B horizons of silicate clay accumulation and 3 of which [Humus Podzols, Podzols (ortstein) and Podzols (orterde)] had B horizons of humus-sesquioxide accumulation. Soils having B horizons like those of Podzolic soils but without Ae horizons were not included in the class.

Through the years since 1955 various changes have been made in concepts and definitions of Podzolic soils in Canada. In 1965, for example specific chemical criteria were introduced specifying the minimum amount of amorphous Fe and Al that must be present in the B horizon of a Podzolic soil (NSSC, 1965). The test, involving the measurement of Fe and Al in acid ammonium oxalate extracts of the B and C horizons, was based on work of Tamm many years previously (Lundblad, 1934). It was revised by Schwertmann (1964) and applied to soils of Canada by McKeague and Day (1966). The introduction of specific chemical criteria reflected a recognition of the fact that soils have a continuum of properties and that without specific criteria, uniformity of classification throughout a large country is not attainable. Specific criteria used for several years in the United States (Soil Survey Staff, 1960) influenced this development in Canada but the criteria were different because the population of soils classified in the United States and Canadian systems differed.

A major change was made in the concept and definition of Podzolic soils in 1968 (NSSC, 1968) when soils having B horizons of clay accumulation and of humus-Al, Fe accumulation were separated at the highest hierarchical level as Luvisolic and Podzolic orders respectively. In addition, soils having podzolic B horizons (humus-Al, Fe accumulation beyond the minimum limit specified) but with no Ae horizon were included in the Podzolic order. Thus by 1968, the original Russian concept of "podzol" (ashy layer) was no longer even a necessary feature of Podzolic soils in the Canadian system.

CURRENT CRITERIA OF PODZOLIC SOILS

Podzolic soils have a "podzolic B" horizon that meets the following morphological and chemical criteria:

1. Morphological: Thickness - 5 cm. or more; Color - black, or 7.5YR or redder, or 10YR near the upper boundary and yellower below, and either a chroma of more than 3 or a color value of 3 or less (Munsell chart); Amorphous material - dark coatings on grains or a silty feel; Texture - coarser than clay; Underlying Bt horizon (clay accumulation) - either none or below a depth of 50 cm.

2. Chemical: Either a) Includes a Bh subhorizon at least 10 cm. thick that contains more than 1% organic carbon, less than 0.3% 0.1M sodium pyrophosphate-extractable Fe, and has a ratio of organic C to pyrophosphate-extractable Fe of 20 or more, or b) Includes a subhorizon at least 5 cm thick that: contains 0.6% or more pyrophosphate-extractable Al and Fe (0.4% for sands), has a ratio of pyrophosphate-extractable Al + Fe to clay (>2μm) of more than 0.05, and has an organic C content of 0.5% or more.

The criteria may seem to be pedantically specific but they were the simplest ones that could be devised to achieve an unambiguous definition of Podzolic soils and hence to permit uniformity of classification throughout the country. They were based upon generalization of properties of hundreds of pedons (the unit of soil; 1 m^3 to about 10 m^3) developed in stable, coarse to medium textured deposits in areas of humid climate and forest or heath vegetation. They were designed to include in the order those soils that satisfied most pedologists' concepts of Podzolic soils while excluding those that did not. The chemical criteria were based upon experimental evidence that pyrophosphate extracts the Al, Fe-organic complex material known to accumulate in the B horizons of Podzolic soils (McKeague, Brydon and Miles, 1971). The specific chemical criteria do not imply that every pedon examined in the field must be sampled and analyzed. Rather they are used as a check to determine the classification of pedons having properties close to the arbitrary boundary lines, through the continuum of soil properties, that separates Podzolic soils from soils of other orders.

From the foregoing material, it is obvious that the concept and definition of Podzolic soils in Canada has changed greatly during the last 50 years. Thus readers of publications dealing with Podzolic soils must consider the concepts of such soils at the times the publications were prepared.

Podzolic soils are subdivided into 3 great groups based upon properties of the B horizon as follows:

Humic Podzol - has a Bh horizon (organic C and Al but little Fe) at least 10 cm thick.
Ferro-Humic Podzol - has a Bhf horizon (at least 5% organic C associated with Al and Fe) at least 10 cm thick.
Humo-Ferric Podzol - has a Bf horizon (less than 5% organic C associated with Al and Fe) at least 5 cm thick.

The great groups are subdivided into a number of subgroups named by placing the following adjectives before the great group names: Orthic (central concept), Ortstein (cemented podzolic B), Placic (thin ironpan), Duric (cemented subsoil), Fragic (fragipan), Luvisolic (horizon of clay accumulation), Sombric (mineral-organic surface horizon), Gleyed. The subgroups are further subdivided into families and the families into series.

The concept of Podzolic soils in Canada is generally consistent with concepts of similar soils in the United States' system but the criteria differ, and the name of the order is Spodosol in the United States' system (Soil Survey Staff, 1976). A Spodosol must have a spodic horizon that satisfies the following criteria among others:

1. The ratio of pyrophosphate-extractable Al + Fe to clay is 0.2 or more (0.05 for a podzolic B in Canada).

2. The ratio of Pyrophosphate- to dithionite-extractable Fe + Al is 0.5 or more (no such criterion for a podzolic B). Thus some Podzolic soils are not Spodosols.

Podzolic soils, as defined in Canada, are similar to Podzols in several western European systems of soil taxonomy (Petersen, 1976). However, in western Europe, Podzols must have an Ae horizon and properties of the B horizon are less strictly defined than in Canada.

PROPERTIES OF PODZOLIC SOILS IN RELATION TO ENVIRONMENTAL FACTORS

The axiom of pedology that soil properties reflect the influence of climate and organisms particularly vegetation, as conditioned by relief, acting over time on soil parent materials applies clearly to soils of the Podzolic order. This is illustrated by presenting site and soil data for 8 pedons from various areas of Canada.

Laurentide Series (Anon., 1973)

This soil might be considered as the archetype of a Podzolic soil. The site in the Montmorency Experimental Forest of Laval University, about 70 km north of Quebec city in the Laurentian uplands of the Canadian Shield at 47°20'N, 71°09'W is representative of a large area in eastern Canada. Vegetation at the well-drained site at an elevation of 750 m is balsam fir-white birch with shrubs, ferns and mosses as ground cover. The stony, sandy glacial till was derived from rocks ranging in composition from granite to gabbro. The mean annual air temperature is 0°C; mean precipitation, 140 cm, 1/3 of which falls as snow; and mean frost-free period about 40 days. An abbreviated profile description follows:

Horizon	Depth cm	Thickness range cm	Description
LFH	8-0	5-15	Dark colored, partly decomposed organic litter
Ae	0-5	3-13	Gray loamy sand, abrupt lower boundary
Bhf	5-17	5-19	Dark reddish brown sandy loam, friable, weak granular
Bf	17-40	17-27	Dark brown loamy sand, friable, weak blocky
BC	40-56	-	Olive brown loamy sand
Cx	56-	-	Olive loamy sand, moderate coarse platy, very firm in place, brittle, some stones; maximum firmness at 150-250cm.

This pedon of the Laurentide series is a well developed Ferro-Humic Podzol as it has a podzolic B horizon with more than 5% organic C associated with complexed (pyrophosphate-extractable) Al and Fe (Table 1). About 80% of the non-silicate iron (dithionite-extractable) in the Bhf horizon is complexed by organic matter. Most of the accessory characteristics typical of Podzolic soils are evident: the acid organic surface layer, the light-colored, strongly leached Ae, the fading of the reddish brown color with depth, the low base saturation, the sandy texture (and high permeability), the low bulk density of the B horizon in relation to that of the parent material, the high pH dependent cation exchange capacity, and the apparent accumulation in the B horizons of products of weathering from the Ae horizon. The dense, brittle C horizon was considered to be a fragipan, a compact horizon that is difficult to distinguish from dense basal

Table 1. Analytical data for Laurentide soil[a]

Horizon	Depth cm	BD g/cm3	sand %	clay %	C %	pH	CEC me/100g	Base Sat. %	ΔCEC me/100g	Fe %t	Fe %d	Fe %p	Al %p	Al %t
Ae	0-5	1.0	76	3	1.4	3.3	0.7	50	3	4.3	0.1	0.02	0.01	6.0
Bhf	5-17	0.75	68	11	12	3.5	6.1	10	40	6.5	3.3	2.6	1.6	6.4
Bf	17-40	1.36	66	6	2.8	4.3	0.9	30	15	5.4	1.5	0.6	1.1	7.6
Cx	60-90	1.84	78	4	0.2	4.7	0.4	70	3	5.3	0.6	0.03	0.2	7.2

[a]These data with the exception of total Fe and Al, are for Canada Soil Survey Committee reference soil samples CSSC 18 to 21 of the Laurentide soil sampled at the site described in Anon. (1973). The methods were as follows (see McKeague, J.A., ed. 1976): bulk density - method 2.23, data courtesy of G. Mehuys and C.R. De Kimpe; sand and clay - method 2.111, pipette; carbon - method 3.611; pH-method 3.11, 0.01 M $CaCl_2$; cation exchange capacity (CEC) - method 3.31, sum of cations extracted by NaCl; base saturation - method 3.31; ΔCEC - the difference between CEC at pH 7, method 3.33, and CEC at the pH of the soil, method 3.31; Fe_d is dithionite-citrate-bicarbonate-extractable Fe (method 3.51); and Fe_p and Al_p are 0.01 M sodium pyrophosphate extractable Fe and Al (method 3.53); Fe_t and Al_t are total Fe and Al values estimated from Anon. (1973).

till. Fragic Ferro-Humic Podzols and other subgroups of Ferro-Humic and Humo-Ferric Podzols occupy vast areas of the Laurentian and Appalachian Highlands.

Whonnock Series (Luttmerding, 1971; McKeague & Sprout, 1975)

This series is representative of a large area of soils at lower and mid elevations in the mountains of humid coastal B.C. It differs from the Laurentide series in lacking a well-developed Ae horizon and in having a strongly cemented subsoil. The site is at 49°21"N, 122°42"W at an elevation of 760 m on a steep lower slope of Burke mountain near Vancouver. Vegetation is flourishing coast forest dominated by Douglas fir. Approximate mean annual climatic data extrapolated from data for weather stations in the area are: total precipitation 250 cm, snow 700 cm, temperature 9°C. The soil is nearly always moist and it does not freeze. The parent material is sandy loam to loam-textured glacial till containing a relatively high proportion of ferro-magnesian minerals, and some volcanic ash may occur near the surface.

A brief description of a profile of Whonnock, a Duric Ferro-Humic Podzol, follows:

Horizon	Depth cm	
L-H	28-0	Undecomposed leaves, twigs and moss at the surface grading to black humic material.
Ahe	0-4	Very dark gray, sandy loam, organic matter rich, friable.
Bhf	4-50	Dark reddish brown grading to brown with depth, sandy loam, friable, common roots.
Bfgj	50-80	Grayish brown, sandy loam, platy, friable to firm, mottled.
BCc	80-150	Olive gray to gray, sandy loam, strongly cemented, very coarse platy, mottled.
C	150-200	Gray, sandy loam, amorphous, very firm, cementation decreases with depth.

Analytical data for the Whonnock pedon (Table 2) are similar in many respects to those for Laurentide. However, appreciable organic complexed Al and Fe occurs in the thin A horizon of Whonnock.

St. Stephen Series (McKeague, Schnitzer and Heringa, 1967)

This series is included as an example of Podzolic soils with a placic horizon, thin ironpan. Placic horizons occur commonly within and below the sola of Podzolic soils in Newfoundland, in some soils of B.C., and in the mineral material underlying some organic deposits in B.C. and Newfoundland. The site at 46°46'N 53°35'W is on an 8% slope about 30 m above sea level on the Avalon Peninsula. Vegetation consists of mosses, shrubs, and grasses. Mean annual temperature is about 5°C and precipitation about 140 cm.

Table 2. Analytical data for Whonnock soil[a]

Horizon	Depth cm	sand %	clay %	C %	pH	Exchangeable Ca me/100g	Exchangeable Mg me/100g	Exchangeable K me/100g	Base Sat. %	ΔCEC me/100g	Fe_d %	Fe_p %	Al_p %
H	5-0	-	-	47	3.5	1.0	0.3	0.3	-	-	-	-	-
Ahe	0-4	46	15	9.2	3.9	0.2	0.07	0.05	50	35	0.8	0.6	1.0
Bhf	4-50	47	17	6.7	4.2	0.1	0.04	0.03	40	38	1.0	0.6	1.8
Bfgj	50-80	50	16	2.5	4.6	-	0.01	0.01	70	17	0.6	0.3	1.3
BCc	80-150	54	14	0.3	5.0	-	-	-	-	6	0.4	0.03	0.3
C	150-200	57	12	0.1	5.5	5.5	1.2	-	100	1	0.4	0.03	0.07

[a]Data from Luttmerding (1971) except for sand and clay courtesy of V.E. Osborne. Methods were the same as those listed in Table 1.

The parent material is stony, gravelly, glacial till derived largely from green sandstones and argillites.

A brief description of a profile of this Placic Ferro-Humic Podzol follows:

Horizon	Depth cm	
FH	15-0	Partly decomposed organic material
Ae	0-4	Pinkish gray silt loam, platy, friable
AB	4-8	Reddish gray gravelly silt loam, subangular block, friable
Bhf	8-22	Reddish black stony gravelly loam, friable
Bhfc	22-23	Black shiny upper layer, brown below, strongly cemented, abrupt wavy boundary
Bfcj	23-35	Dark reddish brown stony, gravelly sand, weakly cemented
Bf	35-58	Dark brown stony gravelly sand, very firm in place.
C	58-100	Olive gray stony gravelly sand, firm in place.

The analytical data (Table 3) shows the relatively high clay content of the upper horizons, due presumably to breakdown of sand and gravel sized argillite fragments. The placic horizon has a much higher iron content than the other B horizons and about 3/4 of the dithionite-extractable Fe is apparently complexed with organic matter.

Morin Series (McKeague, *et al.*, 1973)

This series is typical of a vast area of Podzolic soils in eastern and western Canada developed under less humid conditions than those associated with the 3 series described previously. The site at 45°44'N 74°59'W is in a valley in the Laurentian Highlands on a 4% slope at an elevation of 150 m. The vegetation is mixed forest including red maple, chokecherry and yew with a ground cover of mosses and grasses. The mean annual temperature and precipitation are about 5°C and 105 cm respectively. The parent material is outwash sand derived from Precambrian Shield rocks.

A brief description of a profile of this Orthic Humo-Ferric Podzol follows:

Horizon	Depth cm	
LH	2-0	Black mixture of partly decomposed organic matter and sand
Ae	0-10	Light gray sand, very friable, abrupt wavy boundary
Bf_1	10-20	Dark reddish brown sand; weak fine granular, friable with discontinuous weak cementation
Bf_2	20-40	Strong brown, single grain, very friable

BCcj	40-110	Yellowish brown stratified sand containing some gravel; weakly cemented, thin layers.
C	110-150	Light brownish gray sand, single grain, loose.

*****

Table 3. Analytical data for St. Stephen soil[a]

Horizon	Depth cm	Sand %	Clay %	C %	pH	CEC me/100g	Base Sat. %	ΔCEC me/100g	Fe_d	Fe_p	Al_p
FH	15-0	-	-	48	3.4	-	-	-	.5	.1	.2
Ae	0-4	8	28	6.4	3.4	7.4	46	11	.2	.1	.1
AB	4-8	14	26	3.7	3.8	7.9	49	11	.5	.3	.1
Bhf	8-22	37	27	8.7	4.0	12.9	56	36	1.8	1.5	.6
Bhfc	22-23	-	-	10	-	-	-	-	13.4	9.9	.8
Bfcj	23-35	89	5	3.7	4.1	5.4	46	23	1.5	1.0	.9
Bf	35-50	92	3	2.4	4.3	2.5	56	16	.6	.3	.9
C	50-80	90	3	.3	4.6	1.2	100	3	.4	.05	.2

[a]Analytical methods were the same as those listed for Table 1, except that the CEC determination involved a 5-day equilibration period. This results in somewhat higher CEC and base saturation values than method 3.31.

The analytical data (Table 4) show typical features of sandy Humo-Ferric Podzols, with the exception of the high values for base saturation of surface horizons. This is probably due to minor deposition of dust from a road near to the site.

Uplands Series (McKeague, 1965)

This series is included to represent soils having close to the minimum degree of development of a podzolic B horizon. Such soils occur commonly in association with soils of other orders in many areas of Canada. The site at 45°25'N 75°35'W is at the crest of a low ridge in an area of low-lying alluvial sand at an elevation of 76 m. Vegetation is a mixed forest including pine, maple and cedar with an undergrowth of ferns and grasses. Approximate mean annual temperature and precipitation at the site are 6°C and 85 cm respectively. The soil is classified as an Orthic Humo-Ferric Podzol but it is not a Spodosol in the U.S. taxonomy.

Horizon	Depth cm	
LF	5-0	Mainly undecomposed pine needles
Ae	0-5	Pinkish gray sand, single grain, very friable
Bf	5-20	Strong brown sand, single grain except for a few hard nodules
Bm	20-43	Strong brown sand, single grain except for a few hard nodules
C	43-125	Grayish brown sand, single grain

The data (Table 5) shows the relatively low amount of pyrophosphate-extractable Fe and Al in the podzolic B horizon and the low proportion of pyrophosphate- to dithionite-extractable Fe. However, the pedon meets both the morphological and chemical limits of a Podzolic soil.

Nevers Road

This soil is included as an example of Humic Podzols few of which have been analyzed in Canada. The site at 46° 40'N 64°50'W near the coast of Kent Co. N.B. was on poorly drained outwash overlying sandstone bedrock. The vegetation is mixed forest and mosses. Mean annual temperature and precipitation are approximately 5°C and 105 cm respectively.

Horizon	Depth cm	
LH	11-0	Needles and moss overlying well decomposed organic matter
Ae	0-8	Dark reddish gray sandy loam, amorphous, friable
Bhcj	8-23	Black to dark reddish brown sand, weakly cemented
Bh	23-30	Dark brown to brown sand, amorphous, compact

BCg	30-50	Brown sand, mottled, amorphous
Cg	50-80	Olive gray sand, single grain, loose
Cg	80-90	Light brownish gray sandy loam, amorphous

*****

Table 4. Analytical data for Morin soil[a]

Horizon	Depth cm	Sand %	Clay %	C %	pH	CEC me/100g	Base Sat. %	ACEC me/100g	Fe_d %	Fe_p %	Al_p %
LH	2-0	-	-	24	4.1	23	99	-	0.5	0.3	0.2
Ae	0-10	88	2	0.7	4.4	1.9	89	2	0.1	0.03	0.04
Bf_1	10-20	89	4	4.0	4.2	3.2	62	18	1.2	0.6	1.0
Bf_2	20-40	92	3	1.2	4.7	2.4	94	6	0.6	0.2	0.5
BCcj	40-110	96	2	0.2	5.2	1.9	100	2	0.3	0.05	0.2
C	110-150	98	1	0.1	5.5	1.4	100	-	0.3	0.04	0.1

[a]Analytical methods were the same as those indicated for Table 3.

Table 5. Analytical data for Uplands soil[a]

Horizon	Depth cm	BD	Sand %	Clay %	C %	pH	CEC me/100g	Base Sat. %	ΔCEC me/100g	Fe_d %	Fe_p %	Al_p %
LF	5–0	-	-	-	37	43	-	-	-	-	-	-
Ae	0–5	1.2	90	3	0.8	3.8	1.5	40	2	0.2	0.1	0.03
Bf	5–20	1.0	88	6	1.8	5.0	2.0	90	11	1.0	0.2	0.3
Bm	20–43	1.3	-	-	0.8	5.3	0.8	100	5	-	-	-
C	43–125	1.5	96	2	0.1	5.2	1.0	100	1	0.2	0.05	0.1

[a]Analytical methods were the same as those indicated in Table 3.

*****

The data (Table 6) show very low Fe contents, and accumulation of organic C and Al in the B horizons. The low Fe values are thought to be related in part to a high water table and the associated reducing conditions.

Table 6. Analytical data for Nevers Road soil[a]

Horizon	Depth cm	Sand %	Clay %	C %	pH	CEC me/100g	Base Sat. %	ΔCEC me/100g	Fe_d %	Fe_p %	Al_p %
Ae	0-8	73	7	1.0	3.7	5.2	31	8	0.04	0.01	0.1
Bhcj	8-23	91	5	4.6	3.9	7.0	28	25	0.14	0.01	0.6
Bh	23-30	94	2	1.6	4.2	3.4	35	12	0.16	0.02	0.6
BCg	30-50	-	-	0.6	4.4	2.0	30	6	0.23	0.02	0.3
Cg	50-80	90	3	0.3	4.6	1.4	50	5	0.27	0.02	0.2
Cg	80-90	-	-	-	4.6	2.7	70	3	0.45	0.03	0.1

[a]Analytical methods were the same as those indicated in Table 3.

*****

Tignish series (McKeague, *et al.*, 1973)

This series has features of many soils developed in compact sandy loam to loam textured deposits in the Maritime provinces. It is close to the arbitrary boundary line between Podzolic and Luvisolic soils. The site is at 46° 11'42"N 62°56'35"W on Prince Edward Island on a 1% slope supporting a forest of birch and maple. The parent material is compact glacial till derived from red Permo-Carboniferous sandstones. The soil is classified as a Podzolic Gray Luvisol because it has a podzolic B horizon underlain by a Bt horizon at a depth of less than 50 cm.

Horizon	Depth cm	
Ah	0-10	Dark reddish brown sandy loam, very friable
Ae	10-28	Light brown sandy loam, friable
Bf	28-38	Reddish brown loam, weak granular, very friable
Bt	38-64	Reddish brown loam, weak blocky, friable, thin clay skins on some surfaces
BC	64-120	Reddish brown loam, amorphous, firm, dense
C	120-200	Reddish brown loam, amorphous, firm, dense

The analytical data (Table 7) indicate eluviation of clay, Fe and Al from the Ae and the development of a podzolic B horizon. The underlying Bt horizon is very weakly expressed but both it and the BC horizon contain oriented clay skins as seen in thin section.

Pine (Stonehouse, *et al.*, 1971, Site 6a)

This series represents soils having the color profile of Podzolic soils but lacking the B horizon development required. Such soils are common in sandy deposits in subhumid areas. The site is at 54°48'N 105°15'W, north of Prince Albert, Saskatchewan in a gently undulating outwash plain of the Lac La Ronge Lowland. Vegetation is a stunted forest of jackpine and white spruce with an understory of shrubs and mosses. The approximate mean annual temperature and precipitation are -1°C and 42 cm respectively.

Horizon	Depth cm	
Ae	0-13	Light brownish gray sand, single grain, abrupt boundary
Bm	13-30	Strong brown sand, single grain, loose, gradual boundary
BC	30-50	Yellowish brown sand, single grain, gradual boundary
C	50-90	Light yellowish brown sand, single grain, loose

The analytical data (Table 8) show that, both in C content and in pyrophosphate Fe + Al content, the B horizon falls far short of the minimum limits of a podzolic B horizon. They show, also, a marked accumulation of dithionite-Fe in the Bm horizon relative to that in either the Ae or the C horizon.

The eight soils described illustrate some of the general relationships between soil development and aspects of the soil environment. For example, the Ferro-Humic Podzols occur in more humid environments than the Humo-Ferric Podzols, and the soil that does not have a podzolic B horizon, Pine, occurs at a relatively dry site. However, it must be pointed out that several classes of soil occur within a radius of a few km of all of the sites discussed. Such variability is usually attributable to differences in either local relief or parent material.

*****

Table 7. Analytical data for Tignish soil[a]

Horizon	Depth cm	Sand %	Clay %	C %	pH	CEC me/100g	Base Sat. %	Fe_d %	Fe_p %	Al_p %
Ahe	0-10	-	-	4.0	4.0	5.1	65	1.0	0.4	0.1
Ae	10-28	48	9	0.5	4.0	3.9	36	0.6	0.1	0.1
Bf	28-38	42	16	1.2	4.3	2.4	42	1.9	0.8	0.5
Bt	38-64	41	17	0.1	4.2	2.4	40	1.6	0.1	0.2
BC	64-120	45	14	0.04	4.2	3.0	44	1.5	-	-
C	120-200	45	16	-	4.0	-	-	1.4	-	-

[a]Analytical methods were the same as those indicated in Table 3.

Table 8. Analytical data for Pine soil[a]

Horizon	Depth cm	Sand %	Clay %	C %	pH	CEC me/100g	Fe_d %	Fe_p %	Al_p %
Ae	0-13	95	1.5	0.16	4.6	1.2	0.04	-	-
Bm	13-30	96	2.2	0.15	5.2	2.4	0.21	0.03	0.04
BC	30-50	98	0.7	0.03	5.9	0.9	0.09	0.01	0.02
C	50-90	98	0.6	-	5.7	0.8	0.04	-	-

[a]The analytical data are from Stonehouse *et al.*, (1971). CEC was measured by ammonium acetate, and pH was measured in water.

*****

WEATHERING OF MINERALS IN PODZOLIC SOILS

Clay Mineralogy

Podzolic soils in Canada, as well as elsewhere, show a characteristic distribution of clay minerals in their profiles (Table 9). Commonly, a mixture of mica and chlorite with minor amounts of kaolinite, quartz, feldspars and sometimes amphiboles forms the assemblage of minerals in

Table 9. Layer silicates[c] in the clay fraction of representative Podzolic soils in Canada

Site	Horizon	Smect.	Verm.	Chlor.	Mica	Kaol.	Chlor-Verm.	Mica-Verm.	Mica Smect.
Uplands (0.2-2µm)	Ae	4	-	-	tr.	1	-	-	1[a]
	Bfh	-	1	1	1	1	2	-	-
	C	-	1	2	2	1	1[a]	1[a]	-
Rubicon (0.2-2µm)	Ae	4	-	-	tr.	1	-	-	1[a]
	Bhf	-	2	1	1	1	2	-	-
	Cg_1	-	1	2	2	1	1[a]	1[a]	-
Morin (<2µm)	Ae	4	-	-	-	1	-	-	-
	Bfh[b]	-	-	-	tr.	-	2	-	-
	C	-	2	1	1	1	-	-	-
Laurentide (<2µm)	Ae[b]	2	3	-	-	1	-	-	-
	C	-	-	-	tr.	-	1	-	-
Richibucto (<2µm)	Ae	2	-	-	3	1	-	-	-
	C	-	-	2	2	1	-	-	-
St. Stephen (<2µm)	Ae	2	2	-	1	tr.	-	-	-
	Bhf	-	1	1	2	1	2	-	-
	C	-	-	2	3	1	-	-	-

[a]Regularly interstratified, [b]samples almost totally amorphous to X-rays, [c]mineral abundance: tr < 10%; 1, 0-20%; 2, 20-40%; 3, 40-60%; 4, 60-80%; 5, 80-100%.

the <2μm fraction of the C horizon. Chloritized vermiculite is generally dominant in the B horizon but some mica, orthochlorite and kaolinite may also be present. Smectite, beidellitic in nature and interstratified with some mica layers, is the dominant layer silicate in the Ae horizon. Clay-size quartz and feldspar persist throughout the profile, but clay-size amphiboles are generally absent in the A and B horizons (Franzmeier *et al.*, 1963; Brydon *et al.*, 1968).

The X-ray diffraction patterns of the clay fraction of the Uplands series (McKeague, 1965) discussed previously show the characteristic mineralogy (Figure 1) although the soil has only the minimal expression of podzolic B horizon development. The patterns are interpreted as follows:

<u>C horizon clay</u> - chlorite (14A) and mica (10A) are the dominant layer silicates; kaolinite (contributes to 7A peak), regularly interstratified mica-vermiculite (12A and 24A in glycerated sample), and regularly interstratified chlorite-vermiculite (12A and 24A in heated sample) are minor constituents.

<u>B horizon clay</u> - chloritized vermiculite (14A in glycerated sample shifting to 11A after heating) is the dominant layer silicate; discrete chlorite (14A in heated sample), mica (10A in glycerated sample) and kaolinite (7A in glycerated sample) are minor constituents.

<u>Ae horizon clay</u> - smectite (18A in glycerated sample shifting to 10A after heating) is the dominant layer silicate; interstratified mica-smectite (small peak at 20A in heated sample) and kaolinite (7A) are also present. The broadening of the reflection near 10A in the heated sample indicates a second order reflection of the mica-smectite component (10A) and a reflection of collapsed smectite (9.5A).

Clay-size quartz (4.24A and 3.34A) and feldspar (3.25A and 3.19A) are present in the A, B and C horizons.

The characteristic Podzolic Ae horizon clay mineralogy, expressed by the absence of chlorite and dominance of smectite or sometimes vermiculite (DeKimpe, 1970), holds even in soils having parent materials deficient in layer silicates such as mica. This was demonstrated by Coen and Arnold (1972) for some Spodosols in New York. They attributed the presence of smectite in the Ae horizons to deposition of stratospheric dust. However, they recognized the possibility that the smectite might have weathered from the traces of mica and chlorite in the sand and silt fractions of the parent materials. Synthesis of smectite from solution is not a plausible explanation as the chemical environment of a Podzolic Ae horizon is one in which layer silicates, other than kaolinite, are unstable (Coen and Arnold, 1972; Kittrick, 1973).

The most widely accepted hypothesis of the characteristic clay mineral distribution in Podzolic soils of

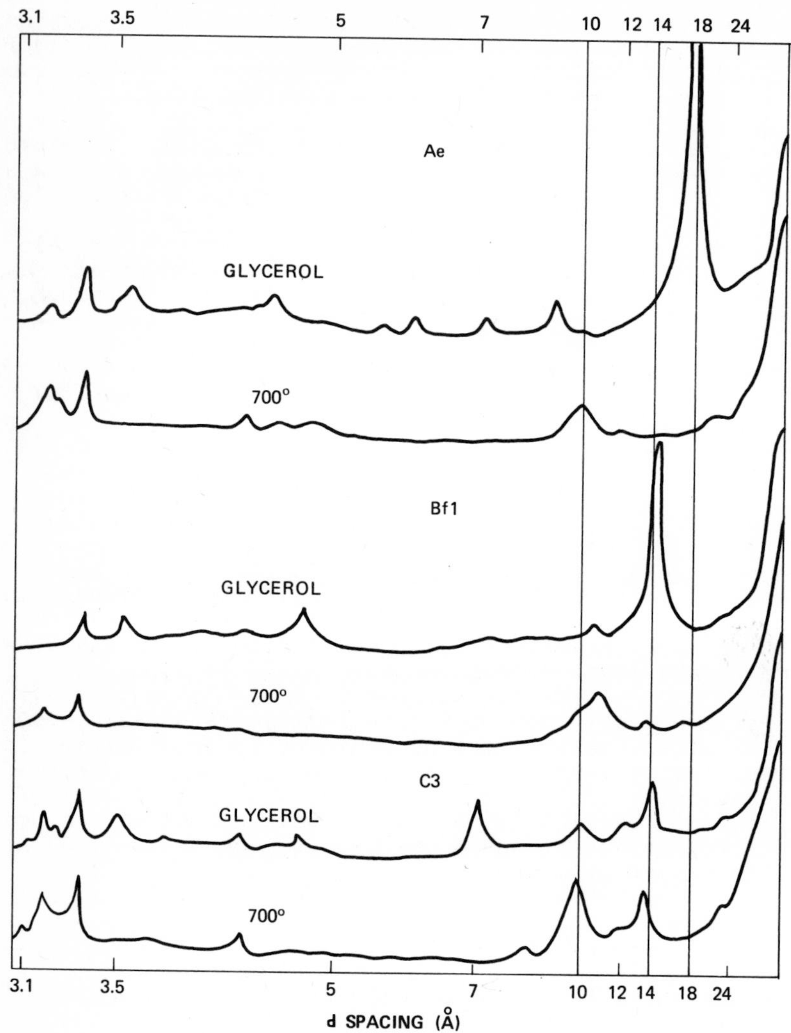

Figure 1 X-ray diffraction patterns of 0.2-2.0 μm clay from three horizons of Uplands soil.

temperate regions is the weathering of micas and chlorites in the C horizon to vermiculite that becomes interlayered with hydrous aluminum in the B horizon. These interlayers are removed in the Ae horizon and the vermiculite expands further to form smectite interstratified with some remaining mica layers. This hypothesis is consistent both with the observation that the distribution of clay minerals in Podzolic soils is relatively independent of the type of parent rock (Sokolova *et al.*, 1971), and with the data for most Podzolic soils. If weathering of sand and silt sized layer silicates is considered, the hypothesis is plausible even for soils, such as those studied by Coen and Arnold (1972) and Laurentide (Anon., 1973) having very low amounts of layer silicate minerals in the clay fraction.

The lack of evidence of kaolinite formation in the Ae horizons of Podzolic soils in temperate regions is commonly attributed to the geological youth of the soils and to the retarding influence on weathering of a cool climate (Sokolova *et al.*, 1971). However, it is surprising that thermodynamically unstable smectite forms and persists in such an environment (Coen and Arnold, 1972). The close association of relatively large amounts of organic matter with the Ae horizon clay (McKeague, 1965; Ross, 1971) may protect the smectite from further alteration.

Silt and Sand Mineralogy

Relatively little information is available on the layer silicate mineralogy of sand and silt fractions of Podzolic soils in Canada. Chlorite and micas occur commonly in these fractions and vermiculite, apparently trioctahedral, occurs in some (McKeague, 1965). Studies of heavy minerals in the sand fractions of Podzolic soils show some evidence of depletion of relatively unstable minerals such as chlorite and pyroxene, and increase of stable minerals such as tourmaline in the Ae horizon with respect to the C (McKeague and Brydon, 1970). Quantitative relationships were obscured in this study by the dominance of microscopically unidentifiable opaque and highly weathered grains. Clear evidence of weathering of sand grains was shown by quartz: feldspar ratios as much as 10 times higher in the Ae than in the associated C horizon of 2 Podzolic soils.

Experimental Weathering Studies

Mineral and organic acids, salt solutions and water have been used in laboratory studies of mineral weathering. Fulvic acids, an alkali and acid soluble, naturally occuring humic substance has been used in some recent studies.

Fulvic acid attack of layer silicates. Fulvic acid (2g/l, pH 2.5) at room temperature attacked chlorites ($<50\mu m$) and micas (1 to $2\mu m$) congruently (Kodama and Schnitzer, 1972; Schnitzer and Kodama, in press). Despite their larger particle size, the chlorites dissolved faster than the micas. The high-Fe species of both minerals dissolved faster than the high Mg species, presumably due to the high affinity of fulvic acid for Fe.

The proportions of the total Mg, Al and Fe dissolved in 2 weeks were 25% for the high-Fe chlorite (thuringite) and 5% for the high-Mg chlorite (leuchtenbergite). Extrapolation of the data indicates complete dissolution of thuringite in 6 months, and of leuchtenbergite in 2 years. For micas, dissolution of total cations in 1 month was 17% for the high-Fe mica (biotite), 9% for the high-Mg mica (phlogopite) and about 2% for muscovite. Extrapolation indicates complete dissolution of biotite and phlogopite in about 1.5 years, and of muscovite in about 5 years.

Sawhney and Voight (1969) also observed congruent structural decomposition of biotite and vermiculite by organic acids. In contrast to the congruent attack by organic acids, the reaction of these minerals with salt solutions resulted in preferential K release and expansion of mica interlayers (Rausell-Colom *et al.*, 1965).

The apparent inconsistency between laboratory studies that show rapid congruent decomposition of layer silicates in organic acid solutions and the persistence of such minerals in the A and B horizons of Podzolic soils may be rationalized, as may the apparent alterations in soils of chlorite and mica to vermiculite.

1. The concentration of organic acids in most soil solutions is probably less than 10% of the 2g/l used in the laboratory. The more dilute natural solution, that includes some cations, may react as a dilute salt solution and effect the hydration of micas and chlorites.

2. Removal of K from soil solution by both leaching and uptake by plants may favor preferential K release from micas and the formation of hydrated layer silicates. In the laboratory experiments, the dissolved K remained in solution.

3. Sesquioxide coatings formed on mineral grains in oxidized Podzolic soils may protect the silicate framework from acid attack (Brydon and Ross, 1966). Such coatings would not develop in these laboratory systems.

Laboratory weathering of chlorites. Treatment of orthochlorites with strong inorganic acids results in congruent dissolution of the minerals (Caillère *et al.*, 1954; Ross, 1969). Recently, however, laboratory experiments have shown that orthochlorites containing sufficient amounts of Fe^{2+} may be altered to vermiculite or smectite products (Ross, 1975; Ross and Kodama, in press). The key to such alteration is strong oxidizing conditions, such as those effected by bromine. The alteration is accompanied by loss of cations in the octahedral sheet in the 2:1 layer, which results in a shift toward a dioctahedral structure of the vermiculite or smectite formed. The latter experimental evidence is consistent with the finding that chlorites weather to vermiculites in some soils (Coffman and Fanning, 1975) and probably in Podzolic soils (Figure 1). Formerly, the absence of trioctahedral vermiculite or smectite in the Ae horizon clay of Podzolic soils was interpreted as indicating that chlorite was destroyed completely. Loss of octahedral cations also occurs during oxidative weathering and expansion of biotite (Farmer *et al.*, 1971; Gilkes *et al.*, 1972). It is clear, therefore, that the absence of trioctahedral vermiculites or smectites can not be accepted as evidence that chlorite, biotite and their expanded products are lost during Podzolic weathering.

CONCEPTS OF GENESIS OF PODZOLIC SOILS

Current concepts of genesis of Podzolic soils are summarized briefly from information in the literature. The framework for this discussion is based upon the classical factors of soil formation: climate, organisms, relief, parent material and time, and their interactions. Some of the key references used are: Stobbe and Wright (1959), Duchaufour (1972), Franzmeier and Whiteside (1963), Bloomfield (1970), Conry *et al.*, (1972), Petersen (1976) and references cited in those publications.

Parent Material

A parent material of coarse texture and moderately high hydraulic conductivity is a favorable medium for the development of Podzolic soils (Lag, 1970). Gravelly or sandy, quartziferous material containing a low percentage of clay and some weatherable Fe- and Al- containing minerals is ideal. A low cation exchange capacity but a sufficient supply of nutrient elements to maintain vigorous plant growth is favorable. The degree of expression of the Ae horizon is presumably related to the rate of release of Fe and Al from weathering minerals as well as to leaching intensity and stability of the soil. Soil materials having an abundant supply of minerals that release Fe and Al at a moderate rate may not have well-expressed Ae horizons, as soluble organic matter may precipitate with Al and Fe close to the surface of the soil. The Whonnock soil discussed herein may be an example.

Climate

In general, Podzolic soil development is favored by a cool, humid climate such that plant remains accumulate at the soil surface and water moves downward in the soil throughout most of the year. Nearly all of the area of Podzolic soils in Canada has this general climate although the range of precipitation and temperature is appreciable and it is reflected to a degree by B horizon development. Increases in effective precipitation and temperature should result in intensified weathering and profile development. However, thermodynamically unstable smectite forms and persists in the Ae horizon of temperate zone Podzolic soils and kaolinite is not known to form as a product of weathering in these soils. In tropical Podzolic soils, on the other hand, kaolinite is commonly the dominant layer silicate (Franco, 1968) and it may be partially depleted in the upper horizons (Andriesse, 1969).

Organisms

Coniferous forest or heath vegetation is commonly thought to favor Podzolic soil development, but well developed Podzolic soils occur also under deciduous forest. De Kimpe and Martel (1976) found more organic-complexed Al and Fe in the upper B horizons of Podzolic soils developed under conifers than in those of similar soils under deciduous vegetation. According to Lag (1971) the undergrowth is more important than the trees in promoting podzol development in Norway. Laboratory studies with drippings from the forest canopy, leaf leachates and extracts indicate that many kinds of vegetation could yield decomposition products effective in podzolisation. The key factor seems to be the persistence of organic litter that decomposes gradually yielding soluble organic acids that dissolve, chelate and transport ions, particularly Fe and Al, from weathering minerals and from secondary sesquioxidic coatings that tend to protect the mineral grains from attack by acids. In addition, growing vegetation depletes the soil solutions of nutrient elements such as K and hence accelerates the release of K from mica and the mica vermiculite smectite transformation. Changes in vegetation as a result of climatic changes, fires and other causes complicate the deduction of relationships between vegetation and Podzolic soil development.

Microorganisms, particularly bacteria and fungi, undoubtedly influence soil development through their role in decomposing organic matter and releasing soluble products that interact with minerals in the soil. Fungi have the dominant role after bases are depleted and acid forest litter accumulates. The soluble organic products probably include simple organic acids as well as the complex polyelectrolyte, fulvic acid (Schnitzer and Khan, 1972) that resists microbial breakdown (Mathur, 1969). Soil fauna, as well as microflora, participate in the breakdown of organic litter (Burges, 1965; Marshall, 1974) and they may mix soil horizons (Langmaid, 1964) and influence the microstructure of podzolic B horizons (Conry *et al.*, 1972).

Relief

Relief affects soil development through its influence on microclimate and vegetation. In sub-humid areas, for example, the relatively wet, depressional sites are the most favorable for Podzolic soil development. Steep slopes subject to slides are unfavorable sites because the surficial material may not be stable for a sufficient time. Protected areas such as those beneath a fallen log or in a microdepression are commonly the sites of maximum expression of Podzolic features. Relatively minor changes in relief can result in the development of other major classes of soil in areas dominated by Podzolic soils. An example is the Uplands catena (McKeague, 1965) in which the sequence downslope is Orthic Humo-Ferric Podzol, Gleyed Ferro-Humic Podzol, Fera Gleysol. The Gleysol has neither a podzolic B horizon nor the typical clay mineralogy profile of Podzolic soils.

Time

The time required for the development of a Podzolic soil depends upon interactions of the factors discussed above. Estimates ranging from a few hundred to a few thousand years have been made (Franzmeier *et al.*, 1963, Soil Survey Staff, 1976). Some of the estimates were made before Podzolic soils were defined rigorously. It seems unlikely that most soils that qualify now as Podzolic formed in less than a few thousand years. Studies of a chronosequence of sandy Podzolic soils in Michigan showed essentially the same typical distribution of clay mineralogy with depth in soils ranging in age from 2000 to 10,000 years (Franzmeier *et al.*, 1963). The rapid development of the apparent weathering pattern discussed previously was supported by results of Gjems (1960) who found smectite to be the dominant clay mineral in the Ae horizon of Podzols in Fennoscandia that were only 300 years old. The smectite appeared to have weathered from biotite.

Interaction of Factors

Although the individual factors discussed above are important, it is the interaction of these factors that is basic to the genesis of Podzolic soils. A conditioning process (Stobbe and Wright, 1959) for the development of Podzolic soils is the depletion of bases by leaching and the development of acidity. This not only accelerates the weathering of minerals but also results in decreased bacterial activity and the accumulation at the surface of acid organic litter. This material is gradually decomposed by fungi and other microbes, releasing organic acids that move downward in the soil solution. Loosening of the surface soil material by frost action, root growth, soil fauna and tree throw is an early process in soil development. In some parent materials, another conditioning process is the translocation of clay.

A generalized view of the development of a Podzolic soil in "conditioned" material follows. Organic matter in solution is leached downward from the surface litter. It reacts with Al and Fe in the mineral soil to form complexes; reduction of ferric iron is not necessarily involved. These Al, Fe-organic complexes continue to move with the descending soil solution and to react with Al and Fe until they reach a critical metal ion content at which they become insoluble. The insoluble complexes deposited on grains or as aggregates in intergranular spaces may react with further Al and Fe ions released by weathering processes. As downward movement of water bearing soluble organic matter continues, the upper part of the mineral soil may be depleted of weatherable minerals and become a light gray colored Ae horizon. If, however, the parent material is rich in weatherable ferromagnesium alumino-silicates, the release of Al and Fe in the uppermost mineral soil may proceed rapidly enough to react with all of the soluble organic matter. In such cases a Podzolic soil without a light colored Ae horizon may develop, such as the Whonnock soil described previously.

In addition to this major process of leaching and complexing, other processes are involved in the development of Podzolic soils (Simonson, 1959; and this volume). The upper sola of some soils is sufficiently acid that Al^{3+} ions unassociated with organic matter could move downward. Reduction of ferric iron promotes Fe movement in some soils but reducing conditions never occur in some well drained Podzolic soils. The possible movement in soils of iron oxide has never been completely discounted. However, complexing of Al and Fe with organic matter seems to be a major process involved in Podzolic soil development.

Working in opposition to the leaching process are processes that return subsurface material to the soil surface. Plants take up soluble substances and deposit them on the surface as litter. Soil fauna may be active in mixing soil material and "tree throw" (Muller and Cline, 1959) may periodically result in mixing of the uppermost half meter of soil.

CALCULATIONS OF HYPOTHETICAL TRANSLOCATION OF ORGANIC MATTER

Al and Fe in a Podzolic Soil

Van Schuylenborgh and Bruggenwert (1965) calculated the capacity of a 10^{-4} M solution of p-hydroxybenzoic acid to translocate Fe and Al in a Podzolic soil. A somewhat similar approach is applied to a consideration of the Laurentide soil discussed herein but a crude estimate of the concentration of organic matter (OM) in solution is made from data in the literature (Table 10). The following estimates and assumptions are made:

1. The average concentration of OM in solution entering the A horizon is 0.1 g/l.

2. The OM in solution has the properties of fulvic acid (FA) as described by Gamble and Schnitzer (1973), with about 8 millimoles (mmol) of carboxyl and 3 mmol of phenolic hydroxyl groups per gram of dry FA.

3. The FA precipitates when about 50% of the carboxyl sites are occupied by metal ions; thus 4 mmol/g. Some of these metal ions may be chelated by adjacent carboxyl and phenolic hydroxyl groups.

4. About 1/4 of these sites are occupied by Al and Fe from the organic litter. Thus about 3 mmol/g of the 4 remain to combine with Al and Fe before precipitation occurs.

5. Three mmol of carboxyl sites complexes an average of 2 mmol of Al or Fe. The extent of hydroxylation and polymerization of the Fe and Al species complexed is not known.

6. None of the FA is lost from the profile with the drainage water.

Table 10. Organic matter reported in solution[a]

Material	Organic matter g/l	Reference
leaf leachates	0.06 to 0.3	Schnitzer, 1952; Schnitzer and DeLong, 1954
leaf extracts	0.05 to 0.06	Malcolm and McCracken, 1968
soil solution, below Ae	1.23	Schnitzer & Desjardins, 1969
soil solution at 10 cm	0.04 to 0.12	Holzhey *et al.*, 1975.
Water extracts of A_0, aerated	0.2 to 1.4	Petersen, 1976
water extracts of A_1, aerated	0.05 to 0.1	Petersen, 1976
soil solution from A_0	0.09 to 1	Petersen, 1976

[a]Additional observations on the composition of leachates from Podzolic soils in Newfoundland were obtained courtesy of A.W.H. Damman and P.K. Heringa.

7. Fifty cm^3 of water per cm^2 of cross section has passed through the soil per year throughout 1×10^4 years.

8. The soil has not been subjected to any major physical disruption since the glaciers receded.

Many of these assumptions are questionable but they provide a tentative basis for considering the changes that might have occurred in a 1 cm^2 column of the Laurentide soil.

1. OM - According to our assumptions, about 50 g was added from solution to the 1 cm^2 column in 10^4 yr. OM found in the 1 cm^2 column is about 4 g (Table 1) In addition to the OM added from solution, a large amount must have been added by roots. Clearly most of the added OM decomposed during the period of soil formation. The mean residence time of OM in one podzolic B horizon was found to be about 600 years and values for organic matter in other soil horizons range from a few hundred to a few thousand years (Schnitzer & Khan, 1973).

2. Translocation of Fe and Al by FA - The 50 g of FA would have a capacity to complex 100 mmol of Fe and Al before precipitating. After precipitating, it would have a capacity to complex an additional 130 mmol (4/3 x 100) of Fe and Al. The apparent loss of Fe and Al from the Ae horizon based upon total Fe and Al values, thickness and bulk density for the Ae and C horizons (Table 1), and not taking into account losses of material from the Ae is about 0.9 mmol of Fe and 2.2 mmol of Al for a total of 3.1 mmol. Corrections for losses of material from the Ae would not double this value. The apparent gains of Fe and Al in the B horizon, not corrected for OM addition, over that in the C are about 2.7 mmol of Fe and 2 mmol of Al for a total of 4.7 mmol. The discrepancy between the apparent losses from the Ae and gains in the B might be due partly to lack of homogeneity of the parent material with depth.

Based upon the amounts extractable by pyrophosphate, the organic-complexed Fe and Al in the B horizon account for about 7.5 mmol of Fe and 18 mmol of Al for a total of 25.5 mmol. Much of this Al and Fe may have come from *in situ* weathering of minerals and combination of FA with the Al and Fe weathering products. Some of it was probably cycled through vegetation and redeposited.

3. Relationship between OM and complexed Al and Fe - The 3.5g of OM in the 1 cm^2 column of B horizon should have, according to the assumptions, a total complexing capacity of 19 mmol. The measured amount of pyrophosphate-extractable, presumably organic-complexed, Al and Fe in the B horizon totalled 25.5 mmol. The approximate agreement of predicted and measured values may be fortuitous as the predicted value is based upon assumption 5 which may be in error. Similar comparisons made on a somewhat different basis for other podzolic B horizons commonly indicated excesses of apparently complexed Fe and Al over the predicted values (McKeague, 1968).

Information Required

The above oversimplified calculations point out the tremendous gap between a conceptual model and a semi-quantitative model of the genesis of a Podzolic soil. They also focus attention on some of the gaps in information essential to the development of a more quantitative model:

1. Information on the quantity and chemical composition of soil leachates in Podzolic soils: a) dripping from the forest canopy, b) immediately below the leaf litter, c) between the Ae and the B horizons, d) below the B horizon, and e) passing through the soil. The estimates made herein of leachate composition were extremely crude.

2. Information on the nature and rate of interaction of dilute organic solutions with soil minerals.

3. Information on the range of compositions and

properties of naturally occurring organic- Al, Fe deposits.

4. Information on the rate of decomposition of organic matter in Podzolic B horizons and the factors that influence this.

5. Quantitative information on the gains and losses of constituents in the horizons of Podzolic soils developed in relatively uniform parent materials (Evans and Adams, 1975).

REFERENCES CITED

Andriesse, J.P., 1969, A study of the environment and characteristics of tropical Podzols in Sarawak (East Malaysia): Geoderma, v. 2, p. 201-227.

Anon., 1973, A visit to Montmorency Forest, Tour guidebook: 4th North Amer. For. Soils Conf., Quebec, 37 p.

Bloomfield, C., 1970, The mechanism of podzolisation: Welsh Soils Discussion Group, Rep. No. 11, p. 112-121.

Brydon, J.E. and Ross, G.J., 1966, Stability of chlorite in dilute acid solutions: Soil Sci. Soc. America Proc., v. 30, p. 740-744.

Brydon, J.E., Kodama, H., and Ross, G.J., 1968, Mineralogy and weathering of the clays in Orthic Podzols and other Podzolic soils in Canada: Trans. 9th Internatl. Congr. of Soil Sci., v. 3, p. 41-51.

Burges, N.A., 1965, Biological processes in the decomposition of organic matter; *in* Hallsworth, E.G., and Crawford, D.V., eds., Experimental Pedology: Butterworths, London, p. 189-198.

Caillère, S., Henin, S., and Esquevin, J., 1954, Transformation experimentale de la chlorite en montmorillonite: Clay Min. Bull., v. 2, p. 166-170.

Coen, G.M. and Arnold, R.W., 1972, Clay mineral genesis of some New York Spodosols: Soil Sci. Soc. America Proc., v. 36, p. 342-350.

Coffman, C.C. and Fanning, D.S., 1975, Maryland soils developed in residuum from chloritic metabasalt having high amounts of vermiculite in sand and silt fractions: Soil Sci. Soc. America Proc., v. 39, p. 723-732.

Conry, M.J., DeConinck, F., Bouma, J., Cammaerta, C. and Diamond, J.J., 1972, Some Brown Podzolic soils in the west and south-west of Ireland: Proc. Roy. Irish Acad., Dublin, v. 72, Sec. B. No. 21, p. 359-402.

DeKimpe, C.R., 1970, Chemical, physical and mineralogical properties of a Podzol soil with fragipan derived from glacial till in the province of Quebec: Can. Jour. Soil Sci., v. 50, p. 317-330.

DeKimpe, C.R. and Martel, Y.A., 1976, Effect of vegetation on the distribution of carbon, iron, and aluminum in the B horizons of northern Appalachian Spodosols: Soil Sci. Soc. America Proc., v. 40, p. 77-80.

Duchaufour, Ph., 1972, Processus de Formation des Sols: Centre de Pedologie, Universite de Nancy, Nancy, France, 182 p.

Evans, L.J. and Adams, W.A., 1975, Quantitative pedological studies on soils derived from Silurian mudstones. V. Redistribution and loss of mobilized constituents: Jour. Soil Sci., v. 26, p. 327-335.

Farmer, V.C., Russell, J.D., McHardy, W.J., Newman, A.C.D., Ahlrichs, J.L. and Rimsaite, J.Y.H., 1971, Evidence for loss of protons and octahedral iron from oxidized biotites and vermiculites: Mineralogical Mag., v. 38, p. 122-137.

Franco, E.P.C., 1968, Some cases of podzolization under tropical conditions in Angola: Trans. 9th Internatl. Congr. of Soil Sci., v. 3, p. 265-273.

Franzmeier, D.P. and Whiteside, E.P., 1963, A chronosequence of Podzols in northern Michigan. I Ecology and description of pedons, and II Physical and chemical properties: Mich. State Univ. Agr. Exp. Sta. Quarterly Bull., v. 46, p. 2-20 and 21-36.

Franzmeier, D.P., Whiteside, E.P. and Mortland, M.M., 1963, A chronosequence of Podsols in northern Michigan. III. Mineralogy, micro-morphology, and net changes occurring during soil formation: Mich. State Univ. Agr. Exp. Sta. Quarterly Bull., v. 46, p. 37-57.

Gamble, D.S. and Schnitzer, M., 1973, The chemistry of fulvic acid and its reactions with metal ions: *in* Singer, P.C., ed., Trace Metals and Metal-Organic Interactions in Natural Waters: Ann Arbor Science, P.O. Box 1425, Ann Arbor Michigan, p. 265-302.

Gilkes, R.J., Young, R.C., and Quirk, J.P., 1972, The oxidation of octahedral iron in biotite: Clays and Clay Min., v. 20, p. 303-315.

Gjems, O., 1960, Some notes on clay minerals in Podzol profiles in Fennoscandia: Clay Min. Bull., v. 4, p. 208-211.

Holzhey, C.S., Daniels, R.B. and Gambe, E.E., 1975, Thick Bh horizons in the North Carolina Coastal Plain, II Physical and chemical properties and rates of organic addition from surface sources: Soil Sci. Soc. America

Proc., v. 39, p. 1182-1187.

Kittrick, J.A., 1973, Mica-Derived vermiculites as unstable intermediates: Clays and Clay Min., v. 21, 479-488.

Kodama, H. and Schnitzer, M., 1972, Dissolution of chlorite minerals by fulvic acid: Can. Jour. Soil Sci., v. 53, p. 240-243.

Lag, J., 1970, Podzol soils with an exceptionally thick bleached horizon: Acta Agric. Scand, v. 20, p. 58-60.

_______, 1971, Some relationships between soil conditions and distribution of different forest vegetation: Acta Agralia Fennica, v. 123, p. 118-125.

Langmaid, K.K., 1964, Some effects of earthworm invasion in virgin podzols: Can. Jour. Soil Sci., v. 44, p. 34-37.

Lundblad, K., 1934, Studies on Podzols and Brown Forest soils I: Soil Sci., v. 37, p. 137-155.

Luttmerding, H.A., 1971, Soil profile descriptions and chemical analyses, CSSS (Western) tour: Soils Division: Kelowna, B.C., B.C. Dept. of Agr., Mineo., 20 p.

Malcolm, R.L. and McCracken, R.J., 1968, Canopy drip: A source of mobile soil organic matter for mobilization of iron and aluminum: Soil Sci. Soc. America Proc., v. 32, p. 834-838.

Marshall, V.G., 1974, Seasonal and vertical distribution of soil fauna in a thinned and urea-fertilized Douglas fir forest: Can. Jour. Soil Sci., v. 54, p. 491-500.

Mathur, S.P., 1969, Microbial use of podzol Bh fulvic acids: Can. Jour. Microbiol., v. 15, p. 677-680.

McKeague, J.A., 1965, Properties and Genesis of three members of the Uplands catena: Can. Jour. Soil Sci., v. 45, p. 63-77.

_______________, 1968, Humic-fulvic acid ratio, Al Fe and C in pyrophosphate extracts as criteria of A and B horizons. Can. Jour. Soil Sci., v. 48, p. 27-35.

_______________, ed., 1976, Manual on soil sampling and methods of analysis: Subcommittee of CSSC on Methods of Analysis: Ottawa, Soil Research Inst., 212 p.

McKeague, J.A. and Brydon,J .E., 1970, Mineralogical properties of the reddish brown soils from the Atlantic Provinces in relation to present materials and pedogenesis: Can. Jour. Soil Sci., v. 50, p. 47-55.

MeKeague, J.A. and Day, J.H., 1966, Dithionite- and oxalate-extractable Fe and Al as aids in differentiating various classes of soils: Can. Jour. Soil Sci., v. 46, p. 13-22.

McKeague, J.A. and Sprout, P.N., 1975, Cemented subsoils (duric horizons) in some soils of British Columbia: Can. Jour. Soil Sci., v. 55, p. 189-203.

McKeague, J.A., Schnitzer, M. and Heringa, P.K., 1967, Properties of an Ironpan Humic Podzol from Newfoundland: Can. Jour. Soil Sci., v. 47, p. 23-32.

McKeague, J.A., Brydon, J.E. and Miles, N.M., 1971, Differentiation of forms of extractable iron and aluminum in soils: Soil Sci. Soc. America Proc., v. 35, p. 33-38.

McKeague, J.A., Dumanski, J., Lajoie, P.G. and Marshall, I.B., 1973, Part 3, Ottawa - Montreal Tour Guide: 4th Intern. Working Meet. Soil Micromorphology, Kingston, Ont. 16 p.

McKeague, J.A., MacDougall, J.I. and Miles, N.M., 1973, Micromorphological, physical, chemical and mineralogical properties of a catena of soils from Prince Edward Island in relation to their classification and genesis: Can. Jour. Soil Sci., v. 53, p. 281-295.

Muir, A., 1961, The Podzol and Podzolic soils: Adv. Agron. v. 13, p. 1-56.

Muller, P.O. and Cline, M.G., 1959, Effect of mechanical soil barriers and soil wetness on rooting of trees and soil mixing by blow down in central New York: Soil Sci., v. 88, p. 107-111.

NSSC (National (Canada) Soil Survey Committee), 1945, Proceedings of National Meetings: Ottawa, Agriculture Canada.

_____, 1955, Proceedings of National Meetings: Ottawa, Agriculture Canada.

_____, 1965, Proceedings of National Meetings: Ottawa, Agriculture Canada.

_____, 1968, Proceedings of National Meetings: Ottawa, Agriculture Canada.

Petersen, L., 1976, Podzols and Podzolization: Copenhagen, DSR Forlag, 293 p.

Rausell-Colom, J.A., Sweatman, T.R., Wells, C.B. and Norrish, K., 1965, Studies on the artificial weathering of mica, *in* Hallsworth, E.G. and Crawford, D.V., eds., Experimental Pedology: London, Butterworths, p. 40-72.

Ross, G.J., 1969, Acid dissolution of chlorites: release of magnesium, iron and aluminum and mode of acid attack: Clays and Clay Min., v. 17, p. 347-354.

__________, 1971, Relation of potassium exchange and fixation to degree of weathering and organic matter content in micaceous clays of Podzol soils: Clays and Clay Min., v. 19, p. 167-174.

__________, 1975, Experimental alteration of chlorites into vermiculites by chemical oxidation: Nature, v. 255, p. 133-134.

Ross, G.J. and Kodama, H., in press, Experimental alteration of a chlorite into a regularly interstratified chlorite-vermiculite by chemical oxidation: Clays and Clay Min.

Sawhney, B.L. and Voigt, G.K., 1969, Chemical and biological weathering in vermiculite from Transvaal: Soil Sci. Soc. America Proc., v. 33, p. 625-629.

Schnitzer, M., 1952, Investigation of properties of cation-enriched leaf extracts and leachates (MSc. Thesis): Montreal, McGill Univ., 92 p.

Schnitzer, M. and DeLong, W.A., 1954, Note on relative capacities of solutions obtained from forest vegetation for mobilization of iron: Can. Jour. Agr. Sci., v. 34, p. 542-543.

Schnitzer, M. and Desjardins, J.G., 1969, Chemical characteristics of a natural soil leachate from a Humic Podzol: Can. Jour. Soil Sci., v. 49, p. 151-158.

Schnitzer, M. and Khan, S.V., 1972, Humic substances in the environment: N.Y., Marcel Dekker, 327 p.

Schnitzer, M. and Kodama, H., in press, The dissolution of micas by fulvic acid: Geoderma.

Schwertmann, U., 1964, The differentiation of iron oxides in soils by a photochemical extraction with acid ammonium oxalate: Z. Pflanzenernahr. Dung. Bodenkunde, v. 105, p. 194-201.

Simonson, R.W., 1959, Outline of a generalized theory of soil genesis: Soil Sci. Soc. America Proc., v. 23, p. 152-156.

Soil Survey Staff, 1960, Soil Classification, a comprehensive system, 7th Approximation: Washington, D.C., U.S. Dept. of Agr., 265 p.

________________, 1976, Soil Taxonomy: A basic system of soil classification for making and interpreting soil surveys, U.S. Dept. of Agric. Handbook 436: Washington, U.S. Government Printing Office, 754 p.

Sokolova, T.H., Targuliyan, V.O. and Smirnova, G.Y.A., 1971,

J.A. McKeague, G.J. Ross and D.S. Gamble

Clay minerals in Al-Fe Humus Podzolic Soils and their role in the formation of the soil profile: Soviet Soil Sci., v. 5, p. 331-341.

Stobbe, P.C. and Wright, J.R., 1959, Modern concepts of the genesis of Podzols: Soil Sci. Soc. America Proc., v. 23, p. 161-164.

Stonehouse, H.B., Acton, D.F. and Ellis, J.G., 1971, Guidebook for a field tour of sandy forested soils in Northern Saskatchewan: Saskatchewan Institute of Pedology, Publ. M19, 92 p.

Van Schuylenborgh, J., and Bruggenwert, M.G.M., 1965, On genesis in temperate humid climate. V. The formation of "Albic" and "spodic" horizon: Neth. Jour. Agric. Sci., v. 13, p. 267-279.

The Pedogenic Profile in the Stratigraphic Section

S. Pawluk

ABSTRACT

The soil profile plays a unique role in the Quaternary sciences since it frequently provides the only record of past environments which existed during the time break within the stratigraphic column. Since soils develop downward into regolith, the surface of which is exposed during any hiatus, the soil morphological features preserved during subsequent burial are the basis for deduction of the pre-existing environments. This is most commonly accomplished by comparison of morphologies of the buried soils with soils forming in present day environments. The validity of this approach is questionable since it is based on the assumptions that existing soils have formed under and are in equilibrium with presently existing environments, that buried soils do not undergo changes after burial, and that the soil morphology is specific to a single genetic pathway.

Since only some of the morphological features relate to specific genetic processes as well as to specific environments, the use of discretion in selecting such features is essential to any interpretation of the pedogenic profile for reconstruction of previous environments.

A concept of soil bodies as open systems within the interface of the biological, geochemical and hydrological cycles provides a basis for such an approach. The interchange and interaction of matter and energy in the soil interface respond to variations in environmental forces and regulate reorganization and transformation of materials within the soil body. An understanding of the dynamics characterizing the pedogenic setting provides a reliable mechanism for the reconstruction of the environments in which it functioned.

INTRODUCTION

The soil profile has always held an important and

unique position in Quaternary sciences. Unlike other features on the earth's surface in which history of our earth is recorded, soil develops downward into the earth's mantle rather than upward from the surface. Thus, the soil frequently provides a record of events written into the upper mantle of the regolith, which under certain conditions is protected against the devastating forces that readily manifest themselves in our changing environment through the course of time.

Therefore, there is good reason as to why researchers so often focus attention on the "buried soil profile" in their attempts to delineate Quaternary history from the stratigraphic column, for often it is the only evidence for events that occurred during the hiatus that marks the break in the stratigraphic section.

At this point it is essential to keep in mind the distinction between relict soils which are of considerable significance as geomorphic markers but of little stratigraphic value, and buried soils which are of considerable stratigraphic significance (Leamy, *et al.*, 1973, p. 723), although some buried soils may have been relict soils at one time.

Unfortunately, those who utilize buried soils for the purpose of recreating historical events related to climatic changes commonly take a simplistic view in making judgments which often reflect shallow, or lack of, training in the discipline of pedology. Such judgments usually stem from a strict adherence to Dokuchaev's classical concept that "soils naturally occur as independent bodies and that their morphology reflects the five factors of soil formation; climate, biota, relief, time and parent material." (Joffe, 1949, p. 17).

The two principal suppositions based on this premise as applied to paleosols are: 1) that soils which form under similar climatic, biotic and drainage conditions during similar time intervals are in themselves similar; and 2) that the degree of differentiation of soil from regolith is a function of time in the absolute sense (that is, that a "well developed" soil must be old and that a "weakly developed" soil must be young). *A priori* reasoning therefore suggests that two soils of similar morphology must, consequently, be developed under similar environmental conditions. The approach followed is one of "indexing" or "fingerprinting", where morphology is described, some routine laboratory analyses are conducted and a search is made of existing soils for possible analogies. Soil analogues thus form the basis for equating similarities in environments. Indexing, as an approach for interpreting paleosols is, in fact, of very common usage.

THE PROBLEM

While buried paleosols may be similar in morphologies

and other characteristics to present day soils, they may not necessarily be analogous in terms of paleo-environmental and regional conditions. An example is the late-Wisconsin Prelate Ferry paleosol of Saskatchewan described by David (1966, p. 685-696). The paleosol, described as a Planosol or Solodized Solonetz, was considered to be analogous to present day Solodized Solonetz of the semi-arid Brown Soil Zone. Since Solonetzic soil formation is primarily a function of groundwater dynamics and chemistry as well as moisture balance, soils with characteristics similar to those described are found widely distributed throughout arid, semi-arid and sub-humid regions of the world. Ruhe (1965, p. 755) among others has pointed out that soils with similar morphologies may develop through different genetic pathways and that all soils to some degree are polygenetic in origin. Black Chernozemic soils with eluviated horizons in the Cypress Hills area of Alberta have been described as developing in ancient Gray Luvisols (Jungerius, 1969, p. 235-246). Yet, soil horizon sequences relatively similar to those described are evident in Eluviated Black Chernozems, Dark Gray Chernozems and Solodic Black Chernozems, each evolving by somewhat different genetic pathways with differences in their associated environment dynamics. Soils with reddish horizons are normally identified with extended periods of warm and hot climates since present day analogues are frequently evident in such environments. Bright reddish soil horizons are often erroneously interpreted as being strongly weathered. Deep reddish paleosols have been described in the Cypress Hills area and near Waterton in southwestern Alberta by some investigators (Westgate, 1972, p.50-61; Wagner, 1966, 141 p.). Yet, interpretations of their data in terms of past climatic events is difficult. According to some concepts of pedogenesis (Cline, 1961, p.442-446) all processes of soil formation are at least potential contributors to the development of every soil but they differ in their rates in different environments. Accordingly, it is at least theoretically possible that the presence or absence of characteristics which reflect a high degree of weathering may indeed not result from the kinds of processes of soil formation in a given set of environmental conditions but rather from the length of time during which the specific processes have been active. On the other hand, if a threshold level of activation energy must first be acquired for a reaction to proceed as suggested by Lavkulich (1969, p. 25-38), then some of the processes active in soil formation may differ in kind, depending upon energy levels governed by the environmental forces. In the latter instance identification of recognized processes within a soil body may reveal a great deal about the environment in which it developed. In reference to the reddish paleosols previously discussed, the question arises as to whether their characteristics primarily reflect a variation in the element of time or rather environments uniquely different from what we experience today, or both. This cannot be resolved on the basis of the two contrasting concepts of soil formation. This same question was previously alluded to by Ruhe (1965, p. 755) in reference to the occurrence of reddish colored soils of the United States.

Further difficulty in evaluating paleosols by comparison to present day analogues arises from the use of morphological features rather than genetic features as the differentiating criteria in modern soil classification schemes of North America. A specific example is the emphasis placed on horizon thickness for the differentiation of soil taxa at several levels of abstraction. This morphological property results in the separation of soils that develop along similar pedogenic pathways into different categories. Soils derived along different genetic pathways are also grouped into the same taxonomic units.

THE PEDOGENIC PROFILE

The "Working Group on the Origin and Nature of Paleosols" of the INQUA Commission on Paleopedology, in their 1970 report recommended not only a need to record morphological features and the three-dimensional aspects of paleosols as well as their classification but also the need for a variety of studies that would provide quantitative evidence of paleopedogenesis that could be applied in combination with the field investigations (Yaalon, 1970). The prime objective in the evaluation of paleosols, it seems, is the elucidation of the dynamics of the soil system prior to and during burial where morphological and other field descriptions provide only the framework for conducting the investigation; in other words, reconstructing the pedogenic profile.

The pedogenic profile may be considered as a conceptual elucidation of all actions and interactions within and among processes contributing to the dynamics of a soil body or to the pedogenic setting in profile view. Just as soil horizons form the building blocks of the pedological profile, soil forming processes may be considered as the building blocks of the pedogenic profile. It is the interaction of the pedogenic profile with the surface mantle of geological material that ultimately results in the differentiation of the pedological profile. While the pedological profile is in part a manifestation of the pedogenic profile, its character also reflects the nature of the material from which it develops. The degree of differentiation observed in the pedological profile is a reflection of the intensity of processes in the pedogenic profile, the resistance of the parent material to alteration and the time of interaction. Therefore it is quite possible for two or more morphologically different soils to form within the same pedogenic setting. Also, it should be understood that processes active in the pedogenic setting are not of essence restricted to the soil body but may extend downward to considerable depth within the surface geological mantle.

Conceptually, the pedogenic setting represents the interface of all reactions that alter the surface of the earth's mantle by cycling of matter and energy through the biosphere, lithosphere and hydrosphere. The soil body itself is a manifestation of the actions and interactions resulting from the cycling of matter and energy. Implications of this concept are several: 1) the dynamics of the hydrological cycle, the "short" carbon cycle, and the geochemical cycle through their interactions at the earth's surface give rise to unique processes responsible for soil reorganization and differentiation (see Figures 1 and 2) components of the soil body as part of such systems exist only in a transient state; 2) as matter cycles through the soil body, energy and matter are removed and new matter and energy are added; 3) as a consequence of 2, reorganization of geological material to a soil may or may not reflect the full impact of the pedogenic forces, depending upon material composition, energy levels, and mean residence time of energy and matter within the soil body; 4) the pedogenic setting remains in a state of dynamic equilibrium with the environment while the soil body only strives towards a steady state.

Placing emphasis on the pedogenic profile rather than on the pedological profile provides an opportunity for recreating paleoenvironmental conditions based on energy relationships rather than material alteration *per se*. Since the nature of cycling of matter and energy through the hydrological, "short" carbon and geochemical cycles relate closely to the total environment, paleopedogenic evidence is most profitably applied to Quaternary studies when used to elucidate the nature of these cycles.

The part of the carbon cycle to which pedogenic studies contribute the most information is the "short" carbon cycle or biological cycle (see Figure 1) where biological agencies interact with the subaerial part of the earth's crust; although shallow subaquous phases in some instances are also included within the realm of pedogenesis. Regardless of the acceptability of previously mentioned concepts relating to thermodynamics of soil forming processes, biological contributions to soil formation are unique to specific environments and are not governed by the classical laws of thermodynamics that relate to geochemical alteration. The transient nature of biological material within the soil body is well demonstrated by many researchers (Paul, 1969, p. 63-76; Nikiforoff, 1959, p. 186-196). As Nikiforoff points out, the amount of organic matter present in any soil is a function of the ratio of the rate of organic matter accumulation/rate of organic matter annihilation where an increase in ratio reflects a build-up of organic matter in soil.

Of what value is the determination of organic matter content and carbon/nitrogen ratio for reconstruction of paleoenvironments? Based on the foregoing premise, total soil biomass relates not only to the rate of biological accumulation but also to the rate of conversion of raw organic matter to partially degraded components, in part humic constituents, and to its ultimate annihilation. So

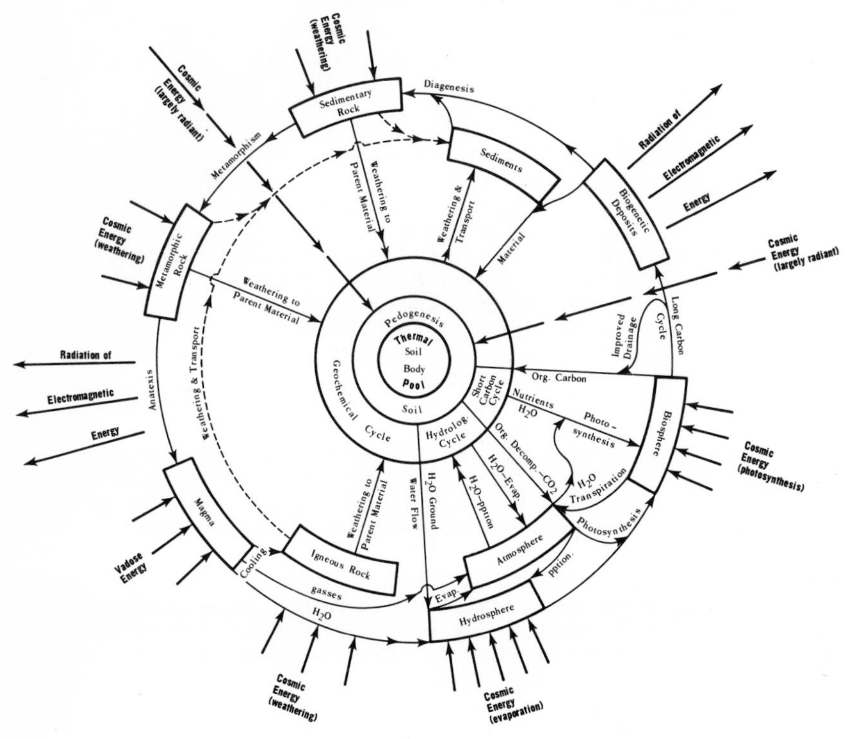

Figure 1 Cycling of matter and energy through the soil body. The soil forms at the interface of the biological (short carbon) cycle, geochemical cycle and hydrological cycle.

that organic matter content in the pedological profile may be quite similar in contrasting environments with very different biological inputs into the pedogenic profile. As pointed out by Paul (1969, p. 63-76), soil organic matter consists of a mixture of materials deposited over thousands of years. These materials are continuously undergoing degradation by organisms which form part of the bulk of the organic matter. Because C/N ratios narrow as decomposition proceeds, values for C/N ratios relate to the degree of decomposition rather than to specific components of the organic matter.

It is well established that the rate of turnover of the different fractions of organic matter is a function of

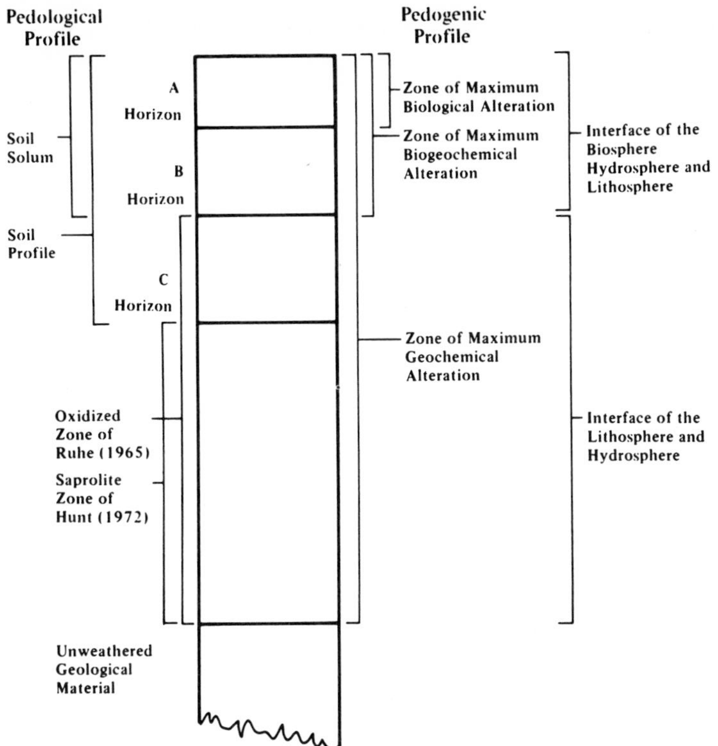

Figure 2 Comparison of the pedogenic profile and the pedological profile. The degree of expression of the pedogenic profile evident in the pedological profile is dependent upon energy levels, material composition and residence time of material within the interface of the biosphere, lithosphere and hydrosphere.

the complexity of their sturctures as well as the environment in which they persist (Stevenson, 1969, p. 470-479; Paul, 1969, p. 63-76). If the assumption can be made that the initial composition of the vascular plants and other organisms making up the bulk of the raw organic matter is relatively similar, then an evaluation of the persistency of organic fractions with different stabilities should provide a useful measure of the environmental conditions under which decomposition takes place.

However, in some instances the picture may be complicated by further degradation after burial. Such degradation by soil microorganisms is likely to selectively remove the more readily decomposable fractions such as loosely bound proteins and carbohydrates (Stevenson, 1969,

p. 470-479) with some selective enrichment of more resistant constituents such as ligno-proteins, from contributions of the microbial tissue to the reservoir of soil organic matter. Investigations of an Orthic Black Chernozemic paleosol buried in a slump bank along the North Saskatchewan River in Edmonton revealed some of the anomalies which may occur as a consequence of selective degradation. The particular soil was dated at *ca.* 2000 BP from a buried tree trunk at the same site. The hexose and pentose contents for the buried soil were 134 and 68 mg/100 g of soil, respectively, as compared to contents of 1000 and 250 mg/100 g respectively for present day Orthic Black Chernozemic soils. Total organic matter content for the buried soil and its present day analogue were 2 and 7 percent, respectively. An attempt was also made to differentiate the paleovegetation on the site at the time of burial according to the procedure proposed by Dormaar (1967, p. 37-45), based on infrared patterns for humic fractions extracted from the soil organic matter. While the slopes between 1800 and 2500 cm^{-1} on the infrared absorption spectra for humic acids extracted from the Ah horizons confirmed the presence of grassland vegetation for both the buried and present day sites, the more pronounced slope for the humic acid extracted from buried soil was believed to reflect the removal of low molecular weight humic constituents through post-burial degradation. This limitation to his technique was recognized by Dormaar. Whether similar trends for specific types of vegetation would be obtained for humic acids in the warmer regions of the world is of interest since rate of organic matter turnover is much more rapid and only the more resistant humic constituents may be expected to persist in the soil. On the basis of the foregoing one may be led to suspect that post-burial degradation of a Chernozemic Ah from the cooler climates of the world would result in development of biochemical characteristics more closely analogous to soils of the warmer climates such as the "Black Cotton" soils of India.

Further, as stated by Paul (1969, p. 63-76), since the various organic degradation products are annihilated at different rates, a specific age for the organic carbon cannot be obtained by radiocarbon dating. Rather, a mean residence time for the various organic fractions formed through the decomposition products is obtained. Organic fractions dated for some present day soils in Saskatchewan showed humic acids as 1308 ± 64 years, fulvic acids as 630 ± 60 years and humins as 240 ± 60 years of age. Mean residence times reported for organic matter in soils of the Canadian Interior Plains ranged from 400 to 2000 years for Chernozems and 250 years for Gray Luvisols. Considerable differences in dates were reported with variable addition of even small amounts of fresh organic matter through minor changes in relief within the same regional environmental setting (Paul, *et al.*, 1964, p. 201-208). Minor additions of selective forms of fresh organic matter during post-burial degradation and selective enrichment of more resistant components can further influence the age of organic carbon in a buried soil solum.

Although fractionation of organic matter from paleosols into components which express a relative degree of degradation and/or polymerization into humic substances, coupled together with C^{14} dating of the various fractions, may provide some evidence as to the original biological processes and their dynamics, the many intervening factors must be accounted for while doing so.

Very useful additional evidence for discerning the kind of biological processes that may have been active in a paleosol can be obtained directly through microscopic study. Micromorphological investigations of suitably prepared soil thin-sections have been used successfully for the identification of paleofeatures produced by soil organisms and in some instances for the actual identification of the organism or portions thereof (Brewer, 1964, p. 253-255). The recognition of fecal pellets and pedotubules in terms of size, shape and composition may be used to identify the soil fauna originally populating the soil and even more important to ascertain the diversity of faunal species which directly reflects environmental constrictions (Brewer and Pawluk, 1975, p. 301-319). Faunal activity can also be recognized through comminution, decomposition and incorporation of plant material into mineral soil. As suggested by Bal (1973, p. 7-10), decomposition and distribution of organic matter in soils are related and therefore can be regarded as an individual entity for which he suggests the term "humon." The humon is defined by Bal as "a collection of macroscopically and/or microscopically observable organic bodies in soil which are characterized by a specific morphology and spatial arrangement; it is a 'natural, three-dimensional, genetic, organic individual' existing in soil."

Since the degradation of organic matter follows different pathways in different soil environments, investigations of the humon reveal the nature of the transient state of organic matter within the soil body as part of the "short" carbon cycle. Bal's proposal for the micromorphological classification of soil organic matter provides a systematic approach to the study of the nature and spatial arrangements of organic components as they relate to pedogenesis. Although his concepts are relatively new, their application to paleosols is exceptionally fruitful. According to the system, original plant components may be recognized through the identification of cellular structures and their decomposition products. Recognition of resistant organic bodies also provides an opportunity for identifying original populations of plants and animals and their associated environments. Identification of sclerotia and spores may indicate the presence of fungi; diatoms suggest wet conditions; phytoliths, pollen grains and relatively resistant chitinous exoskeletons may be used to identify the species of plant or animal organism. Biological processes greatly influence the nature of the surface soil horizon and by definition the zone of maximum biological activity is confined to the A horizon (see Figure 2).

Inorganic components within an actively developing soil profile (*i.e.* where the pedogenic profile and the pedological profile coincide) are also maintained in a transient state. As part of the geochemical cycle erosional processes remove the surface material from the soil body and displace the pedogenic profile downward into new parent geological material while depositional processes form accretionary surfaces that displace the pedogenic profile upward with the lower portion of the soil profile left intact as a relict feature. In situations where residence time is brief as a consequence of a relatively rapid and uniform rate of surface erosion, the pedogenic processes are only weakly evident in the reorganization and differentiation within the pedological profile, horizonation is weakly expressed, thin or lacking and the soil remains a Regosol. In the case of an accretionary surface, a uniform rate of accumulation generally results in excessively thick but weakly expressed surface horizons; variable rates of accumulation result in relatively weakly developed vertical sequences of soil sola within the stratigraphic section, the arrangement of which reflects variability in rates of deposition. An excellent example of the latter is the paleosol developed on an accretionary aeolian surface near Hinton, Alberta and reported on by Dumanski (1970, p. 18). Only on relatively stable land surfaces is the residence time of the soil body sufficiently long so that reorganization of geological material fully reflects the expression of the pedogenic setting. Composition of geological parent material also determines the degree of expression of the pedogenic profile within the pedological profile. In situations where the pedogenic setting is relatively constant within a stable environment, different kinds of soil bodies may be expected to form from parent materials diverse in their characteristics. Podzols on coarse textured acid parent materials, Luvisols on fine textured calcareous parent materials and Regosols on acidic quartzitic sands can all occur in close proximity. While the pedological profiles are quite different the only variation in their pedogenic profiles may be reflected in biological processes related to microenvironmental conditions.

The processes responsible for reorganization of materials within the pedological profile relate to both kind and intensity. While the zones of biological and biogeochemical influence dominate the soil solum and principally establish the pedogenic setting, evidence for geochemical weathering reactions is largely subdued. Dominant expression of geochemical processes extends below the zone of biological influence and encompasses the "oxidized zone" as defined by Ruhe (1965, p. 758) for weathering zones of the temperate regions and the saprolite zone as described by Hunt (1972, p. 152) for the warmer temperate and tropical regions (see Figure 2).

Geochemical alterations, unaffected by biological

agencies, generally follow the laws of thermodynamics for inorganic reactions and therefore more closely reflect the total energy distribution within the surficial mantle. The geochemical weathering profile may be expected to vary in both thickness and intensity along a thermal gradient with minimal expression in arctic regions and maximum expression in tropical environments. Consolidated bedrock and permafrost may result in anomalously thin weathering profiles along such a gradient. Through horizontal segregation of thermal energy as suggested by Nikiforoff (1959, p. 186-196) variations in intensity of processes not only extend laterally but also vertically with intensity highest near the surface and decreasing with depth. If the "collision" theory for reactions as suggested by Lavkulich (1969, p. 31) is applied, this concept suggests that the geochemical alterations below the soil profile are characteristic of specific thermal regimes and therefore provide valuable clues for discerning paleoenvironments. However, according to Cline's (1961, p. 442-446) concept of soil formation, much of this variability may be attributed to the time factor.

The nature of the geochemical weathering profile is not a function of free energy of mineral components and time alone. As with the soil body, the underlying materials in the geochemical weathering zone exist in an open system where hydrological cycling as part of the hydrosphere interphases with the lithosphere. The flux of ionic solutions into and out of the zone of weathering through water movement in such an open system interacts to alter the direction of reactions from that which is predictable for a closed system. Therefore, regardless of the controversy that may exist in energy concepts related to pedogenesis, the sequence of alterations in the geochemical weathering zones along any thermal gradient *in toto* may be expected to be unique to the particular environment irrespective of intensity and/or time. Much of the confusion in the past reflects the emphasis placed on the nature of the weathering materials rather than the dynamics of the system. This conclusion is based on the premise that a series of independent geochemical weathering reactions will not respond to external forces imposed through the introduction of materials and energy as part of an open system in precisely the same manner, but rather in a pattern unique to its environment. Despite the element of time, evidence for weathering patterns should be apparent, at least on the microscale, even in the relatively short period of time since the last glaciation.

Biogeochemical processes relate to the pedogenic profile and significantly influence the reorganization of materials immediately below the zone of maximum biological activity (Hallsworth and Crawford, 1965). Previously much of the alteration and reorganization within this zone has been attributed to geochemical weathering reactions. However, the presence of biological materials and organic constituents interact directly and biological material also serve as catalysts to greatly influence the intensity and direction of reactions during material alteration. Organic decomposition products initiate weathering reactions and combine with weathering products to provide a "sink" for certain constituents that normally enter into reaction to form new geochemical weathering products. Lack of "weathering crusts" and pseudomorphs as weathering products of primary minerals in the soil sola of cool environments probably reflect such complexing action in the chain of weathering reactions. Biochemical interactions reflect the persistence of intermediate organic decomposition products in cool environments and generally may be expected to be more weakly expressed in warm tropical environments where organic matter turnover is much more rapid. Stabilization and formation of soil structure are attributable to polysaccharides and polyurinides and their biochemical interaction with inorganic soil components. The degree of development of subangular blocky structure in B horizons reflects the release and persistence of polysaccharides and polyurinides in the soil system; subangular block structures are more commonly evident in soils of cool temperate regions than in tropical regions. Translocation of iron and aluminum in Podzols have been attributed to the persistence of organic constituents such as polyphenols introduced from decomposition of forest litter into the leaching waters in cool temperate environments (DeLong and Schnitzer, 1955, p. 360-363; Lutwick and DeLong, 1954, p. 203-213). Translocation of colloidal clays has been attributed to various transport mechanisms (Hallsworth and Crawford, 1965, p. 360, 372 and 373). While translocated clay is recognized through similar pedological parameters in a variety of environments, the pedogenic mechanism related to translocation is likely very specific for the prevailing environment. Formation of minerals *in situ* such as lepidocrocite and hematite traditionally has been attributed to variable thermal regimes, but more recently has been demonstrated to reflect specific biological interactions with ionic solutions (McHardy, *et al.*, 1974, p. 471-482). Abundances of such minerals and their genesis in the pedological profile can be related to very specific environmental conditions.

Reorganization and alteration of soil materials through biogeochemical and geochemical processes result in fabric types unique to the pedogenic setting. As for biological material, such fabric types are readily discernible in soil thin-sections. Integration of petrographic and soil micromorphological procedures with X-ray microcamera, electron microprobe and other energy dispersive techniques, while still in its formative stage as a "total package" soil research tool, provides a useful approach to paleopedogenic studies. As an example, the recognition of simple cutans indicate relatively uniform conditions during formation whereas compound cutans suggest changes in conditions during formation. The kind of cutans reveal a great deal of information concerning processes of formation. Inferences related to mode of pedogenesis may be inferred from the characteristics of the various kinds of glabules and their internal fabric. Composition and fabric of crystallaria reflect the nature of the ionic flux through the soil body during formation (Brewer, 1964, p. 229, 281, 282, 292).

Lastly, the hydrological cycle also plays a prominent role in the pedogenic profile within the soil interface. The water regime is responsible for reorganization and transport of colloidal material or plasma and for the exchange of materials in solution between the land surface and the water table. Freezing and thawing as well as wetting and drying also profoundly influence the reorganization of materials within the soil body in a unique and characteristic manner. In Alberta, paleosols complete removal of carbonate, sulfate and chloride salts into the groundwater from the unsaturated zone suggests a leaching environment (precipitation exceeds evaporation) characteristic of groundwater recharge conditions; the presence of secondary lime carbonate enrichment suggests local groundwater discharge; the presence of gypsum rosettes and/or sodium salts suggests groundwater discharge normally in intermediate or regional flow systems of relatively arid environments; post-burial salinization is evident from coincidence of soil horizonation and salt accumulation; and accumulation of material of polygenetic origin may be discerned from the mutual compatibility of the pedogenic processes.

CONCLUSION

The tremendous amount of excellent research information available in the discipline of soil science as yet has not been fully exploited in establishing a model by which information on the Quaternary history could be discerned on a morphogenetic basis. The basis for establishing a model of this type is dependent upon the integration of the dynamics of soil formation with the changing environmental forces. The concept of the pedogenic profile provides this basis.

REFERENCES CITED

Bal, L., 1973, Micromorphological analysis of soils: Soil Survey Institute, Wageningen, The Netherlands, 174 p.

Brewer, R., 1964, Fabric and mineral analyses of soils: Sydney, Wiley, 470 p.

Cline, M.G., 1961, The changing model of soil: Soil Sci. Soc. America Proc., v. 25, p.442-446.

David, P.P., 1966, The Late-Wisconsin Prelate Ferry Paleosol of Saskatchewan: Canadian Jour. Earth Sciences, v. 3, p. 685-695.

DeLong, W.A., and Schnitzer, M., 1955, Investigations on the mobilization and transport of iron in forested soils. I. The capacities of leaf leachates and extracts to react with iron: Soil Sci. Soc. America Proc., v. 19, p. 360-363.

Dormaar, J.F., 1967, Infrared spectra of humic acids from soils formed under grass and trees: Geoderma, v. 1, p. 37-45.

Dumanski, J., 1970, A micromorphological investigation on the genesis of soils developed on calcareous aeolian material (Ph.D. Dissert.): Edmonton, Univ. of Alberta, 143 p.

Hallsworth, E.G., and Crawford, D.V., 1965, Experimental Pedology: London, Butterworths, 414 p.

Hunt, C.B., 1972, Geology of soils: Freeman, 344 p.

Joffe, J.S., 1949, Pedology: Somerville, N.J., Somerset Press, p. 17.

Jungerius, P.D., 1969, Soil evidence of postglacial tree line fluctuations in the Cypress Hills area, Alberta, Canada: Arctic and Alpine Research, v. 1, p. 235-246.

Lavkulich, L.M., 1969, Soil dynamics in the interpretation of paleosols, *in* Pawluk, S., ed., Pedology and Quaternary research: Edmonton, University of Alberta Press, p. 25-38.

Leamy, M.L., Milne, J.D.G, Pullar, W.A., and Bruce, J.G., 1973, Paleopedology and soil stratigraphy in the New Zealand Quaternary succession: N.Z. Jour. Geology and Geophysics, v. 16, p. 723.

Lutwick, L.E. and DeLong, W.A., 1955, Leachates from decomposing leaves. II. Interaction with soil forming minerals: Canadian Jour. Agr. Sci., v. 34, p. 203-213.

McHardy, W.J., Thomson, A.P., and Goodman, B.A., 1974, Formation of iron oxides by decomposition of iron-phenolic chelates: Jour. Soil Sci., v. 25, p. 472-482.

Nikiforoff, C.C., 1959, Reappraisal of the soil: Science, v. 129, p. 186-196.

Paul, E.A., 1969, Characterization and turnover rate of soil humic constituents, *in* Pawluk, S., ed., Pedology and Quaternary Research: Edmonton, University of Alberta Press, p. 63-76.

Paul, E.A., Campbell, C.A., Rennie, D.A., and McCallum, R.J., 1964, Investigations of the dynamics of soil humus utilizing carbon dating techniques: Trans. 8th Internatl. Congress Soil Sci. Bucharest, p. 201-208.

Pawluk, S., and Brewer, R., 1975, Micromorphological, mineralogical and chemical characteristics of some alpine soils and their genetic implications: Can. Jour. Soil Sci., v. 5, p. 415-437.

Ruhe, R.V., 1965, Quaternary paleopedology, *in* Wright, Jr., H.E., and Frey D., eds., The Quaternary of the United States: Princeton University Press, Princeton, N.J., p. 755-764.

Stevenson, F.J., 1969, Pedohumus: accumulation and diagenesis during the Quaternary: Soil Science, v. 107, p. 470-479.

Wagner, W.P., 1966, Correlation of Rocky Mountain and Laurentide glacial chronologies in southwestern Alberta, Canada (Ph.D. Dissert.): Ann Arbor, University of Michigan, 141 p.

Westgate, J.A., 1972, The Cypress Hills, *in* Rutter, N.W. and Christiansen, E.A., eds., Quaternary geology and geomorphology between Winnipeg and the Rocky Mountains: Field Excursion C-22, Internatl. Geological Congress, 24 session, 601 Booth St., Ottawa, 101 p.

Yaalon, D.H., 1970, Paleopedology, an aid in Quaternary stratigraphical investigations: Newsl. Stratigr., v. 1, p. 33-34.

Soil
Stratigraphy

Quaternary Soil Stratigraphy–Concepts, Methods, and Problems

Roger B. Morrison

ABSTRACT

Soil profiles are valuable elements in the geological record. They form during episodes of landscape stability at a given site, when erosion and deposition are negligible. They can be treated as stratigraphic entities, having some attributes like, and others unlike, both lithostratigraphic units and unconformities. They are not diastems; they have thickness and distinctive physical and chemical characteristics. Commonly they provide the best, perhaps the only, information on climatic conditions during the episodes when they form. A soil stratigraphic unit commonly is subject to lateral variation in its physical and chemical characteristics (analogous to facies changes in lithostratigraphic units) caused by changes in parent material, slope, climate, and vegetation, and also in subsequent erosional and secondary modifications.

I now favor using the term "geosol" to designate <u>all</u> soil stratigraphic units, in a less restrictive sense than I proposed a decade ago. In the American Stratigraphic Code "geosol" should replace the term "soil" which is ambiguous because of its varied usage among farmers, engineers, soil scientists, and other specialists (the term "paleosol" generally is restricted to buried soils).

Ancient geosols may be preserved in places at the present land surface as <u>relict geosols</u> (albeit invariably modified by subsequent climatic, erosional, biologic, and other environmental changes)- or they may occur buried beneath younger deposits as <u>buried geosols</u>. A geosol maintains a consistent age and stratigraphic relationship to older and younger deposits (and geomorphic surfaces), subject to the qualifications discussed below. A geosol is younger than the youngest material or surface on which it occurs and is older than the oldest deposit that overlies it. Relict geosols provide no evidence of their minimum age; this must be determined from buried geosols. However, buried geosols commonly show some variation in their age bracket from locality to locality. Some well-known geosols (*e.g.*, the Sangamon and Yarmouth soils and their equivalents in the U.S., and the Paudorf Soil in Austria) have been shown to be in places represented by several distinct soil profiles with intervening sediments. (These places are relatively rare sites of rapid deposition without erosion during the time when the main geosol formed.) Thus, there is need for not one but two or more hierarchies of soil-stratigraphic units. This would recognize the common occurrence of geosol complexes (composites of several pedogenic episodes represented in a single polygenetic soil profile), that locally may be divisible into individual member geosols. The system used for the Czechoslovakian loess sequences provides an instructive model.

Intensive stratigraphic studies of sequences of buried geosols at localities having exceptionally favorable depositional and preservation conditions have demonstrated both in the U.S. and in Europe the presence of many more geosols than were recognized a few years ago. Many of these studies, especially the European ones also supply valuable climatic data from malacologic, botanical, and micropedologic evidence. These studies reinforce my chief conclusions from a decade ago at least for the second half of the Quaternary:

(1) Episodes of relatively rapid soil profile development (geosol forming episodes) alternated with times of much slower or negligible soil profile development. Thus, the concept of "intermittency" of Quaternary soil forming intervals is valid.

(2) The episodes when the stronger geosols formed were considerably shorter than the intervening intervals.

(3) The geosols invariably formed during warm humid intervals, with the degree of profile development corresponding closely to the degree of warmth. The strongest geosols formed during interglacials, moderate to weak ones during interstadials. Negligible profile development took place during pleniglacials. Furthermore, a growing body of paleontologic and oceanographic evidence indicate that the warmer intervals, when the geosols formed, were also wet--the interglacials were the true pluvials of the Quaternary, not only at low but also in middle latitudes. Pleniglacials, on the other hand, had markedly reduced precipitation, though in their earlier parts, increased effective ground moisture.

(4) From the above, it appears that Quaternary geosols are good stratigraphic markers and can be very useful for local and long distance time-stratigraphic division and correlation, provided that studies of individual localities (and also the quality of the stratigraphic record preserved in them) are adequate for resolving problems of composite geosols and possible homotaxis.

INTRODUCTION

My primary purpose in this paper is to outline the concepts and methods of soil stratigraphy (revised from Morrison, 1967), and a few of its problems, from a geologist's viewpoint as contrasted to that of a soil scientist. My secondary purpose is to try to mitigate a very serious problem that now confronts geologists and soil scientists, the problem of effective communication between their respective disciplines (see Appendix 1).

Why should a geologist study soil profiles? Soil profiles commonly are the only records of substantial portions of Quaternary time. They record episodes when erosion and deposition were negligible at a site, so-called "landscape stability episodes." They provide clues to geologic history, particularly climatic history, during intervals of time that generally are not recorded by deposits. Study of soil profiles by modern pedologic, micropedologic, paleontologic, geochemical, and geologic techniques can yield semi-quantitative data on the temperature, precipitation, vegetation, and erosion-drainage conditions that existed while a soil profile was forming.

Another reason why soil profiles are important to geologists is that those that qualify as soil-stratigraphic units (geosols) are excellent key beds for division and correlation of Quaternary stratigraphic sequences. Geosols, particularly buried ones, commonly are distinctive and widespread stratigraphic markers in the sequences. Surface (relict) geosols are useful for distinguishing and correlating various geomorphic surfaces.

BASIC CONCEPTS OF SOIL STRATIGRAPHY

Soil-Stratigraphic Units (Geosols)

The American Stratigraphic Code (Amer. Comm. on Stratigraphic Nomenclature, 1961, Art. 18,) defines a soil stratigraphic unit as "a soil with physical features and stratigraphic relations that permit its consistent recognition and mapping as a stratigraphic unit." The Code further characterizes soil-stratigraphic units as distinct from both rock-stratigraphic and pedologic units; they are "the products of surficial weathering and of the action of organisms at a later time and under ecological conditions independent of those that prevailed while the parent rocks were formed"; also, "a soil-stratigraphic unit may comprise one or more pedologic units or parts of units" because "stratigraphic relations are an essential element in defining a soil-stratigraphic unit but are irrelevant in defining a pedologic unit."

I proposed the term "GEOSOL" (1967; Morrison and Frye, 1965) to replace the term "SOIL" for the basic soil-stratigraphic unit in stratigraphic usage. "Soil" is ambiguous and inappropriate for this usage because of its various meanings to farmers, engineers, soil scientists, and other specialists.[1] The term "paleosol" generally is restricted to buried soil profiles.

I now consider all soil profiles that qualify as soil-stratigraphic units under the American Stratigraphic Code to be geosols. Recognizing the need to demonstrate both mappability and stratigraphic relations (which implies lateral extensiveness in a type area), a geosol can be defined as an assemblage of soil profiles that have similar stratigraphic relations on lithostratigraphic, morphostratigraphic or other stratigraphic evidence. In addition, the soil profiles representing a geosol have physical and chemical pedogenic features that permit recognition, lateral tracing, and mapping as a stratigraphic unit, although specific pedogenic features can vary with changes in soil-forming environment (parent material, climate, vegetation, topography/drainage) from place to place. There are no limitations on degree or type of profile development, except that there must be evidence of pedogenesis. Different geosols range in degree of pedologic development from weak A-C soil profiles to strongly developed profiles many meters thick. An A horizon is not required, nor is a B horizon if an eluvial A horizon and/or a Cca, Ccs, or Csi horizon is present. A geosol may or may not be a "weathering profile."

The various soil profiles comprising (in their lateral extent) a geosol need not be exact time equivalents. The requirement of similar stratigraphic relationship implies some degree of time equivalence, but this requirement must be interpreted quite broadly, particularly for surface (relict) geosols, as will be discussed later. The soil-profile assemblage that represents a geosol is akin to a chrono-catena but includes buried as well as surface (relict) soil profiles that can be demonstrated to have a similar stratigraphic relationship.

The broader, less restrictive redefinition of geosol seems desirable because of the great increase in knowledge of Quaternary stratigraphy in the last dozen years. Modern studies of the "fine stratigraphy" of exceptionally complete and well preserved sequences of Quaternary deposits in Europe and North America have proved the existence of many more soil-stratigraphic units than were known a few years ago. Also, we now know that some geosols occur most commonly as polygenetic soil profiles that record in a single profile the composite effects of several closely following soil-forming episodes, but in places these geosols are "stretched out" as pedocomplexes with several separate soil profiles representing the various pedogenic episodes. Consequently, to accommodate these findings, I am proposing

---

[1] An example is the title of a conference held several years ago: "Engineering properties of sea-floor soils and their geophysical identification."

other types of soil-stratigraphic units of lesser rank than geosol. They are discussed in a later section.

Types of Stratigraphic Occurrence of Geosols

Geosols may occur in two principal ways -- as surface soil profiles, which are called relict geosols, and as buried soil profiles, or buried geosols. Most geosols have both types of occurrence, in landscapes of any appreciable extent.

Relict Geosols. A relict geosol has remained exposed at the land surface after developing most of its diagnostic profile characteristics, without burial by younger deposits, and without enough erosion or subsequent pedogenic activity to destroy or largely transform its original profile. A relict geosol obviously has been exposed to all the surficial processes -- pedogenic, biologic, and geologic -- that have acted since the land surface on which it occurs was formed. It represents a summation of the surficial processes on this surface.

If the original profile is weakly developed, it will be transformed into and masked by any subsequent stronger pedogenesis. On the other hand, once a strongly developed profile has formed, it remains relatively unaffected by subsequent weaker pedogenesis. Clay accumulation is essentially irreversible and so is carbonate accumulation in most semiarid and arid environments. A relict geosol therefore can be said to have developed its diagnostic characteristics in these horizons during the strongest soil-forming episode to which it has been subjected.

Buried Geosols. Buried geosols are sometimes called paleosols or fossil soils. They are seen much more rarely than relict geosols. They are seen in good natural or artificial exposures (but also have been identified in bore-hole cores) at localities where the geosol has been sheltered from erosion prior to its burial. A geosol buried by only a few feet of sediment is removed from the erosive, pedogenic, and biologic agencies of the surficial environment. Secondary changes may occur, but normally these are relatively rare and minor. They include deposition of iron or manganese hydrolozates or calcium carbonate or other salts by ground water; also oxidation of organic matter, and even physical changes such as compaction by loading by sedimentation, by glacial ice, *etc*.

Weakly developed geosols are found only in buried occurrences, except in the case of the later Quaternary (generally late Woodfordian and Holocene) geosols. (Figure 1).

The definitions of "buried soil" and "bisequum" in Soil Taxonomy (Soil Survey Staff, 1976) are far too restrictive and unworkable to soil stratigraphers and should be rejected for reasons discussed in Appendix 1.

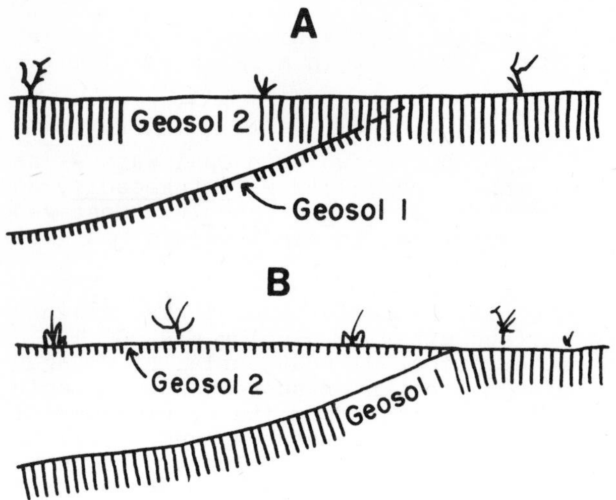

Figure 1 Relict and buried geosol relationships. In A, the weakly developed geosol 1 can be identified only where it is buried beneath a younger deposit. As it approaches close to the landsurface it merges with and is masked by the stronger geosol 2. In B, geosol 1 is the more strongly developed. The weakly developed younger geosol 2 can be identified only where intervening sediments separate it from geosol 1; the younger geosol cannot be distinguished in relict occurrences of geosol 1, whose profile is similar in both buried and relict situations.

SOIL-STRATIGRAPHIC COMPLEXES

Composite (Polygenetic) Geosols

Quite commonly a soil profile that represents a geosol at a given site has characteristics that a skilled pedologist may be able to interpret as the result of more than one episode of pedogenesis. Such a polygenetic soil profile is said to be a composite geosol (Figure 2A). These profiles show only the sequence of pedogenic events, not their duration or the length of any nonpedogenic episodes between them. However, it may be possible to determine the stratigraphic and age relations of the various pedogenic and nonpedogenic events by tracing the soil profile as a geosol from where it is composite to sites where it is a multistory profile as discussed below. Relict occurrences of geosols always are polygenetic to greater or lesser degree but many buried occurrences also are polygenetic. An example is the Paudorf Soil of Austria (see below and Figure 4); another is the Sangamon Soil of the Midwestern U.S., which commonly occurs as a buried composite soil profile, particularly if its cumulic "Late Sangamon" A horizon is included. This portion includes one or more post-Sangamonian soil-forming episodes, represented by the Chapin, Pleasant Grove, and

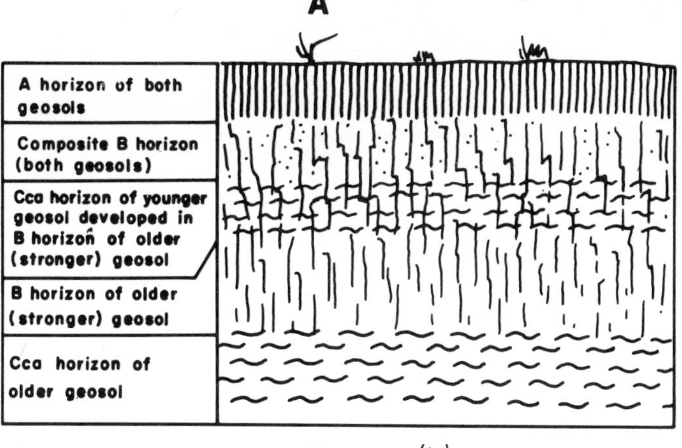

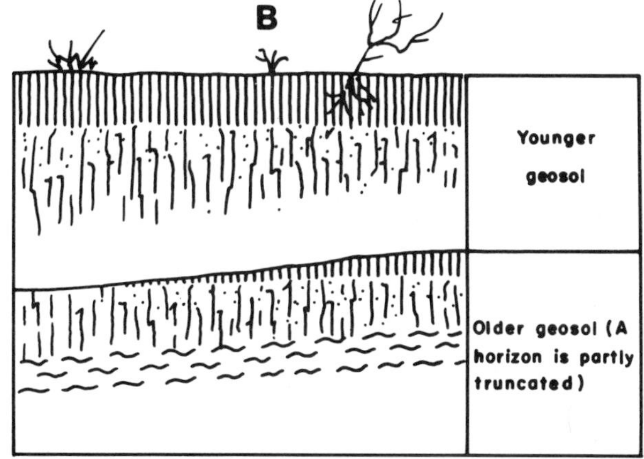

Figure 2 Composite and compounded geosols. Part A diagrams a composite geosol in a relationship commonly found in pedocal areas. Here the profiles of two individual geosols partly overlap each other. Part B shows two compounded geosols whose profiles are separated by a small thickness of intervening sediment.

Farmdale Soils, of early and middle Wisconsinal (Altonian and Farmdalian) age (Willman and Frye, 1970), as determined from tracing to multistory occurrences in Illinois. Also, in some situations a "Sangamon" geosol represents more than one interglacial.

Multistory "Stretched-Out" Soil Profiles; Compounded and Subdivided Geosols

Multistory or "stretched-out" soil profiles (commonly called pedocomplexes by European workers) are those where more than one soil profile occurs in close vertical succession, but separated by enough intervening sediment so that the individual soil profiles do not overlap or overlap so slightly that they can be distinguished from one another. In some cases each soil profile by definition is considered to represent a separate geosol. These occurrences are called compounded geosols (Figure 2B). In other instances, a geosol that usually occurs as a single composite soil profile (and is defined as a geosol on this basis) may be traceable to a few places where it is separated by intervening sediment into two or more soil profiles; such occurrences can be called a subdivided geosol (Figure 3), because here the individual profiles are part of what normally is represented by the single profile, albeit a polygenetic one. The distinction between compounded and subdivided profiles is largely a matter of how the individual geosols are defined in a given area. A sequence of soil profiles that one worker calls a subdivided geosol may be considered by another worker in a different area to be two or more separate geosols. This is akin to the practice with rock-stratigraphic units, where units of one rank (*e.g.*, formation) may be equivalent to units of lower or higher rank (*e.g.*, member) in different areas. A good example of a geosol that locally is subdivided is the Paudorf Soil (Figure 4). For many years this geosol was thought to represent a late-middle Würm interstade. Recently however, it has been determined in its type area (the vicinity of Paudorf-Furth-Aigen, Austria; Fink *et al.*, 1976) to actually represent the whole time-span from the start of the last (Riss/Würm) interglacial to the beginning of the main Würm glacial stade, a time-span at least 10 times longer. Subdivided equivalents of the Paudorf Soil are known at several localities in northeastern Austria and many more are known in Czechoslovakia. A few localities in Czechoslovakia are the most fully subdivided and show the sequence given on the left side of Figure 8. In Austria, Stillfried is the most fully subdivided locality; here most of the soil-stratigraphic units shown on the left side of Figure 8 can be identified.

Several possible systems might be used to designate the individual geosols in a subdivided sequence that normally is represented by a single composite geosol. The system that seems preferable to me is that now used in Czechoslovakia; it is explained in Appendix B. If the system recommended in the American Stratigraphic Code is used, with soils (geosols) designated after place names, perhaps the individual geosols in subdivided occurrences could be called either "the_____ (sub) unit of the _____ geosol" or "the _____ member of the _____ geosol."

How the Stratigraphic Relations of a Geosol are Determined

The usefulness of a geosol as a stratigraphic marker depends on how closely its stratigraphic interval (its maximum and minimum age relative to associated deposits) can be determined within the local Quaternary succession.

Any geosol, relict or buried, obviously is younger than the youngest stratigraphic unit or geomorphic surface on which it developed. By tracing it across a variety of substrates its maximum age can be determined, and also its

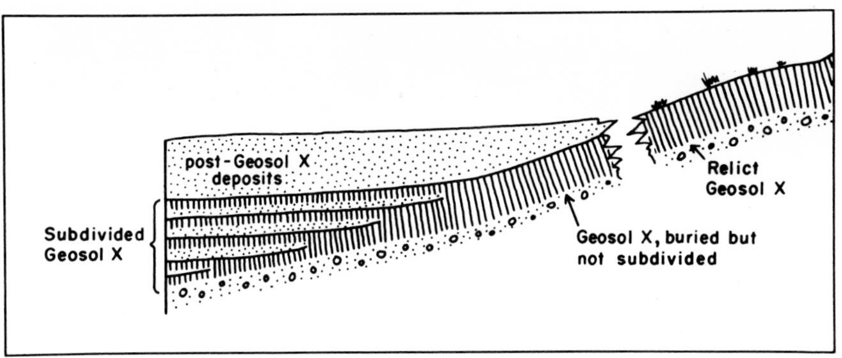

Figure 3 Subdivided and non-subdivided occurrences of a single geosol.

separate identity, by demonstrating the continuity of the geosol as compared with the diversity of underlying surficial or bedrock units, showing that none of them can be parts of the geosol.

The minimum age of a geosol is found by observing the geosol in various buried occurrences, noting the oldest deposit that overlies the geosol. Relict geosols provide at best only an approximate idea of their minimum age: a relict geosol is older than the oldest geomorphic surface or erosional feature that is cut into it.

Thus, to determine the stratigraphic/age relations of a geosol with precision, detailed stratigraphic study is necessary, of an area having a reasonably continuous, well-exposed Quaternary sequence within which the geosol is buried. The stratigraphic/age relations of relict geosols cannot be determined with anything like the precision that they can be for buried geosols at favorable localities. Relict geosols provide so incomplete and blurred evidence of their age that they should always be regarded as low-grade soil-stratigraphic units, inferior to buried geosols. Not only do relict geosols fail to provide good evidence on their minimum age by stratigraphic superposition, but also they usually are composite profiles that represent long spans of time that include various pedogenic episodes alternating with episodes of little or no pedogenesis, especially on landsurfaces older than the last interglacial. For these reasons, the desirability of including relict geosols with soil-stratigraphic units may seem questionable. However, relict geosols are by far the most common type of geosol occurrence; also, they provide evidence for distinguishing between various geomorphic surfaces, albeit imperfectly and with only very crude stratigraphic/age resolution. It is most unfortunate that soil scientists doing research on the age of soils (time factor in pedogenesis) usually confine their studies to relict soils.

Figure 4A The Paudorf Soil. This well-known example of a buried geosol is an important object lesson for soil stratigraphers -- it has many analogues in the U.S. For many years it was believed to be a simple soil profile that formed during a late-middle Würm interstadial, an episode considerably less than 10,000 years long. Recent research (Kukla, 1975; Fink *et al.*, 1976), however, has determined that this is a composite geosol whose development started at the beginning of the last (Riss/Würm) interglacial and ended with the last interstadial of the Würm, a time span of about 100,000 years. Part A shows this geosol at its type locality, Paudorf, Austria, where it developed on loess of the before-last pleniglacial and is buried by loess of the last (main Würm) pleniglacial.

Geosol Facies

Facies changes in geosols are an important concept in stratigraphy. A geosol, being laterally traceable and correlatable, can change considerably in its profile characteristics from place to place (*e.g.*, from one soil series to another) as a result of changes in local environmental factors. The changes in profile characteristics of a geosol that result from changes in these factors (climate, vegetation, topography/drainage, and parent material) are called facies of the geosol, analogous with the usage by

Figure 4B Part B is a closeup showing the argillic B and weak Cca horizons. This profile telescopes all the soil-forming episodes that are represented (where fully subdivided) by the individual profiles shown in the B cycle of Figures 5 and 7 and at the left edge of Figure 8.

geologists of the term "facies" for lateral changes in lithostratigraphic and biostratigraphic units. Geosol facies changes can be on a small or large scale. They can be caused, for example, by variations in soil drainage resulting from minor topographic changes in a farmer's field, from changes in parent material from one geologic terrain to another, and from climatic changes with altitude from lowlands into mountains, as well as over several degrees of latitude.

GEOLOGIC EVIDENCE ON TIME AND CLIMATE IN GEOSOL DEVELOPMENT

Introduction

R.B. Morrison

A decade ago (1967) I proposed the concept that the stronger Quaternary geosols formed intermittently during relatively short and discrete episodes correlative with interglacials and main interstadials, and that the geosol-forming episodes alternated with generally longer times of little or no soil formation, correlative with glacial stadials.[1] Thus, the soil-forming episodes were approximately synchronous, at least at middle latitudes. Because the degree of pedogenesis correlates well with the degree of warmth of the interglacials and interstadials, I also suggested that increase in temperature was the main climatic factor causing an accelerated rate of soil development.

Birkeland and Shroba recently (1974) argued against the idea of intermittency of soil-forming intervals, and postulated that soil development perhaps may have been continuous at a steady rate under climate near that of the present in a particular area. They presumably would explain the numerous soil profiles that can be proven to be stratigraphically discrete as the result of random accidents of local erosion and deposition. In particular, they disputed the thesis that increase in temperature was the primary control for soil development, claiming that there is abundant evidence that soil development took place during times of glaciation in a variety of environments in the western U.S.

I dispute Birkeland and Shroba's arguments, particularly their stratigraphic/chronologic evidence that substantial soil development took place during glacials, and

---

[1] In this paper the terms "interglacial," "glacial," "interstadial" (or interstade), and "glacial stadial" (or glacial stade) are used in the sense of the geologic-climate classification of the 1961 American Stratigraphic Code. In this sense, these climastratigraphic terms refer to physical stratigraphic units that record climatic episodes of broad extent. Although based on the glacial record, they apply not only to glaciated areas but also to areas far removed from those affected by continental and alpine glaciers.

It should be recognized also that the traditional concept of "interglaciation" in the U.S. differs from the European usage of "interglacial." According to modern European usage, an interglacial is a relatively short semi-stable interval of climate as warm or warmer than the Holocene. According to U.S. usage, however, interglaciations such as the Sangamonian and Yarmouthian were long intervals having climatic oscillations of considerable magnitude, including both warm and cold episodes, *i.e.*, several interglacials separated by weak glacials, if the testimony of the deep-sea record is correct. Likewise, the Illinoian, Kansan, and Nebraskan Glaciations actually each include several glaciations (of the marine record) separated by interglacials.

the ancillary implications. It is impractical to answer their arguments in detail in this paper. I will, however, touch on some of the major evidence supporting my conclusions. A large body of data has been published and I can only summarize here that which seems to be most significant and well established.

At the outset, it is very important to recognize two key points: (1) Modern studies of high-resolution stratigraphic sequences have established a multiplicity of Quaternary geosols -- many more than were recognized even 10 years ago. This greatly increases the problems in determining an exact chronology of the soil-forming intervals and in correlating the geosols from one area to another. (2) Nevertheless, the Quaternary sequences at most localities provide only a blurred, partly telescoped, partly eroded soil-stratigraphic record. This is why our knowledge of Quaternary soil development and soil stratigraphy has progressed so slowly and is fraught with so many questions.

Because of the multiplicity of geosols, answers to the questions about the factors of time (duration, multiplicity, synchroneity, *etc.*) and climate (temperature vs precipitation) in soil genesis, and about the relative importance of time and climate, hinge largely upon precise methods of time correlation. I reviewed the problems and methods of time-stratigraphic correlation of Quaternary sequences in an earlier paper (1968). Reliable time correlation of geosols must be based on detailed studies of <u>buried</u> occurrences of the geosols in high-resolution stratigraphic sequences, using a variety of ancillary methods of investigation, perhaps including geochemistry, paleomagnetism, isotopic dating, volcanic-ash chronology, and paleontology. Data on time and climate commonly are intertwined -- for example, proof that a geosol formed during an interglacial demands exact chronologic correlation with a specific interglacial; also, proof that the climate when the geosol formed was interglacial in character requires evidence that this climate was at least as warm as now, from paleontologic data (*e.g.*, from molluscan fauna, pollen or macro-plant flora), or oxygen-isotope or other chemical data.

The Deep-Sea Record

For the most conclusive evidence on the chronology and character of Quaternary climatic changes (which inevitably must be the basis for subdivision of the Quaternary Period), I will begin with what may seem to be a strange detour, to the ocean floors. Information on deep-sea stratigraphy, obtained from cores raised from the floors of the oceans of the world, has revolutionized our understanding of Quaternary stratigraphy, chronology, and climatology. It has established the basic climatic history of the last million or so years -- the number of glacials and interglacials and their comparative duration and the magnitude of temperature changes[1] on a global basis. Sedimentologic, micropaleontologic, paleomagnetic, and oxygen-isotope analyses of the cores has established beyond reasonable doubt the number of glacial cycles during the Brunhes polarity epoch, and their general climatic character. Eight full interglacial-glacial cycles occurred during the Brunhes (*i.e.*, during the ≃ 730,000 years before the Holocene), and probably 17 or 18 glacial cycles during the entire Pleistocene. This compares with only four glacial cycles recognized in the standard classification of the Pleistocene in the midwestern U.S. and 5 glacial cycles (including Donau) in the Alps. Furthermore, Emiliani's (1972) oxygen-isotope determinations from planktonic foraminifera in deep-sea cores from the Caribbean and equatorial Atlantic have established that during about 90% of the last half-million years the near-surface ocean-water temperature was colder than now, and during only about 10% of this time the water was as warm or warmer than now. This means, in short, that there were six interglacials (including the Holocene) in the last half-million years, each lasting an average of less than ten thousand years, which is less than the duration of the present (Holocene) interglacial.

To demonstrate the above, I have chosen to show the oxygen-isotope curve (Figure 5, part A) obtained from a single deep-sea core raised from the Solomon Plateau in the western equatorial Pacific Ocean (Shackleton and Opdyke, 1973). This core is considered by oceanographers to be representative of the temperature data from many dozens of cores taken from ocean floors all over the world, and it also supplies an especially long record. Good additional references are: Be *et al.*, 1976; Broecker and Van Donk, 1970; Emiliani, 1966a, 1966b, 1967, 1970, 1971; Ericson and

---

[1]Information on temperature changes recorded in deep-sea cores is obtained in five principal ways: (1) variations in abundance of the warmth-loving *Globorotalia menardii* group of foraminifera, (2) variations in the coiling direction of two foraminifera, *Globiginerina pachyderma* (left coiling predominates during times of <u>cold</u> climate, and *vice-versa*) and *Globorotalia truncatulinoides* (left coiling dominates during times of <u>warm</u> climate); (3) changes in total carbonate content resulting from variations in abundance of coccoliths, foram tests, and pteropod tests (carbonate content increases with increase in ocean-surface temperature); (4) oxygen-isotope analyses from the planktonic foraminiferan, *Globigerinoides sacculifera* (which is relatively insensitive to temperature changes). The O^{18}/O^{16} ratio changes inversely with temperature of ocean-surface water and directly with the volume of ice stored on land; temperature apparently is the more important factor (see Shackelton, 1969); (5) changes in the >250μm particle-size fraction (in North Pacific Ocean deep-sea cores; the >250μm fraction is believed to have been transported by ice-rafting during glacials, see Kent, Opdyke, and Ewing, 1971).

Wollin, 1968; Hays *et al.*, 1969, 1976; Imbrie and Kipp, 1971; Imbrie *et al.*, 1973; Ruddiman, 1971; Shackleton, 1969; Shackleton and Opdyke, 1973; Kukla and Nakagawa, 1977.

Especially significant in the deep-sea record is the evidence of rapid warming of the surface layer of the oceans after each pleniglacial. The mid-points of these times of rapid warming are called "Terminations" (Broecker and Van Donk, 1970), because they are the most sharply defined boundaries, not only in the oxygen-isotope curves but also in the microfossil curves and on the basis of sedimentologic changes in the cores.

Regarding the better preservation of the stratigraphic record on the ocean floors, the comment of Ewing (1971, p. 572) is appropriate: "The large number of glacial stages indicated may seem surprising, but may perhaps be less so if the problems of estimating the number of glaciations from evidence on continents is contrasted with the problem of estimating the number of glaciations from evidence in deep-sea sediments. The first may be compared in complexity to estimating the number of times the blackboard has been erased; while the second may be compared to finding the number of times the wall has been painted."

The Loess Record Of Central Europe

Next, let us consider the most complete and best-studied soil-stratigraphic records that I know of that are exposed on land.

In the loess regions of central Europe, notably Czechoslovakia and Austria, some remarkable sequences are exposed in brickyard pits. They show stratigraphic detail that is practically unrivaled elsewhere on land and some of them also have a long time range. Dozens of buried soils occur in these sequences. These geosols have been studied intensively in terms of their stratigraphic associations, sedimentology, pedology, micropedology, malacology (snail faunas), and paleomagnetism (Demek and Kukla, 1969; Kukla, 1970, 1975; Fink and Kukla, 1977; Fink *et al.*, 1976). Several sequences include the Brunhes-Matuyama magnetic polarity reversal, about 740,000 years ago. The loess sequences show a cyclicity of deposition and soil formation that is impressive (Figures 5, 6, 7, 8). First, second, and third-order cycles can be distinguished. The first-order cycles represent glacial cycles (each starting with an interglacial and progressing through the following glacial), analogous to the glacial cycles between the various Terminations of the deep-sea record. At the base of each loess cycle is a brownearth soil with an argillic B horizon, formed under mixed deciduous forest and characterized by interglacial snails, *Helix pomatia* and *Helicigona banatica* faunal assemblages, indicating temperate-moist climate comparable to that of the Holocene or warmer and wetter. The base of each interglacial soil, overlying calcareous loess of the preceeding pleniglacial, is an especially

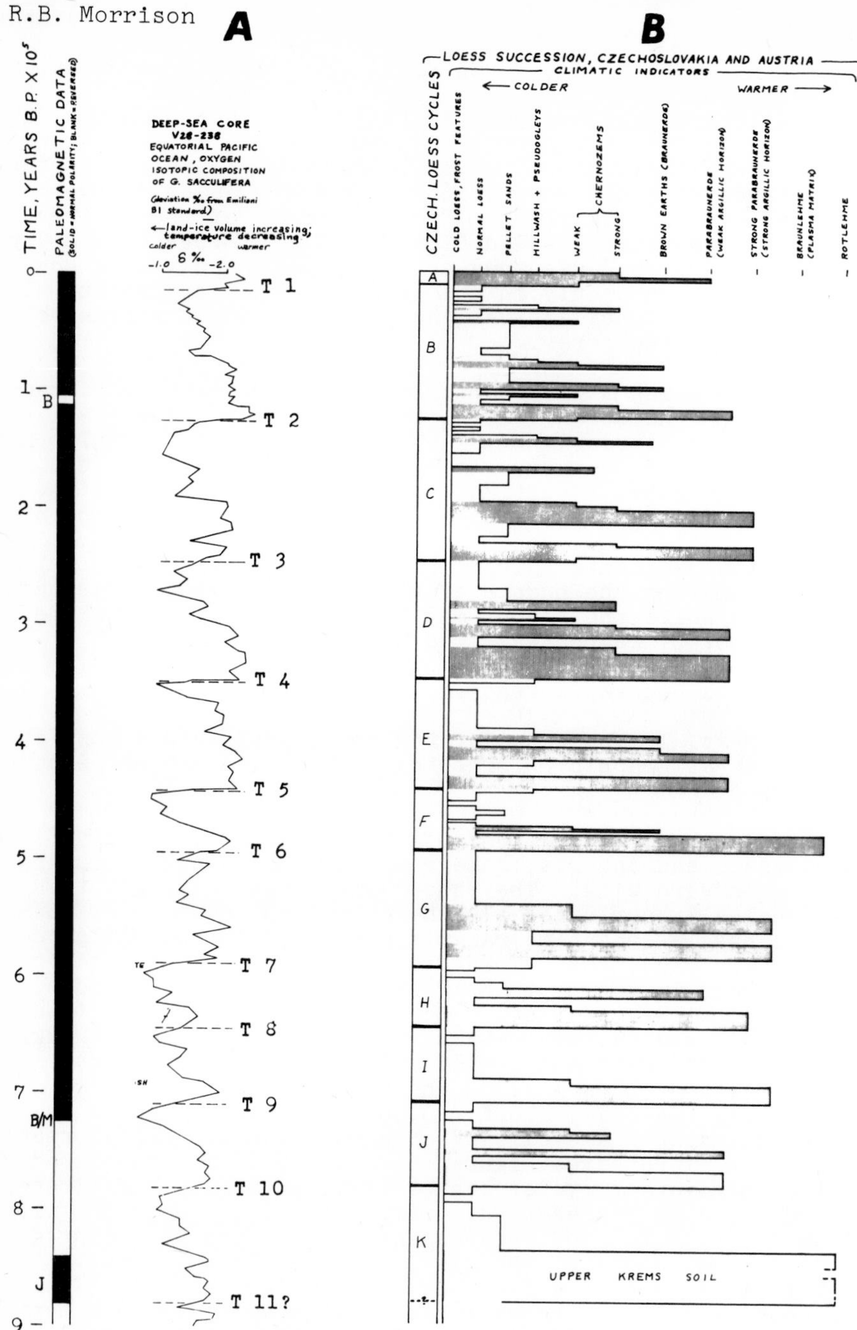

Figure 5 Comparison of the deep-sea and central-European loess records for the last 900,000 years. Vertical scale represents time in (continued Page 93)

Figure 5 (Continued)

hundreds of thousands of years before the present. Part A shows the glacial cycles indicated by the oxygen-isotope "temperature" curve from a deep-sea core from the western equatorial Pacific Ocean. Note the Terminations (T-1 to T-11) marking the start of each interglacial. Part B is the loess curve derived from studies of dozens of loess localities in Czechoslovakia and Austria (Kukla, 1970, 1975). On the horizontal axis are plotted semiquantitative indications of cold to warm climate (increasingly warm conditions to the right) as determined from sedimentologic and pedologic criteria. Note particularly that the time intervals of formation of the various geosols are shaded, and the degree to which these intervals extend to the right indicates the degree of profile development attained. Each loess cycle starts with an interglacial soil. This record shows that significant soil-profile development occurred during only about 25% of the time during the last full glacial cycle (the B cycle) and about 1/3 of the time during the last 700,000 years.

distinctive stratigraphic horizon and is termed a "Marker" ("Leitgrenze" in German; Kukla, 1969, 1970, 1975). Thus the various Markers are analogous to and correlative with the Terminations of the deep-sea record.

Within the Brunhes polarity epoch, the loess sequences show nine interglacials (including the Holocene), represented by forest soils, and several times this many interstadials, represented by grassland and open-forest soils. Figure 5, part B, shows all the stronger geosols (shaded) that formed during the last 900,000 years (and the duration and climatic character of each episode of geosol formation), as determined from detailed studies of buried soils in dozens of loessial sequences in Czechoslovakia and Austria. The vertical axis represents time in hundreds of thousands of years before present. (Note also the paleomagnetic polarity scale and the position of the Brunhes-Matuyama (B/M) polarity reversal.) On the horizontal axis are plotted various semiquantitative indications of temperature from geologic and pedologic data, with colder-climate indications increasing toward the left and warmer-climate indications increasing toward the right. (Comprehensive research on the snail faunas in the loessial sequences in Czechoslovakia and Austria has produced independent semiquantitative evidence on the temperature fluctuations, graphed as a "malacological temperature curve" or "malacofauna curve" (Lozek, 1969a, 1969b; Kukla, 1970), that

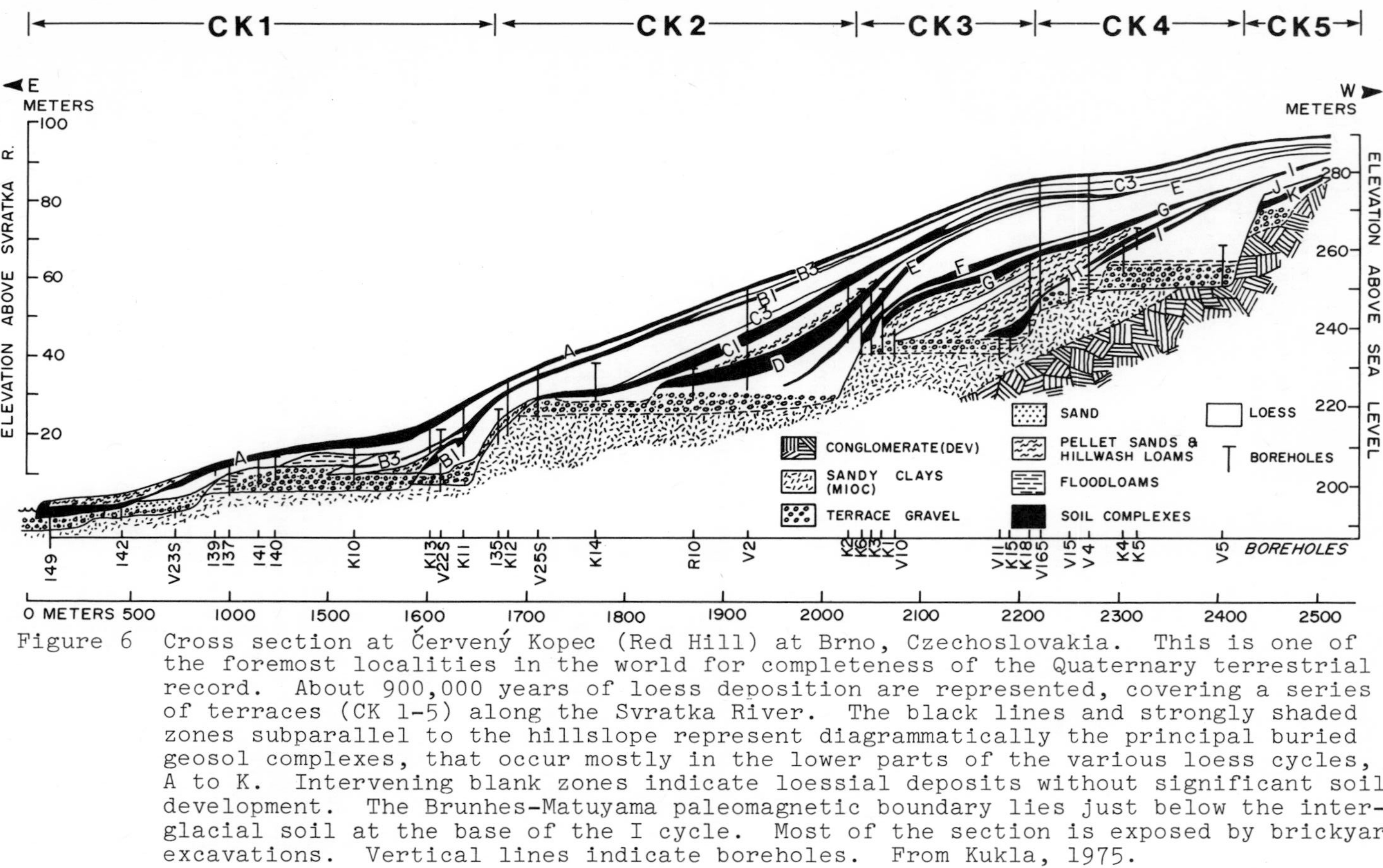

Figure 6 Cross section at Červený Kopec (Red Hill) at Brno, Czechoslovakia. This is one of the foremost localities in the world for completeness of the Quaternary terrestrial record. About 900,000 years of loess deposition are represented, covering a series of terraces (CK 1-5) along the Svratka River. The black lines and strongly shaded zones subparallel to the hillslope represent diagrammatically the principal buried geosol complexes, that occur mostly in the lower parts of the various loess cycles, A to K. Intervening blank zones indicate loessial deposits without significant soil development. The Brunhes-Matuyama paleomagnetic boundary lies just below the interglacial soil at the base of the I cycle. Most of the section is exposed by brickyard excavations. Vertical lines indicate boreholes. From Kukla, 1975.

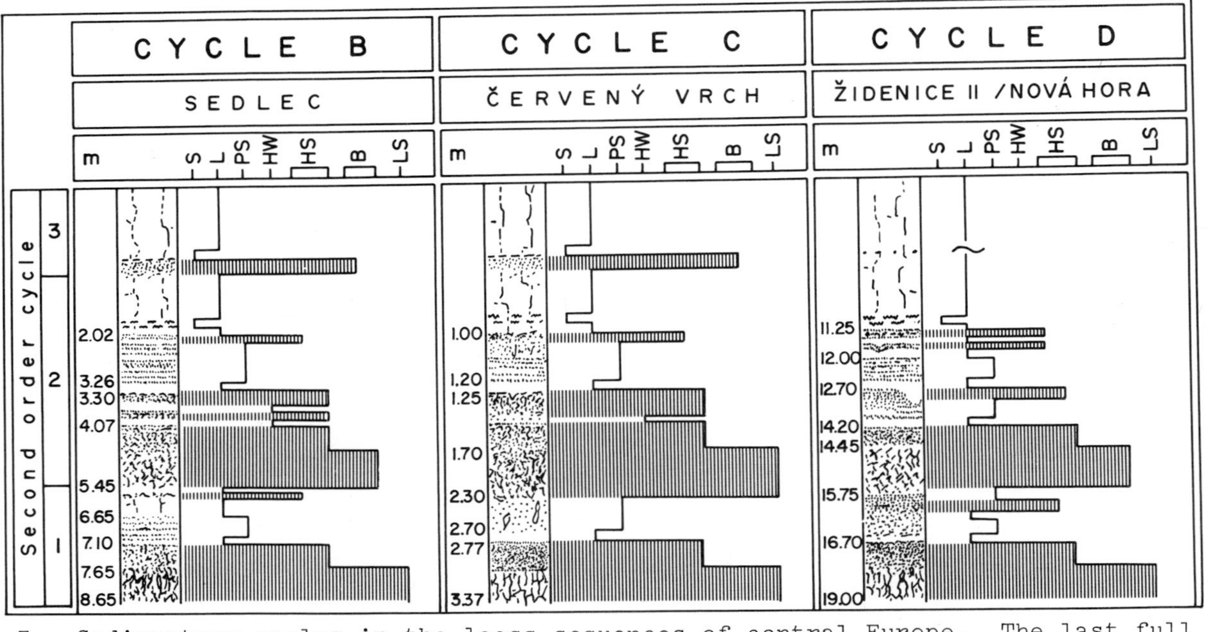

Figure 7 Sedimentary cycles in the loess sequences of central Europe. The last full glacial cycle (B) is compared with the two preceding cycles. To facilitate comparison, the real thickness of the particular layers is not respected, but their relative depth is shown in meters at the left sides of the columns that diagram their sedimentologic and pedologic character. To the right of each column the loess curve is drawn, which indicates semiquantitatively the temperature conditions recorded by the deposits and soil profiles, with progressively warmer conditions to the right. Abbreviations: S = material moved by solifluction, L = loess, PS = pellet sand, HW = hill wash, HS = humic steppe (grassland) soil, B = brownearth, weak or moderate; LS = lessive brownearth with argillic B horizon. The principal geosols are shaded in the loess curves. After Kukla, 1970.

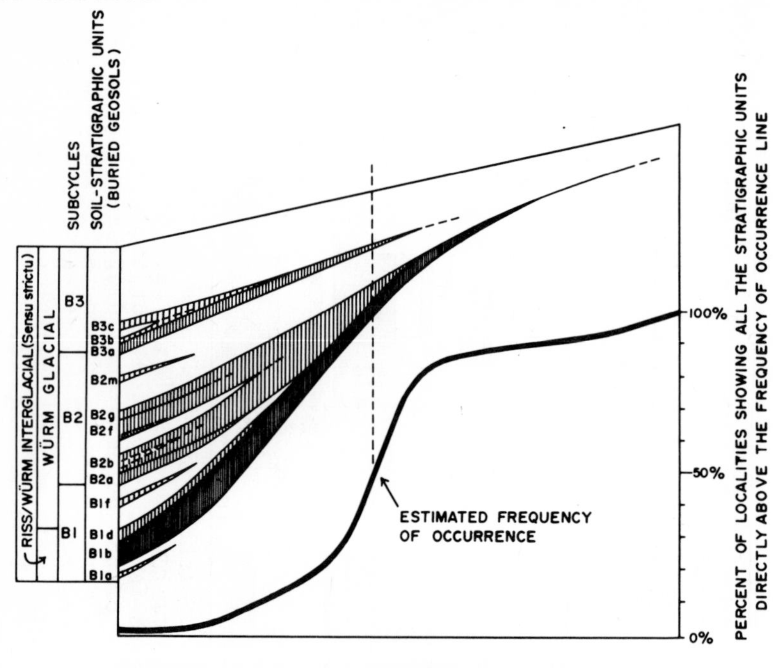

Figure 8 Principal geosols in the B cycle and frequence of their occurrence in loess localities in Czechoslovakia. The left edge of this diagram shows all the geosols that are known from detailed study of dozens of localities. This portion represents the fullest subdivision, separation, and preservation of the individual buried geosols. Progressively to the right are represented the actual observed effects of factors that tend to decrease the number of geosols preserved at a locality. These factors include: (1) compositing of geosols as a result of non-deposition of intervening sediment, and (2) removal of degradation of the soil profile because of erosional truncation, biogenic turbation, and retrograde weathering.

The heavy lower line shows the percentage frequency of occurrence of localities that display the various soil-stratigraphic units that are shown vertically above a given point on this line. (Blank areas above this line represent loess units.) Completely subdivided sequences like that shown at the

(continued Page 97)

Figure 8 (Continued)

left margin are very rare, being seen in less than 2% of the loess localities. In about 50% of the studied localities (represented by the vertical dashed line in the center of the diagram), only the interglacial forest soil B1b, Chernozem B2g, and Brownearth B2a can be recognized. (The Paudorf Soil, Figure 4, is a case with all three geosols composited directly on one another.) (After Kukla, 1975.)

resembles and supports the loess curves shown in Figures 5 and 7.) By these exceptionally detailed studies in a region especially favorable for obtaining the fullest possible soil-stratigraphic record, the various pedogenic episodes can be clearly separated from intervening episodes with virtually no soil development.

The composite, integrated evidence from the multidisciplinary studies of the loess sequences clearly establishes that formation of the stronger soils was not random in time and space: they formed during interglacials and first-order interstadials, which botanical and malacologic evidence shows were warmer and wetter than today (these were the true pluvials of the Pleistocene). Furthermore, for the interstadials, there is a close correlation between the degree of warmth and the degree of soil development. Conversely, negligible soil development took place during the glacial stadials, particularly during the pleniglacials which were cold and dry.[1]

I am not aware of any European workers who report reliable stratigraphic/chronologic evidence of significant soil development during glacial stadials. Among the modern workers who cite evidence of significant soil development during interglacials and interstadials (only) are: Brunnacker, 1957, 1967, 1969, 1974; Brunnacker and Lozek, 1969; Brunnacker *et al.*, 1969a, 1969b, 1971; Demek and Kukla, 1969; Fink, 1974; Fink *et al.*, 1976; Fink and Kukla, 1977; Kukla, 1969, 1970, 1975; Kukla *et al.*, 1972; Lozek, 1969a, 1969b, 1972; Paepe, 1974; Paepe and Vanhoorne, 1967; Rohdenburg and Sabelberg, 1973; Schirmer, 1970, 1974; Schirmer and Streit, 1967; Sibrava, 1972; Turner, 1970; Turner and West, 1968.

---

[1]Pleniglacials were 8° - 10° C cooler than now on land in Europe and 6° - 8° C cooler at the sea surface at latitude 40° N; the global average cooling was 2.3° C. The decrease in precipitation was less regular from place to place, but the general marked dryness of pleniglacials at low to high latitudes is well established from plant floral, mollusk faunal, and sedimentologic evidence.

The U.S. Record

Geologists in the U.S. also are generally in agreement that the stronger geosols formed during interglacials and main interstadials; for example, Willman and Frye, 1970; Reed and Dreeszen, 1965; Scott, 1963, 1965; Morrison, 1964, 1965a, 1965b; Morrison and Frye, 1965; Pierce *et al.*, 1976; Richmond, 1962, 1965, 1970. Here again, I know of no case where significant soil-profile development can surely be correlated with a glacial stadial.

I consider highly questionable several instances in the southwestern U.S. where geosols have been correlated with glacial stadials (Las Cruces, New Mexico; Las Vegas, Nevada; San Pedro Valley, Arizona), given in Birkeland *et al.*, 1971. These correlations are based partly on a few C^{14} dates (the older of which are from carbonate and hence are only minimum dates), partly on finger-matching correlation of the alluvial units and soils (including the probably incorrect assumption that the Bull Lake Glaciation is correlative with the early part of the last glaciation; see Pierce *et al.*, 1976), and partly on the theoretical but mistaken idea that lack of moisture was the chief constraint on soil development in this warm region, especially during interglacials and interstadials, whereas augmented moisture during the glacial stadials favored pedogenesis. (It now is well established that moisture increased during the interglacials and interstadials, and decreased during the glacial stadials, making the glacial stadials the least favorable times for soil development; see footnote in preceeding section.)

Discussion

The close agreement between the loess and deep-sea cycles, the good correlation between them based on paleomagnetic stratigraphy and isotopic dating, and the vast oceanic regions represented by the deep-sea cores, all indicate that the first- and second-order temperature changes during the last half-million years were globally synchronous. The loess record of central Europe (supported by evidence cited from elsewhere) establishes clearly the correlation of the stronger buried geosols with interglacials, which were relatively warm and humid compared to the glacial stadials. Also, the fact that no significant soil development took place during glacial stadials is well established. Inasmuch as the individual interglacial geosols in the fully subdivided loess sequences are buried and not polygenetic, there is no evidence in support of Birkeland and Shroba's suggestion that significant soil formation has gone on at a fairly steady rate, at other times than interglacials and interstadials, and that time alone is the key factor in soil genesis, not climate.

CONCLUSIONS ON TIME AND CLIMATE IN THE FORMATION OF GEOSOLS AND IMPLICATIONS FOR LONG-DISTANCE CHRONO-STRATIGRAPHIC CORRELATION

The modern stratigraphic evidence summarized above reinforces my chief conclusions of a decade ago, at least for the second half of the Quaternary:

(1) Episodes of relatively rapid soil-profile development (geosol-forming episodes) alternated with times of negligible soil-profile development. Thus, the concept of "intermittency of Quaternary geosol-forming episodes" is valid.

(2) The stronger geosols invariably formed during interglacials which were not only as warm or warmer than the Holocene, but also times of increased precipitation. Moderately strong to weak geosols formed during interstadials, with the degree of soil development corresponding closely to the degree of warmth. Negligible profile development took place during the pleniglacials, which were cold and dry.

(3) Because temperature and precipitation both increased during the interglacial and interstadial episodes, these episodes were times of accelerated pedogenesis compared to the glacial stadials. This holds true over regions of wide climatic diversity, regardless of the local climatic constraint-tendency on soil development, whether it be moisture (in arid areas) or temperature (in cool regions).

(4) The interglacial geosol-forming episodes were much shorter than the intervening glacial-interstadial parts of the glacial cycles. In central Europe they comprised less than 15% of the last half-million years. The interstadial geosol-forming episodes likewise generally were shorter than the non-soil-forming intervals of the glacial stadials.

(5) The deep-sea record proves that the interglacials and first-order interstadials were nearly synchronous on a global basis, although they began somewhat earlier and ended somewhat later at low latitudes compared with high latitudes.

(6) From the above, it appears that buried geosols that are not composite and thus represent individual interglacials or first-order interstadials are among the best chronostratigraphic units in Quaternary successions. They are more nearly time parallel than any common sedimentary and biostratigraphic units in Quaternary terrestrial sequences. Many of them also are widespread and distinctive, making them good stratigraphic markers. Thus, such geosols are useful for time-stratigraphic subdivision of sequences, as well as for time-correlation (at least regionally and probably when all the evidence is in, on a continental and intercontinental scale), provided that they can be identified reliably. (Various ancillary stratigraphic inputs should be used wherever possible as aids

to correlation, such as isotopic dating, tephrochronology, and archeologic, paleontologic, and paleomagnetic studies of the associated deposits.)

(7) Relict (surface) geosols, on the other hand, are poor time-stratigraphic units, permitting only rough correlations.

APPENDIX 1

A Geologist's Viewpoint on the U.S. System of Soil Classification

Geologists have been frustrated by recent developments in soil science in the United States. Increasingly, geologists are treating soil profiles as stratigraphic units and using them for subdivision and correlation of Quaternary stratigraphic sequences, as well as for evidence on climatic, erosional, and other changes in the surficial environment. Thus, geologists wish to apply in their work the principles and techniques of soil classification. However, the new U.S. soil classification system, recently codified in *Soil Taxonomy* (Soil Survey Staff, Soil Conservation Service, 1976) has exacerbated the problem of communication between the two sciences (Hunt, 1974). *Soil Taxonomy* is almost impossibly difficult for non-soil scientists to use, for it requires extensive background in soil science and commonly also chemical and mechanical analyses as well as climatic data in order to classify soils. Also, *Soil Taxonomy* suppresses or distorts certain pedologic distinctions which can be important to geologists, by various legalistic and narrow restrictions; for example, the thickness limitations for most epipedons and diagnostic subsurface horizons, the limitation on texture for cambic horizons, and the minimum allowable concentration of $CaCO_3$ for calcic horizons.

In addition, *Soil Taxonomy* is woefully inadequate in its treatment of buried soils, and it ignores the problem of polygenetic soil profiles (which are widespread in both surface and buried situations). The treatment of buried soils is a disaster. On one hand, *Soil Taxonomy* defines a "*bisequum*" as consisting of two successive sequa of diagnostic soil horizons in a vertical sequence of what is called (incorrectly) "a single soil." (A single sequum is defined as an eluvial horizon and its subjacent B horizon, if one is present). Two examples are given of bisequal soils. In one (Pl. 7B) the upper sequum is about 1 foot thick; in the other (pedon 44, p. 570) it is 36 cm (14.3 in) thick. By geologic standards, the lower sequum surely is a buried soil. Yet, *Soil Taxonomy* defines a soil as *buried* only if it is covered by a surface mantle of essentially unaltered "new material" that is at least 50 cm (20 in) thick (or between 30 and 50 cm thick if the surface mantle is at least half the thickness of the diagnostic horizons in the buried soil). Therefore, excluded from consideration as a buried soil are any parts of a soil profile having less than 50 (or 30) cm of "surface mantle."

Also, this "surface mantle" cannot have any developed soil horizons -- to which I object strenuously.

I submit, as a geologist-soil-stratigrapher, that the so-called Fluvent illustrated on page 192 (Fig. 37) of Soil Taxonomy in reality is a four-story pedocomplex with three buried geosols of Holocene age, each marked by an organic-rich A horizon.

A serious deficiency that exists in the system used in the U.S. for describing soil profiles (given in Appendix I of Soil Taxonomy) is that no symbols exist for identifying geologic unconformities[1] in soil profiles. (The Roman Numeral symbols apply, of course, to changes in particle size/lithology of "parent material," which may or may not be related to a geologic unconformity.) Geologic unconformities within soil profiles are much more common than most soil scientists realize -- but generally it takes a geologist skilled in Pleistocene stratigraphy and soil stratigraphy to recognize them.

I suspect that the various deficiencies in the U.S. system of soil classification and description that are cited above have arisen because few soil scientists in the U.S. are trained in Quaternary stratigraphy, geomorphology, and climatology to recognize how useful inputs from these disciplines can be to soil classification and to understanding soil genesis.

APPENDIX 2

The Czechoslovakian System of Designating Soil-Stratigraphic Units

Originally, the buried geosols in the loess sequences of Czechoslovakia were named after type localities. After detailed studies at many loess localities, however, the geosols were found to be so numerous that the practice of naming individual geosols after type localities was abandoned. Instead, the following practice has been adopted:

(1) The primary (first-order) division of the Czech loess succession is into loess cycles. Each loess cycle begins with an interglacial geosol and continues with the deposits and interstadial geosols of the ensuing glacial interval. The various loess cycles are designated by capital letters, starting with A for the

---

[1] Geologic unconformities in soil profiles can be either of two kinds: (1) a diastem, which is the record of a hiatus in deposition without appreciable erosion, and (2) a disconformity, which records a break in deposition marked by erosion, with the strata above and below the break parallel with each other.

Holocene, then B for the last complete interglacial-glacial cycle, C for the next -to-last cycle, and so on.

(2) Second-order divisions within the loess cycles are designated by Arabic numerals. Generally only three divisions of this rank are distinguished.

(3) The individual geosols (as identified at high-resolution fully stretched-out localities) are designated by a lower-case letter following the first two symbols, *e.g.*, B2a (where the small a designates an interstadial geosol).

By this means it is possible to indicate precisely the individual loess cycle (glacial cycle) and second- and third- order divisions in which each geosol occurs. The intention is to use the same designation for geosols believed to be of the same age throughout the loess region. The abbreviated form of designation (outlined above, *e.g.*, B2a) is used in general discussions of the stratigraphy of the loess region. However, correct correlation cannot always be taken for granted at each locality. Therefore, the complete designation at a particular locality has to include the abbreviated locality symbol (*e.g.*, CK for Cerveny Kopec); also, when correlations are made with stratigraphic units outside the loess region, the affix L is used to avoid confusion, thus: L-CK-C3j.

An older and less specific designation is that still used to designate pedocomplexes (indicated by PK) of more than one buried soil within a small stratigraphic interval. The system used in Czechoslovakia starts with the Holocene, thus:

PK0 = all Holocene soils
PK1 = soils of the B3 subcycle
PK2 = soils of the B2 subcycle
PK3 = soils of the B1 subcycle
PK4 = all soils of the C cycle
PK5 = all soils of the D cycle
etc.

These pedocomplexes can be regarded as subdivided/compounded geosols where the individual geosols are separated by intervening sediment, but generally they grade laterally into composite geosols.

REFERENCES CITED

American Commission on Stratigraphic Nomenclature, 1961, Code of Stratigraphic nomenclature: Amer. Assoc. Petroleum Geologists Bull., v. 45, p. 645-665.

Bé, A.W.H., Damuth, J.E., Lott, L., and Free, R., 1976, Late Quaternary climatic record in western equatorial Atlantic sediments, *in* Cline, R.M. and Hays, J.D., eds.,

Investigations of late Quaternary paleoceanology and paleoclimatology: Geol. Soc. Amer. Mem. 145, p. 165-200.

Birkeland, P.W., and Shroba, R.R., 1974, The status of the concept of Quaternary soil-forming intervals in the western United States, *in* Mahaney. W.C., ed., Quaternary environments, Geographical Monographs: Toronto, York University Ser. Geog., p. 241-276.

Birkeland, P.W., Crandell, D.R., and Richmond, G.M., 1971, Status of correlation of Quaternary stratigraphic units in the western conterminous United States: Quaternary Research, v. 1, p. 208-227.

Broecker, Wallace S., and Van Donk, Jan, 1970, Insolation changes, ice volumes, and the O^{18} record in deep-sea cores: Reviews of Geophysics and Space Physics, v. 8, no. 1, p. 169-197.

Brunnacker, Karl, 1957, Die Geschichte der Böden im jüngeren Pleistozän in Bayern: Geologica Bavarica no. 34, 95 p.

________________, 1967, Grundzüge einer Löss- und Bodenstratigraphie am Niederrhein: Eiszeitalter u. Gegenwart, v. 18, p. 142-151.

________________, 1969, Affleurements de loess dans les regions nord-Mediterranneennes: Rev. Geog. Phys. Geol. Dynamique (2), v. 11, p. 325-334.

________________, 1974, Results of Quaternary stratigraphy on the middle and lower courses of the Rhine: in Sibrava, V., ed., Quaternary glaciations in the Northern Hemisphere: IUGS-UNESCO Internatl. Geological Correlation Programme, Project 73/1/24, Rept. 1 (Sept. 15-20, 1973 session in Cologne, FRG), p. 16-33.

Brunnacker, Karl, and Ložek, Vojen, 1969, Löss-Vorkommen in Südost-spanien: Zeitschr. Geomorphologie, n.f., v. 13, p. 297-316.

Brunnacker, K., Basler, D., Ložek, V., Beug, H-J, and Altemuller, H-J., 1969a, Zur Kenntnis der Lösse im Neretva-Tal: N. Jb. Geol. Palaont. Abh., v. 132, p. 127-154.

Brunnacker, Karl, Streit, R., and Schirmer, W., 1969b, Der Aufbau des Quartär-Profils von Kärlich/Neuwieder Becken (Mittelrhein): Mainzer Naturwiss Arch. J. 8, p. 102-133 (Mainz, FRG).

Brunnacker, K., Heller, F., and Lozek, V., 1971, Beiträge zur Stratigraphie des Quartär-Profils von Kärlich am Mittelrhein: Mainzer Naturwiss. Arch. J. 10, p. 77-10 (Mainz, FRG).

Demek, Jaromir, and Kukla, Jiří, 1969, Periglazialzone löss und paläolithikum der Tschechoslowakei: Brno, Institute of Geography, Czechoslovak Academy of Sciences, 158 p.

Emilianai, Cesare, 1966a, Isotopic paleotemperatures: Science, v. 154, no. 3750, p. 851-857.

________________, 1966b, Paleotemperature analysis of Caribbean cores P6304-8 and P6304-9 and a generalized temperature curve for the past 425,000 years: Jour. Geol., v. 74, no. 2, p. 109-126.

________________, 1967, The Pleistocene record of the Atlantic and Pacific Oceanic sediments; correlations with the Alaskan stages by absolute dating; and the age of the last reversal of the geomagnetic field: Progress in Oceanography, v. 4, p. 219-224.

________________, 1970, Pleistocene paleotemperatures: Science, v. 168, p. 822-825.

________________, 1971, The last interglacial-paleotemperatures and chronology: Science, v. 171, p. 571-573.

________________, 1972, Quaternary hypsithermals: Quaternary Research, v. 2, p. 270-273.

Ericson, David B., and Wollin, Goesta, 1968, Pleistocene climates and chronology in deep-sea sediments: Science, v. 162, no. 3859, p. 1227-1234.

Ewing, Maurice, 1971, The late Cenozoic history of the Atlantic basin and its bearing on the cause of the ice ages, *in* Turekian, K.K., ed., The late Cenozoic glacial ages: New Haven, Conn., Yale Univ. Press, p. 565-573.

Fink, J., 1974, Key sites of Quaternary stratigraphy in the Danubian area: *in* Sibrava, V., ed., Quaternary glaciations in the Northern Hemisphere: IUGS-UNESCO Internatl. Geol. Correlation Programme, Project 73/1/24, Rept. 1 (Sept. 15-20, 1973 session in Cologne, FRG), p. 50-68.

Fink, J., and Kukla, G.J., 1977, Pleistocene climates in central Europe; at least 17 interglacials after the Olduvai Event: Quaternary Research, v. 7, p. 363-371.

Fink, J., Fischer, H., Klaus, W., Koci, A., Kohl, H., Kukla, J., Lozek, V., Piffl, L., and Rabeder, G., 1976, Excursion durch den Osterreichischen Teil des nördlichen Alpenvorlandes und des Donauraum zwischen Krems und Wiener Pforte, *in* Fink, J., ed., Kommission für Quartärforschung d. Osterreichischen Akad. d. Wissenshaften, Mit., Bd. 1, 113 p.

Hays, J.D., Imbrie, J., and Shackleton, N.J., 1976, Variations in the Earth's orbit: pacemaker of the ice ages: Science, v. 194, p. 1121-1132.

Hays, J.D., Saito, T., Opdyke, N.D., and Burckle, L.H., 1969, Plio-Pleistocene sediments of the equatorial Pacific: their paleomagnetic, biostratigraphic, and climatic record: Geol. Soc. Amer. Bull. v. 80, p. 1481-1514.

Hunt, C.B., 1974, Soil -- a dirty four-letter word: Geotimes, v. 9, no. 2.

Imbrie, J., and Kipp, N.G., 1971, A new micropaleontological method for quantitative paleoclimatology: application to a late Pleistocene Caribbean core: *in* Turekian, K. K., ed., The Late Cenozoic Glacial Ages: New Haven, Conn., Yale University Press, p. 71-183.

Imbrie, J., van Donk, J., and Kipp, N.G., 1973, Paleoclimatic investigation of a late Pleistocene Caribbean deep-sea core; comparison of isotopic and faunal methods: Quaternary Research, v. 3, p. 10-38.

Kent, D., Opdyke, N.D., and Ewing, M., 1971, Climate change in the North Pacific using ice-rafted detritus as a climatic indicator: Geol. Soc. Amer. Bull., v. 82, p. 2741-2754.

Kukla, J., 1969, Lagerungsverhältnisse und Stratigraphie der Lösse; *in* Demek, J. and Kukla, J., eds., Periglazialzone Löss und Paläolithikum der Tschechoslowakei: Geograph, Inst., Czech. Acad. Sci., Brno, p. 4-18.

_________, 1970, Correlations between loesses and deep-sea sediments: Geologiska Foreningen i Stockholm Forhandlung, v. 92, pt. 2, p. 148-180.

_________, 1975, Loess stratigraphy of central Europe; *in* Butzer, K. and Isaac, G.L., eds., After the Australopithecines: The Hague, Mouton, p. 99-188.

Kukla, G., and Nakagawa, H., 1977, Late-Cenozoic magnetostratigraphy: Quaternary Research, v. 7, p. 283-293.

Kukla, G.J., Matthews, R.K., and Mitchell, J.M., 1972, The end of the present interglacial: Quaternary Research, v. 2, p. 261-269.

Ložek, V., 1969a, Paläontologische Charakteristik der Löss-Serien; *in* Demek, J., and Kukla, J., eds., Periglazialzone Löss und Paläolithikum der Tschechoslowakei: Geograph. Inst., Czech. Acad. Sci., Brno, p. 43-59.

_________, 1969b, Uber die malakozoologische Charakteristik der pleistozänen Warmzeiten mit besonderer Berücksichtigung des letzten Interglazials: Ber. deutsch. Ges.

geol. Wiss., Geol. Paläont., v. 144, p. 439-469.

________, 1972, Holocene interglacial in central Europe and its land snails: Quaternary Research, v. 2, p. 327-334.

Morrison, R.B., 1964, Lake Lahontan; Geology of southern Carson Desert, Nevada: U.S. Geol. Survey Prof. Paper 401, 156 p.

________, 1965a, Quaternary geology of the Great Basin, *in* Wright, H.E., Jr., and Frey, D.G., eds., The Quaternary of the United States: Princeton, New Jersey, Princeton Univ. Press, p. 265-285.

________, 1965b, Lake Bonneville; Quaternary stratigraphy of eastern Jordan Valley south of Salt Lake City, Utah: U.S. Geol. Survey Prof. Paper 477, 80 p.

________, 1967, Principles of Quaternary soil stratigraphy; *in* Morrison, R.B. and Wright, H.E., Jr., eds., Quaternary Soils, Internatl. Assoc. for Quaternary Research (INQUA), VII Congress 1965, Proc. v. 9: Reno, Nevada, Center for Water Resources Research, Desert Research Inst., Univ. Nevada, p. 1-69.

________, 1968, Means of time-stratigraphic division and long-distance correlation of Quaternary successions; *in* Morrison, R.B., and Wright, H.E., eds., Means of correlation of Quaternary successions, Internatl. Assoc. for Quaternary Research (INQUA), VII Congress 1965, Proc., v. 8: Salt Lake City, Utah Univ. Press, p. 1-113.

Morrison, R.B., and Frye, J.C., 1965, Correlation of the middle and late Quaternary successions of the Lake Lahontan, Lake Bonneville, Rocky Mountain (Wasatch Range), Southern Great Plains, and Eastern Midwest areas: Nevada Bur. Mines Rept. 9, 45 p.

Paepe, R., 1974, Correlation of middle Pleistocene deposits with the aid of paleosols in Belgium: *in* Sibrava, V., ed., Quaternary glaciations in the Northern Hemisphere, IUGS-UNESCO Internatl. Geol. Correlation Programme, Project 73/1/24, Rept. 1 (Sept. 15-20, 1973 session in Cologne, FRG), p. 69-78.

Paepe, R., and Vanhoorne, R., 1967, The stratigraphy and paleobotany of the late Pleistocene in Belgium: Service Geologique de Belgique, Mem. 8, 95 p.

Pierce, K.L., Obradovich, J.D., and Friedman, I., 1976, Obsidian hydration dating and correlation of Bull Lake and Pinedale Glaciations near West Yellowstone, Montana: Geol. Soc. Amer. Bull., v. 87, p. 703-710.

Reed, E.C., and Dreeszen, V.H., 1965, Revision of the

classification of the Pleistocene deposits of Nebraska: Nebraska Geol. Survey Bull. 23, 65 p.

Richmond, G.M., 1962, Quaternary stratigraphy of the La Sal Mountains, Utah: U.S. Geol. Survey Prof. Paper 324, 135 p.

______________, 1965, Glaciation of the Rocky Mountains, *in* Wright, H.E., Jr., and Frey, D.G., eds., The Quaternary of the United States: Princeton, New Jersey, Princeton Univ. Press, p. 217-230.

______________, 1970, Comparison of the Quaternary stratigraphy of the Alps and Rocky Mountains: Quaternary Research, v. 1, no. 1, p. 3-28.

Rohdenburg, H., and Sabelberg, U., 1973, Quartäre klimazyklen im westlichen Mediterrangebiet und ihre Auswirkungen auf die Reliefund Bodenentwicklung: Catena, v. 1, p. 71-180.

Ruddiman, W.F., 1971, Pleistocene sedimentation in the equatorial Atlantic: stratigraphy and faunal paleoclimatology: Geol. Soc. Amer. Bull., v. 82, p. 283-302.

Schirmer, Wolfgang, 1970, Das jüngere Pleistozän in der Tongrube Kärlich am Mittlerhein: Mainzer Naturwissenschaftliches Archiv, J. 9, p. 257-284.

__________________, 1974, Mid-Pleistocene gravel aggradations and their cover-loesses in the southern lower Rhine basin: *in* Sibrava, V., ed., Quaternary glaciations in the Northern Hemisphere, IUGS-UNESCO Internatl. Geol. Correlation Programme, Project 73/1/24, Rept. 1 (Sept. 15-20, 1973 session in Cologne, FRG), p. 34-42.

Schirmer, W., and Streit, R., 1967, Die Deckschichten der niederrheinischen Hauptterrasse bei Erklenz: Geol. Inst. Univ. Köln, Sonderveröff, 13, p. 81-94.

Scott, G.R., 1963, Quaternary geology and geomorphic history of the Kassler Quadrangle, Colorado: U.S. Geol. Survey Prof. Paper 421-A, 70 p.

__________, 1965, Nonglacial Quaternary Geology of the southern and middle Rocky Mountains; *in* Wright, H.E. and Frey, D.G., eds., The Quaternary of the United States: Princeton, Princeton Univ. Press, p. 243-254.

Shackleton, N.J., 1969, The last interglacial in the marine and terrestrial records: Proc. Royal Soc. London, B. 174, p. 135-154.

Shackleton, N.J., and Opdyke, N.D., 1973, Oxygen isotope and paleomagnetic stratigraphy of equatorial Pacific core v 28-238: oxygen isotope temperatures and ice volumes on a 10^5 year and 10^6 year scale: Quaternary Research, v. 3, p. 39-55.

Sibrava, Vladimir, 1972, Zur Stellung der Tschechoslowakei im Korrelierungssystem des Pleistozäns in Europa: Anthropozoikum (Sbornik Geologickych Ved), rada A, sv. 8, 218 p. (On the position of Czechoslovakia in the correlation system of the Eruopean Pleistocene; English summary).

Soil Survey Staff, 1976, Soil Taxonomy: A basic system of soil classification for making and interpreting soil surveys, U.S. Dept. Agriculture Handbook No. 436: Washington, U.S. Government Printing Office, 754 p.

Turner, C., 1970, The Middle Pleistocene deposits at Marks Tey, Essex: Philosoph. Trans. Royal Soc. London, v. 257, p. 373-440.

Turner, C., and West, R.G., 1968, The subdivision and zonation of interglacial periods: Eiszeitalter und Gegenwart, B. 19, s. 93-101.

Willman, H.B., and Frye, J.C., 1970, Pleistocene stratigraphy of Illinois: Ill. Geol. Survey Bull. 94, 204 p.

Clay Mineral Weathering and Quaternary Soil Morphology

C. G. Olson, K. L. Brunson, R. V. Ruhe

ABSTRACT

Peorian loess, exposed in southwest Indiana, exhibits weathering zones. No intraloessial buried soils are present below the ground soils. Four clay-mineral zones may be defined on the basis of illite and expandable clay-mineral content. A basal sandy loess, common to Wisconsinan loesses of southwest Indiana, is not discernible in the thick loess area. Based on similar clay-mineral content, the sandy zone in the thinner loess of the ground soils appears to be stratigraphically comparable to the lower two clay-mineral zones of Peorian loess. Correlation cannot be made with the Roxana as has recently been proposed.

There is little morphological evidence for strong soil development in the Farmdale loess. These soils have only an accumulation of organic matter and are leached of carbonates. The dominant clay-mineral suite, a sedimentary feature is kaolinite-chlorite and expandable clays.

The Sangamon paleosol is formed on Illinoian glacial drift, Loveland loess, and is also formed, in part, on Pennsylvanian and Mississippian bedrock in the glacial re-entrant in the south-central part of the state. These buried soils probably developed with a chemistry and clay mineralogy similar to those of the ground soils but were subjected to more intensive or prolonged weathering.

INTRODUCTION

Weathering and soil profiles at four sections in southwestern Indiana were examined and compared on the basis of clay-mineral content and soil morphology.

Clay-mineral zones have been identified in Peorian loess at New Harmony. Illite and expandable clay minerals are dominant. Zoning appears to be a sedimentary phenomena. Ground soils developed in the Peorian loess may reflect this zoning, but pedogenic processes including fragipan formation obscure the primary sedimentation. Correlation of clay-mineral zones from the thick loess at New Harmony to the ground-soil sites in thin loess indicate that the basal sandy zone is part of the Peorian loess and not Roxana.

Soil formation in the Farmdale loess includes only the accumulation of organic matter and leaching of carbonates. In contrast to the Peorian, the Farmdale has greater amounts of kaolinite-chlorite and expandable clay minerals.

The Sangamon paleosol has formed in Illinoian glacial drift, Loveland loess and Paleozoic bedrock. The degree of soil formation is a function of intensity of weathering and the type of material in which it formed. Kaolinite-chlorite and illite are more abundant than in the Farmdale. Chemistry and clay mineralogy of the Sangamon is very similar to that of the overlying ground soils. However, increased abundance of kaolinite-chlorite indicates that weathering of the Sangamon has been more intense or prolonged.

SOILS IN PEORIAN LOESS

Upper Wisconsin Loess

Approximately 10 meters of Peorian loess are exposed in a borrow pit one mile south of the intersection of State Roads 69 and 460 in New Harmony, Indiana (Figure 1). The upper 5 meters are a massive calcareous yellowish brown (10YR hue) loess. It is a fossiliferous silt loam. No identifiable soils, which may be considered buried, are observed within the loess below the ground soils (Figure 2). In Illinois weakly developed soils were reported within the Peorian loess (Willman and Frye, 1970), signifying pauses in loess deposition to allow accumulation of organic carbon. They are considered to be associated with mineral zoning. The Jules Soil of Illinois has been given formal stratigraphic status (Frye *et al.*, 1974a, 1974b). Detailed analysis of "dark-colored bands" in the loess in Iowa indicate minimal differences in the properties of banded and non-banded loess (Daniels *et al.*, 1960; Ruhe, 1969b; Ruhe *et al.*, 1971).

Distinct weathering zones occur within the Peorian loess. Weathering zones are defined on the basis of presence or absence of calcium and magnesium carbonate and the amount and distribution of iron oxide in a matrix. As many as six weathering zones have been described in loess (Ruhe, 1969a). Loess can be assumed to be in a chemically oxidized form containing calcium and magnesium carbonates at the time of transport (Worcester, 1973). The uppermost 5 meters at the New Harmony section correspond to the oxidized and unleached zone. Characteristic of this zone are secondary calcium carbonate concretions (loess kindchen), fossil gastropod shells and uniformly distributed iron oxide, as indicated by the yellowish brown color of the matrix. A deoxidized zone, consisting of alternating thin lamellae of

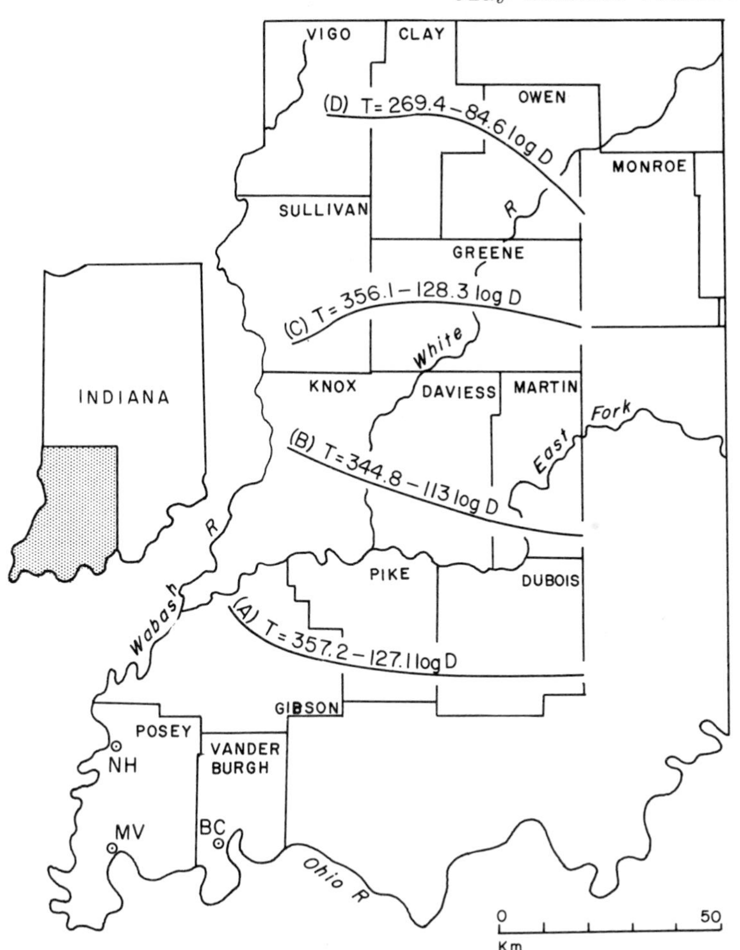

Figure 1 Study area in southwestern Indiana. Traverses A...D from Hall (1973): D, distance in km from Wabash River; T, loess thickness in cm; BC, Bayou Creek section; MV, Mt. Vernon section; NH, New Harmony Section.

oxidized and deoxidized loess, occurs from 5 to 8 meters below datum. The configuration of the lamellae resembles flow lines, suggesting a direct relationship to fluctuating water movements. The remaining 2 meters are deoxidized, but of slightly different nature. Strong mottles, reds (5YR, 7.5YR) and yellowish brown (10YR 6/6), are in a gray (10YR 5/1) matrix. Iron oxide tubules, called pipestems, are prevalent. Properties associated with this zone suggest relict gleying in the subsurface where ground-water zones once stood higher than at present (Ruhe, 1969b).

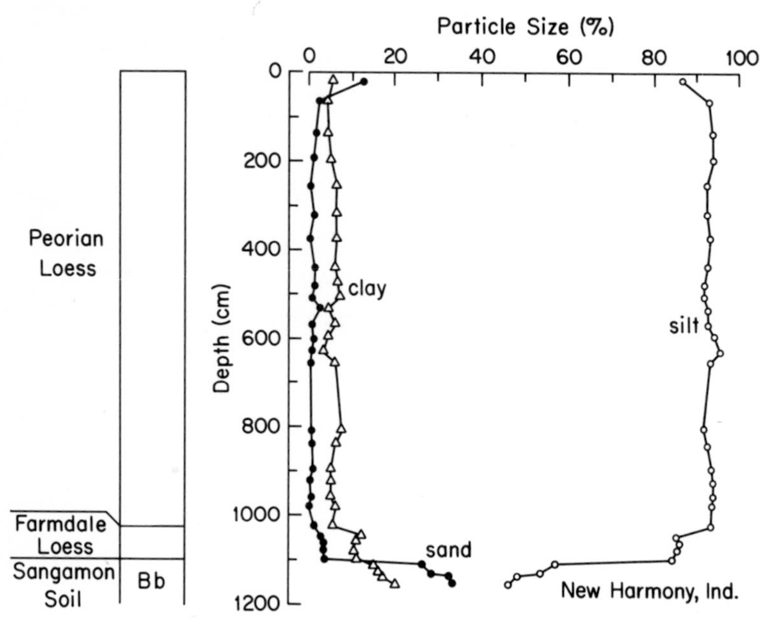

Figure 2 Stratigraphy and particle-size distribution at New Harmony.

A basal sandy loess, common to the Wisconsinan loesses of south-western Indiana, is not discernible at New Harmony (Figure 2). Studies by Hall (1973) and Harlan (1975) refer to a basal increment of a "gritty" or sandy loess, designated the basal Wisconsinan loess. The silty and sandy loess deposits are separated by an increase in the sand fraction from 1 to 7 percent in the silty loess to 7 to 33 percent in the sandy loess (Harlan, 1975). Hall (1976) following Follmer (1970) in Illinois, now assigns the sandy zone to the Roxana loess although upon completion of his work in Indiana, he did not do so (Hall, 1973). Barnhisel *et al.* (1971) reported similar sandy loess deposits, underlying silty loess, in central and eastern Kentucky.

Clay-mineral zones occur in this section. Clay minerals here are defined in the following manner: kaolinite-chlorite has a 7.2Å basal spacing and illite, with a 10Å spacing, does not expand upon glycolation. Montmorillonite has a 17Å spacing when glycolated and includes swelling chlorite and vermiculite. Vermiculite-chlorite mixed-layer clay minerals have a 14Å spacing upon glycolation and include non-expandable vermiculite and chlorite. Illite-montmorillonite mixed-layer clays exhibit a 10Å spacing when heated. Random mixed-layer clay minerals have 10 to 17Å spacings on heated diffractometer traces.

Frye *et al.* (1968) delineated four major clay-mineral zones in the upper Wisconsinan loesses of central and western Illinois. Zone 1, at the base of the loess, is characterized by low illite content; zone II intermediate; zone III high; and zone IV low illite. Zoning was related to source area.

At New Harmony, Indiana, where thick loess is related to the Wabash Valley source, clay-mineral zones are identifiable and may be arranged at the whim of the observer (Figure 3). If illite is used as the index as in Illinois, one arrangement may be an upper zone where illite progressively increases from about 15 to 45 percent from the surface to a depth of about 400 cm. A second zone may be from 450 to 850 cm where illite generally increases from about 30 to 57 percent, and a third zone may be from 900 to 1050 cm where illite generally decreases from about 35 to 25 percent. Note, however, a more pronounced fourfold stratification in the expandable clay minerals. An upper zone to a depth of 300 cm has 55 to 65 percent expandable clay minerals. From 300 to 600 cm the content is 35 to 45 percent. From 650 to 850 cm the content is 15 to 25 percent, and below 900 cm the content is 45 to 60 percent. Illite may be zoned accordingly and from the surface downward is low, intermediate, high, and low content. If this arrangement is used, the second and third zones are inverted in relation to the zones in Illinois (Frye *et al.*, 1968). At New Harmony illite and expandable clay-mineral contents are inversely related. Note further (Figure 3) that other more numerous zones of both constituents may be arranged if desired.

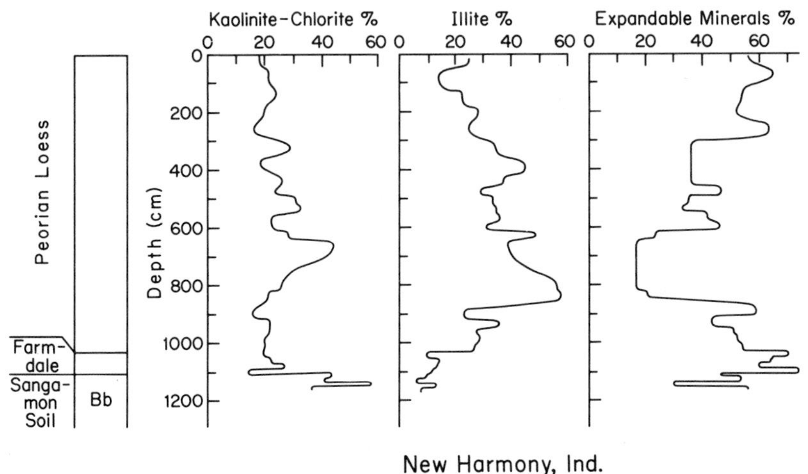

Figure 3 Clay-mineral distribution at New Harmony.

As loess thins away from the Wabash Valley, clay-mineral distributions become complicated by soil formation.

Ground Soils

Ground soils in Peorian loess in southwestern Indiana are dominantly Alfisols, some of which have fragipans (Fragiudalfs). These soils are generally bisequal with an upper sequum of Ap or A2 and B horizons and a lower sequum of A'2 and B'X horizons. The plow layer usually contains the disturbed former A2 horizon of the upper sequum. Note the two eluvial horizons, Ap (A2) and A'2, and the two illuvial horizons, B2 and B'X, shown by the clay-distribution curves of a typical bisequal soil (Figure 4).

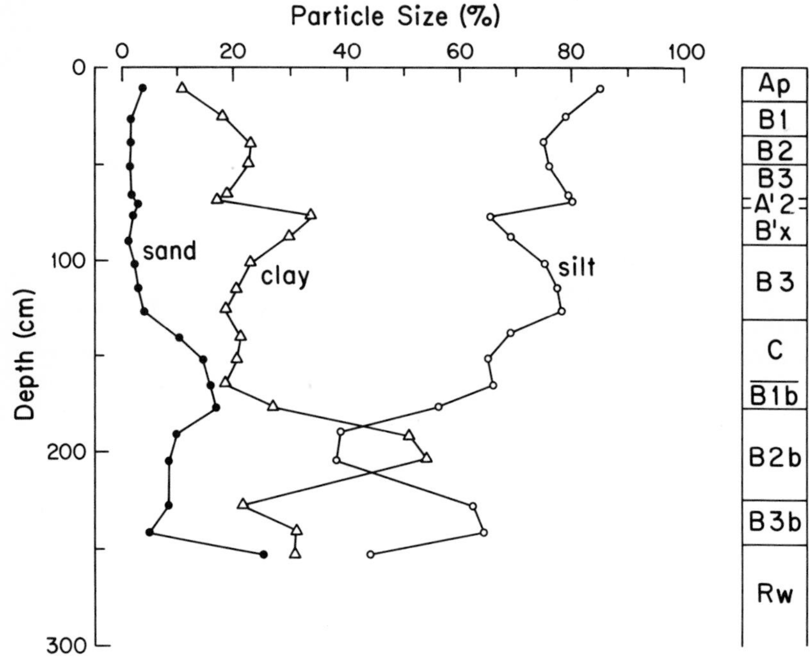

Figure 4 Particle-size distribution for a bisequal ground soil and Sangamon paleosol in Daviess County.

Now compare the clay-mineral distribution in the profile (Figure 5). Kaolinite-chlorite and illite generally decrease downward in the upper sequum while expandable clay minerals generally increase. In the A'2 horizon kaolinite-chlorite sharply increases with a corresponding sharp decrease in expandables. In the fragipan (B'X horizon) and remainder of the lower sequum the expandable minerals are the dominant clay minerals and kaolinite-chlorite and illite are relatively low.

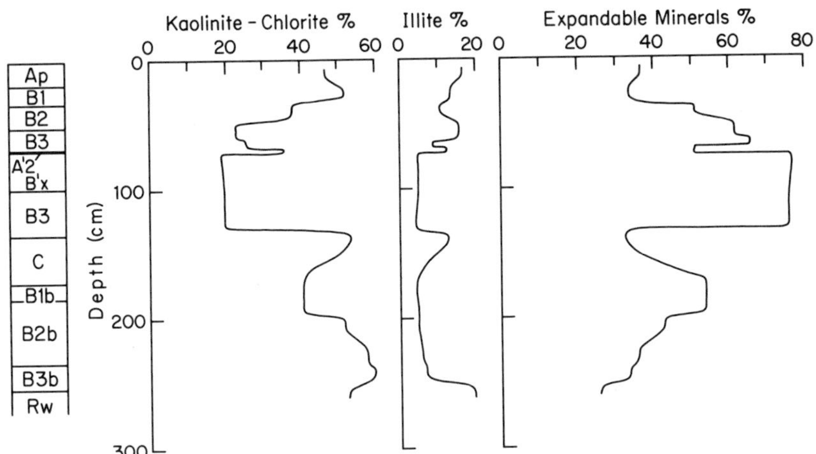

Figure 5   Clay-mineral distribution for the bisequal ground soil and Sangamon paleosol in Daviess County.

In the C horizon (Figure 5) kaolinite-chlorite and illite contents sharply increase with a relative sharp decrease in expandable clay minerals. This stratigraphic change may correspond to the clay-mineral zone in the Peorian loess at New Harmony at a depth of 600 to 900 cm where kaolinite-chlorite and illite are correspondingly high and expandables are relatively low (Figure 3). If this correlation is correct, the bisequum of the ground soil is formed in the equivalent of the upper two clay-mineral zones at New Harmony where the thickness of these zones is about 600 cm. At the ground-soil site the equivalent thickness is 135 cm. The lower two clay-mineral zones in the Peorian loess at New Harmony are 425 cm thick and the equivalent thickness at the ground-soil site is 35 cm. These thinning relations are similar to those described in Iowa (Ruhe, 1969a) and Illinois (Kleiss, 1973) where the younger increments of Wisconsin (Peorian) loess are the greater part of the thickness of the loess at progressively greater distances from the source area.

Note further that the sand content progressively increases in the C horizon (Figure 4) where the clay-mineral composition changes sharply (Figure 5). Note also that the sand content progressively increases into the upper part of the underlying buried soil (Sangamon Soil). This local mixing phenomena is common where loess has been deposited on a sandier material.

If the correlation of clay-mineral zones from New Harmony to the ground-soil site is correct, the sandy zone in the loess at the latter sites is part of the Peorian loess, and there is no need to consider the sandy zone as Roxana loess as recently proposed (Hall, 1976).

SOIL IN FARMDALE LOESS

The Farmdale loess is stratigraphically below the Peorian loess in southwestern Indiana. The term Farmdale is used here for the following reasons: (1) Regardless of the numerous changes in nomenclature in Illinois (Frye and Willman, 1960; Willman and Frye, 1970; Willman *et al.*, 1975), the term Farmdale loess has had long priority and was applied specifically to the New Harmony section (Leighton and Willman, 1950) described in this paper. See discussion on loesses of midwestern United States (Ruhe, 1976). (2) The Farmdale loess is formally recognized as a member of the Atherton Formation in the Pleistocene stratigraphy in Indiana (Wayne, 1963; Shaver *et al.*, 1970).

In southwestern Indiana the dark reddish brown (7.5YR) hue and higher bulk density of the Farmdale loess contrast sharply with those of the overlying Peorian loess. Bits of organic carbon are uniformly distributed throughout. Nearly 30 cm of Farmdale are partially exposed at New Harmony, Indiana (Figure 2) and in the section at Mt. Vernon (Figure 6). The Farmdale in the Bayou Creek section along the Ohio River is generally thicker than that of the other localities, varying from 1 to 1.5 m (Figure 7).

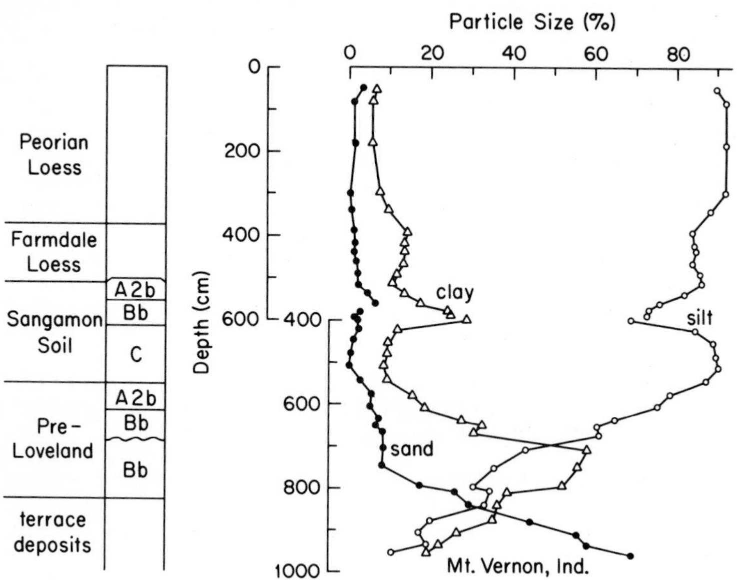

Figure 6 Stratigraphy and particle-size distribution at Mt. Vernon. Composite section of hillslope boring 0 to 600 cm matched to cut 400 to 1000 cm.

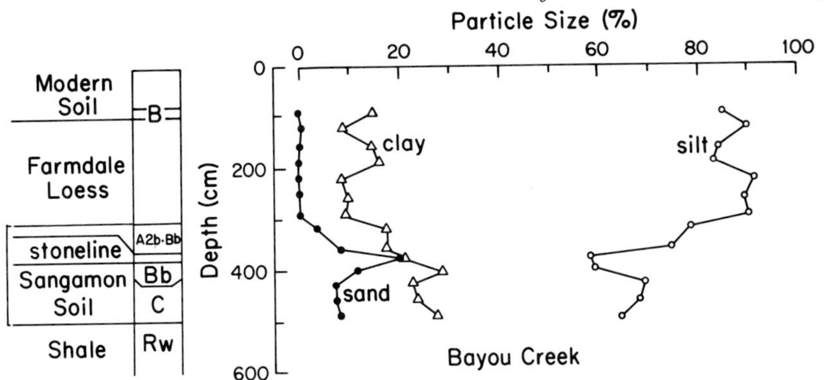

Figure 7 Stratigraphy and particle-size distribution at Bayou Creek.

A discontinuous, dark-colored band is detectable at the Bayou Creek section. It is a localized accumulation of organic carbon, mainly charcoal flakes. Similar bands have not been determined in the Farmdale at New Harmony or Mt. Vernon. The charcoal will be radiocarbon dated.

Near Mt. Vernon, Indiana (SW 1/4, SW 1/4, Sec. 14, T75, R14W), Farmdale loess is not exposed in a roadcut that contains two well-developed paleosols (Figure 6). This loess was penetrated in an auger boring on the hillslope above the face of the cut. In the boring the Farmdale loess directly overlies the Sangamon soil, which is the uppermost buried soil that is exposed in the roadcut. In a second roadcut 85 m to the south and behind a residence, 5 m of Peorian loess are exposed. This loess is calcareous and fossiliferous. A core taken near the base of this cut penetrated 6 m of coarse silt and sand, but did not penetrate Farmdale on older deposits. Apparently dune sand and loess of Peorian age drape down the flank of the hill to the north that contains the Farmdale loess and older deposits.

Soil development in the Farmdale loess consists essentially of an accumulation of organic matter and leaching of carbonates. There is a slight increase but no discernible accumulation of clay in the profiles (Figures 2, 6, 7) and no soil structure is morphologically evident. Sand increases slightly but not significantly in the soil or loess. In fact, a sandy phase occurs in the overlying Peorian loess that drapes down the flank of the hill at Mt. Vernon (Figure 8).

Clay mineralogy of the Farmdale loess differs from the overlying Peorian loess (Figures 3, 9, 10). In general, the Farmdale loess has greater amounts of kaolinite-chlorite and expandable clay minerals whereas the Peorian loess has greater illite content. The different clay-mineral suite

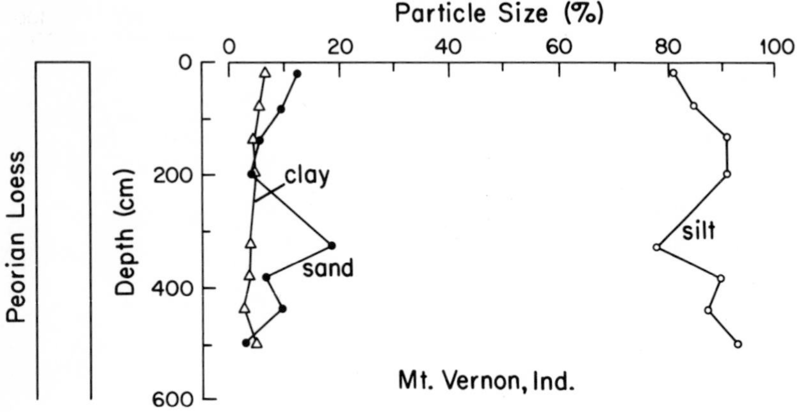

Figure 8 Particle-size distribution for the Peorian loess on hillslope flank at Mt. Vernon.

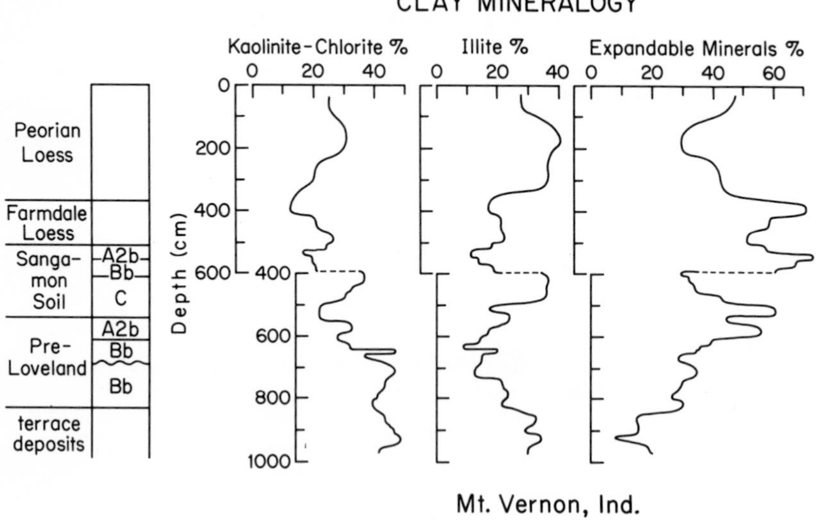

Figure 9 Clay-mineral distribution at Mt. Vernon. (*cf* Figure 6)

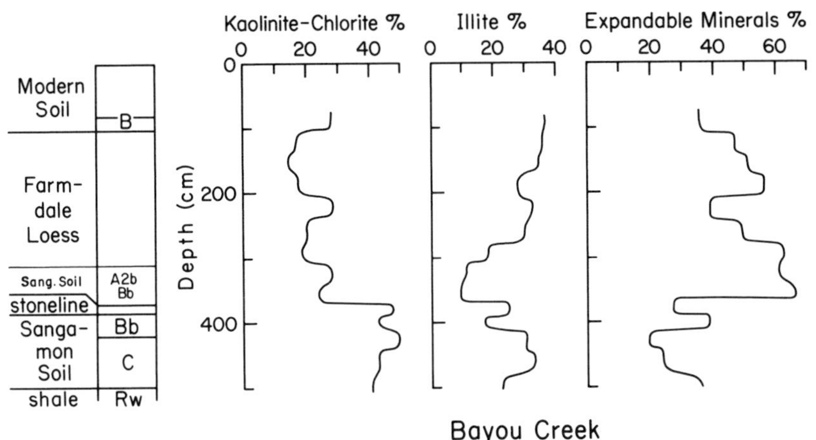

Figure 10 Clay-mineral distribution at Bayou Creek.

in the Farmdale loess is a sedimentation feature like the clay-mineral zones in the Peorian loess and cannot be attributed to soil formation.

It must be noted that if the Roxana loess of Illinois occurs in Indiana, it must underlie the Farmdale soil. No identification of such a loess is possible in the New Harmony, Mt. Vernon, and Bayou Creek sections. Instead the Farmdale directly overlies the Sangamon paleosol at all sites.

SANGAMON PALEOSOL

As previously reported (Ruhe *et al.*, 1974), the Sangamon paleosol is formed on Illinoian glacial drift and Pennsylvanian and Mississippian bedrock in southwestern Indiana. A Sangamon soil formed in Loveland loess is reported in this paper.

Exclusive of the buried soils on Loveland loess, the paleosol is typically a composite soil with its upper part formed in pedisediment and its lower part developed in Illinoian glacial drift or weathered bedrock (Ruhe *et al.*, 1974). This phenomenon is similar to the development of the Late Sangamon paleosol on pediments in the Kansan till of Iowa (Ruhe, 1960, 1969a).

Soils in Weathered Bedrock and Drift

The horizon sequence in the best preserved paleosols is similar to that of the ground soil. Depending upon whether the substrata is glacial drift or bedrock, the sequence is A2b, B1b, B2b, B3b, and Cb or Rb (Ruhe *et al.*,

1974). The A2b horizon is generally yellowish brown or brown. Texture is usually loam and structure may be massive to weak subangular blocky. Brown, strong brown, and reddish brown colors are characteristic of the Bb horizons. The B2b horizons of the Sangamon paleosol usually have a heavier texture and stronger structure than the B horizon of the ground soil (Figure 4).

The chemistry of the Sangamon paleosol is complicated by the fact that it has been buried beneath Wisconsinan loess. Weathering of various intensities, including formation of fragipans, has progressed throughout the thickness of the cover loess (Ruhe *et al.*, 1974). The pH in the paleosol deviates little from neutrality. The organic carbon content is about 0.2 percent, and the base saturation is high. The paleosol has been resaturated by the translocation of bases, and perhaps organic carbon, from above.

At Bayou Creek the Sangamon profile is interrupted by a stoneline (Figure 7). The stoneline is composed of fragments of the underlying weathered bedrock and erratics from 0.6 to 5.0 cm in diameter. Above the stoneline is pedisediment in which the upper part of the solum is formed. A well-developed red Bb horizon with strong mottling lies below. The Cb horizon is gradational to a shale substratum. Particle-size analysis indicates a sand increase in the mixing zone of the Farmdale loess above the Sangamon soil. Clay content also increases substantially through the Bb horizon (Figure 7). Overall, there is a marked increase in kaolinite-chlorite and illite, while expandable clay minerals decrease in the Sangamon soil, as compared with the overlying Farmdale (Figure 10).

Similarities between the ground soil and Sangamon paleosol are substantiated by the dominant clay-mineral suites (Figure 5) except that more intense weathering is indicated for the paleosol. Kaolinite is the dominant clay mineral in the Bb horizon. The increased abundance of kaolinite, compared with its abundance in the overlying ground soil, suggests a similar but more intense or prolonged weathering regime.

Soils in Loveland Loess

At Mt. Vernon the Sangamon paleosol, developed in the Loveland loess, consists of the A2b, Bb and Cb horizons (Figure 6). The A2b horizon is weakly calcareous, secondarily enriched from the overlying calcareous Peorian loess. Sand content increases slightly in the A2b horizon. Texture and structure increase to a maximum in the red Bb horizon (Figure 6). Clay skins along ped surfaces are common. The Cb horizon is lighter in color and is more massive. Kaolinite and illite are abundant with lesser amounts of expandable clay minerals in the paleosolum

PRE-LOVELAND PALEOSOL

A second, older or pre-Loveland paleosol is exposed at Mt. Vernon (Figure 6). The A2b and Bb horizons are developed above a convoluted lag accumulation of chert gravel. A heavily textured and mottled Bb horizon extends downward from the lag deposit. The Bb horizon terminates in stratified deposits. Mottling and pedogenic features disappear within 117 cm. The stratified material, high in percent sand and presumably of fluvial origin, is an oxidized deposit, laced with deoxidized stringers. Topographic maps indicate probable terrace remains.

With the exception of weathering intensity, clay-mineral suites in the pre-Loveland paleosol and in the Sangamon soil developed in Loveland loess, are similar (Figure 9). Expandable clay minerals dominate the eluvial horizons. Illuvial horizons are enriched in kaolinite-chlorite and illite.

SUMMARY

(1) No buried soils are identified in Peorian loess below ground soils in southwestern Indiana. Thick loess exhibits weathering zones.

(2) Numerous clay-mineral zones can be identified in the Peorian at New Harmony. Tentative correlation of these zones to the ground-soil site places the sandy zones of the ground soils in the Peorian and not the Roxana as recently proposed.

(3) The Farmdale loess contains no morphological evidence of soil structure. Soil development only consists of an accumulation of organic matter and is leached of carbonates.

(4) Similarities between the ground soil and Sangamon paleosol are illustrated by the clay-mineral suites with the exception of weathering intensity.

REFERENCES CITED

Barnhisel, R.I., Bailey, H.H., and Matondang, S., 1971, Loess distribution in central and eastern Kentucky: Soil Sci. Soc. America Proc., v. 35, p. 483-486.

Daniels, R.B., Handy, R.L., and Simonson, G.H., 1960, Dark-colored bands in the thick loess of western Iowa: Jour. Geology, v. 68, p. 450-458.

Follmer, L.R., 1970, Soil distribution and stratigraphy in the Mollic Albaqualf region of Illinois, (Ph.D. Dissert.): Champaign-Urbana, Univ. Illinois.

Frye, J.C. and Willman, H.B., 1960, Classification of the Wisconsinan stage in the Lake Michigan glacial lobe: Illinois Geol. Survey Circ. 285, 16 p.

Frye, J.C., Glass, H.D., and Willman, H.B., 1968, Mineral zonation of Woodfordian loesses of Illinois: Illinois Geol. Survey Circ. 427, 44 p.

Frye, J.C., Follmer, L.R., Glass, H.D., Masters, J.M., Willman, H.B., 1974a, Earliest Wisconsinan sediments and soils: Illinois Geol. Survey Circ. 485, 12 p.

Frye, J.C., Leonard, A.B., Willman, H.B., Glass, H.D., and Follmer, L.R., 1974b, The late Woodfordian Jules Soil and associated molluscan faunas: Illinois Geol. Survey Circ. 486, 11 p.

Hall, R.D., 1973, Sedimentation and alteration of loess in southwestern Indiana (Ph.D. Dissert.): Bloomington, Indiana Univ.

__________, 1976, Sedimentation of Wisconsinan loess and related surficial materials in southwestern Indiana: Geol. Soc. Amer., 10th Ann. Mtg., North-Central Sec., Abstract with Programs, v. 8, no. 4, p. 480.

Harlan, P.W., 1975, Soil formation as influenced by loess thickness and soil drainage in southwest Indiana (Ph.D. Dissert.): Lafayette, Indiana, Purdue Univ.

Kleiss, H.J., 1973, Loess distribution along the Illinois soil-development sequence: Soil Sci., v. 115, p. 194-198.

Leighton, M.M. and Willman, H.B., 1950, Loess formations of the Mississippi Valley: Jour. Geology, v. 58, no. 6, p. 599-623.

Ruhe, R.V., 1960, Elements of the soil landscape: Trans. 7th Internatl. Cong. Soil Sci., Madison Wisconsin, v. 4, p. 165-170.

__________, 1969a. Quaternary landscapes in Iowa: Ames, Iowa State Univ. Press, 255 p.

__________, 1969b, Application of pedology to Quaternary research, *in* Pawluk, S., ed., Pedology and Quaternary research: Edmonton, Univ. Alberta Press, p. 1-23.

__________, 1976, Stratigraphy of mid-continent loess, USA, *in* Mahaney, W.C., ed., Quaternary Stratigraphy of North America: Stroudsburg, Pa., Dowden, Hutchinson and Ross, p. 197-211.

Ruhe, R.V., Miller, G.A., and Vreeken, W.J., 1971 Paleosols, loess sedimentation, and soil stratigraphy, *in* Yaalon, D.H., ed., Paleopedology: Origin, nature, and dating of paleosols: Jerusalem, Israel Univ. Press, p. 41-60.

Ruhe, R.V., Hall, R.D., and Canepa, A.P., 1974, Sangamon paleosols of southwestern Indiana, USA: Geoderma,

v. 12, p. 191-200.

Shaver, R.H., Burger, A.M., Gates, G.R., Gray, H.H., Hutchinson, H.C., Keller, S.J., Patton, J.B., Rexroad, C.B., Smith, N.M., Wayne, W.J., and Wier, C.E., 1970, Compendium of rock-unit stratigraphy in Indiana: Indiana Geol. Survey Bull. 43, 229 p.

Wayne, W.J., 1963, Pleistocene formations in Indiana: Indiana Geol. Survey Bull. 25, 85 p.

Willman, H.B. and Frye, J.C., 1970, Pleistocene stratigraphy of Illinois: Illinois Geol. Survey Bull. 94, 204 p.

Willman, H.B., Atherton, E., Buschback, T.C., Collinson, C., Frye, J.C., Hopkins, M.E., and Lineback, J.A., 1975, Handbook of Illinois stratigraphy: Illinois Geol. Survey Bull. 95, 261 p.

Worcester, B.K., 1973, Soil genesis on the stable primary divides of the southwestern Iowan loess province, (Ph.D. Dissert.): Ames, Iowa State Univ.

The Sangamon Soil in its Type Area - A Review

L. R. Follmer

ABSTRACT

The type area of the Sangamon Soil is an indefinite region in central Illinois. The soil was first mentioned by Worthen in his report on Sangamon County in 1873, but it was named by Leverett in 1898 on the basis of generalized descriptions of many wells in northwestern Sangamon County that were reported by Worthen and supported by observations of Leverett and others.

Concepts of the Sangamon Soil were reasonably well understood by 1898, as indicated by Leverett in his paper introducing the Sangamon as the "weathering zone between the loess [Wisconsinan] and the Illinoian till sheet ... found from central Ohio westward to southeastern Iowa, *i.e.*, to the limits of the Illinoian till sheet." Leverett recognized a topographic relation between "the black soil" in nearly level areas and "a leached and slightly reddened till surface" on slopes. As the term "soil" was used at that time, it was essentially equivalent to the A horizon of modern pedologic nomenclature, but Leverett and his contemporaries agreed that a zone with color alteration and leaching was present under the "soil." The relation of the weathered zone to the soil was first described in the Midwest in 1916 by Kay, who defined five zones of a weathering profile in which the "soil" was the top zone. In 1928, Norton and Smith proposed that major soil horizons be designated A, B, C, D, and E. This was the first time soil horizon designations were clearly related to Kay's weathering zones. Leighton and MacClintock in 1930 assigned numbers to the zones and reviewed the processes of weathering in relation to the development of each zone.

Although the concepts of buried soils that developed from 1896 to 1930 were to a large extent based on studies of the Sangamon Soil, none of the work was performed in the type locality of the Sangamon Soil nor has any been done since. However, many studies have been made within a region that could be called the type area -- central Illinois.

At least four described sections in central Illinois can be considered reference sections for the Sangamon Soil. Two sections, Farm Creek and Effingham, were described by Leighton and have been reinterpreted several times. The primary controversy at Farm Creek has been the "Farmdale problem," whereas at Effingham it has been the "gumbotil and accretion-gley dilemma." The other two sections, Chapin and Rochester, were described in 1970 by Willman and Frye, who considered them paratype sections for the Sangamon Soil. These four sections display several types of Sangamon Soil and the common variations in the loess that overlies the Sangamon Soil in Illinois. The many variations of the Sangamon Soil and disagreements about their genesis have caused several deeply rooted controversies.

When the Sangamonian Interglacial Stage ended is an unresolved question that bears on the present concept of the Sangamon Soil. That Leverett intended the "Sangamon" to include the cooling interval is indicated by his statement, "The aspect of the flora [in the soil] is decidedly boreal." Radiocarbon determinations on "the black soil" horizon within the type area yield a Farmdalian age (approximately 24,000 years BP). World-wide evidence now makes the Farmdalian or its equivalent an interstade separated from the Sangamonian by a stade, or glacial substage. Therefore, the black soil correlated to the Sangamon by Leverett is the Farmdale Soil. His Sangamon Soil is time transgressive and contains several merged soils of Wisconsinan, Sangamonian, and possibly late Illinoian age. However, where the early Wisconsinan deposits are less than about two feet thick overlying weathered Illinoian till, they are still considered a part of the Sangamon Soil as originally interpreted by Leverett.

One of the conclusions drawn by the INQUA Commission on Paleopedology in 1970 was "...profiles should be traced laterally in the landscape to determine their spatial variation"(Yaalon, 1971). This has not been done for the Sangamon Soil in its type area because the work done on the soil was of a more general nature than the standards recommended by the Commission. The spatial variation of soil characteristics can be best represented by a type transect that would include the range of drainage and topographic conditions common to the soil-stratigraphic unit. In a special sense this would be a paleocatena for paleosols. An effort should be made to find complete soil profiles that represent a conformable sequence of materials containing no erosion surfaces. A distinction between the eroded and uneroded profiles is essential for an accurate characterization of the soil. A type Sangamon Soil transect would resolve many of the previous controversies and would provide a useful reference for comparisons with Sangamon profiles in other areas.

INTRODUCTION

The purpose of this review is to summarize the previous studies on the Sangamon Soil in central Illinois and demonstrate that additional work needs to be done in order to better understand the Sangamon Soil and its implications to time and Pleistocene stratigraphy. The review must begin with the concept of buried soils, as the Sangamon was one of the first recognized buried soils in the Midwest. A considerable amount of emphasis is placed on the early concepts of the Sangamon Soil and the evaluation of the concepts from 1898 to 1930. The latter part of the review is devoted to a summary of the information available on the Sangamon Soil in its type area and a discussion of four important reference sections in the area. As a conclusion, recommendations are made for future studies on the Sangamon Soil, including the establishment of a type Sangamon Soil transect.

CONCEPTS OF BURIED SOILS IN ILLINOIS BEFORE 1898

The first buried soil in Illinois that was consistently recognized in its correct stratigraphic position was the Sangamon Soil. It was first recognized as a soil by Worthen in the fifth volume of his report to the Illinois General Assembly in 1873. Worthen was commissioned by Governor W. H. Bissell in 1858 to make a geological and mineralogical survey of Illinois (Worthen, 1866). He was given the charge to determine the "geological formations within the state... search for and examine all [mineral resources] obtain chemical analysis on the undetermined substances [and to determine the] relative elevations and depressions of the different parts of the state."

In the eight volumes prepared by Worthen and his assistants, the geology and paleontology of Illinois was discussed including a detailed account for each county. In Volume 1, Worthen makes no reference to a buried soil but states that "the trunks and branches of coniferous trees, ..., are quite common in the blue clays at the base of the drift; and in the brown clays above..." (1866, p. 38). Worthen assumed there was only one drift that had some variations -- true drift and modified drift. He thought that modified drift appeared "to have been subjected to a partial sifting process" which created a sequence of strata composed of true drift (till) and stratified deposits. The importance of water in his thinking on glacial processes even carried over into his explanation on the origin of loess. He considered loess to be a water deposit from a large body of water or a "chain of lakes." Therefore, within his model of glacial theory, buried soils were not possible within the drift.

In Volume 3, Worthen (1868, p. 86) described a reddish clay that occurs below 3 feet of "soil and subsoil." The reddish clay is 14 feet thick and overlies sand and gravel. His following paragraph contains a discussion on the "blue mud" at the base of the drift which is "composed, in good part, of vegetable matter, consisting of leaves and partly decayed wood, embedded in a muddy sediment." The blue mud may be an alluvial deposit but it is no doubt contemporaneous with a soil environment which is pre-drift in age. The reddish clay in the near subsurface was occasionally mentioned in Worthen's subsequent work but was never recognized by him to be a zone of weathering or related to a soil. However, Worthen (1868, p. 124) noted that "the upper part [was composed] of the common reddish-brown clay, so generally characteristic of this formation [drift]."

Not until 1870 in the Adams County report in Volume 4 did Worthen mention a buried soil. It was "the first point in the State where a bed resembling the surface soil was observed below the Drift." Although this coal mine shaft at Coatsburg was sunk in 1859, "no public notice was made of it at the time, as it was then supposed to be merely a local phenomenon that might not be verified elsewhere." The black soil described in the log was considered to be the "Post Tertiary soil" by Worthen (1870, p. 47). Underlying the black soil, 26 feet of clay was described that was interpreted to be older than "the Drift proper." Present day correlations of the black soil and clay are most likely the Yarmouth Soil and the Banner Formation of Kansan age.

Subsequent discoveries of buried organic soils were recorded by Worthen and his assistants in the county reports that followed the Adams County report in 1870. These soils were typically described as black "mould," muck or soil by drillers and "the forest bed" by early geologists. The coal shaft sunk in Bloomington, McLean County, reported by Bannister (1870, p. 178) eventually received special attention (Geikie, 1877 and Horberg, 1953) because it encountered two beds of black mould. Worthen apparently did not see the significance of the beds because he maintained that a buried soil could only be found at the base of the true drift and not within it.

In 1873 Worthen came to the conclusion in his Sangamon County report that there are two buried soils in the Quaternary deposits, one below the "boulder clay" and the other above the "boulder clay" and beneath the loess (Worthen, 1873, p. 307). The soil above the "boulder clay" was later named the Sangamon Soil by Leverett (1898). In the Sangamon County report, Worthen (1873) presented a generalized sequence of "beds" described by a well driller to occur in the northwest part of Sangamon County and the adjoining portion of Menard County. The sequence is:

No. 1	Soil	(1 to 2 1/2 ft)	30-75 cm
No. 2	Yellow clay	(3 ft)	90 cm
No. 3	Whitish gray jointed clay with shells	(5 to 8 ft)	1.5-2.5 m
No. 4	Black muck with fragments of wood	(3 to 8 ft)	90 cm-2.5m
No. 5	Bluish colored boulder clay	(8 to 10 ft)	2.5-3 m

No. 6	Gray hard pan, very hard	(2 ft)	60 cm
No. 7	Soft blue clay without boulders	(20 to 40 ft)	6-12 m

The present interpretation of these beds based on the recent work of Bergstrom, *et al.* (1976) is:

No. 1 A horizon of Modern Soil in Peoria Loess
No. 2 B horizon of Modern Soil in Peoria Loess
No. 3 C horizon in calcareous Peoria Loess
No. 4 Organic horizon of the Farmdale Soil, includes the Roxana Silt
No. 5 Bg (gley) horizon of Sangamon Soil in accretion gley and/or Illinoian till
No. 6 Unaltered calcareous Illinoian till
No. 7 Unknown, Illinoian lacustrine deposit or Yarmouthian accretion-gley

EARLY CONCEPTS OF THE SANGAMON SOIL

In the years between 1873 and 1898 considerable progress was made on resolving many of the complexities of the Quaternary. The idea of multiple glaciations separated by interglaciations, characterized by episodes of nonglacial erosion and weathering of the surficial materials, had been largely accepted. The number of glacial and interglacial stages grew as many of the midwestern states were conducting their geological surveys in a manner similar to Worthen's survey of Illinois. The U.S. Geological Survey contributed to the progress with a program directed toward the study of the glacial formations of the Midwest. Their greatest contribution to the Quaternary studies in Illinois came from Frank Leverett. Leverett began his work with the U.S. Geological Survey in 1886 and started his endless journeys of field work which he essentially continued up to two years beyond his retirement in 1929.

Leverett and his contemporaries spent a considerable amount of time crossing back and forth from Minnesota to Ohio studying outcrops and landforms. In the process they assembled an enormous collection of information by interviewing local people, particularly water well drillers and coal mine operators. Leverett's focus on Illinois was primarily during the years from 1886 to 1898 and culminated in 1899 with U.S.G.S. Monograph 38, *The Illinois Glacial Lobe*.

During the course of his work, Leverett discovered that one soil occurred above and another soil occurred below a formation of glacial deposits that was named the Illinoian till sheet (Chamberlin, 1896). In 1897 at the meeting of the Iowa Academy of Sciences, Leverett gave these soils formal status by naming them the "Sangamon soil" and the "Yarmouth soil," respectively (Leverett, 1898a and 1898b). By 1898, the concept of the Sangamon Soil was reasonably well understood as indicated in Leverett's paper introducing the Sangamon as "the weathering zone between the [Wisconsinan] loess and the Illinoian till sheet...found from central Ohio westward to southeastern Iowa, *i.e.* to the limits of the Illinoian till sheet" (p. 75). The first use of the term "Sangamon soil" by Leverett in 1898 restricted it to the black soil, muck or peat that contains remains of coniferous wood occurring at the base of the loess. The purpose of naming the Sangamon Soil was to formalize a term so that a geologic interval of time could be named, "the Sangamon interglacial stage," to separate the Illinoian and Wisconsinan glaciations.

Recognition of a soil naturally precedes the recognition of particular soil characteristics. The relation of the black muck or peat beds to the underlying "blue clay" and/or leached horizons was not fully understood in 1898. The terms soil and subsoil were often used to infer that they were two beds and not genetically related as in a soil profile. For example, Leverett (1898a, p. 78) used the phrase "there are exposures where the subsoil beneath the Sangamon soil...," which appears to make a clear distinction between the soil and the material immediately underlying the soil called the subsoil. But in the same paper he made the statement "It [the buried soil] has a deep black color to a depth of ten or twelve inches, beneath which it assumes a greenish-yellow color, such as presented by sub-soils beneath poorly drained regions." This statement implies that the subsoil is a genetic part of a soil as it is understood today.

Whether Leverett meant to include the subsoil as a part of the soil is not clear. The apparent dual usage of soil came with the fact that a weathered zone was recognized on the Illinoian deposits whether the "black soil" was present or not (Leverett, 1898a, p. 77). He probably meant this to mean that a catena relationship existed and therefore, the weathered zone had to be included as a part of the soil-time unit. This concept was in parallel with concepts of the modern soil and the other buried soils known at the time. The previous concept that soil is the organic-rich zone, the A or O horizon of modern terminology, continued to be used.

The relation of a leached horizon to the soil was also used in a dual sense. Leverett (1898a) commonly referred to a zone of leaching and weathering beneath the "Sangamon soil" but also inferred that this zone is the Sangamon Soil where the "black soil" is absent. The uncertainty of the relationship between the Sangamon Soil and some of its features carried into Monograph 38 (Leverett, 1899). His chapter title, "The Sangamon Soil and Weathered Zone," implies some uncertainty or was he just trying to be explicit? In the opening remarks of this chapter (p. 125) he stated that the "Sangamon interglacial stage" is "a period marked by leaching and oxidation of the Illinoian drift, of peat and soil accumulation, and of erosion." The leaching process is not necessarily related to soil formation from this statement but he goes on to say later that "the leaching, therefore took place prior to the loess deposition in connection with the development of the soil." A two member catena, "the black soil" and "the slightly reddened till surface," was confirmed in this chapter by Leverett (1899, p. 126) with his statement "These two phases seem to be mutually exclusive."

To draw any specific conclusions on the Sangamon Soil from Leverett's description is difficult. The primary significance of the soil is what it meant to the resolution of the glacial stratigraphy in the Midwest. His model of the Sangamon consisted of two types:

(1) The organic-rich zone of peat, muck, or black soil that occurs in poorly drained positions and has no genetic subsoil.

(2) The "reddish soil" or oxidized zone in till that occurs on "undulatory tracts" and generally does not have an overlay of "black muck," a dark colored A horizon.

In the framework of thinking at the time, Leverett had a correct grasp of glacial stratigraphy and a rudimentary understanding of soil stratigraphy. But he lacked a technical vocabulary to express soil concepts. Several factors concerning soils seemed to give him serious problems. The main problem stemmed from his inconsistency in recognizing the Sangamon Soil. A subordinate problem was equating different soil horizons. He correlated the A and O horizons of the wet member of his catena to the B horizon of the oxidized, well drained member. In the present context, these two members are at opposite ends of a catena which contains several intermediate members. Therefore, the main problem was that he generally did not recognize as soils the somewhat poorly drained and poorly drained, mineral soil profiles. However, the realization that the organic horizon had a relation to the oxidized portion of the Illinoian deposits in surrounding areas with relief led to their correlation as a stratigraphic feature, "The Sangamon."

Perhaps Leverett's most astute observation was that the type or organic matter in the "black soil," particularly the coniferous wood, is not characteristic of the climax interglacial conditions, but of "the close of that stage when glacial conditions were being inaugurated." Probably all of the woody deposits that Leverett observed below the loess in central Illinois is post-Sangamonian by present definition, but was defined as the Sangamon Soil by Leverett.

EVOLUTION OF THE GENERAL CONCEPTS OF THE SANGAMON SOIL

The evolution of concepts of the Sangamon Soil after 1899 paralleled the evolving concepts of the Yarmouth Soil and to a large degree the Modern Soil. Much of the progress made in Iowa from the studies on the Yarmouth Soil was applied to the Sangamon Soil in Iowa and Illinois (Leighton and MacClintock, 1962). The concepts of Modern Soil formation in the United States were still in their infancy up to the 1920's. Before about 1923 American soil scientists were laboring with the problems of classification and mapping of poorly defined, ground surface soils and were not making much progress on concepts of soil formation.

The term "soil" was used by geologists and pedologists prior to 1923 in a restricted sense to mean an organic-enriched layer or ground surface "accumulation." The term subsoil had no particular meaning other than it was the next underlying layer. Weathering was normally related to oxidation and leaching processes. The effects of reduction or gleying in a poorly drained environment were essentially not recognized or were poorly understood.

The poorly drained, gleyed material on the Kansan and Illinoian deposits received much attention in the years between 1898 and 1920 (Leighton and MacClintock, 1962). The gleyed deposits were referred to as "gumbo" by Leverett and earlier workers and were thought to be of sedimentary origin. Kay (1916) proposed the name "gumbotil" to replace the term gumbo in the "super-drift" position because gumbo was a common term which had been used in many different ways. He claimed that gumbotil was chiefly the result of chemical weathering and defined it to be "a gray to dark-colored, thoroughly leached, non-laminated, deoxidized clay, very sticky and breaking with a starch-like fracture when wet, very hard and tenacious when dry, and which is, chiefly, the result of weathering of drift." Kay also added that "the name is intended to suggest the nature of the material and its origin."

Four years later Kay and Pearce (1920) published a paper entitled: "The Origin of Gumbotil." Their insight into the chemical soil-forming factors was a giant step forward in the understanding of soil genesis as well as the origin of gumbotil. However, their theory contains several flaws. The major flaw is that they assumed that an oxidizing environment was required to promote leaching and other processes in the formation of gumbotil. This is in contradiction to the definition of gumbotil given earlier by Kay that it is a "deoxidized clay."

Kay and Pearce explained in detail the horizonation that is typically found in exposures containing gumbotil. The organic-rich layer at the top of the sequence was referred to as "soil," but was greatly de-emphasized. Gumbotil occurred under the "soil" or was frequently placed at the top of the sequence if the "soil" was not recognized. Under the gumbotil they described the oxidized and leached zone which overlays the oxidized and unleached zone. At the base of complete sections, the unweathered till occurs which they referred to as the unoxidized and unleached zone.

Gumbotil as viewed by Kay and Pearce was strongly weathered till, but they did not discount that "wind action, freezing and thawing, burrowing of animals, slope wash and other factors may have contributed to the formation of gumbotil."

In 1923, C.F. Marbut of the U.S. Department of Agriculture presented a series of lectures at the University of Illinois which had a large impact on the understanding of soil genesis, including buried soils. He introduced to the United States the Russian school of thinking on concepts of soil genesis, morphology, and classification. The fundamentals of modern pedology were established during the 1920's largely due to Marbut and his translation of a book by Glinka into English in 1927. The concept of a soil profile was introduced and was defined as a sequence of three horizons -- A, B, and C. The A is the organic-enriched surface layer informally called "topsoil." The B is the genetic layer under the A and quite naturally assumed the informal designation, "the subsoil." The C horizon is the "parent material" from which the A and B horizons develop, but only if one geologic material can be demonstrated in the profile. The A and B horizons have to meet certain criteria for recognition, but the C horizon does not. The C horizon merely underlies the A or B horizon. The lack of specific criteria for the definition of the C horizon frequently causes communication problems that still exist today.

Marbut introduced and developed many new ideas about soils (Leighton, 1958) and was challenged by Leighton and MacClintock (1962). Pleistocene geologists in Illinois, according to Leighton (1958), accepted the soil profile concept but chose to call it a profile of weathering in order to distinguish "the differences in the objectives, scope and application of geology and pedology."

Norton and Smith (1928), soil scientists from the University of Illinois, were the first to clearly relate soil horizons to the weathering zones of Kay. The definitions of the soil horizons were not settled at the time, so they described profiles from south-central Illinois to explain their proposal for standardizing the soil horizon concepts. Without recognizing it as such, they were the first to make soil-stratigraphic correlations between a well drained, oxidized Sangamon Soil and a poorly drained, gleyed Sangamon Soil. However, their examples were thin loess-over-till profiles, and they did not recognize the buried Sangamon Soil beneath the loess. They were under the impression that the profiles had developed in drift. Five major horizons were designated:

A -- the topsoil, the horizon of biotic accumulation and eluviation;
B -- the subsoil, the horizon of illuviation and disintegration;
C -- the horizon of oxidation, disintegration and leaching;
D -- the horizon of oxidation and slight decomposition, calcareous; and

E -- the calcareous, unaltered parent material.

Norton and Smith later discovered they had not been consistent with the use of soil horizon designations and made some corrections (Smith and Norton, 1929). In making the changes, they acknowledged Leighton and MacClintock as contributing to cooperative studies that led to some adjustments so that the soil horizons occupy the same positions in both profiles. The important change that they made was to adjust the lower boundary of the B down to the top of the zone of iron accumulation, a feature of the oxidized and leached horizon.

In the following year Leighton and MacClintock (1930) published their classic paper on the "Weathered Zones of the Drift-Sheets of Illinois." They acknowledged the "active collaboration" with Smith and Norton but did not reference their work. Leighton and MacClintock reached a very important point in the understanding of the Sangamon Soil. They recognized a type of catena; the gumbotil profile in poorly drained areas, the siltil profile in well drained areas, and the mesotil profile in intermediate areas. They did not call them types of Sangamon Soil but weathering profiles on Illinoian drift. They used the term Sangamon only in a time-stratigraphic sense.

Leighton and MacClintock summarized the views of Kay and Pearce on the factors of gumbotil (soil) genesis and added explanations on the origin of the "horizons" (zones) and on rates of development of soil profiles. Leighton and MacClintock continued with the assumption that oxidizing conditions are dominant in the origin of gumbotil and never changed their interpretations on this fact (Leighton and MacClintock, 1962, p. 271). They recognized the thin loess that Norton and Smith misidentified in south-central Illinois, adjusted the upper two horizons downward to exclude the loess, dropped Norton and Smith's horizon designations and numbered them downward, 1 through 5 (Leighton and MacClintock, 1930).

In 1931, the stratigraphic position of the Sangamon Soil was adjusted when Leighton (1931) reinterpreted the loess-like silt described at the "Farm Creek exposure" (Leighton, 1926) to be the Late Sangamon loess. This exposure was considered by Leighton to be a "type Pleistocene section." The inference that can be drawn from Leighton (1931) is that the Sangamon Soil transgresses from interglacial to glacial conditions and consists of two parts: 1) Illinoian gumbotil in the lower part and 2) a youthful soil profile formed in the Late Sangamon loess which may have developed during the "Iowan," the first glacial stage of the "Wisconsin." The two part idea was expressed earlier by Leighton (1926) but in more uncertain terms and the "Iowan" in 1926 was considered to be a separate glacial stage. After 1931 no significant modification of the two part concept of the Sangamon Soil was made for about 20 years. Then Leighton eliminated the "upper Sangamon" by changing the name of the Late Sangamon loess to the Farmdale loess (Wascher *et al.*, 1948) and placing it into the "Wisconsin" Stage (Leighton and Willman, 1950). During the 1940's, Leighton *et al.* came to the realization that loess was a deposit related to glacial conditions. But the Sangamon peat described by Leverett (1899) at the "Farm Creek exposure" overlies the Farmdale loess. Therefore, by placing the peat and Farmdale loess into the Wisconsinan the peat bed containing the boreal remains (coniferous wood) was removed from the Sangamon Soil as conceived by Leverett. This eliminated most, if not all, peat or muck deposits in this stratigraphic position in central Illinois from the Sangamonian time-stratigraphic unit because of the organic soil over loess relationship. The organic soil is now known to be the Farmdale Soil.

The Sangamon Soil at the "type Pleistocene section" during the 1950's was restricted to the non-organic portion of the "profile of weathering." This led to a false conclusion, which was commonly drawn after field examination of similar exposures, that the Sangamon Soil did not (or appear to) have an A horizon. Recognition of a buried A horizon at this time required identifiable organic remains, otherwise the A horizon, if present, was misidentified as gumbotil or a deposit overlying the gumbotil. Criteria for establishing the presence or absence of a buried A horizon is a point for discussion that has not been resolved in the literature on buried soils. The problem is caused by post-burial alteration referred to as pedometamorphism (Gerasimov, 1971) or diagenesis (Valentine and Dalrymple, 1976), which causes the loss of soil characteristics.

The geo-pedologic solution to the apparent absence of the A horizon in many buried soils and the cause of the time-transgressive nature of the Sangamon Soil in central Illinois first came to light in a concept proposed by Thorp, Johnson, and Reed (1951). They reasoned that "the first Wisconsin loess seemed to have collected very slowly on the old Sangamon soil in Nebraska and Kansas..., soil formation kept pace with deposition, ..." and produced an "over-thickened A horizon...." They also suggested that this loess may correlate to the Farmdale loess in Illinois. This concept explains how an organic rich horizon can develop upwards as loess is being deposited. It also explains why some layers of gumbotil (the gleyed zone) appear too thick (Simonson, 1954). The most significant proposition Simonson (1954) offered is, "If deposition of the loess were slow and continuous, the B horizon (of a gumbotil profile) could 'grow' upward, first into the former A horizon and later into the loess." Simonson was the first soil scientist to recognize gumbotil as a part of a soil (1941), first to classify the buried soils in pedologic nomenclature and first to present results of chemical and particle size analyses of buried soils (1941, 1954). Recently Ruhe (1974) discussed the distribution of the Sangamon paleosol [*sic*] in the Midwest, its environment of formation, and the classification of the various types of Sangamon Soil profiles in the current pedological terminology.

Allen (1959) performed petrographic studies in 1932 on the weathering zones defined by Leighton. He found a reduction of heavy minerals passing upwards from horizon 5 to horizon 1. Significant amounts of feldspars and ferromagnesian minerals were lost or altered in horizons 1 and 2. Allen was the first to establish that the fine fraction of gumbotil (the A and B horizons) is chiefly composed of clay minerals of the montmorillonite [*sic*] group, although Kay and Pearce (1920) were aware of the colloidal properties of the fine fraction.

The last and most controversial change in the concept of the Sangamon Soil occurred in 1960 when Frye, Shaffer, Willman and Ekblaw published the paper, "Accretion-gley and the Gumbotil Dilemma." They criticized the dualism of the empirical and genetic definition of gumbotil and suggested that gumbotil be restricted to the truly *in situ*, gleyed soil. They reviewed the term gley, a product of reduction in a wet environment, and developed the concept of "accretion-gley," a product of "slowly accumulating deposits of surficial clay" in a wet soil environment. In their criticism of gumbotil they contended that there are five genetic processes that can meet the empirical definition of gumbotil: 1) glaciofluvial deposition before soil formation, 2) slope wash accretion during soil formation, 3) eolian accretion during soil formation, 4) combination of accretion and *in situ* gleying during soil formation, and 5) *in situ* gleying during soil formation.

The concept of gleyed Sangamon Soil profiles was discussed much earlier by Thorp, Johnson, and Reed (1951) but was rejected by Leighton and MacClintock (1962) and other gumbotil advocates because it was in contradiction to their concept that oxidizing conditions are required for gumbotil formation.

Very shortly after their gumbotil dilemma paper, Frye *et al.* (1960b) presented the evidence to support their interpretation of accretion-gley. To support their interpretations they used: 1) field relationships, 2) petrographic analyses of the light and heavy minerals, 3) clay mineral analysis, 4) grain size analysis, and 5) size and composition of pebbles. This was the first broad spectrum analysis of the physical features of the Sangamon Soil in Illinois but they (Frye *et al.*, 1960a) did not describe any soil profiles. They modified the standard U.S. Department of Agriculture horizon designations and used them in generalized sketches illustrating soil-stratigraphy of two accretion-gley profiles. A probable misinterpretation was introduced with the use of their "BG-zone." In their sketch, they showed the "B-zone" in the till passing under the accretion-gley and becoming the "BG-zone." They stated in the definition that "a BG-zone commonly is a secondary modification of a primarily developed B-zone." Willman *et al.* (1966) clarified the meaning with the statement that a "BG-zone is a gleyed B-zone." This concept along with the sketch implied that an oxidized soil existed before the accretion-gley formed. From a soil-geomorphic point of view, it is unlikely that the B horizon of a soil in a depression would be transformed from "better drained" to "poorer drained" with the addition of accreted material. They have confused the iron accumulation zone under a gley horizon in a B3 or C1 position with the B2 horizon of a somewhat poorly drained soil.

The significant conclusions drawn by Frye *et al.* (1960b) are: 1) the degree of mineral decomposition in accretion-gley profiles is less than in the *in situ* profiles and much less than ascribed to the gumbotil; and 2) the term gumbotil is not a good scientific term and "should be used only in a general sense to refer to those plastic and sticky surficial clays resting on till." Leighton and MacClintock (1962) refuted much of the work of Frye *et al.* but acknowledged that some deposits are accretion-gleys, and by quoting Trowbridge stated: "gumbotil as originally defined can be distinguished readily enough from other clays."

Frye and Willman (1963) countered by commenting on what they considered to be archetypical gumbotil sections that "At every reported exposure that we have recently examined the 'gumbotil' is accretion-gley." Part of the explanation of the dilemma can be derived from their diagram of the Hipple School Sections which shows a cross section of a Sangamon Soil catena. One section passes through the accretion-gley member and the other section is in the moderately well drained member of an apparent toposequence (catena). An *in situ*, poorly drained, gleyed profile must conceptually exist between the two catena members described by Frye and Willman, and could be called gumbotil.

However, Willman *et al.* (1966) did not approve of differentiating a poorly drained, *in situ* soil from the better drained, *in situ* soils because they did not consider it practical. They critically reviewed the concept of gumbotil again, presented a wealth of new mineralogical data and using pedological terminology, described 12 sections, 11 of which contained the Sangamon Soil. Then they concluded that the material called gumbotil is accretion-gley.

Frye and Willman succeeded in replacing the concept of gumbotil with the concept of accretion-gley in Illinois but this exchange may create a new dilemma. Accreted material in a soil environment is generally distinguishable from *in situ* material in the field. The identification can be supported with laboratory analyses, but the distinction is often difficult. As a result accretion-gley and *in situ* gley are commonly confused because they are both gleyed materials. A probability exists that all gleyed soils will be called accretion-gley by some geologists, and thus, a new dilemma may develop because *in situ* gleyed Sangamon Soil profiles are common in central Illinois (Bergstrom *et al.*, 1976).

The current concept of the Sangamon Soil in Illinois was described by Willman and Frye (1970). They described two paratype sections and several other typical occurrences of the Sangamon Soil. The following features of the Sangamon Soil can be derived from their discussion.

(1) It underlies Wisconsin-age deposits and has developed on Sangamonian, Illinoian or older deposits.

(2) It is composed of two genetic types:

a. Accretion-gley- a gray to blue-gray massive clay, with some pebbles and sand, formed by accretion under poor drainage and gleying conditions; may have sharp contact with the till below, gradational upper boundary;

b. *In situ* - a red-brown to sometimes dark gray profile developed in place under moderately good drainage to sometimes poor drainage on till or other glacial deposits; has a distinctive clayey B2-zone, typically has Mn-Fe pellets and staining; gradational lower boundary, sharp upper boundary.

(3) It is leached of carbonates to about 1.8 to 3 m (6 to 10 ft) depth of leaching slightly diminishes northward, leached below the till contact in some accretion-gley profiles.

(4) It is strongly developed:

a. *In situ* profiles are more strongly developed in the south than the north;

b. Accretion-gley profiles are similar throughout Illinois.

(5) It is not restricted to the Sangamonian time interval:

a. Soil formation started in some localities during the Illinoian Stage;

b. Early Wisconsinan sediments, to some extent, were incorporated into the gleyed profiles.

The Sangamon Soil is time-transgressive into the Wisconsinan Stage. Therefore, the recognition of a Sangamon Soil profile in a sequence of deposits does not necessarily establish that the beginning of Wisconsinan time is marked by the top of the soil. The Wisconsinan time boundary commonly lies within the A horizon of the Sangamon Soil and has been determined in Illinois by detailed grain-size analyses (Follmer, 1970, summarized in Johnson *et al.*, 1972) or by mineralogical analysis (Frye *et al.*, 1974). The beginning of Wisconsinan time has been estimated by Frye *et al.* to be about 75,000 years ago. Studies in Iowa (Ruhe, 1976) and in Indiana (Kapp and Gooding, 1964) suggest that the Wisconsinan begins at a younger age. Future work may well show that this time boundary may need to be adjusted younger if studies in the type area can resolve some absolute dates in the range of 50,000 to 75,000 years BP.

The fundamental question that arises at this point concerns the basis of assigning the Wisconsinan-Sangamonian boundary in the continental record. Should the beginning of the Wisconsinan be based on the first Wisconsinan sediments on the Sangamon Soil in its type area (Frye *et al.*, 1974) or based on the introduction of, or some measure of, boreal remains into the Sangamon Soil profile? Kapp and Gooding (1964) included the upper boreal pollen zones in the Sangamon Soil because of the observed field relation of the soil with overlying sediments and because of dead radiocarbon dates ($>$41,000 yrs BP) obtained from the top of the soil identified as the Sangamon Soil.

A study in south-central Illinois by Gruber (1972a and b) used the increase of boreal pollen for establishing the Sangamonian-Wisconsinan boundary, but was reinterpreted by Frye and Willman (1973) to be a warm-cool climatic transition within the Wisconsinan. Gruger studied the pollen profiles in lacustrine deposits at three sites and correlated the time zones only on the basis of pollen. Voss (1939) studied pollen profiles in the peat deposits thought to be Sangamonian at two sites in central Illinois, but the peat was later correlated to the "Farm Creek intraglacial" (Farmdalian Substage) of the Wisconsinan by Leighton (cited in Kapp and Gooding, 1964).

Finite age determinations are needed in the type area of the Sangamon Soil where a complete section of early Wisconsinan sediments can be documented to overlie the soil. Identification of the type of organic remains in the soil and at the base of the Wisconsinan sediments is needed to answer the fundamental question on the age. A possibility exists that the "first sediment" criteria will be compatible with the "first boreal remains" criteria. In the near future, advances in radiocarbon technology will make the age determination possible.

CENTRAL ILLINOIS QUATERNARY STRATIGRAPHY

The type area of the Sangamon Soil is an indefinite region in central Illinois. The foundation for the present classification of the Quaternary stratigraphy in Illinois was established by Willman and Frye (1970). Johnson (1976) reviewed the work of Willman and Frye, added the results of recent work in Illinois and discussed some of the current problems with regional stratigraphic correlations. The history of the classification of the Wisconsinan Stage has been reviewed in detail by White (1973).

The stratigraphic concepts of the Quaternary in Illinois are divided into categories which are part of a scheme of multiple classification described by Willman and Frye (1970). Figure 1 illustrates the time-stratigraphic, rock-

TIME		ROCK			SOIL
STAGE	SUBSTAGE	eolian	till and related deposits	glaciofluvial and other deposits	
HOLOCENE				Grayslake Peat; Cahokia Alluvium; Peyton Colluvium	Modern
WISCONSINAN	VALDERAN	Parkland Sand; Peoria Loess	Richland Loess; a	Equality Formation; Henry Formation; a; a; a	
	TWOCREEKAN	Parkland Sand; Peoria Loess	Richland Loess; a	Equality Formation; Henry Formation; a; a; a	
	WOODFORDIAN	Parkland Sand; Peoria Loess	Richland Loess; Wedron Fm. {till members and interbedded water-laid deposits}; Morton Loess	Equality Formation; Henry Formation; a; a; a	Jules
	FARMDALIAN	a; Robein Silt	Robein Silt	Equality Formation; Henry Formation; a; a; a	Farmdale
	ALTONIAN	a; Roxana Silt: Meadow Loess M., McDonough Loess M., Markham Silt M.	a; a	Equality Formation; Henry Formation; a; a; a	Pleasant Grove; Chapin
SANGAMONIAN			Berry Clay M.		Sangamon
ILLINOIAN	JUBILEEAN	a; Loveland Silt	Teneriffe Silt; Glasford Formation {till members and interbedded water-laid deposits}	a; Pearl Formation; a; a; a	unnamed
	MONICAN	a; Loveland Silt	Teneriffe Silt; Glasford Formation {till members and interbedded water-laid deposits}	a; Pearl Formation; a; a; a	Pike
	LIMAN	a; Loveland Silt	Teneriffe Silt; Glasford Formation; Petersburg Silt	a; Pearl Formation; a; a; a	
YARMOUTHIAN			Banner Formation: Lierle Clay M.		Yarmouth
KANSAN		a; a	a; Banner Formation {till members and other glacial deposits}	a; a; a; a; a	
AFTONIAN			Enion Formation: a		Afton
NEBRASKAN		a; a	a; Enion Formation {tills and other glacial deposits}	a; a; a; a; a	

a Stratigraphic position of lithologic equivalents not known to occur or are not differentiated.

Figure 1 Quaternary stratigraphy of central Illinois. Modified from Willman and Frye (1970).

stratigraphic and soil-stratigraphic terminology currently used in Illinois and is modified from the stratigraphic classification presented by Willman and Frye (1970). Most of the units currently recognized in Illinois occur in central Illinois. In comparison to rock units recognized by Johnson (1976) for the whole state, only five formations in the Wisconsinan and two in the Nebraskan are not present in central Illinois.

All of the time-stratigraphic units of the Quaternary are shown in Figure 1. The rock-stratigraphic units are listed in columns according to their origin. Members of most formations are not shown to simplify the figure. The purpose of this organization is to demonstrate the decreasing degree of rock-stratigraphic differentiation in the older units. Figure 1 clearly shows that a greater degree of lithologic and genetic differentiation is attained in the younger deposits. Therefore, the degree of generalization increases with depth, simply because of the diminishing amount of information available.

All the soil-stratigraphic units currently recognized in the state occur in central Illinois. The relation of the soil units to the rock units as shown in Figure 1 will help explain some of the problems concerning the age and position of the Sangamon Soil. Several rock units occurring only at the surface, eolian Parkland Sand, Equality Formation, Grayslake Peat, Cahokia Alluvium and Peyton Colluvium do not have recognized equivalents in the subsurface in Illinois. Of equal importance, note that the Sangamonian Berry Clay Member of the Glasford Formation does not have an equivalent in the Holocene. The Berry Clay is the rock-stratigraphic name of the Sangamonian accretion-gley. Modern accretion-gleys are common in depressions and alluvial positions but they have not been defined as a rock-stratigraphic unit. A typical modern accretion-gley in current pedologic terminology is the Cumulic Haplaquoll.

The previous discussion on the time-transgressive nature of the Sangamon Soil established that the soil formation continued into the Wisconsinan Stage in central Illinois. The Roxana Silt (formerly called the Farmdale loess) began accumulating in central Illinois during the Altonian Substage of the Wisconsinan. The first deposit, Markham Silt Member, was incorporated into the "growing" soil. The top of this zone (horizon) was interpreted to be the top of the Chapin Soil (Frye *et al.*, 1974). In gleyed positions the whole Roxana where about 1.5 m (5 ft) thick or less was incorporated into the Sangamon Soil (Bunker Hill Section, Willman *et al.*, 1966; and Willman and Frye, 1970). In other localities where Roxana Silt had not been recognized, it was included in the Sangamon paleosol [*sic*] (Follmer, 1970) or Berry Clay (Johnson *et al.*, 1971).

Similar problems of differentiation exist between Berry Clay and other rock units of Illinoian age. Glaciolacustrine deposits analogous to the Equality Formation have not been formally differentiated in the Illinoian. Such deposits occur in Sangamon County and many other localities (Bergstrom *et al.*, 1976). Generally these deposits when recognized have been included in the Teneriffe Silt (Willman and Frye, 1970) which was defined to include eolian, glaciofluvial and related deposits. At its type section, the Teneriffe is probably a loess and if differentiated from the glaciofluvial deposits, the eolian silt should retain the name Teneriffe (1977, H.B. Willman, pers. comm.). The glaciolacustrine deposits have been misidentified as accretion-gley in some instances by Frye *et al.*, (1960a, Sangamon County exposure) and possibly by Willman and Frye (1970) at the Rochester section. Problems with differentiating other types of Illinoian deposits from Berry Clay are not known but are certainly possible.

REFERENCE SECTIONS OF THE SANGAMON SOIL

Many profiles of the Sangamon Soil have been described in Illinois since Leverett (1898a and b) named the soil. Most of the descriptions are skeletal and only describe the general appearances. The recognition of a "soil" was sufficient for the purposes of much of the work in the early days because glacial stratigraphy was the major objective. Therefore, little attention was given to the actual characteristics of the soil.

A summary of the locations of all known published descriptions of the Sangamon Soil in central Illinois is given in Table 1. A few important sections are included where the Sangamon was identified but not described. Thesis work is not included. The descriptions were judged against present pedological standards for describing a soil profile and rated on a scale from 0 to 3. The lowest value (0) indicates that the Sangamon Soil was recognized by name but not described. The next value (1) means that the Sangamon was recognized by name and some general appearances were described. The value (2) means that a semi-detailed description was made which described some of the characteristics of the major soil horizons. The highest value (3) is given to the detailed descriptions which described the major horizons and some of the standard features of the minor (sub) horizons, such as the B1, B2 and B3 or a near equivalent.

The locations of the sites of description are shown in Figure 2. The type area of the Sangamon Soil has never been designated in the past. Therefore, the area shown on Figure 2 will be designated the type area for the purposes of this review so that four important reference sections are included within the area. Also shown on Figure 2 is the location of all radiocarbon dates determined on samples collected in the Sangamon Soil or immediately overlying a horizon of a described Sangamon Soil.

Four sections (Farm Creek, Effingham, Chapin and Rochester) can be considered reference sections for the

Table 1. References Containing Descriptions of the Sangamon Soil in the Type Area

County Sec-Twp-R	Section name or number	Reference	Description class[a]
Cass			
11-18N-11W	Cottonwood School	Willman and Frye, 1970............	(1)
Christian			
36-13N-2W	Taylorville Dam..	Johnson, 1964....	(3)
DeWitt			
15-21N-1E	Campbells Hump...	Willman and Frye, 1970............	(1)
15-21N-1E	#24..............	Horberg, 1953....	(1)
Effingham			
6-7N-6E	Effingham........	Willman, Glass and Frye, 1966...	(2)
6-7N-6E	Effingham........	Brophy, 1959.....	(2)
6-7N-6E	Effingham........	Simonson, 1954...	(2)
6-7N-6E	Effingham........	Leighton and MacClintock, 1930	(2)
4-7N-5E	Funkhouser.......	Willman, Glass and Frye, 1966...	(2+)
Fayette			
30-6N-1E	Hickory Ridge....	Jacobs and Lineback, 1969...	(1)
21-8N-1W	Liberty Creek....	Jacobs and Lineback, 1969...	(1)
17-7N-2E	Linn Creek.......	Jacobs and Lineback, 1969...	(1)
31-6N-1W	Mulberry Grove...	Jacobs and Lineback, 1969...	(1)
26-9N-1W	Ramsey Creek.....	Jacobs and Lineback, 1969...	(1)
7-8N-1E	Ramsey West......	Jacobs and Lineback, 1969...	(1)
24-6N-2E	#18..............	MacClintock, 1929	(0)
16-6N-1E	#19..............	MacClintock, 1929	(0)
4-8N-1E	#21 and #22......	MacClintock, 1929	(0)
Fulton			
33-7N-5E	#48..............	Wanless, 1957....	(1)
26-6N-3E	#51..............	Wanless, 1957....	(1)
27-6N-3E	#52..............	Wanless, 1957....	(1)
8-5N-4E	#54..............	Wanless, 1957....	(1)
16-4N-3E	#58..............	Wanless, 1957....	(1)
26-4N-3E	#60..............	Wanless, 1957....	(1)
30-4N-3E	#61..............	Wanless, 1957....	(1)
32-4N-3E	#62..............	Wanless, 1957....	(1)
8-3N-2E	#66..............	Wanless, 1957....	(1)
35-8N-4E	#69..............	Wanless, 1957....	(1)

(Continued Next Page)

Table 1 (Continued)

County Sec-Twp-R	Section name or number	Reference	Description class[a]
Fulton			
32-4N-3E	Enion............	Willman and Frye, 1970............	(1)
21-5N-3E	Lewistown........	Willman and Frye, 1970............	(0)
8-7N-3E	Hipple School....	Frye and Willman, 1965a...........	(2)
36-8N-2E	Fairview.........	Willman, Glass and Frye, 1966...	(0)
31-8N-3E	Flamingo.........	Brophy, 1959.....	(3)
2-6N-3E	Country Club.....	Brophy, 1959.....	(3)
33-7N-5E	Breeds...........	Brophy, 1959.....	(3)
Green			
27-12N-13W	Hillview.........	Frye, Glass and Willman, 1962....	(1)
Logan			
32-19N-2W	Sugar Grove School	Johnson, 1964....	(3)
McLean			
32-25N-1W	Danvers..........	Frye and Willman, 1965b...........	(1)
32-25N-1W	#22..............	Horberg, 1953....	(1)
Macon			
8-16N-3E	#26..............	Horberg, 1953....	(1)
Macoupin			
14-11N-8W	#44..............	Ball, 1952.......	(1)
27-10N-7W	#47..............	Ball, 1952.......	(1)
22-11N-8W	#48..............	Ball, 1952.......	(1)
13-12N-7W	#50..............	Ball, 1952.......	(1)
18-12N-7W	#51..............	Ball, 1952.......	(1)
34-10N-7W	#53..............	Ball, 1952.......	(1)
1-7N-7W	Sawyerville......	Frye and Willman, 1963............	(1)
28-7N-7W	Bunker Hill......	Willman, Glass and Frye, 1966...	(1)
Madison			
20-3N-8W	Pleasant Grove School...........	Willman and Frye, 1970............	(1)
11-5N-10W	Reliance Whiting Quarry..........	Frye and Willman, 1965b...........	(2)
17-6N-8W	Dorsey..........	Frye and Willman, 1963............	(1)
Menard			
23-18N-7W	Petersburg......	Willman and Frye, 1970............	(0)

(Continued Next Page)

Table 1 (Continued)

County Sec-Twp-R	Section name or number	Reference	Description class[a]
Menard			
26-18N-7W	Petersburg Dam #1.................	Johnson, 1964....	(3)
36-18N-6W	Unnamed...........	Shaw and Savage, 1913.............	(0)
22-18N-5W	Unnamed...........	Shaw and Savage, 1913.............	(0)
Montgomery			
23-7N-4W	Panama - A........	Willman, Glass and Frye, 1966.......	(2)
35-9N-5W	#25...............	MacClintock, 1929	(0)
32-9N-4W	#26...............	MacClintock, 1929	(0)
16-8N-4W	#27...............	MacClintock, 1929	(0)
Morgan			
35-13N-11W	#30...............	MacClintock, 1929	(0)
8-15N-11W	Chapin............	Willman and Frye, 1970.............	(2)
9-15N-11W	Jacksonville NW...	Frye, Follmer, Glass, Masters and Willman, 1974	(2)
3-16N-10W	Literberry........	Frye and Willman, 1965a...........	(2)
9-13N-8W	Waverly...........	Frye and Willman, 1963.............	(1)
Moultrie			
13-13N-4E	#27...............	Horberg, 1953....	(1)
Peoria			
7-10N-7E	Jubilee College...	Willman and Frye, 1970.............	(1)
31-7N-6E	Tindall School....	Willman and Frye, 1970.............	(1)
Sangamon			
34-15N-4W	Rochester.........	Willman and Frye, 1970.............	(1)
4-14N-4W	New City..........	Frye and Willman, 1965b...........	(2)
16-15N-4W	Rochester West....	Frye and Willman, 1965b...........	(0)
5-14N-4W	unnamed...........	Frye, Shaffer, Willman and Ekblaw, 1960.............	(1)
29-16N-3W	Allis-Chalmers....	Johnson, 1964....	(3)
28-17N-4W	Bice School.......	Johnson, 1964....	(1)
23-17N-6W	Georgetown School.	Johnson, 1964....	(1)
14-15N-3W	Roby..............	Johnson, 1964....	(2)
19-17N-4W	Sherman...........	Johnson, 1964....	(1)

(Coninued Next Page)

Table 1 (Continued)

County Sec-Twp-R	Section name or number	Reference	Description class[a]
Sangamon			
12-17N-4W	Unnamed..........	Shaw and Savage, 1913............	(0)
Scott			
16-13N-11W	#32..............	MacClintock, 1929	(0)
Shelby			
8-11N-4E	Shelbyville Dam...	Johnson, Glass, Gross, and Moran, 1971............	(2)
8-11N-4E	Shelbyville Borrow Pit..............	Johnson, Glass, Gross, and Moran, 1971............	(2)
21-10N-1E	Oconee...........	Leighton and MacClintock, 1930	(2)
Tazewell			
30-26N-3W	Farm Creek........	Leverett, 1899...	(1)
30-26N-3W	Farm Creek........	Leighton, 1926...	(2)
30-26N-3W	Farm Creek (#20)..	Horberg, 1953....	(1)
30-26N-3W	Farm Creek........	Willman and Frye, 1970............	(1)
33-26N-4W	#19..............	Horberg, 1953....	(1)
8-24N-2W	#23..............	Horberg, 1953....	(1)
Woodford			
20-28N-2W	Richland Creek....	Frye, Glass and Willman, 1962....	(1)
20-28N-2W	#18..............	Horberg, 1953....	(1)
3-25N-1W	#21..............	Horberg, 1953....	(1)
9-26N-E	Secor............	Frye and Willman, 1965b...........	(1)

[a]Relative assessment of the morphologic description of the Sangamon Soil compared to present pedologic standards. Class criteria:

(0) named or weathering zone recognized but not described
(1) skeletal; general appearances described
(2) semi-detailed; major horizons described
(3) detailed; major and minor horizons described.

Figure 2 Location of reference sections, Sangamon Soil descriptions and radiocarbon dates in central Illinois.

Sangamon Soil. These sections have received the most attention in the literature; they display four common types of Sangamon Soil and contain the common variations in the loess that overlies the Sangamon in Illinois. However, they are geographically separated from each other (Figure 2). The Farm Creek was, in effect, a reference section for the Sangamon peat of Leverett (1899), a "type Pleistocene section" to Leighton (1926), the type section of the Farmdale loess of Leighton (Wascher *et al.*, 1948) and is presently the type section for the Farmdalian Substage and the Robein Silt (Willman and Frye, 1970).

The Effingham Section was, in effect, the type gumbotil section for Leighton and MacClintock (1930), a Sangamon Soil reference section for Simonson (1954) and Brophy (1959), and a reference section for the accretion-gley (Berry Clay) for Willman *et al.* (1966) and Willman and Frye (1970).

The Chapin and Rochester Sections have been proposed as paratype sections for the Sangamon Soil and the Sangamonian Stage by Willman and Frye (1970). The Chapin Section has been previously discussed by Frye and Willman (1965a) and Willman *et al.* (1966). The Rochester Section has been discussed by Frye *et al.* (1960) and Frye and Willman (1963 and 1965b).

The exposures along Farm Creek were first discussed by Leverett (1899). In one exposure he described 10 feet of Iowan loess resting on a deeply leached and weathered zone at the top of the Illinoian till. In another exposure, the railroad cut, he found a peat within a silt sequence in the Iowan position. He called the peat Sangamon and questioned whether the loess-like silt under the peat should be called Iowan. Leverett offered a two-phase concept for the origin of the Sangamon Soil. The first phase was the weathering on the Illinoian till and the second was the formation of the peat. This explains the apparent dualism in his expression "The Sangamon soil and weathered zone." Leverett was aware that the wood found in the Sangamon peat was coniferous and interpreted that "The aspects of the flora is decidedly boreal." He then concluded that these vegetal remains represented the closing phase of the interglacial stage "when glacial conditions were being inaugurated."

The Farm Creek exposures were studied again by Leighton (1926) and Willman and Frye (1970). The correlation of their sections to the Farm Creek Sections described by Leverett are shown in Figure 3. The thickness of the units shown in Figure 3 are adjusted by 30 cm (1 ft) to maximize similarity. The three Farm Creek Sections were intended to be measurements of the same section but show some obvious differences which illustrates the difficulty in remeasuring Pleistocene sections. The railroad cut section at Farm Creek is about one-half mile away, but is included because it has the stratigraphic features that can be easily correlated to the features described by Leighton and by Willman and Frye. The major stratigraphic features, except

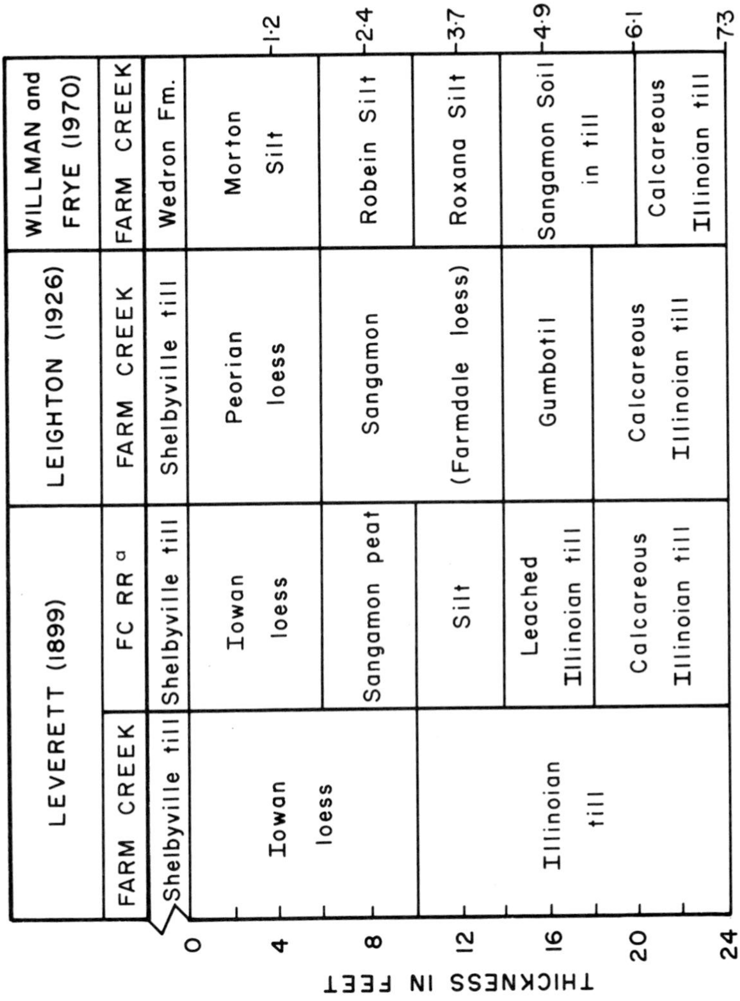

LEVERETT (1899)		LEIGHTON (1926)	WILLMAN and FRYE (1970)
FARM CREEK	FC RR [a]	FARM CREEK	FARM CREEK
Shelbyville till	Shelbyville till	Shelbyville till	Wedron Fm.
Iowan loess	Iowan loess	Peorian loess	Morton Silt
	Sangamon peat	Sangamon	Robein Silt
Illinoian till	Silt	(Farmdale loess)	Roxana Silt
	Leached Illinoian till	Gumbotil	Sangamon Soil in till
	Calcareous Illinoian till	Calcareous Illinoian till	Calcareous Illinoian till

THICKNESS IN FEET: 0, 4, 8, 12, 16, 20, 24

Right-hand scale: 1·2, 2·4, 3·7, 4·9, 6·1, 7·3

[a] Railroad cut near Farm Creek Section

Figure 3 Interpretation of Farm Creek Section stratigraphy.

for Leverett's Farm Creek Section, are in good agreement; however, the names and time interpretations have changed.

Leighton (1926) described the top one to one and one-half feet of the "Sangamon" as "old soil, dark with flakes of carbon, some...wood..., loessial in texture, non calcareous." The next zone down, 2.1 to 2.4 m (7 to 8 ft) thick, he described as a "loess-like silt," leached to about 1.6 m (5 1/2 ft) at the east side of the exposure where the silt has brown and yellow colors but only leached to about 15 cm (6 in) below the "old soil" at the west side. The 15 cm (6 in) is described as "greenish loess" and overlies a bluish-gray calcareous zone, presumably 1.2 to 1.5 m (4 to 5 ft) thick which contains "scattered small pebbles in lower 3 feet." Leighton (Wascher *et al.*, 1948) changed the name of the loess-like silt to the Farmdale loess after his 1926 descriptions. Under the silt, he described 1.2 m (4 ft) of gumbotil as "brown with reddish specks on each side, brownish to brick red at top with bluish spots on west side." He also noted that it is "tenacious, [has a] hackly fracture, [and contains] siliceous pebbles, mostly under 3/4-inch, ...and grades downwards into very calcareous till..."

The interpretations of the Farm Creek Section, drawn by Leighton (1926) were largely in agreement with Leverett. Important interpretations made by Leighton were 1) the silt under the "Sangamon peat" is loess, 2) the original color of this loess was probably a "brown to grayish yellow color" and was altered to the "bluish-gray," and 3) the section displayed a catena, or local conditions in which "the subsurface drainage of an oxidizing character" existed within a short distance to a "nearly stagnant drainage of either unoxidizing or deoxidizing character." Leighton apparently lost sight of this catena concept in his 1930 paper and virtually denied the existence of the process of "deoxidation" in 1962.

Willman and Frye (1970) reinterpreted the Farm Creek Section using current concepts of the Sangamon Soil and rock-stratigraphy. Farmdale loess had been renamed the Roxana Silt (Frye and Willman, 1960). The Roxana is largely loess and has a colluviated silt with some sand at the base (Willman and Frye, 1970). The Roxana is the first Wisconsinan deposit in central Illinois (Frye *et al.*, 1974). The Sangamon peat of Leverett, known for many years as the Farmdale Silt, was renamed Robein Silt (Willman and Frye, 1970) because of confusion caused by the multiple use of the name Farmdale. The name has been retained for the Farmdale Soil and Farmdalian Substage. The radiocarbon ages of two wood samples taken from the Robein Silt at the Farm Creek Section are given in Table 2 along with all other radiocarbon determinations on organic materials historically associated with the Sangamon Soil. Most of the dates are remarkably similar, ranging from about 20,000 to 30,000 years BP. A few determinations are in the greater than 30,000 years BP range and were interpreted to be from the

Robein Silt or Illinoian or older deposits. The Fayette County, Pittsburg Basin, series was included in the table because Gruger (1972 a, b) concluded that all of his samples stratigraphically overlie a pollen zone he interpreted to be Sangamonian in age.

The description of the Sangamon Soil at the Farm Creek Section by Willman and Frye (1970) is skeletal. Essentially only the general appearances are described, although they recognized the B2 horizon as being "thinner than typical." They described the soil as "till, leached, brown with some streaks and splotches of red-brown, tough, clayey: ..." and is overlain by 1.1 m (3.5 ft) of Roxana Silt and 1.4 m (4.5 ft) of Robein Silt."

The Effingham Section was first described by Leighton and MacClintock (1930). They referred to a "weathered zone on the Illinoian drift-sheet" and never used the term Sangamon Soil. They proposed a type of catena which is made up of three members: gumbotil, mesotil, and siltil. This catena was really a physiographic separation between "poorly drained areas," "partially drained areas" and "well drained areas," respectively. The Effingham Section was used by them to characterize the gumbotil profile which occurs on the broad flat, poorly drained areas in the Illinoian drift region of southern and western Illinois.

The Sangamon Soil description by Leighton and MacClintock at Effingham is semi-detailed. The major horizons are described by color, texture, consistence, and few other features that appear to be related to soil structure. Subdivisions are described but appear to be more related to material boundaries than soil horizon boundaries. Their example "profile" is overlain by 85 cm (2 ft, 10 in) of "soil and loess." Horizon 1 is the "Fossil soil," meaning ancient or buried soil. Horizon 2 is the gumbotil. Horizon 3 is the leached and oxidized, little altered till. Horizon 4 is the calcareous, oxidized till and Horizon 5 is the unweathered, unaltered, blue-gray till.

Figure 4 compares the horizons of the Sangamon Soil profile described by Leighton and MacClintock to the zones (horizons) described by Simonson (1954), Brophy (1959) and Willman *et al.* (1966) at the Effingham Section. The original designations of the authors are retained. The measurements at this section were also intended to be at the same location but were probably separated by some distance less than 30 m (100 ft). The separation between the measured sections is not known. Lateral variations can explain the difference in thickness but not the interpretation of each horizon or zone. The Effingham Section was the first place where the gumbotil was directly equated to accretion-gley (Willman *et al.*, 1966) although the equivalency was pointed out earlier by Frye *et al.* (1960) and Frye and Willman (1963).

Simonson (1954) interpreted the Sangamon Soil at

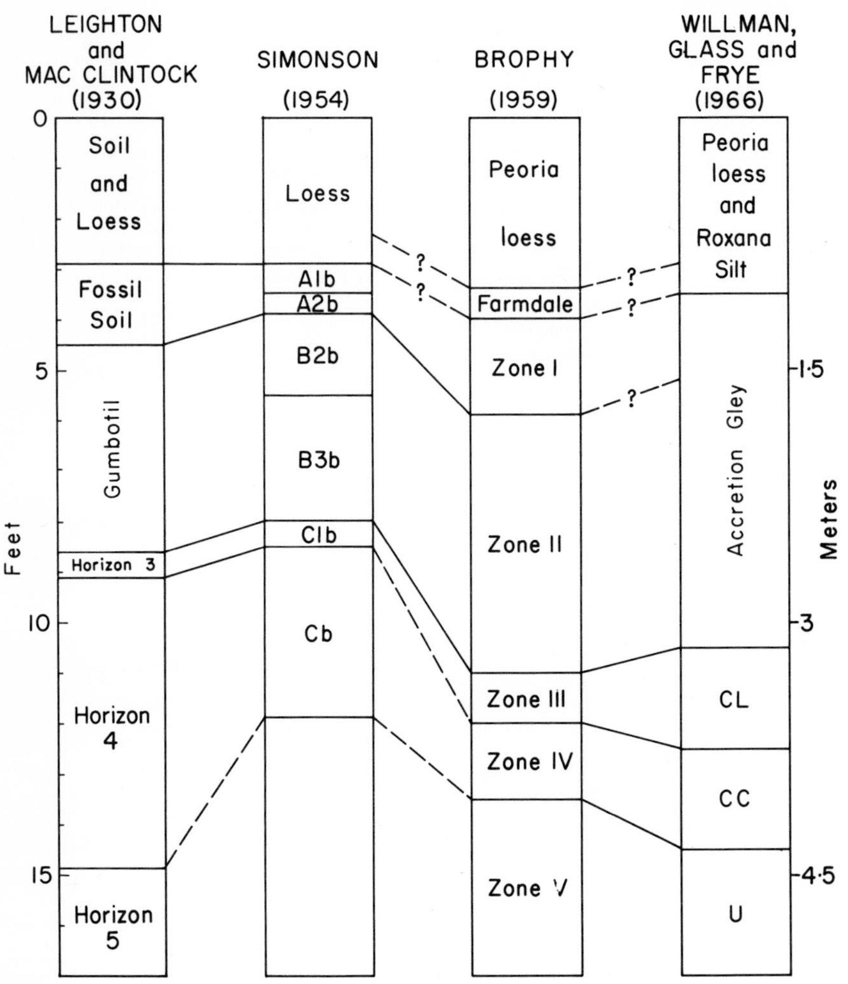

Figure 4 Comparison of the horizons of the Sangamon soil (Leighton and MacClintock) to the zones (horizons) described by Simonson (1954), Brophy (1959) and Willman *et al.* (1966) at the Effingham Section.

Table 2. Radiocarbon dates from weathered horizons in or superjacent to the Sangamon Soil in its type area.

County Sec-Twp-R	Locality or Section	Unit	Laboratory Number[a]	Date Years BP	References[b]
Cass					
24-18N-11W....	Virginia NW[c]........	Robein Silt..	ISGS-122..	24,980 ± 420	Rubin and Alexander, 1960
25-17N-9W.....	Ashland[c]...........	Robein Silt..	ISGS-123..	21,080 ± 370	Leighton, 1965
22-17N-10W....	Virginia...........	Roxana Silt..	W-526.....	29,000 ± 1200	Kleiss and Fehrenbacher, 1973 Coleman, 1974
DeWitt					
21-21N-5E.....	Farmer City.........	Robein Silt..	I-2517....	21,950 ± 500	Kempton, DuMontelle and Glass, 1971
Fayette					
3-5N-1W.......	Pittsburg Basin.....	Equality Fm..	ISGS-47...	21,370 ± 810	Kim, 1970
3-5N-1W.......	Pittsburg Basin.....	Equality Fm..	ISGS-53...	24,200 ± 1900	Coleman, 1972, 1973
3-5N-1W.......	Pittsburg Basin.....	Equality Fm..	ISGS-65...	24,200 ± 800	Gruger, 1972a,b
3-5N-1W.......	Pittsburg Basin.....	Equality Fm..	ISGS-9....	22,300.......	
3-5N-1W.......	Pittsburg Basin.....	Equality Fm..	ISGS-67...	34,000 ± 1200	
3-5N-1W.......	Pittsburg Basin.....	Equality Fm..	ISGS-10...	27,000.......	
3-5N-1W.......	Pittsburg Basin.....	Equality Fm..	ISGS-5....	22,000.......	
3-5N-1W.......	Pittsburg Basin.....	Equality Fm..	ISGS-22...	40,000.......	
3-5N-1W.......	Pittsburg Basin.....	Equality Fm..	ISGS-71...	37,200 ± 900	
3-5N-1W.......	Pittsburg Basin.....	Equality Fm..	ISGS-11...	38,100 ± 1000	
3-5N-1W......	Pittsburg Basin.....	Equality Fm..	ISGS-13...	40,000.......	

Fulton					
33-4N-3E.....	Enion Terrace......	Robein Silt..	W-745.....	23,500 ± 400	Rubin and Alexander, 1960
13-6N-4E......	Buckheart Strip Mine	Robein Silt..	W-849.....	23,700 ± 500	Frye and Willman 1960, 1963
13-6N-4E......	Buckheart Strip Mine	Robein Silt..	W-853.....	25,500 ± 600	Leighton, 1965
McLean					
32-25N-1W.....	Danvers...........	Robein Silt..	W-406.....	26,150 ± 700	Rubin and Alexander, 1958
32-25N-1W.....	Danvers...........	Robein Silt..	ISGS-12...	23,900 ± 200	Leighton, 1965
20-22N-4E.....	Le Roy............	Robein Silt..	I-2220....	27,200 + 1000 - 900	Kim, 1970 Kempton, Dumontelle and Glass, 1971
5-21N-5E......	Le Roy SE[c]........	Robein Silt..	I-2785....	24,600 ± 750	
1-25N-2E......	Lake Bloomington Spillway..........	Illinoian or older........	ISGS-16...	40,000.......	
Macon					
13-16N-1W.....	Niantic...........	Robein Silt..	ISGS-21...	25,500 ± 600	Kim, 1970
13-16N-1W.....	Niantic...........	Robein Silt..	ISGS-25...	33,000.......	Gruger, 1972a
34-17N-1W.....	Niantic NW[c].......	Robein Silt..	ISGS-135..	31,400.......	Coleman, 1974
Menard					
24-18N-5W.....	Fancy Prairie......	Robein Silt..	ISGS-90...	24,450 ± 280	Coleman, 1973, 1974
9-17N-6W......	Rock Creek[c]........	silt.........	ISGS-110..	20,740 ± 720	Miller, 1973
Sangamon					
23-16N-4W.....	Clear Lake[c].......	Robein Silt..	ISGS-107..	22,020 ± 300	Coleman, 1973, 1974
22-16N-4W.....	Sangamon River Valley............	silt.........	ISGS-102..	24,640 ± 430	Miller, 1973
9-15N-4W......	South Fork[c].......	silt.........	ISGS-108	22,150 ± 330	
28-16N-4W.....	Sugar Creek[c]......	silt.........	ISGS-99	22,700 ± 1100	

Table 2 (Continued)

County Sec-Twp-R	Locality or Section	Unit	Laboratory Number[a]	Date Years BP	References[b]
28-16N-4W.....	Sugar Creek[c].......	sand.........	ISGS-118..	29,140 ± 270	
Shelby					
8-11N-4E......	Shelbyville Bor.Pit	Robein Silt..	ISGS-26...	20,000 ± 200	Kim, 1970
8-11N-4E......	Shelbyville Bor.Pit	Robein Silt..	ISGS-32...	21,300 ± 500	Johnson, Glass, Gross and Moran, 1971
8-11N-4E......	Shelbyville Bor.Pit	Berry Clay...	ISGS-46...	21,400 ± 1000	Coleman, 1972
Tazewell					
30-26N-3W.....	Farm Creek.........	Robein Silt..	W-68......	22,900 ± 900	Suess, 1954
30-26N-3W.....	Farm Creek.........	Robein Silt..	W-69......	25,100 ± 800	Leighton, 1965
Woodford					
1-25N-1E......	Six Mile Creek Dam.	Robein Silt..	I-2218....	26,500 + 1000 - 900	Kempton, DuMontelle and Glass, 1971
1-25N-1E......	Six Mile Creek Dam.	Illinoian or older........	I-2219....	39,000.......	

[a] I - Teledyne Isotopes, Westwood, New Jersey; ISGS - Illinois State Geological Survey, Urbana Illinois; W - U.S. Geological Survey, Reston, Virginia.

[b] Grouped by county to eliminate duplication. References before 1970 are listed in Willman and Frye (1970, Table 1).

[c] Name assigned for this paper. Previous references did not assign a geographic name.

Effingham to be a planosol and described the soil using soil horizon designations. His description is semi-detailed and is almost the same as Leighton's. Although he used sub-horizons (A1, A2, *etc.*), he only described color, texture, consistence, and a few other features. No quantifying expressions were used nor was any pedological structure interpreted. Simonson's purpose in his 1954 paper was to draw attention to buried soils in the area of the Mississippi River drainage and to explain their significance to the interpretations of Pleistocene geology.

Brophy (1959) redescribed the Effingham Section using weathering zone designations. His description is similar to Simonson's but included Munsell color notations, quantified the amount of mottling and described zone II as structureless. Brophy's purpose was to study weathering and the heavy mineral ratios in four "Sangamon weathering profiles" and not to specially describe the soil. Brophy established that there was significant depletion of hornblende, illite, and chlorite in Zones I and II and only a slight depletion in Zones III and IV. His results on the degree of weathering were in agreement with the work in Iowa by Ruhe (1956) and he concluded that the weathering ratio method is a promising technique for "evaluating profile maturity, especially in buried profiles where standard pedological tests of maturity may not work." His reference to "standard pedological tests" means the assessment of the degree of development of soil structure and may include the degree of soil horizon differentiation. The stronger the expression of soil structure or the greater the contrast between the A and B horizons, the more mature the soil is interpreted to be. But as Brophy noted, this test often fails in some buried soils, particularly buried poorly drained soils.

The garnet to hornblende ratio determined by Brophy in the Effingham profile shows a marked discontinuity in the upper part of Zone II, but he did not make any comment on this feature. It happens to occur where his medium silt fraction changes from a higher value, similar to the loess above to a lower value, similar to the till derived material below. Loess or loess derived material appears to be significantly contributing to Zone I, but Brophy did not recognize it.

Willman *et al.* (1966) recognized that local slope wash, which may include some loess, was a component present at the Effingham Section but expanded the interpreted accretionary zone to include all the gleyed deposits which they called accretion-gley. The site they sampled could have been in such a position. The grain size data they presented are not comparable with Brophy's which supports the conclusion that the sampling sites were not at the same place. The purpose of the study by Willman *et al.* (1966) was to perform a total mineralogical analysis of four Sangamon Soil profiles and two Yarmouth Soil profiles for general characterization. At the Effingham Section, they collected six bulk samples, three from the gley and three from below the gley, which is not a sufficient number nor the right type of samples to distinguish a lithologic discontinuity with much certainty. However, the upper sample in the accretion-gley does have less sand and more silt than the underlying sample. They described the Sangamon Soil in general terms but the major horizons can be identified within their description. They noted the presence of pebbles and the krotovinas throughout the profile. Krotovinas are caused principally by burrowing animals which contribute to mixing and the confused soil horizon characteristics. Interpretations made by Leighton and MacClintock and Simonson and Brophy on the Effingham Sections are in general agreement and contrast with the interpretations made by Willman *et al.* (1966). The differences arise from the genetic interpretations of the origin of gumbotil and accretion-gley. The former is interpreted to be a product of intense weathering under oxidizing conditions and the latter is interpreted to be the product of accretion and weathering under reducing conditions.

The Chapin Section was first described in 1965 for the purpose of demonstrating a complete sequence of the overlying Roxana Silt to the INQUA field conference (Frye and Willman, 1965a). The Sangamon Soil at this section is a well drained, *in situ* soil developed in Illinoian till. They described the Sangamon Soil in reasonable detail, including color, texture, structure, and several other features. They recognized sub-horizons of the B but did not describe them completely. They described the Sangamon 2 m (6 1/2 ft) thick and overlying 30 cm (1 ft) of weakly calcareous, oxidized till. The top horizon is described as an A or B1 but they did not comment on whether the A is missing or not identified.

Willman *et al.* (1966) described the Chapin Section again and sampled the Sangamon Soil for their mineralogical investigation. Their description of the Sangamon Soil included only the thickness of the B1 or A, B2 and B3 horizons. Then in 1970 Willman and Frye selected the Chapin Section to be a paratype section of the Sangamon Soil, representing the *in situ* profile developed in Illinoian till. Their description of the Sangamon is essentially the same as the 1965 description. Willman and Frye considered the Sangamon to be better developed here than in some localities because younger Illinoian deposits are missing, making the interval of soil formation longer. The Sangamon is overlain by the Markham Silt, the oldest member of the Roxana Silt. The profile is leached to 2 m (6 1/2 ft) and has a distinctive red-brown, clayey B2 horizon with iron-manganese concretions and stains.

The Rochester Section was first described in a general way by Frye *et al.* (1960). The profile was described by defining the major zones of weathering and using the zone designations on a generalized sketch of the section. The Rochester Section at this time was considered a typical exposure of the Sangamonian accretion-gley. The purpose of the 1960 study was to establish the concept of accretion-gley and not to describe the Sangamon Soil. The Rochester Section was discussed again by Frye and Willman (1963) but not redescribed. The INQUA field conference also stopped at the Rochester Section, and a skeletal description of the Sangamon Soil was presented (Frye and Willman, 1965b). Then in 1970 the Rochester Section was chosen to be a paratype section of the Sangamon Soil, serving as a representative section of the accretion-gley profile (Willman and Frye, 1970). The accretion-gley was named Berry Clay at this section. Their Sangamon Soil description is skeletal. General features of color, texture, and structure are noted, and only two zones are recognized. The upper zone appears to be the result of post burial alteration and was interpreted to be evidence of Farmdale Soil formation. They stated that the Roxana has been truncated and Peoria Loess directly overlies the accretion-gley.

Willman and Frye (1970) also considered the Rochester Section to be a paratype section of the Sangamonian Stage. However, they stated that "the interval of soil formation exceeded the time span of the Sangamonian Stage." This statement can be interpreted in two ways. In the context of their statement, it means that soil formation began before the end of the Illinoian at the Rochester Section. But they also stated that "early Wisconsinan sediments" were incorporated into the "gleyed material" of the Sangamon Soil after the end of Sangamonian time. The fact that the Roxana has been truncated at the section indicates that an erosion surface overlies the accretion-gley and suggests that some of the gley has also been removed by erosion. Therefore, it is not clear from their discussion whether the Rochester Section contains a complete sequence of Sangamonian (Berry Clay) deposits or whether the Sangamon Soil profile is complete. However, the top of the Sangamon Soil at Rochester has been arbitrarily defined as the Sangamonian-Wisconsinan boundary even though in a regional context the Sangamon Soil may be incomplete.

A similar problem of interpreting the Sangamonian and the Sangamon Soil exists at the Chapin Section. The A and B1 horizons of the Sangamon Soil were not differentiated and were described as having a combined thickness of 6 inches. This means that the A horizon is probably missing because a "normal" Sangamon Soil in this position is a podzolic type of soil (Ultisol, Ruhe, 1974) and should have an A horizon 8 to 12 inches thick and a B1 horizon 25 to 45 cm (10 to 18 in) thick. Therefore, an important part of the Sangamon profile appears to be missing or has been misidentified. The uncertainty of the probable erosion surfaces at the Rochester and Chapin Sections greatly diminishes the possibility of deriving a more precise definition of the Sangamonian interval and the beginning of the Wisconsinan Stage at those sections. New reference sections should be established that describe the common types of Sangamon Soil profiles and record the complete sediment record free from erosion surfaces.

CONCLUSIONS

The major concepts of the origin and stratigraphic position of the Sangamon Soil in Illinois have evolved into a reasonably clear picture in the 80 years since the introduction of the Sangamon Soil by Leverett (1898a). However, some of the details remain to be resolved. The details pertaining to the Sangamon Soil and its age have become increasingly important as more precise correlations to other areas, particularly the oceanic record, are being attempted.

The need for more precise information has always been recognized. Leighton initially went to the Farm Creek Section in 1926 because he thought a "detailed examination" was needed. Even after the great amount of work Leighton accomplished, he described the need for a comprehensive study of the weathering profiles (Leighton and MacClintock, 1962) and made recommendations that the "Farm Creek Section should be opened up" and studied again (1965). In the most recent work, Willman and Frye (1970) thought that paratype sections of two types of Sangamon Soil profiles were needed because none had existed before. However, it has been shown in this review that more work remains to be done.

For a general summary of the present status of the Sangamon Soil in its type area, the following conclusions can be drawn:

(1) It has generally been consistently recognized in stratigraphic sections.

(2) It has been successfully used as a general marker bed to separate Wisconsinan and Illinoian deposits.

(3) The mineralogy of the profile has been satisfactorily characterized by major horizons.

(4) The parent material uniformity has not been specifically tested at the reference sections, except for the work by Brophy (1959).

(5) The soil morphology has not been studied in sufficient detail.

(6) A catena or toposequence has not been studied or adequately described. A type transect should be established.

(7) The contributions of early Wisconsinan loess and Illinoian lacustrine deposits in gleyed profiles have not been fully assessed.

(8) The age of the Sangamon Soil has not been precisely defined.

(9) The paratype sections of the Sangamon Soil do not appear to contain complete profiles on which the Sangamonian should be defined.

One of the conclusions drawn by the INQUA Commission on Paleopedology (Yaalon, 1971) was "...profiles should be traced laterally in the landscape to determine their spatial variation." This has not been done for the Sangamon Soil in its type area because the work done on the soil was of a more general nature than the standards recommended by the commission. The spatial variations of the Sangamon Soil can be best represented by a type transect that would include the range of drainage and topographic conditions common to the soil-stratigraphic unit. In a special sense this would be a paleocatena. At least four members of a Sangamon Soil catena should be described: 1) a very poorly drained, organic rich accretion-gley profile, 2) a poorly drained, *in situ* profile, 3) an imperfectly drained profile, and 4) a well drained profile. The *in situ* profiles should be developed in till and a complete study of the morphology, mineralogy, and chemical parameters of each profile should be attempted. An effort should be made to find complete soil profiles that represent a conformable sequence of materials containing no erosion surfaces. A distinction between the eroded and uneroded profiles is essential for an accurate characterization of each profile. A type Sangamon Soil transect would resolve many of the previous controversies and would provide a useful reference for comparisons with Sangamon profiles in other areas.

ACKNOWLEDGMENTS

The manuscript was critically reviewed by H.D. Glass, C.S. Hunt, W.H. Johnson, J.P. Kempton and E.D. McKay who made many constructive suggestions. Special recognition is due to the members of the Illinois State Geological Survey for their willingness to participate in discussions on Pleistocene problems. Appreciation is extended to Barbara Roby for drafting the figures and to Mary McGuire for typing the manuscript. I greatly appreciate the assistance provided by W.C. Mahaney who encouraged the undertaking of this review and extended his patience for the completion of the manuscript.

REFERENCES CITED

Allen, V.T., 1959, Gumbotil and interglacial clays: Geol. Soc. Amer. Bull., v. 70, no. 11, p. 1483-1486.

Ball, J.R., 1952, Geology and mineral resources of the Carlinville Quadrangle: Illinois Geol. Survey Bull. 77, 110 p.

Bannister, H.M., 1870, Geology of Tazewell, McLean, Logan and Mason Counties, *in* Worthen, A.H., Geology and palaeontology: Geol. Survey of Illinois, Vol. IV, p. 176-189.

Bergstrom, R.E., Piskin, K., and Follmer, L.R., 1976, Geology for planning in the Springfield-Decatur region, Illinois: Illinois Geol. Survey Circ. 497, 76 p.

Brophy, J.A., 1959, Heavy mineral ratios of Sangamon weathering profiles in Illinois: Illinois Geol. Survey Circ. 273, 22 p.

Chamberlin, T.C., 1896 (Editorial): Jour. Geol., v. 4, p. 872-876.

Coleman, D.D., 1972, Illinois State Geological Survey radiocarbon dates III: Radiocarbon, v. 14, p. 149-154.

__________, 1973, Illinois State Geological Survey radiocarbon dates IV: Radiocarbon, v. 15, p. 75-85.

__________, 1974, Illinois State Geological Survey radiocarbon dates V: Radiocarbon, v. 16, p. 105-117.

Follmer, L.R., 1970, Soil distribution and stratigraphy in the Mollic Albaqualf region of Illinois (Ph.D. Dissert.): Urbana, Univ. Illinois, 155 p.

Frye, J.C., and Willman, H.B., 1960, Classification of the Wisconsinan Stage in the Lake Michigan glacial lobe: Illinois Geol. Survey Circ. 285, 16 p.

__________, 1963, Loess stratigraphy, Wisconsinan classification and accretion-gleys in central western Illinois: (Midwestern Sec.) Friends of the Pleistocene, 14th Ann. Mtg., Illinois Geol. Survey Guidebook Ser. 5, 37 p.

__________, 1965a, *in* Schultz, C.B., and Smith, H.T.U., eds., (Illinois part of) Guidebook for field conference C--Upper Mississippi Valley: Lincoln, Nebr., Nebraska Acad. Sci., Internat. Assoc. Quaternary Research 7th Cong., p. 81-110; Illinois Geol. Survey Reprint 1966-B (supplemental data, J.P. Kempton and H.D. Glass, p. C-S1-C-S11), 41 p.

__________, 1965b, Illinois, *in* Schultz, C.B., and Smith, H.T.U., eds., Guidebook for field conference G--Great Lakes-Ohio River Valley: Lincoln, Nebr, Nebraska Acad. Sci., Internat. Assoc. Quaternary Research 7th Cong., p. 5-26; Illinois Geol. Survey Reprint 1966-B (supplemental data, H.D. Glass, p. G-S1-G-S4), 26 p.

__________, 1973, Wisconsinan climatic history interpreted from Lake Michigan Lobe deposits and soils, *in* Black, R.F., Goldthwait, R.P., and Willman, H.B., eds., The Wisconsinan Stage: Geol. Soc. Amer. Mem. 136, p. 135-152.

Frye, J.C. *et al.*, 1960, Accretion-gley and the gumbotil dilemma: Amer. Jour. Sci., v. 258, no. 3, p. 185-190.

Frye, J.C. *et al.*, 1960, Gumbotil, accretion-gley, and the weathering profile: Illinois Geol. Survey Circ. 295, 39 p.

Frye, J.C. *et al.*, 1962, Stratigraphy and mineralogy of the Wisconsinan loesses of Illinois: Illinois Geol. Survey Circ. 334, 55 p.

Frye, J.C. *et al.*, 1974, Earliest Wisconsinan sediments and soils: Illinois Geol. Survey Circ. 485, 12 p.

Geikie, James, 1877, The great ice age (2nd ed.): London, 624 p.

Gerasimov, I.D., 1971, Nature and originality of paleosols, *in* Yallon, D.W., eds., Paleopoedology: Jerusalem, Israel, Israel Univ. Press, p. 15-27.

Gruger, E., 1972a, Pollen and seed studies of Wisconsinan vegetation in Illinois: Geol. Soc. Amer. Bull., v. 83, no. 9, p. 2715-2734.

__________, 1972b, Late Quaternary vegetation development in south-central Illinois: Quaternary Research, v. 2, no. 2, p. 217-231.

Horberg, C.L., 1953, Pleistocene deposits below the Wisconsin drift in northeastern Illinois: Illinois Geol. Survey Rept. Inv. 165, 61 p.

Jacobs, A.M., and Lineback, J.A., 1969, Glacial geology of the Vandalia, Illinois, region: Illinois Geol. Survey Circ. 442, 24 p.

Johnson, W.H., 1964, Stratigraphy and petrography of Illinoian and Kansan drift in central Illinois: Illinois Geol. Survey Circ. 378, 38 p.

______________, 1976, Quaternary stratigraphy in Illinois: Status and current problems, *in* Mahaney, W.C., ed., Quaternary stratigraphy of North America: Stroudsburg, Penn., Dowden, Hutchinson and Ross, p. 161-196.

Johnson, W.H., *et al.*, 1971, Glacial drift of the Shelbyville Moraine at Shelbyville, Illinois: Illinois Geol. Survey Circ. 459, 22 p.

Johnson, W.H., *et al.*, 1972, Pleistocene stratigraphy of east-central Illinois: Illinois Geol. Survey Guidebook Ser. 9, 97 p.

Kapp, R.O., and Gooding, A.M., 1964, Pleistocene vegetation studies in the Whitewater Basin, southeastern Indiana: Jour. Geol., v. 72, p. 307-326.

Kay, G.F., 1916, Gumbotil, a new term in Pleistocene geology: Science, new ser., v. 44, p. 637-638.

Kay, G.F., and Pearce, J.N., 1920, The origin of gumbotil: Jour. Geol., v. 28, no. 2, p. 89-125.

Kempton, J.P., *et al.*, 1971, Subsurface stratigraphy of the Woodfordian tills in the McLean County region, Illinois, *in* Goldthwaite, R.P., Forsyth, J.L., Gross, D. L., and Pessl, F., Jr., eds., Till, a symposium: Columbus, Ohio State Univ. Press, p. 217-233.

Kim, S.M., 1970, Illinois State Geological Survey radiocarbon dates II: Radiocarbon, v. 12, no. 2, p. 503-508.

Kleiss, H.J., and Fehrenbacher, J.B., 1973, Loess distribution as revealed by mineral variations: Soil Sci. Soc. America Proc., v. 37, p. 291-295.

Leighton, M.M., 1926, A notable type Pleistocene section: the Farm Creek exposure near Peoria, Illinois: Jour. Geol., v. 34, no. 2, p. 167-174.

______________, 1931, The Peorian Loess and the classification of the glacial drift sheets of the Mississippi Valley: Jour. Geol., v. 39, p. 45-53.

______________, 1958, Principles and viewpoints in formulating the stratigraphic classifications of the Pleistocene: Jour. Geol., v. 66, no. 6, p. 700-709.

______________, 1965, The stratigraphic succession of Wisconsin loess in the Upper Mississippi River Valley: Jour. Geol., v. 73, no. 2, p. 323-345.

Leighton, M.M., and MacClintock, P., 1930, Weathered zones of the drift-sheets of Illinois: Jour. Geol., v. 38, no. 1, p. 28-53.

______________, 1962, The weathered mantle of glacial tills beneath original surfaces in north-central United States: Jour. Geol., v. 70, no. 3, p. 267-293.

Leighton, M.M., and Willman, H.B., 1950, Loess formations of the Mississippi Valley: Jour. Geol., v. 58, no. 6, p. 599-623.

Leverett, F., 1898a, The weathered zone (Sangamon) between the Iowan Loess and Illinoian till sheet: Jour. Geol., v. 6, p. 171-181.

______________, 1898b, The weathered zone (Yarmouth) between the Illinoian and Kansan till sheets: Iowa Acad. Sci. Proc., v. 5, p. 81-86.

______________, 1899, The Illinois glacial lobe: U.S. Geol. Survey Monogr. 38, 817 p.

MacClintock, P., 1929, I. Physiographic divisions of the area covered by the Illinoian drift-sheet in southern

Illinois. II. Recent discoveries of pre-Illinoian drift in southern Illinois: Illinois Geol. Survey Rept. Inv. 19, 57 p.

Miller, J.A., 1973, Quaternary history of the Sangamon River drainage system, central Illinois: Ill. State Museum, Rep. Inv. 27, 36 p.

Norton, E.A., and Smith, R.S., 1928, Horizon designations: Amer. Soil Surv. Assoc. Bull. 9, p. 83-86.

Rubin, M., and Alexander, C., 1958, U.S. Geological Survey radiocarbon dates IV: Science, v. 127, no. 3313, p. 1476-1487.

________, 1960, U.S. Geological Survey radiocarbon dates V: Amer. Jour. Sci., Radiocarbon Supp., v. 2, p. 129-185.

Ruhe, R.V., 1956, Geomorphic surfaces and the nature of soils: Soil Sci., v. 82, p. 441-455.

__________, 1974, Sangamon paleosols and Quaternary environments in midwestern United States, *in* Mahaney, W.C., ed., Quaternary Environments: Proceedings of a Symposium, Geographical monographs no. 5: Toronto, York University Ser. Geog., p. 153-167.

__________, 1976, Stratigraphy of mid-continent loess, U.S.A., *in* Mahaney, W.C., ed., Quaternary Stratigraphy of North America: Stroudsburg, Penn., Dowden, Hutchinson and Ross, p. 197-211.

Shaw, E.W., and Savage, T.E., 1913, Description of the Tallula-Springfield quadrangles: Washington, D.C., U.S. Geol. Survey Geol. Atlas Folio No. 188, 12 p.

Simonson, R.W., 1941, Studies of buried soils formed from till in Iowa: Soil Sci. Soc. Amer. Proc., v. 6, p. 373-381.

______________, 1954, Identification and interpretation of buried soils: Amer. Jour. Sci., v. 252, p. 703-732.

Smith R.S., and Norton, E.A., 1929, The structural anatomy of the soil profiles: Amer. Soil Surv. Assoc. Bull. 10, p. 35-39.

Suess, H.E., 1954, U.S. Geological Survey radiocarbon dates I: Science, v. 120, no. 3117, p. 467-473.

Thorp, J., *et al.*, 1951, Some post-Pliocene buried soils of the central United States: Jour. Soil Sci., v. 2, p. 1-19.

Valentine, K.W.G., and Dalrymple, J.B., 1976, Quaternary buried paleosols: A critical review: Quaternary

Research, v. 6, no. 2, p. 209-222.

Voss, J., 1939, Forests of the Yarmouth and Sangamon interglacial periods in Illinois: Ecology, v. 20, no. 4, p. 517-528.

Wanless, H.R., 1957, Geology and mineral resources of the Beardstown, Glasford, Havana, and Vermont Quadrangles: Illinois Geol. Survey Bull. 82, 233 p.

Wascher, H.L., *et al.*, 1948, Loess in the southern Mississippi Valley - Identification and distribution of the loess sheets: Soil Sci. Soc. Amer. Proc., 1947, v. 12, p. 389-399.

White, G.W., 1973, History of investigation and classification of Wisconsinan drift in north-central United States, *in* Black, R.F., Goldthwaite, R.P., and Willman, H.B., eds., The Wisconsinan Stage: Geol. Soc. Amer. Mem. 136, p. 3-34.

Willman, H.B., and Frye, J.C., 1970, Pleistocene stratigraphy of Illinois: Illinois Geol. Survey Bull. 94, 204 p.

Willman, H.B., *et al.*, 1966, Mineralogy of glacial tills and their weathering profiles in Illinois. Part II - Weathering profiles: Illinois Geol. Survey Circ. 400, 76 p.

Worthen, A.H., 1866, Geology: Geol. Survey of Illinois, v. I, 504 p.

____________, 1868, Geology and palaeontology: Geol. Survey of Illinois, v. III, 574 p.

____________, 1870, Geology and palaeontology: Geol. Survey of Illinois, v. IV, 508 p.

____________, 1873, Geology and palaeontology: Geol. Survey of Illinois, v. V, 619 p.

Yaalon, D.H., ed., 1971, Criteria for the recognition and classification of paleosols, A report of the working group on the origin and nature of paleosols, INQUA Commission on paleopedology, 1970, *in* Yaalon, D.H., ed., Palaeopedology: Jerusalem, Israel University Press, p. 153-158.

The Brussels Formation—A Stratigraphic Re-appraisal

W. H. Allen, Jr., R. A. Ward

ABSTRACT

The Brussels Formation, originally described in Illinois and Missouri, has recently been rejected as a stratigraphic unit because of insufficient evidence regarding its relationship to other Illinoian or Wisconsinan deposits. As such, the materials which had been given a rather definitive formational status were subdivided into three separate units of Wisconsinan or Illinoian Age. The Brussels nomenclature was discontinued for two specific reasons; there was, at that time, no conclusive evidence of a Sangamon Paleosol or Roxana Silt overlying the sediments.

Admittedly, the original discussion regarding this formation was weak from the standpoint of not giving a standard section, but from a stratigraphic point of view, it was developed on a very strong concept that was well documented in the field. Rubey, in 1952, held that the Brussels Formation represented sediments of fluvio-lacustrine origin which were deposited behind an early Illinoian ice dam that blocked the flow of the Mississippi River at St. Louis, Missouri.

Based on stratigraphic evidence just north of St. Louis in St. Charles County, it would appear that the oldest and most extensive Illinoian glacial advance (the westernmost extension of the Lake Michigan Lobe) turned northwest as it entered the Mississippi River Valley at the juncture of the Illinois and Mississippi Rivers and extended at least 10 miles up river. The stratigraphy as defined at the Arrowhead Park composite section includes the following Illinoian and Wisconsinan units respectively: Petersburg Silt - Loveland Silt, Kellerville Till Member - Brussels Formation, Pike Soil, Teneriffe Silt, Sangamon Soil, Roxana Silt, Robein Silt, Farmdale Soil, and Peoria Loess.

The Brussels Formation at the Arrowhead Park Site consists of a sequence of varved sediments that alternate between silts and clays. The material is highly calcareous, but contains no fossils. Moving away from the terminal edge of the Illinoian ice, the sediments generally tend to coarsen and reflect more of the local source material. This is true in the Dardenne and Cuivre Rivers as well as up valley in the Mississippi and the Illinois River Systems. The thickness of the sediments which are related to the Illinoian Brussels Formation seldom exceeds 8 to 10 feet. A well developed Sangamon Paleosol immediately overlies the varved sediments. Pedogenic development obscures the upper part of the varve sequence but within the zone where the varves are well defined, it is estimated that over 800 alternating layers of dark and light deposits can be distinguished.

At the Arrowhead Park locality, the Sangamon Soil can be traced from the northwest where it lies over the varved sediments, to the southeast where it separates the Teneriffe Silt from the Roxana Silt. The Teneriffe Silt is bounded on the base by a weakly developed paleosol (Pike Soil) in the top of the Kellerville Till Member of the Glasford Formation.

Overlying the Sangamon Soil is a moderately thick sequence of Wisconsin loess. Both the Roxana Silt and the Peoria Loess can be distinguished with a faint indication of a Farmdale Soil, developed in Robein Silt, lying between the two aforementioned units.

On the basis of rock and soil stratigraphy, it would appear that the decision to reject the use of the Brussels nomenclature was somewhat premature.

INTRODUCTION

Recent work in Mt. Charles County, Missouri, northwest of St. Louis, has revealed an interesting section of Pleistocene deposits preserved within a high-level terrace remnant. The terrace is located on the southern edge of the Mississippi River Valley and is the site of the Arrowhead Industrial Park. The site is immediately across the valley from the Brussels and Hardin quadrangles (Calhoun County) located in bordering Illinois (Figure 1).

Rubey (1952) reported on the geology of Hardin and Brussels qudrangles and as part of his work proposed the term "Brussels Formation" for a thick sequence of thinly laminated silts that are found in well-defined terraces north of the city of St. Louis, Missouri. These materials were considered to be of early-Illinoian age. Locally, significant amounts of sand and gravel were included within this stratigraphic unit. Rubey (1952) described the materials as fluvio-lacustrine and attributed their deposition to the damming by an early Illinoian glacier which impinged on the west bank of the Mississippi River in the vicinity of St. Louis, thereby causing the development of Glacial "Lake Brussels." Rubey (1952) did not identify a Sangamon

Soil above the Brussels deposits. However, Leighton and Brophy (1961) indicated that on the western edge of Calhoun County, Illinois (Hardin and Brussels quadrangles) excellent exposures of the Sangamon Soil were found, developed within the "Brussels Formation."

Willman and Frye (1970, p. 127) chose to drop the "Brussels Formation" from the stratigraphic lexicon of Illinois for the following reason: "Because there is no evidence of Sangamon Soil and Roxana Silt on the sediments, the Illinoian age is questionable and the deposits are tentatively assigned to the Equality Formation." The Equality is restricted to deposits of Wisconsinan age, and represents lake deposits overlain by loess or Holocene sediments. Elsewhere Willman and Frye (1970, p. 72-74) stated: "The Brussels Terrace sediments in Calhoun, Jersey, and Greene Counties...are largely lacustrine and are tentatively assigned to the Teneriffe Silt."

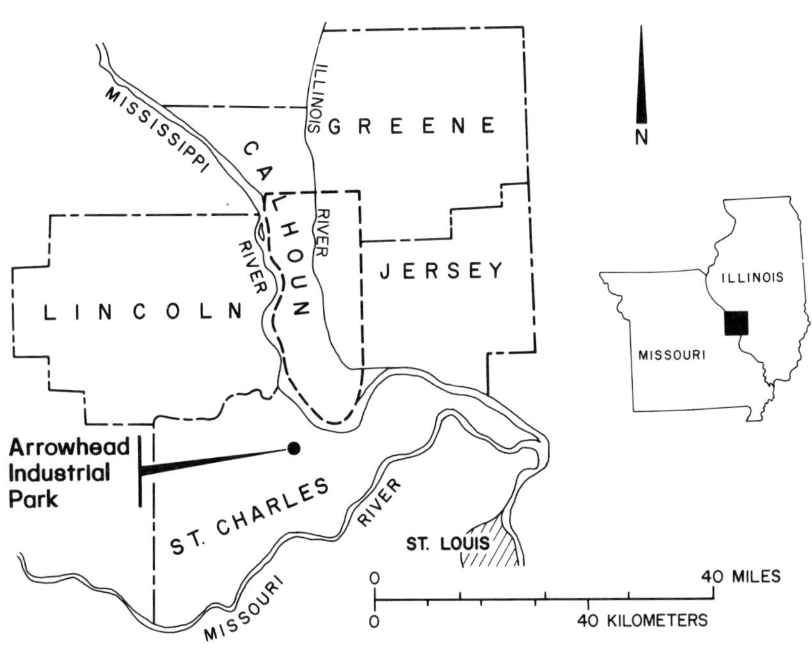

Figure 1 Location map of area of investigation. Hardin and Brussels quadrangles studed by Rubey (1952) shown in heavy dashed outline.

W.H. Allen, Jr. and R.A. Ward

In this paper we will offer evidence to show that a Sangamon soil does exist above, and is sometimes developed into, a varved silt-clay sequence at the Arrowhead Industrial Park in St. Charles County Missouri. The varved unit is separate from the Teneriffe Silt which overlies it, and is not included as part of this overlying unit. Rubey (1952) referred to the varved sequence as the "Brussels Formation" in his investigations in Hardin and Brussels quadrangles. We are reintroducting the term "Brussels" as the Brussels Member of the Glasford Formation and limiting it to the varved silt-clay sequence present at the Arrowhead Industrial Park. Rubey's stratigraphic control was not as definitive as is presented herein, thereby permitting him to be more general and include a wider variety of materials within his formational boundaries.

Additionally we are recognizing the following Pleistocene stratigraphic terminology currently used for the Illinoian stage of Illinois investigators: Petersburg Silt, Kellerville Till Member of the Glasford Formation, Pike Soil, and Teneriffe Silt.

AREA OF INVESTIGATION

The terrace remnant on which Arrowhead Industrial Park site is located lies within St. Charles County, Missouri. St. Charles County (Figure 1) is at the junction of the Missouri and the Mississippi Rivers. One third of the county is floodplain and the remainder is classified as upland loess covered glacial plain and heavily dissected nonglaciated topography typical of the Ozark Dome.

The terrace site (Figure 2) is 1.8 km (1.1 mi) northwest of the town of St. Peters, bounded on the south by the Norfork and Western Railroad tracks, and crossed by Missouri State Highway 79. Similar terrace remnants are also located east of St. Peters and along the Dardenne, Gelleau, Peruque, and other smaller creeks within St. Charles County. The site lies in the SE1/4 of Section 23 and NW1/4 of Section 26, T. 47N., R. 3E.

The highest surface elevation is slightly greater than 155 m (510 ft) AMSL and its outer edge lies at approximately 146 m (480 ft) AMSL. The main surface of the terrace is relatively level, with no more than 1 m difference in elevation. Small streams have dissected the flat on either end to a depth of 10 to 12 m; however, slumping and vegetative cover conceal the nature of the stratigraphic relationships.

STRATIGRAPHIC FRAMEWORK

The Division of Geology and Land Survey is in the process of establishing a Pleistocene stratigraphic framework for Missouri. Terminology in use by the Illinois Geological Survey for western Illinois is being heavily relied upon in this report. The stratigraphic section shown in Table 1 represents that which is present at the Arrowhead Industrial

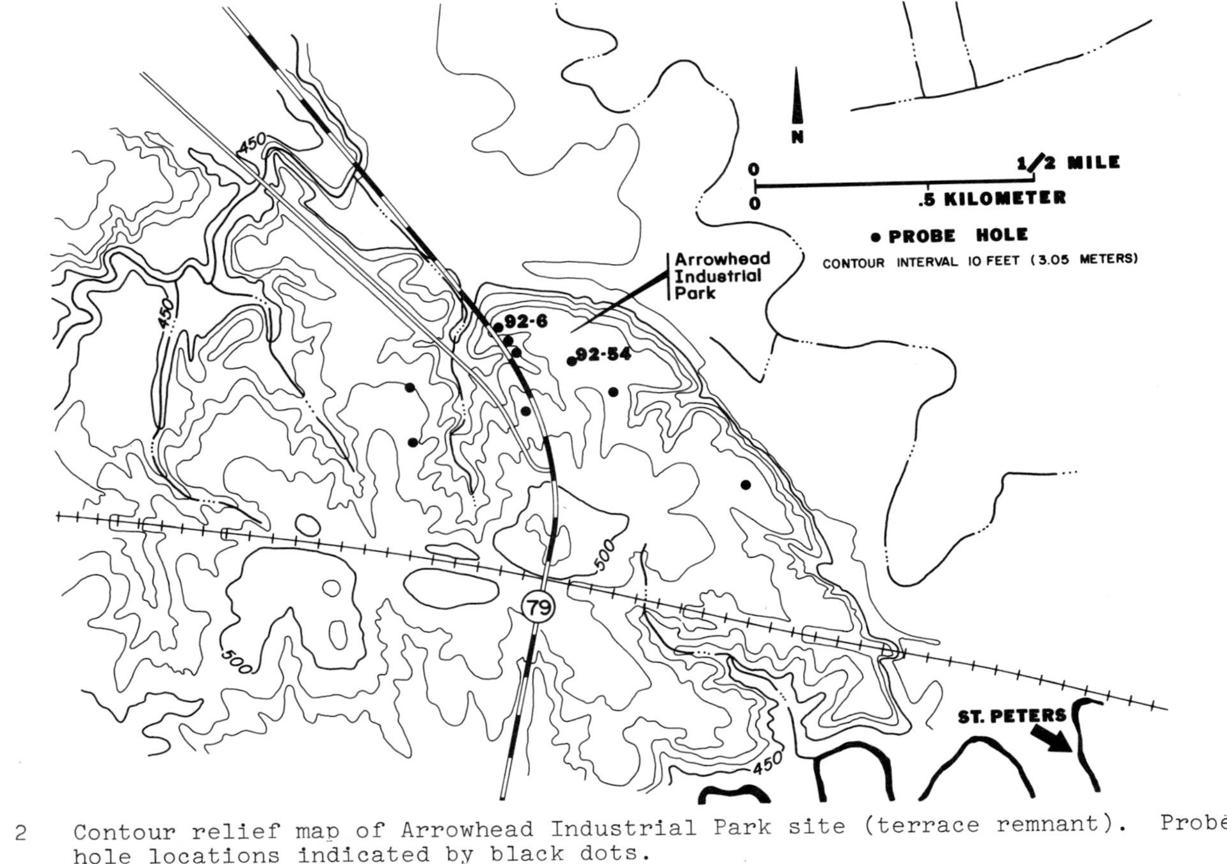

Figure 2 Contour relief map of Arrowhead Industrial Park site (terrace remnant). Probe hole locations indicated by black dots.

Table 1. Arrowhead Industrial Park Composite Stratigraphic Section.

Stage	Substage	Unit	Soil
HOLOCENE STAGE			Modern Soil
WISCONSINAN STAGE	Woodfordian Substage	Peoria Loess	
	Farmdalian Substage		"incipient Farmdale Soil"
	Altonian Substage	Farmdale Loess	
SANGAMONIAN STAGE			Sangamon Soil
ILLINOIAN STAGE	Monican and Jubileean Substage	Teneriffe Silt	
	Liman Substage	Glasford Fm. Kellerville Till Mbr.	Pike Soil Brussels Mbr.
		Petersburg Silt	
YARMOUTHIAN STAGE			Yarmouth Soil
KANSAN STAGE		Kansan "Outwash"?	

Park site, and is being used by the Division of Geology and Land Survey.

Arrowhead Industrial Park Composite Section

All observations discussed in this paper have been made from nine exploratory holes (Figure 2) probed into the terrace to a maximum depth of 15.85 m (52 ft). Each core has been described in detail with particle-size (pipette method) and pH data obtained for selected holes.

KANSAN STAGE?

Hole 92-6 (Figure 3) lies along the northwestern edge of the terrace. At a depth of 12.8 m (42 ft) outwash material was encountered that extends to 13.7 m (45 ft), where refusal occurred. While the particle-size data would support a till interpretation, the sample appears to be outwash of Kansan (?) age. This unit is unnamed at the present time.

YARMOUTHIAN STAGE?

In hole 92-6 (Figure 3) and three others, silty clay with angular to subangular blocky structure was encountered in the lower 1.5 to 3.5 m. The color is 10YR 6/1 (gray) with 10YR 5/6 (yellowish-brown) mottles. Some limonite nodules were noted in several cores within this stratigraphic horizon. The particle size as well as the pH data, would indicate that the material has been pedogenically altered. This unit appears to be the stratigraphic equivalent of the Lierle Clay Member as defined by Willman and Frye (1970).

ILLINOIAN STAGE

Petersburg Silt

This unit is most commonly in contact with the Yarmouth Soil at the Arrowhead Industrial Park site; however, in two sample cores it is in contact with residuum on bedrock. Either the Kellerville Till Member or Brussels Member overlies the unit locally. A stone line occurs at the top of the Petersburg Silt at certain locations on the terrace.

In hole 92-6 (Figure 3), the Petersburg Silt lies above the Yarmouth Soil and below a 2.5 m thick sequence of silty-clay varves (Brussels Member). The separation is marked by a strong increase in the sand percentage which occurs at the base of the varves. In hole 92-54 (Figure 4) the Petersburg Silt lies below a till. The use of the name Petersburg Silt to describe that silt beneath the silty-clay varves, beyond the glacial-till boundary, is at variance with Illinois usage (Willman and Frye, 1970). It is their practice to use the name Loveland Silt beyond the till boundary. However, in this case the lateral facies equivalent of the Kellerville Till Member is a varved sequence of materials (Brussels Member) that represents deposition in a lacustrine environment and is tied directly to the glacial regime. Therefore, a glacially associated environment of deposition is the basis for usage of the name Petersburg Silt beyond the glacial-till boundary in this case.

In a probe hole located along Dardenne Creek 9.5 m from the mouth, the Petersburg Silt was recognized below a truncated Sangamon Soil and is the lateral facies equivalent of the varved clays.

Glasford Formation

A detailed discussion of the Glasford Formation is given in Willman and Frye (1970, p. 52). Presently, the only recognized members in Missouri are the Kellerville Till and Brussels Members.

Kellerville Till Member

Willman and Frye (1970, p. 55) named the Kellerville Till for roadcut exposures near the town of Kellerville,

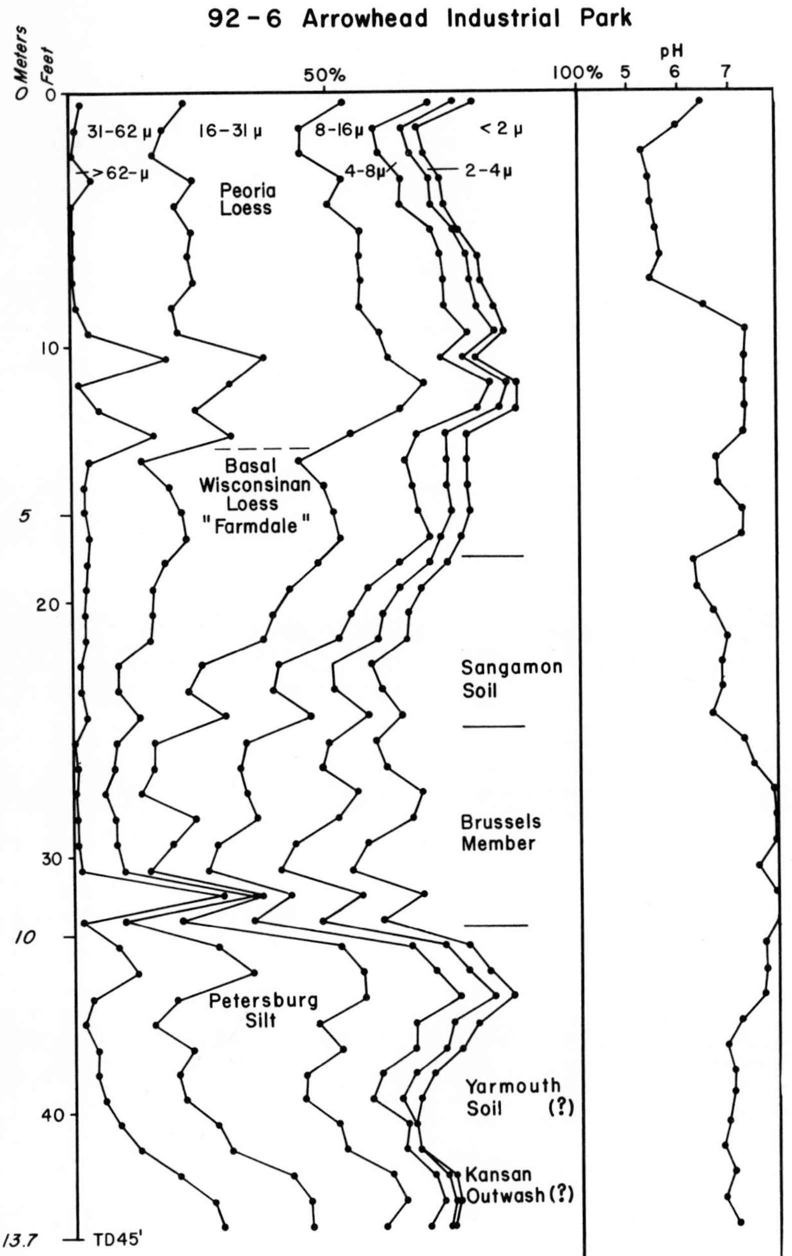

Figure 3 Probe hole 92-6, Arrowhead size analysis, pH, and

Color	Structure	other remarks
10YR5/4 yellowish brown		
10YR4/4 dark yellowish brown with 10YR6/2 light brownish gray mottles	subangular blocky	weak mottling
10YR4/4 dark yellowish brown		
10YR4/6 dark yellowish brown	massive	
10YR5/4 dark yellowish brown with 10YR5/8 yellowish brown mottles		mottling not uniform
10YR5/4 yellowish brown	weak subangular blocky	
10YR5/4 yellowish brown with 10YR5/6 yellowish brown mottles	angular blocky	
10YR6/6 brownish yellow and 10YR6/2 light brownish gray	laminated	
10YR5/3 brown with 10YR5/6 yellowish brown mottles	massive	
10YR5/4 yellowish brown		
2.5YR6/2 pale red	granular	scattered rock fragments

Industrial Park site,
descriptive data, illustrated.

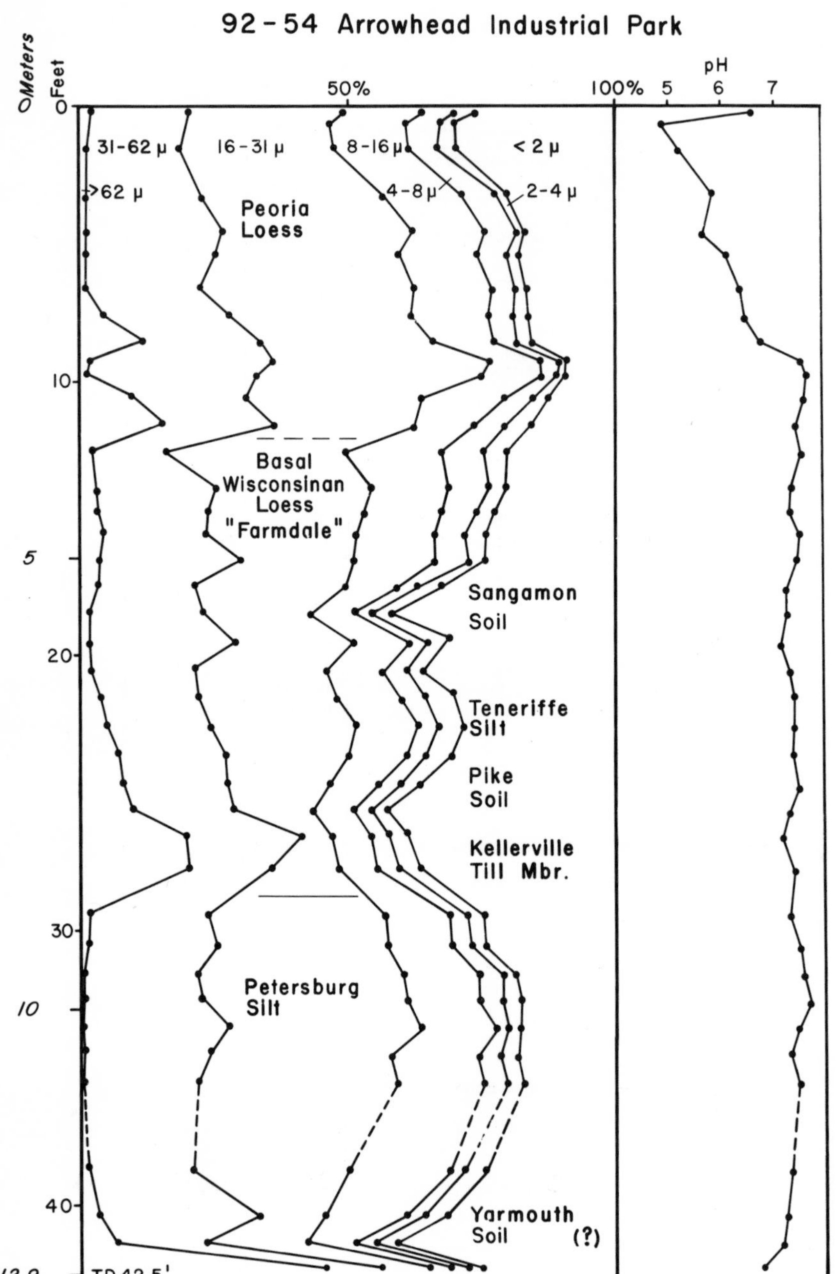

Figure 4 Probe hole 92-54, Arrowhead size analysis, pH, and

Color	Structure	other remarks
10YR5/4 yellowish brown	subangular blocky	
10YR4/4 dark yellowish brown with 10YR6/2 light brownish gray mottles		
10YR4/4 dark yellowish brown	massive weak pseudo platy massive	minor Mn lined rootlets
10YR4/3 brown with 10YR5/6 yellowish brown mottles		
10YR4/3 brown	weak pseudo platy	
10YR4/4 dark yellowish brown	massive	
10YR4/4 dark yellowish brown with 10YR6/2 light brownish gray and 10YR5/6 yellowish brown mottles		
10YR5/3 brown with 10YR6/3 pale brown mottles	subangular blocky	
10YR5/6 yellowish brown with 5YR6/1 light gray and 7.5YR6/4 light brown mottles	angular blocky to massive	
10YR5/4 yellowish brown with 5YR6/1 light gray and 7.5YR6/4 light brown mottles	weak subangular blocky	
10YR5/4 yellowish brown	platy	
10YR4/4 yellowish brown to 10YR5/6 dark yellowish brown	massive	partially oxidized and leached
7.5YR5/6 strong brown	platy	Mn staining
7.5YR5/4 brown	massive	
5YR4/6 yellowish red		Mn vienlets throughout

Industrial Park site,
descriptive data, illustrated.

Adams County, Illinois. At the Arrowhead Industrial Park site the till is bounded at the base by the Petersburg Silt or older deposits, and at the top by the Teneriffe Silt. The unit is laterally equivalent to the Brussels Member. Kellerville Till occurs in hole 92-54 (Figure 4) and five others at the Arrowhead Industrial Park site. Its upper limit is the top of the Pike Soil developed in the till. The Pike Soil is a moderate to strongly developed paleosol that may have been partially eroded. The Pike Soil and the overlying Sangamon Soil samples registered pH values of 7 or greater from this hole.

The Kellerville Till Member lies at a lower elevation on the southeastern end of the site. This is due in part to the absence of the Petersburg Silt beneath it (Figure 5). Till thickness averages 2 m, which is less than has been reported by Goodfield (1965) and others for the more easterly St. Louis area, and areas adjacent in Illinois.

The occurrence of the Kellerville Till at Arrowhead Industrial Park constitutes the westernmost location known to be reported in Missouri.

Brussels Member (New)

The Brussels Member of the Glasford Formation was formerly named in a report by Rubey (1952) for the town of Brussels in Calhoun County, Illinois. As a formation, it was rejected by the Illinois Geological Survey because of insufficient stratigraphic evidence for its proper placement. We are reintroducing the name at the "member" level (Table 1), aware of its relatively limited extent. In addition, it is not included with the Teneriffe Silt as indicated by Willman and Frye (1970, p. 74).

The Brussels Member consists of a maximum 8 ft (2.5 m) thick sequence at Arrowhead Industrial Park of silty-clay varves as can be seen from hole 92-6 (Figure 3). A sharp increase in sand is recorded in the lower meter. The unit changes compositionally upvalley to a silt and finally to a fluvially derived sand and gravel composed predominantly of detritus from residuum as reported by Rubey, 1952 (p. 83-84). Work by the authors confirm this observation for the St. Charles County area. The material is generally calcareous, but no fossils have been found in the sections observed at Arrowhead Industrial Park. Rubey (1952) did report aquatic pulmonates from some sites in Illinois.

The upper limit for the Brussels Member in hole 92-6 appears to be in the B3 horizon of the Sangamon Soil. However, to the east of this location the Teneriffe Silt overlies the Brussels Member and the Sangamon Soil is developed on the Teneriffe Silt. This relationship can be seen in Figure 5. Whether or not the Pike Soil (developed on Kellerville Till) is present on the Brussels Member could not be determined with certainty. The Brussels Member is bounded at the base by the Petersburg Silt.

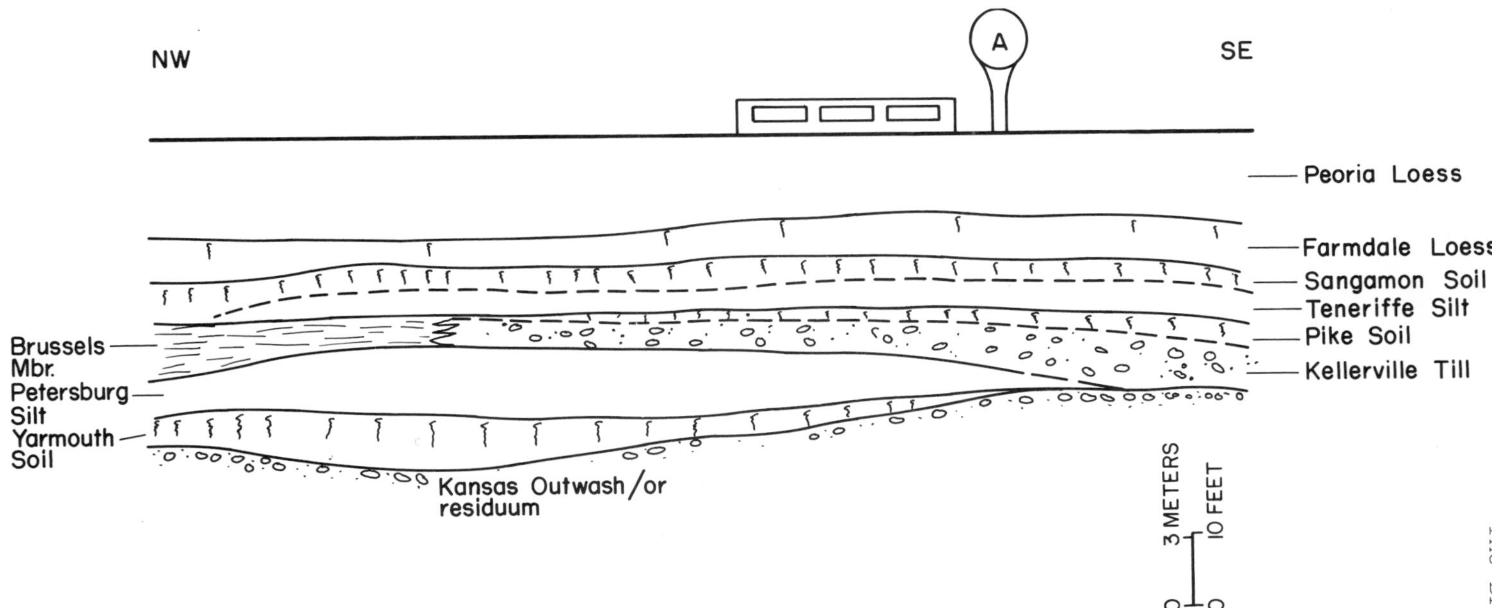

Figure 5 Diagrammatic northwest-southeast cross-section of Pleistocene deposits at Arrowhead Industrial Park.

The Brussels Member is considered to represent a fluvio lacustrine deposit resulting from damming of the Mississippi River by an early Illinoian glacier. The Kellerville Till is the deposit left behind by this glacial advance. Estimates by H.M. Groves of the Division of Geology and Land Survey (pers. comm., 1974) indicate that there are approximately 800 varves present, as determined from hole 92-6. This confirms data reported by Rubey (1952) on the number of varves occurring in various places in Hardin and Brussels quadrangles, Illinois to the northeast (Figure 1).

The Arrowhead Industrial Park site is proposed as a reference section for the Brussels Member, as the section is not exposed at the surface at this location.

Teneriffe Silt

Defined by Willman and Frye (1970, p. 60), this unit rests on the Pike Soil developed in the Kellerville Till throughout much of the area (Figure 5). Elsewhere, the Teneriffe Silt rests on the Brussels Member with uncertainty as to the presence of the Pike Soil at the base. At Arrowhead Industrial Park it has an average thickness of 2.1 m (7 ft). It is slightly calcareous where not modified by paleosolic development, and has a color of 10YR 5/4 (yellowish-brown) to 7.5YR 6/6 (reddish-yellow).

In this locality the Teneriffe Silt appears to range from the late Liman Substage through the Monican and Jubileean Substages of the Illinoian Stage.

SANGAMON STAGE

The presence of the Sangamon Soil and its stratigraphic position is critical to this discussion. At the Arrowhead Industrial Park site the paleosol is an *in situ* profile developed on the Teneriffe Silt and on the Brussels Member (presence of Teneriffe uncertain in hole 92-6). In hole 92-6 (Figure 3) where the Sangamon soil is developed into the Brussels Member, it is 2.4 m (8 ft) thick. The entire sequence is leached of carbonates and a strong structure is noted in the "B" horizon of the profile. The maximum clay content is 40.5%. The color is 10YR 5/4 (yellowish-brown) with mottles. The Pike Soil may be represented in the lower part of the profile.

In hole 92-54 (Figure 4) the Sangamon Soil is developed in the Teneriffe Silt. The profile is 1.5 m (5 ft) thick, with colors similar to those noted in hole 92-6 (yellowish-brown). At this point the materials show a slight reduction in pH values from the overlying Wisconsin loess, but it never goes below 7.2. A multiple profile is suspected, not to be interpreted as Sangamon on Pike, with a maximum clay accumulation (41.7%) occurring at 5.6 m (18.5 ft) and a secondary clay bulge (35.5%) at 6.2 m (20.5 ft). The lower member of this two-story profile probably represents an incipient paleosol that developed sometime during the

Jubileean Substage.

The Sangamon Soil is noted in all the holes probed on the Arrowhead Industrial Park site. It is the most consistent and best stratigraphic marker. The paleosol shows strong evidence of having been partially truncated at some locations.

WISCONSINAN STAGE

The loess that composes much of the section at the Industrial Park is relatively thin compared to that found on the upland bluff at Elm Point. The Elm Point section (Figure 6) is located 14 km (8.7 mi) to the east. The maximum loess thickness at the Arrowhead Industrial Park site is 5.8 m (19 ft), and some of the sections exhibit as little as 4 m (13 ft) of deposit. The loess has been very influential in masking the older depositional surface, resulting in the present flat form of the terrace.

The Altonian and Woodfordian Substages are considered to be present at Arrowhead Industrial Park, with some question regarding the Farmdalian Substage.

Altonian Substage

Farmdale Loess. This unit is bounded at the base by the Sangamon Soil and at the top by the incipient Farmdale Soil. A semantic controversy exists regarding priority of names for this unit. However, there is no controversy over the fact that there is a sequence of material that lies above the Sangamon Soil and below the Peoria Loess. In 1899, Leverett described a loessial unit that lay beneath a peat and above the top of the Illinoian till 0.5 mi (0.8 km) east of the Farm Creek locality, Illinois. It was overlain by Peorian Loess and underlain by Illinoian "gumbotil." He later renamed it the Farmdale in a personal communication quoted by Wascher, Humbert, and Cady (1948, p. 390). Leighton and Willman (1950) called it a pro-Wisconsin Loess (Farmdale); they indicated in diagrammatic sections that it was deposited in the Wisconsinan Stage. Frye and Willman (1960) subdivided the Farmdale Loess into Farmdale Silt, at the top, and Roxana Silt below, from further studies made at the Farm Creek site in Tazewell County. In 1970, Willman and Frye replaced the Farmdale Silt with the name "Robein Silt," and restricted the use of the Farmdale to the Farmdalian Substage and the Farmdale Soil. Name changes, as such, have historically been strenuously opposed, and it is felt by the authors that valid prior usage should have precedence over subsequent designations. Consequently, the basal Wisconsinan Loess is herein tentatively designated as Farmdale Loess and correlated with the Roxana Silt of Illinois as defined by Willman and Frye in 1970.

At Arrowhead Industrial Park the Farmdale Loess is between 1 to 2 m thick. It is a silt loam and has a color 10YR 4/4 (dark yellowish-brown). It frequently exhibits a

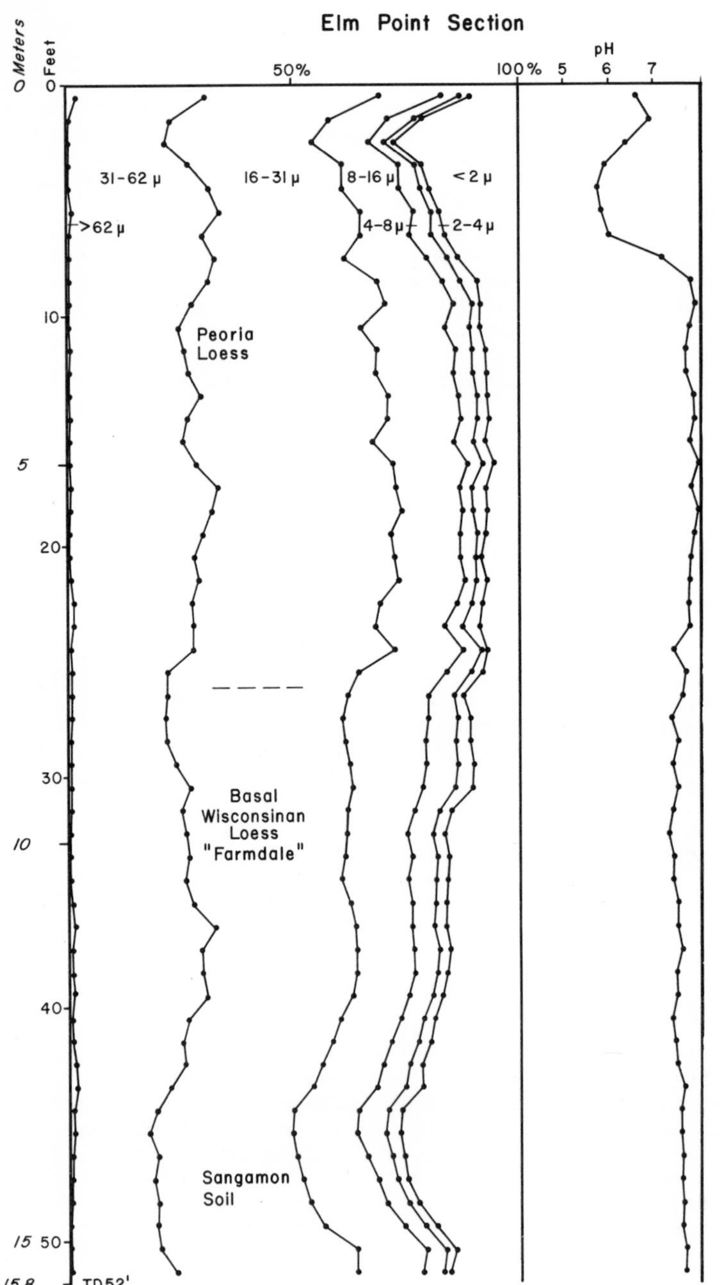

Figure 6 Probe hole 92-15, Elm Point section, 8.7 km (14 mi) east of Arrowhead Industrial Park si

Color	Structure	other remarks
10YR4/3 brown	massive	
7.5YR4/4 dark brown	weak subangular blocky	
10YR4/3 brown	massive	wormtube voids
10YR4/4 dark yellowish brown with 10YR6/3 pale brown mottles	massive	mottling weak
10YR5/4 yellowish brown with 10YR5/6 yellowish brown and 10YR5/8 yellowish brown mottles	pseudo platy	
	massive	
10YR5/6 yellowish brown		
7.5YR5/4 brown with 10YR5/6 strong brown mottles	weak subangular blocky	mottling weak
	massive	
7.5YR5/4 brown with 5YR5/4 reddish brown mottles	weak subangular blocky	
7.5YR5/6 strong brown	massive	

Size analysis, pH, and descriptive data indicated.

pseudo-platy structure and is almost always calcareous. Along the bluff at Elm Point (Figure 6) the Loess maintains similar color and structural characteristics to the Industrial Park section but is thicker, commonly exceeding 5 meters.

Farmdalian Substage

The Farmdalian Substage is not clearly understood in the area of concern and therefore will not be considered in this report. It is thought that an incipient paleosol does exist, separating the Altonian and the Woodfordian Substages.

Peoria Loess. At Arrowhead Industrial Park this loess is almost always leached of carbonates. It has a well-developed modern soil profile and a rather consistent color of 10YR 5/4 (yellowish-brown) to 10YR 4/4 (dark yellowish-brown). It seldom exceeds 4.5 m (14.75 ft) in thickness; however, along the bluffs it may be found to be in excess of 9 m (29.5 ft).

It is bounded on the base by the "Farmdale Loess" and its top is the modern ground surface.

CONCLUSIONS

(1) The Brussels Member of the Glasford Formation is herein reestablished as a valid stratigraphic unit and restricted in its definition to be a lateral facies equivalent of the Kellerville Till Member. It is overlain by the Teneriffe Silt and/or the Sangamon Soil and underlain by the Petersburg Silt. It is a fluviolacustrine deposit that resulted from blockage of the Mississippi River at its confluence with the Missouri River. The blockage lasted for approximately 800 years, and was the result of an early Illinoian age glacial advance into the area. The Kellerville Till Member was deposited during this glacial advance. It is a mappable unit in the subsurface. At the present time the ability to identify it is dependent on the recognition of the Teneriffe Silt and/or the Sangamon Soil above it and the relatively constant characteristics of the Petersburg Silt below it.

(2) The Kellerville Till Member exists farther west than had previously been thought. This allows for the extension of western Illinois Pleistocene terminology into eastern Missouri (see Table 1) as related to the Illinoian Stage.

(3) The Pike Soil, known to be developed on the Kellerville Till, is a soil-stratigraphic unit which has been observed at the Arrowhead Industrial Park site, thus extending the name usage into the state.

ACKNOWLEDGMENTS

We wish to extend our appreciation to all those who have aided in the preparation of the paper, and to those who have offered their criticism.

REFERENCES CITED

Frye, J.D., and Willman, H.B., 1960, Classification of the Wisconsinan Stage in the Lake Michigan glacial lobe: Illinois Geol. Survey Circ. 285, 16 p.

Goodfield, A.G., 1965, Pleistocene and surficial geology of the city of St. Louis and the adjacent St. Louis County, Missouri (Ph.D. Dissert.): Urbana, Univ. of Illinois, 207 p.

Leighton, M.M., 1926, A notable type Pleistocene section: Jour. Geology, v. 34, p. 167-177.

Leighton, M.M., and Brophy, J.A., 1961, Illinoian glaciation in Illinois: Jour. Geology, v. 69, p. 1-31.

Leighton, M.M., and Willman, H.B., 1950, Loess formations of the Mississippi Valley: Jour. Geology, v. 58, p. 599-623.

Leverett, F., 1899, The Illinois glacial lake: U.S. Geol. Survey, Monograph 38.

Rubey, W.W., 1952, Geology and mineral resources of the Hardin and Brussels Quadrangles: U.S. Geol. Survey Prof. Paper 218, 179 p.

Wascher, H.L., Humbert, R.P., and Cady, J.G., 1947, Loess in the southern Mississippi Valley: Identification and distribution of the loess sheets: Soil Sci. Soc. America Proc., v. 12, p. 389-399.

Willman, H.B, *et al.*, 1975, Handbook of Illinois stratigraphy: Ill. Geol. Survey Bull. 95, 261 p.

Willman, H.B., and Frye, J.C., 1970, Pleistocene stratigraphy of Illinois: Ill. Geol. Survey Bull. 94.

Quaternary Soil-Geomorphic Relationships, Southeastern Mojave Desert, California and Arizona

Roy J. Shlemon

ABSTRACT

The approximate age of soils and related landforms in a 10,000 sq km area of the southeastern Mojave Desert is determined by correlation to newly available radiometric and paleomagnetic dates for sediments along the lower Colorado River. Four distinct classes and ages of alluvial deposits flank 15 major mountain ranges centered around the proposed Sundesert Nuclear Generating Station near Blythe, California. Geomorphic surfaces are differentiated by relative surface dissection (preservation of divides), and development of desert pavement and varnish (patina). Diagnostic characteristics of corresponding soils are the thickness of a stone-free (A_2) vesicular horizon below desert pavement, the presence of cambic and argillic horizons, and the depth and morphology of carbonate (calcic) accumulations.

Major soil orders are Entisols and Aridisols. Torripsamments and Camborthids occur on "post-pluvial" surfaces less than about 15,000 years old. Calciorthids and some Camborthids typify low and intermediate geomorphic levels less than about 100,000 years old. Haplargids may be on surfaces possibly as old as 500,000 years. Incipient Paleargids are less than about 700,000 years old based on correlation of parent material to sediments underlying the Bishop Tuff (Coxcomb Mountains) and the Brunhes-Matuyama paleomagnetic boundary (Colorado River terrace deposits). Paleargids with strong K-horizons are probably at least 1,000,000 years old. Formation of Aridisols in the southeastern Mojave Desert is thus perhaps an order of magnitude slower than for comparable soils on well-dated sediments in southern New Mexico.

INTRODUCTION

The geological literature abounds with studies of landform evolution and geomorphic processes operative in warm deserts. This is particularly true for the Mojave Desert of southeastern California and western Arizona (Cooke and Warren, 1973). Unfortunately, however, the absolute ages of most desert landforms remain enigmatic, for few deposits older than about Holocene age are susceptible to dating by radiocarbon or other radiometric techniques. Thus most landforms and underlying sediments by necessity are usually deemed "early," "middle," or "late" Pleistocene in age.

With few exceptions, the age of desert soils is similarly unknown. In general, most profiles are Aridisols some certainly developing under environments of the past, as indicated by relative profile development and certain morphological characteristics, but their geological age is yet speculative.

Most information about desert profiles in the southwestern United States has come mainly from the pioneering soil-geomorphic studies in southern New Mexico, primarily by Ruhe (1967), Gile (1966, 1968, 1974), Gile and Hawley (1966, 1968), Gile *et al.* (1965, 1966, 1970) and Smith and Buol (1968). For the Mojave Desert in general, desert soil morphology was described early by Nikiforoff (1937), and Buol and Yessilsoy (1964); and most recently, as related to Quaternary investigations in Nevada and Arizona, by Nettleton *et al.* (1975).

Unfortunately, few data have been available to assess the age of alluvial landforms and related soils for that portion of the southeastern Mojave Desert, California and Arizona bordering the lower Colorado River between, approximately, the Whipple Mountains on the north and Yuma, Arizona, on the south (Figure 1). Now, however, recent investigations for a nuclear power plant siting have shed much light on desert soil-geomorphic relationships. These studies, following requirements of the Nuclear Regulatory Commission (NRC) have sought to date Quaternary sediments and related geomorphic surfaces over an approximately 10,000 sq km area of the southeastern Mojave Desert in order to ascertain the last movement of any fault which might be identified in geological mapping. Most of the studies have not yet been formally published in well-known geological or pedological journals. They are, however, available as appendices in various Early Site Review Reports (ESRR), and Preliminary Safety Analysis Reports (PSAR); see, for example, Bull (1974a; 1974b), Ku (1975), Kukla and Opdyke (1973), Lee and Bell (1975), Murray *et al.* (1976), and Shlemon and Purcell (1976). These nuclear site-related studies have now yielded new information regarding the Quaternary stratigraphy, soils, and landforms in this portion of the Mojave Desert, their relative ages and, in some cases, their "absolute" ages based on radiometric and magnetic dating.

This paper therefore summarizes key aspects of these soil-geomorphic studies in the southeastern Mojave Desert, California and Arizona. Particularly emphasized are data

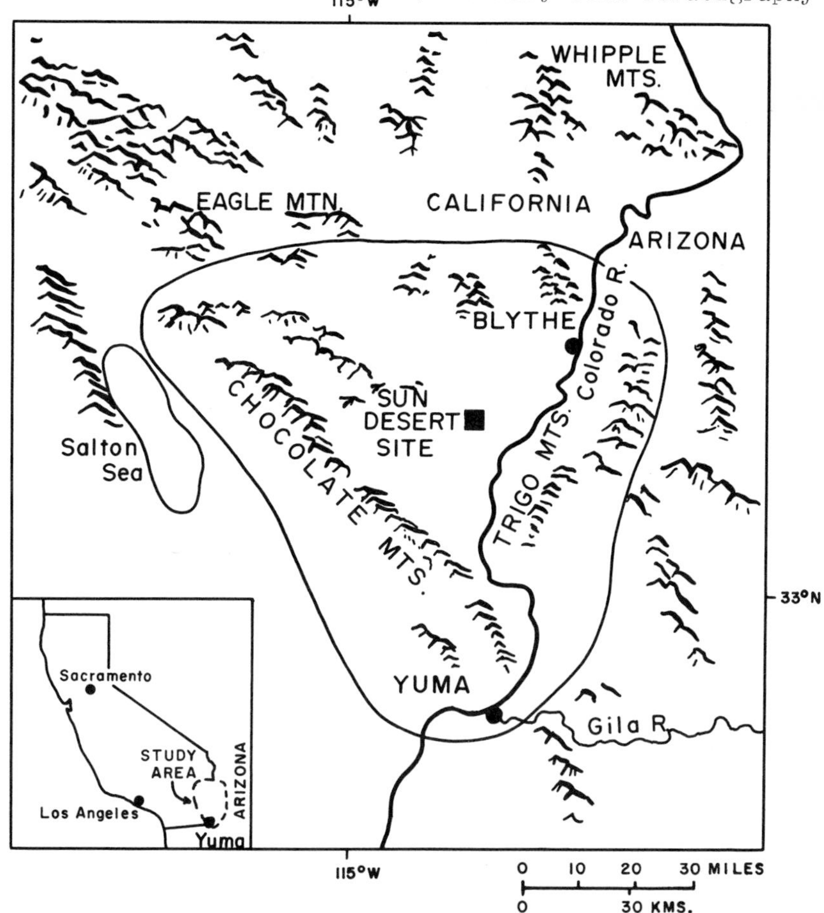

Figure 1 Reconnaissance area, soil-geomorphic investigations, southeastern Mojave Desert, California and Arizona.

from siting studies for the Sundesert Nuclear Generating Station near Blythe, California (San Diego Gas and Electric Co., 1976). Pointed out are the broad alluvial sequences in this area and geomorphic criteria useful for their separation in reconnaissance, and the approximate ages of related geomorphic surfaces and the soils forming upon them. Reviewed briefly also is the probable rate of Aridisol formation in this area compared with the well-studied soils on similar flights of geomorphic surfaces in the semi-arid terrain of southern New Mexico (Gile, 1974, 1977; Gile and Hawley, 1966, 1968; Gile *et al.*, 1966, 1970).

GEOMORPHIC SETTING

The area studied is approximately triangular-shaped,

covering some 10,000 sq km in California and Arizona, centered around the proposed Sundesert Nuclear Generating Station, about 32 km southwest of Blythe, California (Figure 1). Fifteen major mountain ranges are encompassed within the study area. The entire region is within the Mojave Desert, a subdivision of the Sonoran physiographic province.

A great variety of lithologies make up the core of the several mountain systems in this region. These rocks, reflected in the Quaternary alluvium, range in age from Mesozoic granitic intrusives to diverse pre-Cretaceous metamorphic and metasediments with some younger volcanics. These contrasting lithologies weather differently, develop distinctive surfaces, and thus affect the geometry, morphology, and photographic tonal expression of Quaternary landforms and soils identified in these studies (Figure 2).

Figure 2 Representative desert alluvial landforms in the southeastern Mojave Desert. Metamorphic bedrock in mountain range giving rise to dissected high- and intermediate-level piedmont fan systems with well-patinated desert pavement. Holocene active washes typified by dispersed vegetation; adjacent surfaces by bar and channel topography.

The climate, reflected in the landforms, is arid to locally semi-arid with precipitation ranging from approximately 5 or 6 cm in the basins to about 25 cm at higher elevations. Vegetation on Quaternary fan systems is confined mainly to riparian cottonwood and phreatophytes along active washes and dispersed Ocotillo and Creosote bush *(Larrea Sp.)* on modern alluvial fans. Older geomorphic surfaces, characterized by well-developed desert pavement, typically are barren of vegetation except for sparse Creosote. Soils in this portion of the southeastern Mojave are unmapped except for irrigated agricultural land and adjacent terraces (Torrifluvents and Torriorthents) in the Palo Verde Valley along the lower Colorado River (Elam, 1974; Figure 3) and reconnaissance of piedmont fan surfaces around proposed nuclear facilities near Parker and Yuma, Arizona (Bull, 1974a; 1974b).

Figure 3 Irrigated Torrifluvents on Colorado River floodplain, Palo Verde Valley, California (left); Torriorthents and Haplargids on dissected Colorado River terrace deposits, Arizona (right).

Field Procedures

Because of the large area covered by the soil-geomorphic survey, many data were obtained by interpreting high altitude (scale 1:125,000), U.S. Air Force aerial photographs, and color small-scale Skylab and ERTS imagery (Shlemon and Purcell, 1976). Where available, recent large-scale photography was also used; as well as field checks by four-wheel drive vehicle, by helicopter and by light aircraft. Additionally, soils within about a 10 km radius of the Sundesert Nuclear site (Figure 1) were described from 24 trenches specifically excavated for that purpose, and from natural exposures in lower, intermediate, and high-level geomorphic surfaces primarily bordering the Mule and Palo Verde Mountains (Figure 4). These surfaces, directly traceable to several dated fluvial terraces of the Colorado River, permit dating and correlating of Quaternary soils and landforms over a wide area in the southeastern Mojave Desert.

Figure 4 Excavation for soil description, intermediate-level alluvial fan, east flank, Palo Verde Mountains, California.

Desert Alluvial Systems

The alluvial landforms in the study area are primarily piedmont and pediment fans ranging in age from undissected Holocene deposits to remnant late-Tertiary fanglomerates. As applicable to the southeastern Mojave Desert, these have been described by Denny (1965, 1967), Hadley (1967), Hamilton (1964), Lustig (1966), Melton (1965), Metzger *et al.* (1973), Olmsted *et al.* (1973), and Royse and Barsch (1971). These alluvial landforms in the southeastern Mojave Desert appear to have been remarkably stable throughout the Quaternary (Bull, 1964; Shlemon and Purcell, 1976), though tectonism in the western Mojave is well known (Denny, 1967; Hooke, 1967, 1972). It appears geomorphic equilibrium and surface form have been controlled mostly by regional climatic change which affected vegetation, discharge, sediment yield, and a host of minor geomorphic variables. In essence, Quaternary climatic change in this area gave rise to epochs of alluviation preceded and followed by relative landscape stability and soil formation.

Regionally, landform evolution, sedimentation, and pedogenesis in the southeastern Mojave Desert is generally associated with major epochs of Pleistocene pluviality. Whether or not this climatic change can be equated to mid-continental glaciations or interglaciations is still debatable. However, an approximate chronology correlative to Wisconsin and at least two or more Illinoian or earlier stages is borne out by radiometrically-dated multiple lake levels throughout the Great Basin and Mojave Desert (Morrison, 1964, 1965) and in some cases by abrupt sedimentological changes in cores from closed lake basins, interpreted as indicative of climatic change (Smith, 1968). In any event, carbonate depth and morphology in Aridisols greater than about 100,000 years old attest to past climates with effective soil moisture greater than the present.

Classes of Quaternary Alluvium

Four distinct classes of alluvial forms and underlying sediments flank almost all fifteen major mountain ranges in the southeastern Mojave study area (Table 1). These are designated, from younger to older, "Q1" through "Q4." These alluvial forms are differentiated primarily by their surface form and secondarily by their relative soil profile development.

With respect to surface form, two distinct aspects of piedmont fan and piedmont terrace morphology permit their discrimination: (1) relative dissection of fan surface or conversely, preservation of divides, and (2) the amount of desert pavement and its appearance on imagery indicated by reflectance (albedo) of desert varnish (patina).

Relative Dissection of Alluvial Surfaces

Each major periodic change in geomorphic equilibrium, whether induced by local tectonism or controlled by regional climatic fluctuation, resulted in channels being incised into bedrock and existing deposits, and new sediments being carried basinward away from the mountain fronts. Consequently, the older landforms were progressively abandoned as surfaces of transport and eventually dissected. In some cases, the oldest, and now topographically highest surfaces, have little if any original divides remaining (Figure 5). In contrast, the younger, topographically lower surfaces are characterized by progressively better preserved, wider divides.

Reflectance of Desert Pavement

The vast majority of alluvial fans older than about Holocene-age, are capped by desert pavement, a veneer of gravels derived from adjacent mountain ranges. Where the gravels are high in iron and manganese, emanating from volcanic and metamorphic terrane, a distinctive varnished coating or patina renders these surfaces readily discernible both in the field and on aerial photographs (Figure 5).

Table 1. Designation of Alluvial Classes and Approximate Ages, Southeastern Mojave Desert, California and Arizona.

This Study		Lee and Bell, 1975		Bull, 1974a		Hamilton, 1964	
Estimated Ages (yrs)	Alluvial Unit	Estimated Ages (yrs)	Alluvial Unit	Estimated Ages (yrs)	Alluvial Unit	Estimated Ages (yrs)	Alluvial Unit
<15,000	Q_1	0	Q_{al}	0	Q_4		Q_{a4}
~15,000-200,000	Q_2	0-10,000	Qf	2,000-11,000	Q_3	(None Given)	Q_{a3}
~100,000-500,000	Q_3	30,000-100,000	Qfc	11,000-200,000	Q_2		Q_{a2}
>500,000	Q_4	500,000	QTfc Tf, QTfa	500,000-1,500,000	Q_1		Q_{a1}

Figure 5 Representative dissected fan of alluvial class Q_3, approximately 500,000 to 700,000 years old. Only narrow, pavemented drainage divides preserved; southern Coxcomb Mountains.

The relative amount and micro-relief of pavemented surfaces is an excellent geomorphic indicator of landform and soil age.

The origin of desert pavement is not completely understood, with hypotheses ranging from deflation of fine-grained interspersed particles by wind and water (Clements, 1952; Lowdermilk and Sundling, 1950; Lustig, 1966) to mechanical processes causing movement of pebbles to the surface (Cooke, 1970; Denny, 1965, 1967; Springer, 1958). Both hypotheses are applicable in the southeastern Mojave Desert as borne out by inspection of natural cuts in arroyos and trenches specifically excavated for soil description.

The time required for desert pavement formation varies greatly from place to place. In the southeastern Mojave, trails and tracks cut in the pavement during military maneuvers of early World War II show no evidence of "healing." Yet Cooke (1970) has observed incipient renewal of artificially removed pavement in as little as four years. In any event, desert pavement forms only on surfaces that are relatively stable as well as endowed with gravels.

Normally the desert pavement is underlain by 2 to about 8 cm of stone-free vesicular silt, especially on the higher and older geomorphic surfaces. The stone-free silty unit, commonly designated A_2 or A_3 eluvial horizons, owes its origin mostly to mechanical disturbance, partially because of wetting or drying, or because of freezing and thawing. As observed by Springer (1958) in western Nevada, repeated wetting and drying of montmorillonite and other expandable-lattice clays is an effective method to move gravels upward in the solum; this process being a function of time as well as of climate.

Another useful indicator to date relatively and to correlate geomorphic surfaces in the southeastern Mojave Desert is the dark-colored varnish or patina characteristics of many desert pavements. The iron and manganese film comprising most desert varnish in part is derived from the rock itself and in part from solution emanating from adjacent weathered soil and debris (Engle and Sharp, 1958; Hooke *et al.*, 1969; Potter and Rossman, 1977). The desert varnish is confined almost exclusively to iron and manganese-rich rocks, and is almost devoid on surfaces characterized by quartzite cobbles or limestone fragments.

The time required for desert varnish formation is still not precisely known. Incipient varnish seems to form in only a few thousand years (Engle and Sharp, 1958; Hooke *et al.*, 1969), but well-patinated surfaces may take tens of thousands of years to develop (Blackwelder, 1948). Varnished pavement greater than about 15,000 years old has likely been subjected to pluviality during the latest Pleistocene, and this paleoclimatic factor may have accelerated the varnishing process.

In brief, the formation and preservation of desert pavement and varnish are excellent indicators for identifying, correlating, and dating relatively intermediate- and high-level geomorphic surfaces throughout the southeastern Mojave Desert. Active fan surfaces, washes and floodplains with little or no pavemented surfaces are probably less than about 15,000 years old (alluvial class Q1; Table 1). Based on limited radiometric and paleomagnetic dates, described in the following section, those well-patinated surfaces with smooth pavement are greater than at least 15,000 years old and more likely 35,000 to about 200,000 years (alluvial class Q2). Locally, some pavement has been partially disrupted by rare surface flow which gives rise to bar and channel topography. Nevertheless, even in these areas the ubiquitous presence of varnished rocks, though transported slightly from their original place of patination, and the presence of strongly developed soils, attest that these surfaces have remained essentially stable since cessation of a previous epoch of alluviation. Those surfaces with obvious deep bar and channel topography, and with none or at best an incipient pavement and varnish, are likely "post-Wisconsin" in age. Higher, and older geomorphic surfaces with well-developed, smooth desert pavement are forming on "pre-Wisconsin" alluvium (Class Q3), and range in age from about 100,000 to 500,000 years old. In some cases, highest remnant drainage divides reflect the aggregate of weathering for the past 500,000 to perhaps 1,000,000 years (alluvial class Q4).

ALLUVIAL SOILS OF THE SOUTHEASTERN MOJAVE DESERT

Description of 22 soil profiles from toposequences of alluvial fans within the southeastern Mojave Desert area indicates that the vast majority of soils are Aridisols, either Orthids or Argids (Shlemon, 1977). Three morphological characteristics of these soils have proven useful for regional correlation and for determining their relative pedological and geological age: (1) the presence and thickness of silty eluvial "A_2" or "A_3" horizons; (2) an argillic horizon usually with 7.5YR or 5YR hues and chromas; and (3) the depth and morphology of calcium carbonate horizons. These three characteristics typify most pre-Holocene alluvial soils in the southeastern Mojave Desert. Thus, the rate of Aridisol development, giving rise to these morphological characteristics, may be ascertained more precisely by correlating alluvial fans and geomorphic surfaces to Quaternary sediments dated by radiometric and magnetic techniques.

Torripsamments and Camborthids

As shown diagrammetrically in Figure 6, Torripsamments and Camborthids typify Holocene-age (post-pluvial) surfaces of active fluvial or eolian transport in the southeastern Mojave Desert. Typically, the surface is still sufficiently unstable so that neither desert pavement nor a stone-free vesicular silt have yet formed. Both surface and subsurface horizons are normally 7.5YR to 10YR in color, with cambic horizons limited to soils on distal segments of alluvial fan derived mainly from more easily weatherable metamorphic and volcanic rocks. Not uncommonly, disseminated lime occurs at a depth of about 60 cm (stage 1 development of Gile *et al.*, 1966), the surface epipedon either non-calcareous or slightly calcareous owing to the continual influx of dust. The alluvial fans and basin deposits giving rise to Torripsamments and Camborthids are traceable directly into Colorado River deposits underlying the present floodplain in the Palo Verde Valley dated by radiocarbon as of Holocene age (Metzger *et al.*, 1973; Olmsted *et al.*, 1973).

Calciorthids

Calciorthids are intimately mixed with some Camborthids on low, intermediate geomorphic surfaces, usually slightly above the present active surfaces of transport. A_2 or A_3 horizons are normally less than 5 cm thick; 7.5YR colors typify the cambic horizon. Carbonate nodules to about 5 mm in diameter occur in sandy matrices, and limey rinds to about 3 mm thick coat the bases of pebbles at a depth of about 90 to 120 cm. Where demonstrably pedogenic and not related to buried Cca horizons, this depth of carbonate accumulation suggests, with other geomorphic data, that the soils have been subject to at least one pluvial epoch. Camborthids and Calciorthids, however, are less than about

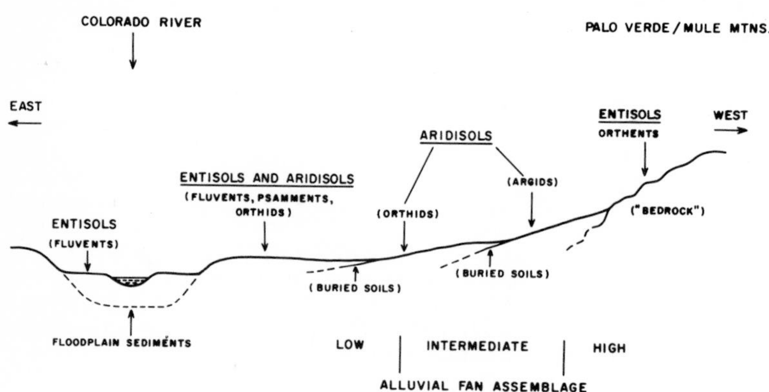

Figure 6 Generalized soil-geomorphic associations showing major soil orders and suborders typical of piedmont fans and pediment terrace, southeastern Mojave Desert, California and Arizona.

100,000 years old, based on tracing their underlying parent material to sediments of that age dated by thorium-uranium-protactinium series assay of vertebrate remains (Ku, 1975). In essence, both radiometric dating of soil parent material and geomorphic position indicate a probable "pre-Wisconsin" yet "post-Sangamon" age for Camborthids and Calciorthids in this part of the Mojave Desert.

Haplargids

Haplargids characterize most intermediate-level geomorphic surfaces in the southeastern Mojave Desert. Desert pavement is well-developed with fan remnants sufficiently high above local base level and dissected such that surface flow is derived wholly from contemporary rainfall.

Vesicular horizons underlying the desert pavement typically are about 5 to 10 cm thick, this increasing with elevation and age of fan surface. Argillic horizons are recognizable in the field by a slight increase in clay, reddish-colors (5YR) and presence of illuvial coatings. Whether the clay is wholly translocated or formed in place is unknown, a problem characteristic of argillic horizons in desert soils throughout the southwestern United States (Nettleton *et al.*, 1975). These horizons (Bca) are, however, within the present depth of wetting, for they are typified by stage I carbonates (Gile *et al.*, 1966).

The morphology and depth of the Haplargid Cca horizon is also indicative of age. Nodular carbonates are abundant occupying perhaps 25 percent of surface area, especially on fan sediments derived from metamorphic and volcanic rocks; or downwind from playa lakes, sources of carbonate dust. Lime coatings on pebble bases are 10 to 15 mm thick (stage II) with a dendritic staining pattern on pebble surfaces as well. Calcic horizons are often 140 to 160 cm deep, far below the present depth of wetting in this truly arid terrane (Arkley, 1963).

Paleargids

Paleargids occur on the highest-intermediate level and highest fan remnants and drainage divides in the southeastern Mojave Desert. Desert pavement and varnish are very strongly developed, for except for eolian dust, these geomorphic surfaces have probably not received new sediments during perhaps the last 200,000 years. Landscape dissection is such that the only water available for transport across these surfaces is from local rainfall and not from present drainages originating in adjacent mountain terrain.

Stone-free vesicular horizons are 10 to 12 cm thick; argillic horizons are distinct with colors in the 5YR range. Illuvial clay is also present except for parent material almost wholly quartzitic where derived from old, reworked Colorado River deposits.

Carbonate depth and morphology also attest to the antiquity of these soils. Carbonate nodules occupy approximately 50 to 60 percent of pore space at a depth of about 150 cm. Incipient K horizons (stage III development) are common (Gile *et al.*, 1965, 1966). These Paleargids, however, are less than about 700,000 years old, for they occur on sediments along the lower Colorado River with positive magnetic polarization ascribed to the Brunhes Normal Epoch (Kukla and Opdyke, 1975), and on fans in the Coxcomb Mountains (Shlemon, 1976) containing the approximately 700,000 year old Bishop Tuff, an ash deposit widespread in eastern California (Merriam and Bischoff, 1975).

On highest fan remnants, topographically above surfaces dated as about 700,000 years old, are remnants of well-developed Paleargids. Elevations above local base level are often in the order of tens of meters, such that fan surfaces are strongly dissected and in most cases completely cut off from contemporary deposition except for eolian dust. Desert pavement may be preserved on the wider fan remnants. Elsewhere the surfaces have been stripped so that only a K horizon is preserved, with petrocalcic rubble mantling the adjacent slopes (Figure 7). These high remnants and associated Paleargids, characterized by argillic, Cca and K horizons (stage III and IV development) are best preserved immediately east of the Chocolate Mountains of California (Figure 1), and on drainage divides within the Trigo Mountains of Arizona (Shlemon and Purcell, 1976). Some of these ancient pedons have plugged carbonate horizons more than two meters thick (Figure 8). The absolute age of these Paleargids and their underlying parent material (alluvial class Q4) is unknown. However, based on their geomorphic position, it appears they may be in the order of at least one million years old.

Figure 7 Paleargid (K-horizon) capping proximal remnant of high-level alluvial fan, east side Mule Mountains, California.

Figure 8 Paleargid (K-horizon) more than two meters thick; early (?) Pleistocene fan system, east flanks Chocolate Mountains, California.

RATE OF ARIDISOL FORMATION

The radiometrically and magnetically-dated sediments provide an approximate "calibration" for the age and hence rate of Aridisol formation in the southeastern Mojave Desert. Gile and Hawley (1969) and Gile (1975, 1977) report stage I and II carbonate development occurring in Aridisols within Holocene time in southern New Mexico. In the Mojave, in contrast, soils of this age ("post-pluvial") are Torripsamments, Torrifluvents or, at best, Camborthids with only fine disseminated lime.

Haplargids in the southeastern Mojave Desert also appear to be geologically much older than their equivalent profiles in New Mexico based on limited radiometric dating of parent materials in both areas (Gile and Hawley, 1968; Gile *et al.*, 1970; and Gile, 1975, 1977). Whereas Haplargids have formed in the last 10,000 years in southern New Mexico, these soils probably range in age between 15,000 and 100,000 years old in the southeastern Mojave Desert.

Incipient Paleargids in the Mojave Desert require at least 700,000 years to form, at least two to three times longer than that reported by Gile *et al.* (1970) for southern New Mexico or by Nettleton *et al.* (1975) for other selected localities in the southwestern United States. Extremely well-developed Paleargids with thick K horizons (laminar) on the highest drainage divides within the southeastern Mojave probably require at least 1,000,000 years to form (Bull, 1974a; Shlemon and Purcell, 1976). Thus the rate of Aridisol formation in this area is probably at least an order of magnitude slower than for comparable profile development in southern New Mexico.

SUMMARY AND CONCLUSIONS

Reconnaissance soil-geomorphic mapping and other studies related to nuclear power plant siting provide new information about the age of alluvial landforms and soils in the southeastern Mojave Desert, California and Arizona.

Four distinct classes of alluvium, designated from younger to older as "Q1" through "Q4" flank some fifteen mountain ranges in a 10,000 sq km area centered around the proposed Sundesert Nuclear Generating Station near Blythe, California. These piedmont and pediment fans range in age from undissected Holocene deposits to remnant late-Tertiary fanglomerates. Regionally, alluviation appears to be climatically controlled, episodically preceded and followed by relative landscape stability and soil formation.

Geomorphic surfaces are differentiated by (1) elevation and increasing surface dissection (preservation of divides); and (2) relative development of desert pavement and varnish (patina). Soil characteristics most applicable for regional correlation and age determination are the (1) presence and thickness of an eluvial (A_2), stone-free vesicular zone below desert pavement; (2) presence of cambic and locally argillic horizons; and (3) depth and morphology of carbonate accumulations.

Torripsamments and Camborthids occur on active surfaces deemed "post-pluvial" or less than about 15,000 years old. Calciorthids and some Camborthids occurring on low, intermediate-level geomorphic surfaces are typified by eluvial horizons generally less than 5 cm thick, cambic horizons and stage I carbonate development. Depth of carbonate below the present average wetting front and geomorphic position suggest a "pre-Wisconsin" soil age. These profiles are, however, less than about 100,000 years old based on correlation with radiometrically dated sediments along the lower Colorado River.

Haplargids in southeastern Mojave Desert occur on intermediate-level geomorphic surfaces. Desert pavement and an underlying silty vesicular (eluvial) horizon characterize these landforms and soils. Cca horizons have stage II carbonate development. From geomorphic position and relative profile development, these soils are greater than about 100,000, and possibly as old as 500,000 years.

Paleargids occur on high-intermediate and highest levels in the southeastern Mojave Desert. Desert pavement is limited to narrow divides. Stone-free eluvial, argillic, and strongly developed (stages III and IV) carbonate horizons are characteristic. Incipient Paleargids are somewhat less than about 700,000 years old based on correlating parent material to sediments containing the 700,000-year-old Bishop Tuff, and to those with positive paleomagnetic polarity (Brunhes Normal Epoch).

Strongly developed Paleargids on highest remnant surfaces have Cca and K horizons to two meters thick. This profile development and geomorphic position well above 700,000-year old dated sediments suggests that these Paleargids may be in the order of at least one million years old.

The rate of Aridisol formation in the southeastern Mojave Desert is substantially slower than that reported for similar flights of geomorphic surfaces in southern New Mexico. Whereas Haplargids have formed within Holocene time in New Mexico, these same soils are greater than 15,000 and possibly up to 100,000 years old in the southeastern Mojave.

Paleargids, similarly, apparently require longer to form. Surfaces stable for about 700,000 years have produced only incipient Paleargids; similar profiles in New Mexico may be as young as Holocene age.

ACKNOWLEDGMENTS

Field work was supported by Fugro, Inc., (Long Beach),

on behalf of San Diego Gas and Electric Company. Greatly acknowledged is the field assistance of G. Lee, and C. Purcell (Fugro); the commentary of J. Bell (Fugro) and Professor W. Bull (University of Arizona); and manuscript review of W.C. Mahaney (York University).

REFERENCES CITED

Arkley, R.J., 1963, Calculation of carbonate and water movement from soil climatic data: Soil Sci., v. 96, no. 4, p. 239-248.

Blackwelder, E., 1948, Historical significance of desert lacquer: Geol. Soc. America Bull., v. 55, p. 1937.

Bull, W.B., 1964, Geomorphology of segmented alluvial fans in western Fresno County, California: U.S. Geol. Survey Prof. Paper 532-F, p. 89-128.

__________, 1974a, Geomorphic tectonic analysis of the Vidal region: *in* Woodward-McNeill and Assoc., Vidal Nuclear Gen. Station, Units 1 and 2, Appendix 2.5B (Geology and Seismology), Southern California Co., Rosemead, Calif., 66 p.

__________, 1974b, Reconnaissance of the Colorado River terraces near the Yuma Dual Purpose Nuclear Plant: *in* Woodward-McNeill and Assoc., Geotechnical Investigations, Yuma Dual-Purpose Nuclear Plant, Yuma, Ariz., Appendix F, pt. 2, 18 p.

Buol, S.W., and Yessilsoy, M.S., 1964, A genesis study of the Mojave sandy loam profile: Soil Sci. Soc. America Proc., v. 28, p. 254-256.

Clements, T.F., 1952, Wind blown rocks and trails on the Little Bonnie Claire Playa, Nye County, Nevada: Jour. Sed. Pet., v. 22, p. 182-186.

Cooke, R.U., 1970, Stone pavements in deserts: Assoc. Amer. Geog. Annals, v. 60, p. 560-577.

Cooke, R.U., and Warren, A., 1973, Geomorphology in deserts: Berkeley, Calif., Univ. California Press, 374 p.

Denny, C.S., 1965, Alluvial fans in the Death Valley region, California and Nevada: U.S. Geol. Survey Prof. Paper 466, 62 p.

__________, 1967, Fans and pediments: Amer. Jour. Sci., v. 62, p. 81-105.

Elam, N.E., 1974, Soil survey-Palo Verde area, California: U.S. Dept. Agric., Soil Cons. Ser., Washington, 37 p.

Engle, E.G., and Sharp, R.P., 1958, Chemical data on desert varnish: Geol. Soc. America Bull., v. 69, p. 487-518.

Gile, L.H., 1966, Cambic and certain non-cambic horizons in desert soils of southern New Mexico: Soil Sci. Soc. America Proc., v. 30, p. 773-781.

__________, 1968, Morphology of an argillic horizon in desert soils of southern New Mexico: Soil Sci., v. 106, p. 6-15.

__________, 1974, Holocene soils and Holocene-geomorphic relations in an arid region of southern New Mexico: Amer. Quaternary Assoc. (Abs.), 3rd Biennial Meet: Madison, Univ. Wisconsin, p. 30-39.

__________, 1975, Holocene soils and soil-geomorphic relations in an arid region of southern New Mexico: Quaternary Res., v. 5, no. 3, p. 321-360.

__________, 1977, Holocene soils and soil-geomorphic relations in a semi-arid region of southern New Mexico: Quaternary Res., v. 7, no. 1, p. 112-132.

Gile, L.H., and Hawley, J.W., 1966, Periodic sedimentation and soil formation on an alluvial fan piedmont in southern New Mexico: Soil Sci. Soc. America Proc., v. 53, p. 261-268.

__________, 1968, Age and comparative development of desert soils at the Gardner Spring Radiocarbon Site, New Mexico: Soil Sci. Soc. America, Proc., v. 32, p. 709-716.

Gile, L.H., Peterson, F.F., and Grossman, R.B., 1965, The K-Horizon--a master horizon of $CaCO_3$ accumulation: Soil Sci., v. 99, p. 74-82.

__________, 1966, Morphological and genetic sequences of carbonate accumulation in desert soils: Soil Sci., v. 101, p. 347-360.

Gile, L.H., Hawley, J.W., and Grossman, R.B., 1970, Distribution and genesis of soils and geomorphic surfaces in a desert region of southern New Mexico: Guidebook, Soil-geomorphology field conference: University Park, New Mexico, Soil Sci. Soc. America, 156 p.

Hadley, R.F., 1967, Pediments and pediment-forming processes: Jour. Geol. Educ., v. 15, p. 83-89.

Hamilton, W., 1964, Geologic map of the Big Maria Mountains Northeast Quadrangle, Riverside County, California and Yuma County, Arizona: U.S. Geol. Survey Quad Map GQ-350.

Hooke, R. LeB., 1967, Processes on arid-region alluvial fans: Jour. Geol., v. 75, p. 438-460.

______________, 1972, Geomorphic evidence for late-Wisconsin and Holocene tectonic deformation, Death Valley, California: Geol. Soc. America Bull., v. 53, p. 2073-2098.

Hooke, R. LeB., Yang, H-Y, and Weiblen, T.W., 1969, Desert varnish: an electron probe study: Jour. Geol., v. 77, p. 275-288.

Ku, Teh-Lung, 1975, Age dating of mammoth tusk by Th-230/U-234 and Pa-231/U-235: *in* Fugro, Inc., Early Site Review Rept., Sundesert Nuclear Power Proj., San Diego Gas and Electric Co., Appendix 2.5-J, Addendum 3-b, 2 p.

Kukla, G., and Opdyke, N., 1975, Preliminary report on magnetostratigraphic study of sediments near Blythe, California and Parker Valley, Arizona: *in* San Diego Gas and Electric Co., Early Site Review Rept., Sundesert Nuclear Power Proj., Appendix 2.5-B, 10 p.

Lee, G., and Bell, J., 1975, Depositional and geomorphic history of the lower Colorado River: *in* San Diego Gas and Electric Co. Early Site Review Rept., Sundesert Nuclear Power Proj., Appendix 2.5-D, 19 p.

Lowdermilk, W.C., and Sundling, H.L., 1950, Erosion pavement, its formation and significance: Trans. Amer. Geophysical Union, v. 31, p. 96-100.

Lustig, L.K., 1966, The geomorphic and paleoclimatic significance of alluvial deposits in southern Arizona: Jour. Geol., v. 74, p. 95-102.

Melton, M.A., 1965, The geomorphic and paleoclimatic significance of alluvial deposits in southern Arizona: Jour. Geol., v. 73, p. 1-38.

Merriam, R., and Bischoff, J.L., 1975, Bishop Ash: a widespread volcanic ash extended to southern California: Jour. Sed. Pet., v. 45, no. 1, p. 207-211.

Metzger, D.G., Loeltz, O.J., and Irelan, B., 1973, Geohydrology of the Parker-Blythe-Cibola area, Arizona and California: U.S. Geol. Survey Prof. Paper 486-G, 130 p.

Murray, K., Bell, J., and Crowe, B., 1976, Stratigraphy and structure of the Orocopia, Chocolate and Cargo Muchacho mountains, southeastern California: *in* San Diego Gas and Electric Co., Early Site Review Rept., Sundesert Nuclear Power Proj., Appendix 2.5-L.

Morrison, R.B., 1964, Lake Lahontan: geology of southern Carson Desert, Nevada: U.S. Geol. Survey Prof. Paper 401, 156 p.

______________, 1965, Quaternary geology of the Great Basin, *in* Wright, H.E., and Frey, D.G., eds., The Quaternary of the United States: Princeton, N.J., Princeton Univ. Press, p. 265-285.

Nettleton, W.D., Witty, J.E., Nelson, R.E., and Hawley, J.W., 1975, Genesis of argillic horizons in soils of desert areas of the southwestern United States: Soil Sci. Soc. America Proc., v. 39, no. 5, p. 919-926.

Nikiforoff, C.C., 1937, General trends of the desert type of soil formation: Soil Sci., v. 43, p. 105-131.

Olmsted, F.H., Loeltz, O.J., and Irelan, B., 1973, Geohydrology of the Yuma area, Arizona and California: U.S. Geol. Survey Prof. Paper 486-H, 227 p.

Potter, R.M., and Rossman, G.R., 1977, Desert varnish: the importance of clay minerals: Science, v. 196, no. 4297, p. 1446-1448.

Royse, C.F., and Barsch, D., 1971, Terrace and pediment-terraces in the Southwest: an interpretation: Geol. Soc. America Bull., v. 82, p. 3177-3182.

Ruhe, R.V., 1967, Geomorphic surfaces and surficial deposits in southern New Mexico: New Mexico Inst. Mining and Tech., Memoir 18, 66 p.

San Diego Gas and Electric Company, 1976, Early Site Review Report, Sundesert nuclear power project, Geology and Seismology, Appendix 2.5-B (with amendments), San Diego, Calif.

Shlemon, R.J., 1976, Quaternary soil stratigraphy, southeastern Mojave Desert, California and Arizona, *in* Mahaney, W.C., ed., Quaternary Soils Symposium, Abstracts-with-Program: Toronto, York University, p. 53-56.

______________, 1977, Soils of the Sundesert Nuclear Generating Station, near Blythe, California: Fugro, Inc., for San Diego Gas and Electric Co., 56 p.

Shlemon, R.J., and Purcell, C.W., 1976, Geomorphic reconnaissance, southeastern Mojave Desert California and Arizona: *in* San Diego Gas and Electric Co., Early Site Review Rept. Sundesert Nuclear Power Proj., Appendix 2.5-M, 28 p.

Smith, B.R., and Buol, S.W., 1968, Genesis and relative weathering studies in three semi-arid soils: Soil Sci. Soc. America Proc., v. 32, p. 261-265.

Smith, G.I., 1968, Late-Quaternary geologic and climate history of Searles Lake, southeastern California, *in* Morrison, R.B., and Wright, H.E., eds., Means of correlation of Quaternary successions, Proc., VII Congr., Internatl. Assoc. Quaternary Res., v. 8: Salt Lake City, Univ. Utah Press, p. 293-210.

Springer, M.E., 1958, Desert pavement and vesicular layer of some desert soils in the desert of the Lahontan Basin, Nevada: Soil Sci. Soc. America Proc., v. 22, p. 63-66.

Palaeosol Studies in Western Canada

J. F. Dormaar

ABSTRACT

While about 120 references refer directly or indirectly to palaeosols in Canada, only about 26 of them are pertinent to the three Prairie Provinces. Geographically, they relate to the Brandon Sand Hills in Manitoba, the Prelate and Saskatoon areas in Saskatchewan, and the Cypress Hills, various river valleys, and regions along the Eastern Slopes of the Rocky Mountains in Alberta. The palaeosols have generally either distinct black or distinct red horizons. Although soil micromorphology is a powerful tool in the study of palaeosols, those of the three Prairie Provinces have been studied with a variety of other methods. Close cooperation between archaeologist, pleistocene geologist, and soil scientist is desirable for further significant contributions to the environmental history of the Prairies.

INTRODUCTION

Palaeosol and soil are two concepts that have defied precise definition. Often, the definition for soil carries an agricultural bias. Within the context of the study of palaeosols, Joffe's definition (1949) of a soil seems appropriate: "The soil is a natural body of mineral and organic constituents, differentiated into horizons of variable depth, which differs from the material below in morphology, physical make-up, chemical properties, and biological characteristics."

A palaeosol is, by definition (Yaalon, 1971), a soil formed in a landscape of the past, *i.e.*, the former soil-forming reaction was either altered as a result of a change in the external environmental conditions or was interrupted by burial. Bos and Sevink (1975) noted, however, that this definition was not very clear, as both landscape and past were not defined.

Palaeosols, and in this paper palaeosols are always

considered to be the buried variety, are of interest in pleistocene stratigraphy because of their importance as stratigraphic markers and as keys to past environments. This presentation surveys some of the major palaeosol studies carried out in the three Prairie Provinces.

MANITOBA

The Brandon Sand Hills

Several dune areas in western Canada have been studied in detail by David (1968, 1971). In the Brandon Sand Hills (Figure 1a), where most sand dunes were stabilized by vegetation, research was concentrated on the morphological development of the dunes, sedimentary properties of the dune sands, and chronology of past dune activities. Radiocarbon dating of humus-rich soil horizons in the Brookdale Road Section revealed the most complete aeolian stratigraphic and chronological record in the area.

The palaeosols developed at the surface of former sand dunes and morphologically resemble the local surficial soils developed on the stabilized dunes, inasmuch as they are all characterized by feeble profile development. This particular dune section was formed during six periods of dune activity that preserved six layers of dune sands and five palaeosols.

For dating, the samples were cleaned of all root hairs and the humic matter was extracted from the samples by flotation in distilled water. Since the soil samples were not treated chemically before dating, the radiocarbon ages of the soils represented the Mean Residence Time (M.R.T.) of the total soil humus of the sampled soil layers at the time of burial of the soils. The date of organic carbon from a surface Ah horizon reflects the age of the organic matter being added daily together with the age of the organic matter synthesized and resynthesized over several thousand years. The organic matter in a palaeosol already has a high M.R.T. at the time of its burial. Also, such organic matter, particularly the young 'active' fraction (Paul, 1970) may continue to decompose after burial. Therefore, the radiocarbon date obtained actually represents the time since burial plus an increment due to these factors and must be corrected (Turchenek *et al.*, 1974). The following corrected ages in C^{14} years BP were found: 3,680 ± 180, 2,150 ± 150, 1,510 ± 150, 920 ± 140, and 430 ± 130. At these times, important fluctuations of the regional climate either corresponded to or preceded the dune formation and stabilization sequence. These dates correspond roughly to the end of early Sub-boreal, the beginning of sub-Atlantic and Scandic, and within Neo-Atlantic and Pacific II (Bryson and Wendland, 1967).

This study (David, pers. commun., Jan. 9, 1970) was mainly concerned with the postglacial aeolian activities in areas where sand dunes occurred as interpreted from the

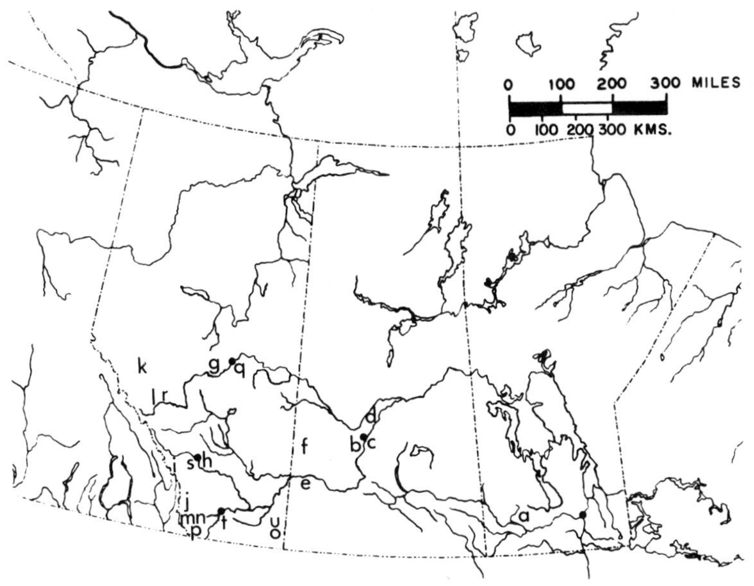

Figure 1 Location of the palaeosols discussed in the text.

a. Brandon Sand Hills
b,c,d. Saskatoon area
e. Prelate Ferry
f. 200 km north of Prelate Ferry
g. West of Edmonton
h. Fish Creek, south of Calgary
i. Marmot Creek Basin
j. Gap in Livingstone Range
k. Hinton
l. Upper North Saskatchewan River area
m. Crowsnest Lake
n. Bellevue
o. Cypress Hills plateau
p. Cloudy Ridge north of Waterton Lakes National Park
q. East of Edmonton
r. South of Nordegg
s. West of Calgary
t. Lethbridge area
u. North of Cypress Hills

presence of the palaeosols and from the dating of the organic carbon derived from them. This use of palaeosols was certainly more geological than pedological, even though some interpretation of the environment related to their formation is always derived from the morphology of the palaeosol.

SASKATCHEWAN

Saskatoon Area

Postglacial palaeosols developed in lacustrine and

aeolian sediments occurring on terraces and in banks of the South Saskatchewan River near Saskatoon (Figures 1b, 1c, 1d) have been described (Turchenek, 1971; Turchenek *et al.*, 1974). At each of three sites a sequence of two palaeosols was studied: a lower soil of immediate postglacial age that was buried by aeolian sands and an upper one, located within the sand dunes, that marked a period of stability of the dunes.

Pedogenic interpretations were based on macro- and micromorphological observations, bulk densities, and particle size distributions. Micromorphology and the determination of calcite and dolomite distributions were particularly useful in differentiating sedimentary, pedogenic, and post-burial alteration processes in the buried soil profiles.

Studies of the organic matter suggested that it had changed markedly after burial. The organic carbon contents and the soil colors did not follow the relationship between percent organic matter and Munsell renotation value for Saskatchewan soils developed on till deposits. A Munsell value of 3 corresponds to an organic matter content of about 5 percent in till soils (Shields *et al.*, 1968). Some of the buried Ah horizons had organic matter contents considerably lower than 5 percent. Similarly, the E_4/E_6 ratios of the humic acids reflected possible post-burial changes by being somewhat higher than those of Saskatchewan soils generally.

The lower palaeosol at two of the sites had similar radiocarbon ages [8,100 ± 120 yr BP (Figure 1d) and 8,160 ± 125 yr BP (Figure 1c), while the age of the lower palaeosol at the third site was 9,940 ± 160 yr BP (Figure 1b)]. The ages of the upper palaeosols differed widely (7,070 ± 115, 3,150 ± 80, and 7,640 ± 150 yr BP, respectively). These ages date more closely the beginning of deposition of the overlying materials rather than the end of sedimentation of the underlying materials.

The general sequence of events along the South Saskatchewan River which followed glaciation included the deposition of lacustrine or alluvial deposits, formation of terraces in these deposits, the deposition of loess in some areas, soil formation, and subsequent burial by aeolian sands. Climatic change was inferred to be the cause of the dune and soil sequences.

Prelate Ferry Area

Further upstream on the South Saskatchewan River near Prelate Ferry (Figure 1e), a late-Wisconsin palaeosol with complex A, B, and C horizons was identified by David (1966). The palaeosol (20,000 ± 850 yr BP), a poorly drained upland soil or possibly a solod, developed in till, is buried by about 37 m of sediments comprising two till sheets and three beds of stratified drift. Further detailed examination

(David, 1969) revealed that there are two palaeosols, a lower one, the Prelate Ferry palaeosol proper, and an upper one, that developed in the colluvial deposits that buried the lower soil. The nonglacial Prelate Ferry interval during which the lower palaeosol formed had a local climate similar to that of today. The beginning of the interval is not known, but it ended about 20,000 yr BP when the last major ice advance occurred in the area.

Christiansen (1965) reported a soil, buried beneath a few meters of till, about 200 km north of the Prelate Ferry palaeosol exposure (Figure 1f). The dating of 21,000 ± 800 yr BP suggested that this soil was also formed during the Prelate Ferry interval.

ALBERTA

A variety of palaeosols have been found in Alberta. Fortunately, several of these are now being studied in more detail and, in time, a synthesis may be possible.

Brunisolic Soils

Several palaeosols, widely separated geographically and belonging to the Brunisolic Order, occur in strata beneath Mazama tephra. The Mazama tephra deposit is an excellent stratigraphic marker, since it was deposited about 6,700 yr BP.

(1) A degraded Eutric Brunisol on aeolian sand west of Edmonton (Pawluk and Dumanski, 1970) (Figure 1g). It consisted of an Aej with a Bfj over a C. This soil has its present day analogue in the Eutric Brunisols developed under pine forest in the sand dune area southwest of Grande Prairie, Alberta.

(2) An Alpine or Orthic Eutric Brunisol south of Calgary (Figure 1h). It consisted of an Ah with a Bm over a C. The surface soil was a Cumulic Rego Black Chernozemic soil. The area is being subjected to an archaeological survey at present (Reeves, pers. commun.).

(3) An Eutric Brunisol on till in Marmot Creek Basin (Figure 1i) about 72 km east of Banff (Beke and Pawluk, 1971). The section contained four geologic deposits: colluvium, mixed colluvium and tephra, tephra, and till. The palaeosol consisted of a Bm with a BC over a C. The surface soil developed in the colluvium was an Orthic Humo-Ferric Podzol.

(4) A Degraded Alpine Eutric Brunisol on aeolian sand in the Gap area (Figure 1j) where the Oldmand River comes through the Livingstone Range (Reeves and Dormaar, 1972). It consisted of an Ah, an AB, a Btj, and a BC over a C. A subalpine to alpine vegetation and a cold, wet climate were inferred. It was suggested that this represented a depression of the tree line, by about

600 m at this point. The surface soil consisted of an Orthic Eutric Brunisol. Examination of this section revealed three successive human occupation sites. Charcoal from the living floor associated with a buried Ah horizon above the Mazama tephra dated 6,060 ± 140 yr BP from the floor associated with a buried Ah horizon immediately below the Mazama tephra dated 6,720 ± 140 yr BP, and from the floor associated with the buried Brunisol dated 8,000 ± 150 yr BP.

Other Red Palaeosols

Several other palaeosols have been recognized on the basis of red horizons:

(1) Dumanski and Pawluk (1971) when examining soils developed in postglacial aeolian deposits on the foothills region of Hinton (Figure 1k) were faced with a classification problem. A thick, distinct yellowish brown to yellowish red 'palaeo' B horizon (2,730 ± 100 yr BP) appeared very near the surface at the eastern extremity of the area but became buried progressively deeper at the western extremity of the area. When this 'palaeo' B was within the zone of active pedogenesis, internal clay translocation within this horizon resulted in many cases in development of a weakly expressed Ae, Bt horizon sequence. The soil profile that would normally be classified as a Regosol was now, with the presence within its solum of a horizon sequence with Luvisolic character classified within the Luvisolic Order. When the 'palaeo' B was present below the solum, it remained a separate entity and did not interfere with the classification of the soil above it.

Since the 'palaeo' B is incidental, the solum in which it is found could be classified as if it, or the horizon sequence formed from it, was not present (Dormaar, pers. commun. to Dumanski, Nov. 10, 1971) or it could possibly be handled in the concept of soil 'intergrade' (Stobbe, 1952). The main underlying cause of the confusion is the fact that the presence of a palaeo horizon cannot be indicated within the present System of Soil Classification for Canada (Dumanski, pers. commun., Dec. 7, 1971).

Bos and Sevink (1975) proposed a system based on the differentiation between gradational and pedomorphic features. The concept of pedogenetic horizons was abandoned in favor of the recognition, or failure to recognize, pedomorphic features at a certain depth and over a certain vertical extent.

(2) Vertical sections along the Upper North Saskatchewan River, adjacent to and inside the front range of the Rocky Mountains (Figure 1) showed one or more distinct brown to reddish brown horizons (Dormaar and Lutwick, 1975). As indicated by tephra stratigraphy of the

area (Westgate and Dreimanis, 1967), these bands and the alluvial/aeolian materials in which they occurred were deposited post 6,600 yr BP (Mazama tephra) to some time after 2,500 yr BP (Bridge River tephra). Based on such criteria as time for water droplet penetration, magnetic susceptibility, and crystallization of Fe oxides, together with present conditions and vegetation of the fans and floodplain in that area, it was concluded that these reddish brown horizons were fire-affected. Many of the colored horizons contained charcoal and sometimes the remains of bog vegetation. The organic debris in the bogs before a fire was continually being enriched with silt blowing from the floodplain. The colored horizons were thus considered to be a combination of accessed aeolian matter reddened in the layer of burning debris and some mineral matter reddened by conducted heat.

(3) The Crowsnest Pass area (Figure 1m,n) is presently being subjected to detailed archaeological investigations (Reeves, 1974a). Several well developed palaeosols have also been found.

(a) An Alpine Eutric Brunisol on fan gravels was present in an excavation at the east end of Crowsnest Lake (Figure 1m). It consisted of an Ah, with a Bm over a C. Time diagnostic projectile points found in the Brunisol indicated a date of *ca.* 7,950 to 7,450 yr BP. The surface soil was an Orthic Black Chernozemic soil which was being transferred towards a Dark Gray Chernozemic soil by poplar invasion. Through the presence of other palaeosols, mainly Regosols, texture of the deposits, and artifact assemblages, the site has become of major value and significance to the prehistoric archaeology of the Crowsnest Pass.

(b) During the construction of a new sewer system in Bellevue (Figure 1n), about 30 km east of the previous site, a thin (about 25 cm), yet complete, Gleyed Bisequa Gray Luvisol was found. It had the following horizon sequence: Ah, Aej, Bf, ABg, Btg, and Cg. Mazama tephra were present above this soil. A bison skull present in the Ah has been dated at 9,860 ± 320 yr BP (Reeves, pers. commun., Feb. 10, 1975). This date supports the partial reconstruction of mean frontal vegetation zones during late-glacial time by Bryson and Wendland (1967) who showed western Boreal Forest in this region.

(4) Jungerius (1966, 1967) described a palaeosol developed in the uppermost 1 m of the Cypress Hills Formation (Figure 1o) on the Cypress Hills Plateau. It was commonly covered by Quaternary aeolian deposits. No A horizon was recognizable but it was readily discernible

because of its reddish B horizon. Westgate *et al.* (1972) indicated that lithologic data justified the use of the term 'palaeosol' and demonstrated that sediments in the B horizon were relatively strongly weathered. It was thought that this soil developed during the long weathering interval that preceded the Elkwater glaciation (Laurentide). After this interval, the Green Lake drift was deposited (probably between 24,000 to 20,000 yr BP) onto the Elkwater drift. However, the age of these drift sheets are not well established because of the scarcity of C^{14} dates. The Plateau itself was unglaciated. Jungerius (1966) suggested that, to allow for the well developed B horizon and the absence of $CaCO_3$, the climate must have been warm and sub-humid, the rainfall having a marked seasonal distribution. Westgate *et al.* (1972) also deciphered two phases of periglacial activity probably superimposed during the Green Lake glaciation.

(5) Wagner (1966) described a palaesol (Cloudy Ridge soil north of Waterton Lakes National Park) under what he believed to be the oldest mountain till in the area (Figure 1p). No A horizon was recognizable. The B horizon was 2.4 m thick and contained strongly weathered material. The Cca horizon was 4.6 m thick and contained highly cemented till. Lack of radiometric control poses a problem. However, the older till was tentatively placed as Illinoian age, the palaeosol as Sangamonian age, and the overlying till as late Farmdalian-early Woodfordian age.

Both the Cypress Hills palaeosol and the Cloudy Ridge palaeosol are being subjected to detailed analyses at present (Dudas, pers. commun., Aug. 20, 1975).

Black Palaeosols

At several locations, one or more black Ah horizons have also been found under the Mazama tephra deposit.

(1) Pawluk and Dumanski (1970) identified a weakly developed Black Chernozemic soil in alluvium along the North Saskatchewan River east of Edmonton (Figure 1q).

(2) Along the North Saskatchewan River south of Nordegg (Figure 1r), charcoal was collected from an Ah horizon 40 cm below a tephra deposit, which was presumed to be Mazama or older, and dated at 8,030 ± 200 yr BP. Reeves (1974b) commented that the analysis provided a date on a period of soil formation and surface stability in the loess deposition sequence in the area. The period of stability, dated in Pinedale IV times (Reeves and Dormaar, 1972), perhaps indicates decreased river flow and silt transport during time of maximum ice advance in the headwaters.

(3) Harrison (1973) identified a 20-cm thick Ah horizon

(8,400 ± 150 yr BP) west of Calgary (Figure 1s).

(4) Around Lethbridge (Figure 1t) up to 25-cm thick Ah horizons may be found between the varves of glacial Lake Lethbridge and the Mazama tephra deposit. One such site has been described by Horberg (1952).

(5) The most conspicuous unit in a postglacial succession exposed at several sites north of the Cypress Hills (Figure 1u) is the brownish-gray Ah horizon of an Orthic Humic Gleysol (Westgate *et al.*, 1972; Dormaar, 1973). The soil was dated at 10,230 ± 150 yr BP.

Several of these black Ah horizons are at present being studied in more detail by Rutter (pers. commun., 1976).

METHODOLOGY

Generally, the same methods used in pedology are also used in the study of palaeosols. These include both morphological and chemical approaches. However, some methods may be useful for some palaeosols but not for others. Also, problems in interpretation of analytical results arise mainly because of changes which have possibly occurred in the soils since burial.

Organic matter is probably the component most susceptible to change after burial of a soil (Stevenson, 1969; Paul, 1970). The extent to which humus has been preserved will depend upon such factors as the change in environment which preceded the new cycle of sedimentation, the circumstances under which the soil was buried, and the activities of living organisms in the buried sediment. After burial, the source of raw plant material for humus synthesis is cut off and the humus is exposed to successive cycles of biological attack with concomitant changes in chemical composition. Indeed, Cook (1970) showed that buried Ah horizons have a rich bacterial and actinomycete flora. Further modifications of the organic matter will also be brought about through chemical processes.

The most profitable approach, from a chemical point, will be to concentrate on those methods in which those pedologic features that are resistant to alteration are studied. Color, for example, may be meaningless for identifying a soil as a particular species because of secondary carbonation, but it is often meaningful for recognizing the palaeosol in the first place. A more intensive study of the chemistry of color in soils is needed to establish its durable component. The use of the ratio of the absorbance of humic materials in solution measured at 465 and 665 nm (E_4/E_6 ratio) may be significant in this context (Paul, 1970). Waxes and lipids are particularly resistant to decomposition (Stevenson, 1969); Dormaar (1964) showed that morphologically similar soils contained significantly different amounts of alcohol-benzene extractable organic matter.

A variety of techniques have been used successfully on a number of palaeosols in western Canada.

Opal phytoliths are minute bodies of isotropic silica that have been precipitated in the leaf epidermal cells of grasses and, when found in soils, can be used as indicators of the vegetative history of the site (Dormaar and Lutwick, 1969). Counts of opal phytoliths in the coarse fraction have been helpful in explaining the genesis of a number of palaeosols from along various river valleys in southern Alberta.

The pattern of original carbonate distribution may be preserved. This depends on the mode of groundwater flow through the palaeosol. David (1966, 1969) compared a weighted mean carbonate content of the soil horizon with the average carbonate content of the parent material. Weighting was done by taking the carbonate content of a particular horizon multiplied by the average thickness of that horizon and dividing it by the total thickness of all horizons so considered.

The character of the infrared absorption spectra of alkali-soluble, acid-insoluble soil organic matter (humic acids) has been helpful in differentiating forested soils from grassland soils (Dormaar and Lutwick, 1969; Reeves and Dormaar, 1972). Although it is often difficult to establish in the field the origin of the B horizon of truncated palaeosols, the infrared absorption spectra have also been useful in distinguishing between Chernozemic and Gleysolic B horizons (Dormaar, 1973).

The study of fire-affected soils required the use of several techniques (Dormaar and Lutwick, 1975):

(1) The amounts of well-crystallized Fe oxides, which are estimated as the difference between Fe extractable with dithionite-citrate-bicarbonate and Fe extractable with acid ammonium oxalate in the dark, increase with pyrolysis of the soil.

(2) The magnetic susceptibility of the soil increases with pyrolysis. In fact, the ignition of any Fe oxide in the presence of organic materials increases the magnetic susceptibility of the sample (Oades, 1963).

(3) For samples subjected to temperatures at 600 to 700°C, differential thermal analysis is worth pursuing. Legitimate interpretation of differential thermal patterns depends on the validity of the assumption that samples so analyzed and compared are mineralogically similar and differ only in their exposure to heat (Dudas, pers. commun., April 9, 1974).

EPILOGUE

About 120 papers refer, directly or indirectly, to

palaeosols in Canada. Of these, 26 were pertinent to the three Prairie Provinces and almost all of these have been used in this assessment. Often palaeosols represent only short periods of soil-forming activity. However, several of the more complete palaeosols are now being studied in detail. Close cooperation between archaeologists, pleistocene geologists, and soil scientists is desirable for further significant contributions to the environmental history of the Prairies.

REFERENCES CITED

Beke, G.J., and Pawluk, S., 1971, The pedogenic significance of volcanic ash layers in the soils of an east slopes (Alberta) watershed basin: Can. Jour. Earth Sci., v. 8, p. 664-675.

Bos, R.H.G., and Sevink, J., 1975, Introduction of gradational and pedomorphic features in descriptions of soils. A discussion of the soil horizon concept with special reference to paleosols: Jour. Soil Sci., v. 26, p. 223-233.

Bryson, R.A., and Wendland, W.M., 1967, Tentative climatic patterns for some late-glacial and postglacial episodes in central North America, *in* Mayer-Oakes, W.J., ed., Life, Land and Water: Winnipeg, Univ. Manitoba, p. 271-298.

Christiansen, E.A., 1965, Geology and groundwater resources, Kindersley area (72-N), Saskatchewan: Saskatchewan Res. Council, Geol. Div., Rept. No. 7, 72 p.

Cook, C.D., 1970, Significance of microorganisms in palaeosols, parent materials, and groundwater, *in* Pawluk, S., ed., Pedology and Quaternary Research: Edmonton, Univ. Alberta Press, p. 53-59.

David, P.P., 1966, The late-Wisconsin Prelate Ferry paleosol of Saskatchewan: Can. Jour. Earth Sci., v. 3, p. 685-696.

__________, 1968, Geomorphology, stratigraphy, chronology, and migration of sand dunes in Manitoba and Saskatchewan, in Report of Activities: Canada Geol. Survey, Paper 68-1, Pt. A, p. 155-157.

__________, 1969, A reappraisal of the Prelate Ferry paleosol: Unpubl. Report prepared for the 19th Midwest Friends of the Pleistocene Field Conference, Saskatchewan and Alberta, 34 p.

__________, 1971, The Brookdale road section and its significance in the chronological studies of dune activities in the Brandon Sand Hills of Manitoba, *in* Turnock, A. C., ed., Geoscience Studies in Manitoba: Geol. Assoc. Canada Spec. Paper 9, p. 293-299.

Dormaar, J.F., 1964, Fractionation of the humus of some Chernozemic soils of southern Alberta: Can. Jour. Soil Sci., v. 44, p. 232-236.

______________, 1973, A diagnostic technique to differentiate between buried Gleysolic and Chernozemic B horizons: Boreas, v. 2, p. 13-16.

Dormaar, J.F., and Lutwick, L.E., 1969, Infrared spectra of humic acids and opal phytoliths as indicators of palaeosols: Can. Jour. Soil Sci., v. 49, p. 29-37.

______________, 1975, Pyrogenic evidence in paleosols along the North Saskatchewan River in the Rocky Mountains of Alberta: Can. Jour. Earth Sci., v. 12, p. 1238-1244.

Dumanski, J., and Pawluk, S., 1971, Unique soils of the foothills region, Hinton, Alberta: Can. Jour. Soil Sci., v. 51, p. 351-362.

Harrison, J.E., 1973, Dated paleosol from below Mazama (?) tephra, in Report of Activities: Canadian Geol. Survey, Paper 73-1, Pt. B, p. 127-128.

Horberg, L., 1952, Pleistocene drift sheets in the Lethbridge region, Alberta, Canada: Jour. Geol., v. 60, p. 303-330.

Joffe, J.S., 1949, Pedology: New Brunswick, N.J., Pedology Publications, 662 p.

Jungerius, P.D., 1966, Age and origin of the Cypress Hills plateau surface in Alberta: Geographical Bull., v. 8, p. 307-318.

________________, 1967, The influence of pleistocene climatic changes on the development of the polygenetic pediments in the Cypress Hills area, Alberta: Geographical Bull., v. 9, p. 218-231.

Oades, J.M., 1963, The nature and distribution of iron compounds in soils: Soils Fert., v. 26, p. 69-80.

Paul, E.A., 1970, Characterization and turnover rate of soil humic constituents, *in* Pawluk, S., ed., Pedology and Quaternary Research: Edmonton, Univ. Alberta Press, p. 63-76.

Pawluk, S., and Dumanski, J., 1970, Notes from the field tour held in conjunction with the Symposium, May 14, 1969, *in* Pawluk, S., ed., Pedology and Quaternary Research: Edmonton, Univ. Alberta Press, p. 187-218.

Reeves, B.O.K., 1974a, Crowsnest Pass archaeological project 1973 salvage excavations and survey, Paper 2, Preliminary Report 1973: Nat. Museum Man, Mercury Series, Paper 24, 90 p.

______________, 1974b, GSC-1944. North Saskatchewan River, site FcQa-8, *in* Lowdon, J.A., Wilmeth, R., and Blake, Jr., W., eds., Canada Geological Survey Radio Carbon Dates XIV, Canada Geol. Survey, Paper 74-7, p. 6.

Reeves, B.O.K., and Dormaar, J.F., 1972, A partial Holocene pedological and archaeological record from the southern Alberta Rocky Mountains: Arctic Alpine Res., v. 4, p. 325-336.

Shields, J.A., Paul, E.A., St. Arnaud, R.J., and Head, W.K., 1968, Spectrophotometric measurement of soil color and its relationship to moisture and soil organic matter: Can. Jour. Soil Sci., v. 48, p. 271-280.

Stevenson, F.J., 1969, Pedohumus: Accumulation and diagenesis during the Quaternary: Soil Sci., v. 107, p. 470-479.

Stobbe, P.C., 1952, The morphology and genesis of the Grey Brown Podzolic and related soils of Eastern Canada: Soil Sci. Soc. America Proc. v. 16, p. 81-84.

Turchenek, L.W., 1971, A study of paleosols in the Saskatoon area of Saskatchewan [M.Sc. thesis]: Saskatoon, Saskatchewan Univ., 160 p.

Turchenek, L.W., St. Arnaud, R.J., and Christiansen, E.A., 1974, A study of paleosols in the Saskatoon area of Saskatchewan: Can. Jour. Earth Sci., v. 7, p. 905-915.

Wagner, W.P., 1966, Correlation of Rocky Mountain and Laurentide glacial chronologies in southwestern Alberta, Canada [Ph.D. dissert.]: Ann Arbor, Michigan Univ., 141 p.

Westgate, J.A., and Dreimanis, A., 1967, Volcanic ash layers of recent age at Banff National Park, Alberta, Canada: Can. Jour. Earth Sci., v. 4, p. 155-161.

Westgate, J.A., Bonnichsen, R., Schweger, C., and Dormaar, J.F., 1972, The Cypress Hills, *in* Rutter, N.W., and Christiansen, E.A., eds., Quaternary Geology and Geomorphology between Winnipeg and the Rocky Mountains: Guidebook Field Excursion C-22 of the 24th Intern. Geol. Congr., p. 50-62.

Yaalon, D.H., 1971, Paleopedology: Origin, Nature and Dating of Paleosols: Jerusalem, Israel Univ. Press, 350 p.

Late-Quaternary Stratigraphy and Soils in the Wind River Mountains, Western Wyoming

W. C. Mahaney

ABSTRACT

Glacial and periglacial deposits of late-Quaternary age in the Wind River Mountains, Wyoming, are assigned relative and absolute ages using radiocarbon dating, topographic position, surface morphology, weathering features, lichenometry, and soil development. Geologic-climatic units, defined on the basis of deposits consisting primarily of till and mass-wasted debris, carry distinctive soil profiles representative of the age of the unit. Five soil-stratigraphic units which developed following periods of glaciation are recognized and named, from youngest to oldest, "post-Gannett Peak soil," post-Audubon soil," "post-Early Neoglacial soil," "post-Pinedale soil," and "post-Bull Lake soil." Deposits are differentiated largely upon differences in field morphology and the degree of soil development as indicated by texture and organic profiles, clay-mineral assemblages, and selected soil-chemical parameters. Soil profiles were sampled on representative moraines and a stone-banked lobe composed largely of granitic and gneissic debris. Soil facies change along an altitudinal gradient from Entisols and Alfisols in the alpine and subalpine ecosystems to Aridisols and Mollisols in the adjacent intermontane basins. All profiles are situated on high and well-drained topographic positions. These soil-stratigraphic units are traceable along the Wind River Range in major drainages. Correlative pedons are found in adjacent areas, such as, the Teton Range, Gros Ventre Mountains, Big Horn Mountains, and Front Range in Colorado.

INTRODUCTION

Despite a century of Quaternary geological research in the mountains of western Wyoming, relatively few publications have emphasized the importance of soils in separating deposits in the glacial and periglacial sequence. Early research stressed the physiography (Hayden, 1863; Fenneman, 1931; Atwood, 1940), and glacial history (Blackwelder, 1915);

more recent studies have refined the glacial history (Moss, 1951; Richmond, 1965, 1976; Currey, 1974). Recent pedological investigations by Miller and Birkeland (1974) and Mahaney and Fahey (1976) have generated new questions related to the age of the older Bull Lake and pre-Bull Lake deposits, to the correlation of type sections in the Wind River Range with Yellowstone National Park, to the age of the controversial Type Temple Lake deposits in the southern part of the Range, and to the separation of deposits of Neoglacial age. The principle objectives of this paper are to briefly summarize the surficial geology, to analyze the soils which have stratigraphic importance, and to discuss problems of age determination.

FIELD AREA

The Wind River Range, a broad northwest-southeast trending anticlinal uplift (Figures 1 and 2) forms part of the Central Rocky Mountain Province as defined by Fenneman (1931). The Range is approximately 190 km (120 mi) in length and 60 km (40 mi) wide, extending from South Pass to the Gros Ventre Mountains.

The anticlinal uplift of Tertiary age exposed a core of Precambrian granite, gneiss and schist (Branson and Branson, 1941; Baker, 1944; Worl, 1968). Rocks of Paleozoic and Mesozoic age flank the eastern slope, while Tertiary conglomerates, shales, tuffs, and lava are common on the western slope (Bradley, 1964).

Along the crest of the Range in the vicinity of the Titcomb Basin (Figures 3 and 4), a complex of igneous and metamorphic rocks with fine, medium, coarse, and porphyroblastic textures prevail. Glacial and periglacial deposits contain clasts of dioritic gneiss, gabbro-dioritic gneiss, quartz-monzonite gneiss, granitic gneiss, granite, amphibolitic schist, and quartz-syenitic gneiss. Linearity of major valleys suggests the importance of fault systems and shear zones in controlling drainage in the area.

Topography

In 1811, members of the Overland Astorian Expedition described the Wind River Range as consisting of three parallel ranges (Irving, 1836). The first description of the range as consisting of "vast masses of black rock [granodiorite ?], almost destitue of wood, and covered in many places with snow" was made by Robert Stuart, who as leader of the return Overland Astorian Expedition of 1812, traversed the western slopes of the Range (via Boulder Lake, Figure 2). After more than two decades of exploration by itinerant fur trappers, who for the most part, did not keep detailed journals (DeVoto, 1964), the central part of the Range was explored by Capt. B.L.E. Bonneville who climbed Gannett Peak and described the headwaters of the Green River in 1833 (Figure 1) (Irving, 1836). More detailed descriptions of the Wind River and Green River drainages were produced by Osborne Russell in 1837, but remained unpublished until 1914 (see Haines, 1965). The central core of the Range was explored by J.C. Fremont in the early 1840's, and visited briefly by the Hayden expedition in 1870 (Bartlett, 1962). That modern glaciers exist in the Wind Rivers, as well as adjacent ranges, seems to have gone unnoticed until Clarence King described them in 1869 (King, 1870).

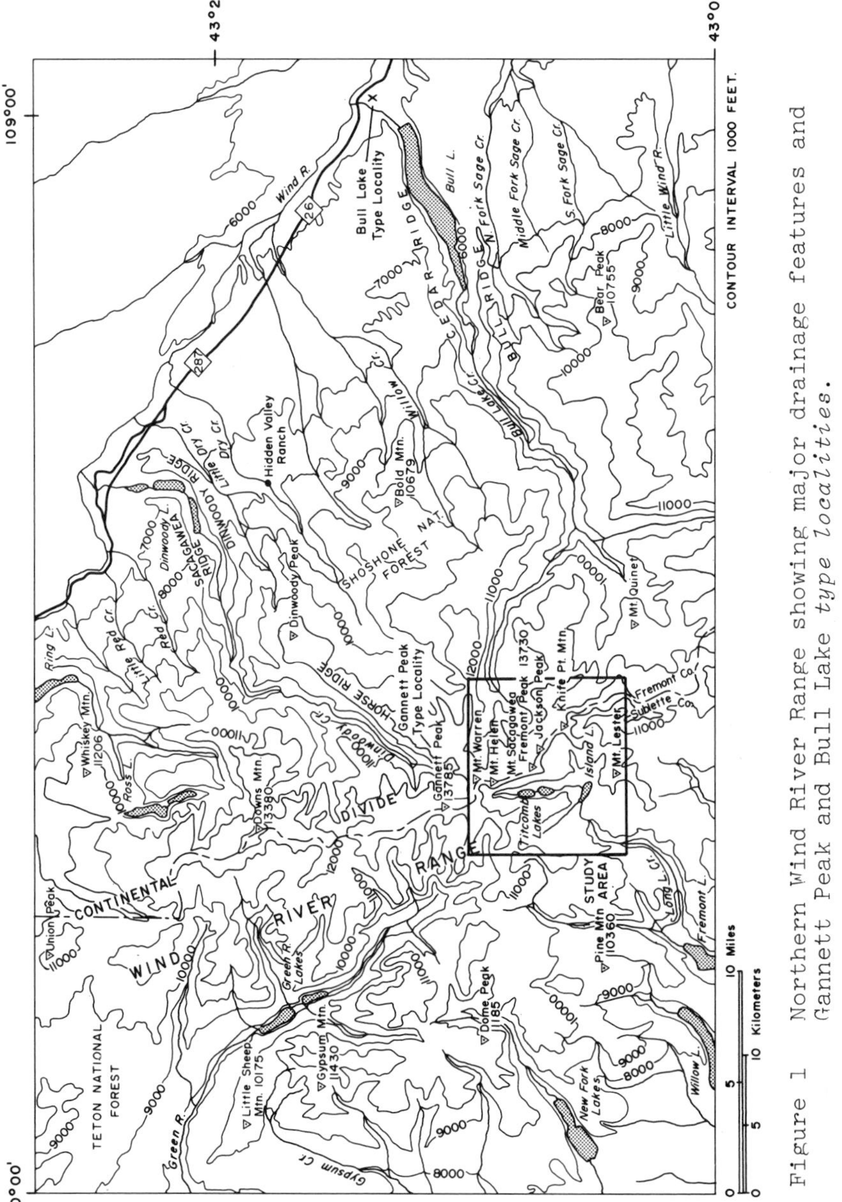

Figure 1 Northern Wind River Range showing major drainage features and Gannett Peak and Bull Lake *type localities*.

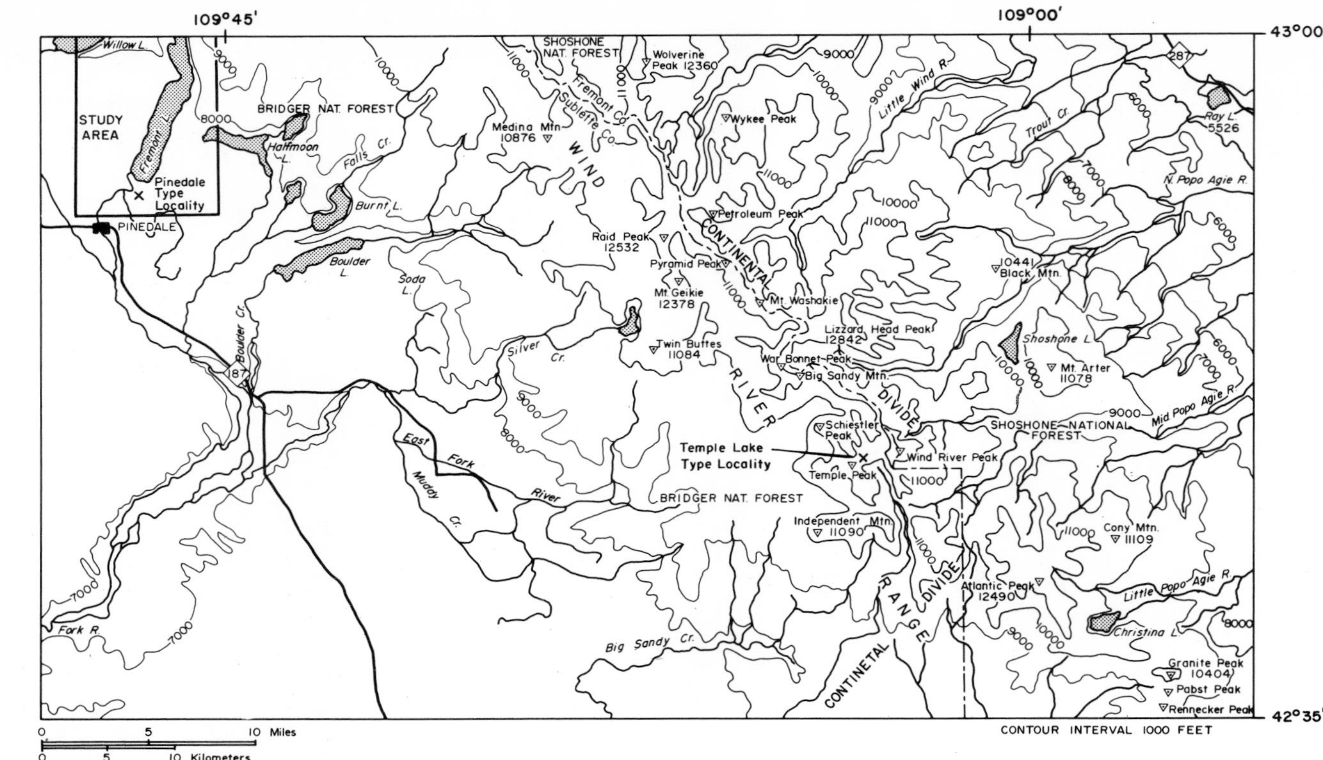

Figure 2 Southern Wind River Range showing major drainage features, Pinedale *Type Locality* of Blackwelder (1915) and Temple Lake *Type Locality*.

The Titcomb Basin (Figures 1, 3 and 4), located 40 km (25 mi) northeast of Pinedale, Wyoming, is representative of glaciated basins along the west flank of the Wind River Range (*cf.* Cirque of the Towers, Temple Lake Cirque, Mammoth Glacier Cirque, and Indian Basin). On both flanks of the Range glaciers are generally limited to cirque basins above 3300 m (11,000 ft). Glaciers are more common in cirques with a northeast orientation, and bear many similarities with glaciers of the Ural-type in the Front Range of Colorado which have formed in response to the drifting of snow by prevailing westerly winds (Outcalt and MacPhail, 1965). In many cases, modern glaciers occupy portions of cirque floors originally formed by Pleistocene and late-Holocene ice.

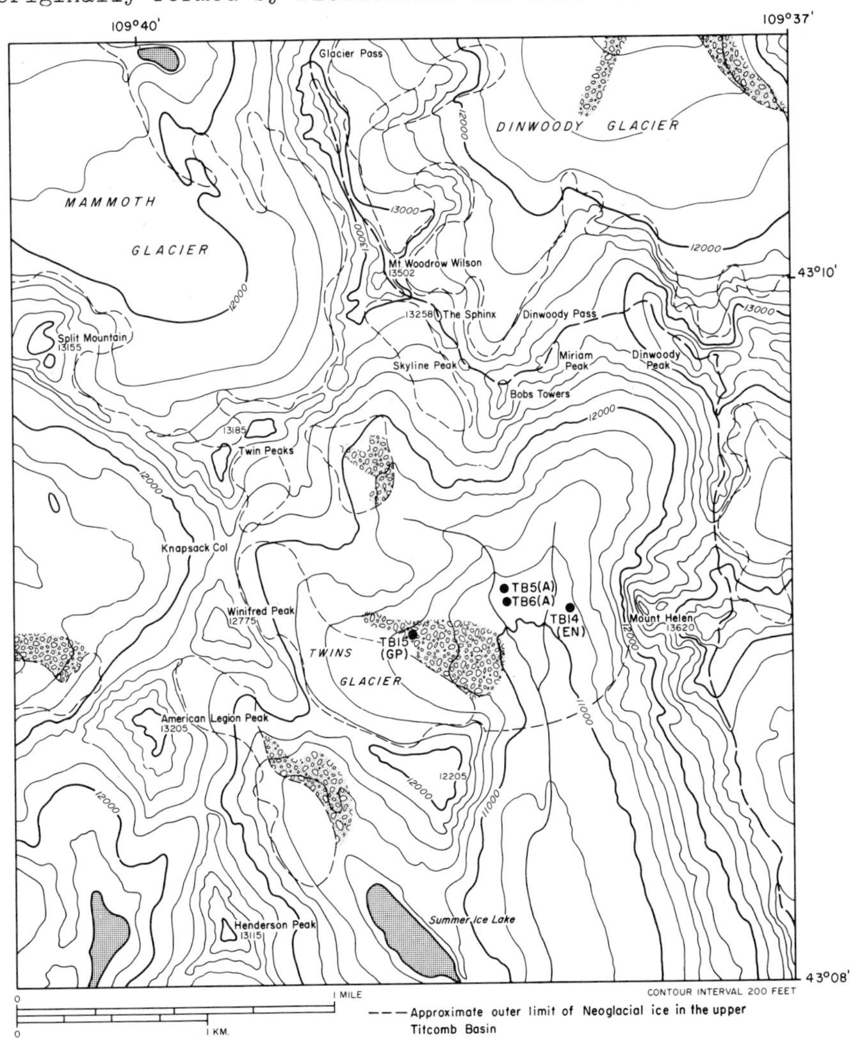

Figure 3 Upper Titcomb Basin showing sites on Neoglacial deposits: Gannett Peak (GP), Audubon (A) and Early Neoglacial (EN).

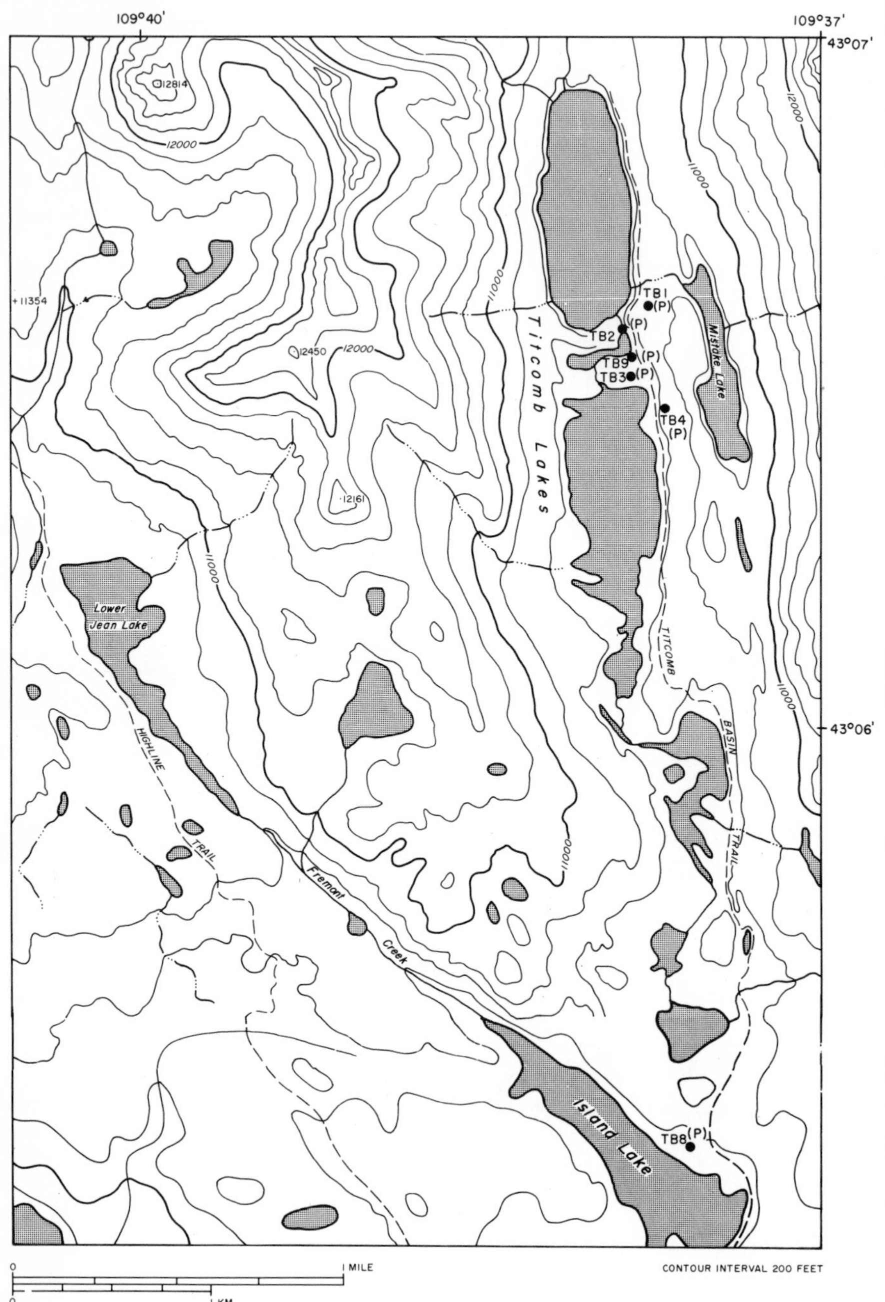

Figure 4 Lower Titcomb Basin showing sites on late-Pleistocene/early-Holocene deposits: Pinedale (P).

Twin Peaks (4020 m), the Sphinx (4041 m), Dinwoody Peak (4115 m), Mount Warren (4185 m) and Mount Helen (4152 m) dominate the upper Titcomb Basin along with rugged and sharp-crested arêtes which drop away to the Basin floor at 3300 m. Relative relief in some adjacent valley systems is more than 1000 m. The Lower Titcomb Basin (Figure 4) has an average elevation of approximately 3200 m, and is a classic U-shaped glacial valley containing, in its upper end, a well-preserved sequence of end moraines and striated bedrock. The steep rugged flanks of Mount Sacajawea (4145 m), and Fremont Peak (4190 m), enclose the basin on the east; the somewhat more subdued ridge from Henderson Peak in the north to Island Lake in the south, flanks the west side of the Basin. Valley walls are commonly 800 m high, and mantled with talus in places; valley floors are typically flat, and sparsely covered with glacial and mass-wasted material. A high proportion of the total area of the valley system consists of scoured bedrock surfaces.

During the Pleistocene valley glaciers extended through deep canyons into several intermontane basins forming large terminal moraines (Figure 5). Large lakes up to 250 m deep are found where drainage has been impounded by moraines.

Climate

Climatic data are summarized in Table 1 from stations along the eastern (Dubois and Lander), and western (Pinedale) flanks of the Range. The average annual temperature, mean seasonal extremes, and absolute maximum and minimum temperatures are calculated using mean winter and summer lapse rates determined in the Colorado Rockies (Barry and Chorley, 1970). This gives an average annual temperature for the Titcomb Basin of -3.5°C which is similar to values computed for the Grand Teton Glacier at 3200 m (Mahaney, 1975). The mean annual precipitation for the alpine zone is unknown, but must be significantly higher than the average snowfall reported for the Basin stations. The wind is predominantly northwest most of the year, and the windiest months of the year are from October through February.

The alpine area is typified by short and cool to cold summer temperatures, with frequent thunderstorm activity. Winter is cold and windy with frequent cloud cover. The subalpine and montane zones are cool in summer with frequent thunderstorms, while winter is cold and cloudy. The intermontane basins experience summer thunderstorm activity comparable to that of higher elevation, but with reduced effect as a result of higher temperatures and rates of evaporation. Winters in the basins are cold with high wind activity.

Precipitation regimes differ somewhat from the east to the west flanks of the Range (Baker, 1944). On the west slope, the precipitation pattern is uniform throughout the year, while on the east slope somewhat higher precipitation is received in May. The east slope is typified by a relatively higher summer rainfall than on the western slope.

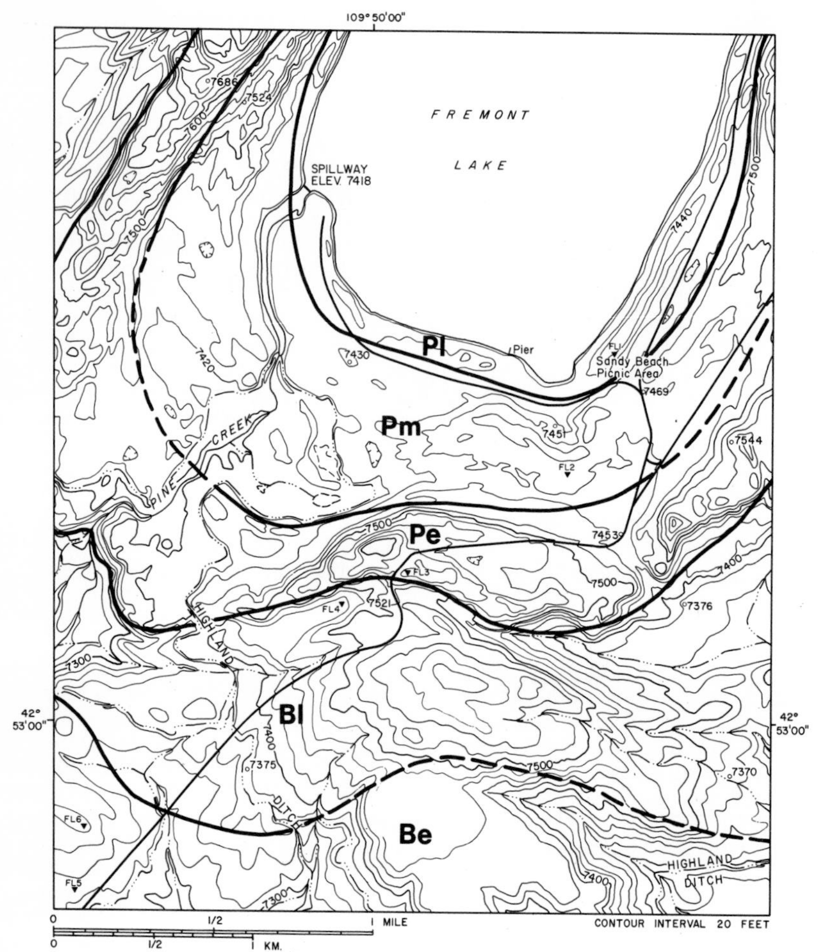

Figure 5 Glacial deposits at the Pinedale *Type Locality* (modified from Richmond 1965; 1974): late Pinedale (Pl), middle Pinedale (Pm), early Pinedale (Pe, late Bull Lake (Bl) and early Bull Lake (Be).

The Range is well-named as chinooks descend the east slope toward Lander and Dubois producing high winds. In winter months they often raise the temperature several degrees.

Vegetation

The alpine tundra, located above the timberline at 3000 m, is dominated by perennial sedges, grasses, and herbaceous plants. Vegetation complexes range from kobresia meadow, dryas, and sedge-grass, to willow-sedge stand types. The timberline zone is composed almost exclusively of white-bark pine *(Pinus albicaulis)*, limber pine *(P. flexilis)*,

Table 1. Climatic data for stations in Southwestern Wyoming.

Station	Location	Elevation m (ft)	Temperature					Mean number of days		Precipitation	
			Average °C (°F)	Mean January °C (°F)	Mean July °C (°F)	Absolute Minimum °C (°F)	Absolute Maximum °C (°F)	>90°	<32°	Average total cm (in.)	Average snowfall cm (in.)
Pinedale[a]	42°52'N, 109°51'W	2188 (7180)	1.8 (35.3)	-10.9 (12.4)	15.8 (60.5)	-20.1 (-4.2)	34.4 (94)	2	267	23.1 (9.11)	157 (61.8)
Dubois[a]	43°33'N, 109°37'W	2109 (6917)	4.7 (40.4)	-5.9 (21.3)	15.8 (60.5)	-42.2 (-44)	33.3 (92)	1	249	23.7 (9.35)	88.4 (34.8)
Lander[a]	42°49'N, 108°44'W	1695 (5563)	6.9 (44.4)	-5.4 (22.2)	21.6 (70.9)	-34.4 (-30)	38.3 (101)	22	187	34.5 (13.6)	238.3 (93.8)
Titcomb[b] Basin	43°07'N, 109°38'W	3200 (10,500)	-3.5 (25.7)	-15.1 (4.8)	9.6 (49.3)	-44.8 (-48.7)	28.2 (82.8)				

[a]Excerpted from U.S. Dept. Commerce (1965); data 1951-1960.

[b]Temperature determined from average lapse rates at Colorado Springs and Pikes Peak, Colo., where the mean winter lapse rate is 4.1°C/km, and the mean summer lapse rate is 6.2°C/km (Barry and Chorley, 1970, p. 42).

and subalpine fir *(Abies lasiocarpa)* (Reed, 1976). Much of the timberline vegetation is procumbent, growing in niches sheltered from the prevailing westerly winds.

Below the timberline a zone of subalpine forest of spruce *(Picea engelmannii)* and fir *(A. lasiocarpa)* dominates. The lower limit of the subalpine forest merges at 2750 m with the montane forests of lodgepole pine *(P. contorta)* that contain some aspen *(Populus tremuloides)* and Douglas Fir *(Pseudotsuga menziesii)*. Douglas Fir and aspen are more common species on crystalline rocks (Reed, 1971). Ponderosa pine *(P. ponderosa)*, so prevalent in the Southern Rocky Mountains, is curiously absent in the Wind Rivers. Below the lower limit of trees sagebrush *(Artemisia tridentata)* dominates.

METHODS

Soil morphological descriptions follow the Soil Survey Staff (1951, 1960), and Birkeland (1974), while the particle size analysis is based on the Westworth scale (Folk, 1968). Particle size data are determined from dry sieving for coarse grade-sizes (64 mm-63μ ; Day, 1965), and on sedimentation for fine grade-sizes ($<63\mu$; Bouyoucos, 1962). An air-dried oriented clay sample of the $<1.95\mu$ grade-size was X-rayed following procedures established by Whittig (1965). Clay minerals were analyzed on a Toshiba ADG-301H diffractometer with Ni-filtered CuKα radiation.

The soil pH was determined from a 1:1 soil paste by glass electrode. Extractable hydrogen was measured by the Triethanolamine method (Olsen and Dean, 1965); extractable bases and total cation exchange capacity by the ammonium acetate method of Peech, *et al.* (1947), and Schollenberger and Simon (1945). Total nitrogen was measured by the Kjeldahl method of Bremner (1965), and organic matter by the Walkley and Black (1934) method. Total soluble salts were determined by the electrical conductivity method outlined by Bower and Wilcox (1965). Free iron oxide was determined by extraction following procedures established by Mehra and Jackson (1960).

SURFICIAL GEOLOGY

Glacial and periglacial deposits in the Titcomb Basin and Fremont Lake areas are differentiated on the basis of topographic position, deposit morphology, weathering features (Tables 2 and 3), and lichen characteristics (Table 4). These relative age-dating methods have been defined by Birkeland (1973) and Mahaney (1973). Neoglacial deposits are separable into three advances, named from youngest to oldest, Gannett Peak (300-100 BP), Audubon (2,000-1,000 BP), and Early (5,000-3,000 BP).[1] Late-Pleistocene-Early Holocent deposits are separable as surficial units of Pinedale (~10,000 BP), and Bull Lake (~100,000 BP) glaciations.

---

[1] (See Footnote Next Page)

Table 2. Weathering data for sites in the Alpine area, Wind River Range, Wyoming.

Site	Age	Weathering ratios[a]			Weathering rinds			Weathering pits		
		%Fresh	%Wx	n	Average Maximum (mm)	Average Minimum (mm)	n	Average depth (mm)	Average width (mm)	n
TB15	Gannett Peak	100	nil	100	nil	nil	50	nil	nil	50
TB5	Audubon	100	nil	100	1.8	nil	50	5.5	38.1	100
TB6	Audubon	100	nil	100	2.9	nil	50	5.0	39.5	50
TB14	Early Neoglacial	100	nil	100	5.2	nil	50	7.3	41.7	50
TB1	Pinedale	98	2	100	26.2	1.1	25	12.1	120.8	100
TB2	Pinedale	97	3	100	...	...	...	17.1	67.2	100
TB3	Pinedale	99	1	100	29.9	1.6	25	11.2	19.8	50
TB4	Pinedale	98	2	100	15.1	1	25	10.8	43.6	50
TB8	Pinedale	99	1	100	15.1	2	25	...	...	...

[a]Weathered-fresh differentiation based on surface state of the boulder as generally unweathered (*e.g.* fresh) or weathered (Wx), cavernous and rotten.

(Footnote to Page 232)

[1]Temple Lake moraines at the type locality in the Wind River Range are referred to in the literature as both post- and pre-Altithermal (Birkeland, 1973; Currey, 1974). Since there is considerable controversy over the age of Temple Lake units, the name Temple Lake will not be used here. The earliest Neoglacial units are referred to as "Early."

Table 3. Weathering data for sites near Fremont Lake, Wind River Range, Wyoming.

Site	Age	Weathering ratios[a]			Weathering rinds			Weathering pits		
		%Fresh	%Wx	n	Average Maximum (mm)	Average Minimum (mm)	n	Average Depth (mm)	Average Width (mm)	n
FL1	Pinedale	98	2	100	12.6	2.4	50	12.0	41.6	100
FL2	Pinedale	92	8	100	10.3	0.8	50	11.8	55.5	100
FL3	Pinedale	91	9	100	14.7	2.3	50	16.9	40.0	100
FL4	Bull Lake	65	35	100	25.4	8.4	50	18.6	43.4	100
FL5	Bull Lake	68	32	100	18.1	4.5	50	19.7	50.6	50
FL6	Bull Lake	68	32	100	19.0	6.1	50	21.0	58.0	100

[a]Weathered-fresh differentiation based on surface state of the boulder as generally unweathered (*e.g.* fresh) or weathered (Wx), cavernous and rotten.

Table 4. Lichen characteristics[a] for Neoglacial deposits in the alpine zone, Wind River Range, Wyoming.

Site	Age	*Rhizocarpon geographicum*	*Lecanora thomsonii*	*Lecanora aspicilia*	*Lecidea atrobrunnea*	% Cover
TB15	Gannett Peak	15	5	nil	nil	1
TB5	Audubon	60	65	145	130	20-30
TB6	Audubon	55	75	123	110	15-25
TB14	Early Neoglacial	127	117	143	235	75+

[a]Lichen measurements of the maximum diameter in mm.

Gannett Peak Advance

Deposits of the youngest Neoglacial advance are sharp- and multiple-crested, steep-sided with slopes of 20-35°, bouldery, and nearly devoid of soil and vegetation cover. Located adjacent to the terminii of small cirque glaciers in the Upper Titcomb Basin, moraines of Gannett Peak age are often ice-cored, and lack patterned ground features. Weathering ratios indicate that boulders and cobbles are 100 percent fresh[1], and weathering rinds and pits are nil (Table 2).

The diameters of *Rhizocarpon geographicum* and *Lecanora thomsonii* are 15 and 7 mm, respectively, and lichen cover is 1 percent. Maximum diameters are somewhat smaller than those reported for Gannett Peak deposits in the Front Range, Colorado (Mahaney, 1974) (*cf.* 18 and 25 mm, respectively). This would indicate that either Gannett Peak moraines in the Wind River Range are somewhat younger than in the Front Range, or that lower rainfall in the Wind River Range, relative to the Front Range, has produced lower lichen growth velocities. While precipitation data are not available for the alpine zone in the Wind River Range, there seems to be a consensus among researchers that the Front Range is wetter than the Wind River Range (Birkeland and Shroba, 1974).

Audubon Advance

Deposits of Audubon[2] age lie adjacent to younger moraines of Gannett Peak age, and are generally not separable on the basis of position. Deposits are usually single-crested, steep-sided with slopes of 15-25°, and lack ice cores. Poorly sorted polygonal ground features reach 1.5m in diameter. Boulders and cobbles are largely unweathered (Table 2). Weathering ratios give 100 percent fresh stones; weathering rinds average from 2 to 3 mm, and 40 to 50 percent of stones do not have any rind; and weathering pits have a depth ranging from 5.0 to 5.5 mm, and widths of 38 to 39 mm.

Lichen measurements indicate that 15 to 20 percent of rock surfaces are covered with a number of crustose lichens. As shown in Table 4, maximum diameters of *R. geographicum* and *L. thomsonii* reach 65 and 75 mm, respectively. Moreover two new species -- *Lecanora aspicilia* and *Lecidea atrobrunnea* -- become common on rock surfaces. Maximum diameters of the four dominant crustose lichens are smaller (*cf.* Front Range, Colorado), a result apparently of differences in age or climate.

Early Neoglacial Advance

---

[1]A weathered boulder is defined as one where more than half the surface is weathered to a depth greater than the average crystal size (*e.g.* fretting ratio of Sharp, 1969).

[2]Deposits of Audubon age described here are correlative with those in the Teton Range described by Mahaney (1975); and in the Front Range (Mahaney, 1974).

Ground moraine formed during this advance is located ~1.5 km from cirque headwalls, and is morphologically less prominent when compared with younger deposits of Gannett Peak and Audubon age. Sorted stone polygons up to 3.0 m in diameter are common on deposit surfaces. Stones on Early Neoglacial surfaces have a state of weathering that is clearly different from younger deposits (Table 2). While it is not possible to separate deposits of Neoglacial age by use of weathering ratios, the data indicate that deposit differentiation can be achieved by use of weathering rind and weathering pit measurements. Stones on Early Neoglacial surfaces have rinds of ~5 mm thickness, and pits up to 7.3 mm deep and 39 mm wide. The data indicate that while rinds thicken and pits deepen with time, pit width does not change appreciably.

Lichen cover on the oldest Neoglacial deposits is approximately 75 percent, and maximum diameters of *R. geographicum* thalli reach 127 mm (Table 4). *R. geographicum* thalli are more prevalent than on younger deposits and exist along with *L. thomsonii*, *L. atrobrunnea* and *L. aspicilia*. The data indicate that, while the ratios of the maximum diameters of *R. geographicum* and *L. thomsonii* are most useful in separating deposits of the three Neoglacial advances, *L. aspicilia* is useful in differentiating Gannett Peak from Audubon deposits only; and *L. atrobrunnea*, although relatively slow in colonizing deposits, may be used to differentiate deposits of Early Neoglacial from Audubon age.

Pinedale Glaciation

Moraines of the Pinedale Glaciation enclose Fremont Lake (Figure 5) at the Pinedale Type Locality as defined by Blackwelder (1915). Lateral moraines thin-out at 2,600 m (8,500 ft), while ground moraine is fairly extensive in places to elevations above 2,750 m (9,000 ft). Deposits of Pinedale age are generally hummocky and bouldery with numerous undrained depressions and poorly integrated drainage.

Weathering ratios indicate slight differences in weathering between Pinedale deposits in the alpine (Table 2, TB1-TB4 sites), krummholz (Table 2; TB8 site), and sagebrush vegetation zones near Fremont Lake (Table 3; FL1-FL3 sites). The data indicate a slightly higher percentage of weathered stones near Fremont Lake suggesting that the Pinedale moraines in the lower basin are older than those in the alpine and subalpine vegetation zones. The moraines enclosing Fremont Lake have similar weathering characteristics (Table 3) suggesting that there is little difference in age between the early and late stades of the Pinedale Glaciation as defined by Richmond, (1965, 1974). However, it may be that weathering methods are not sensitive enough to differentiate deposits at the stadial level.

Bull Lake Glaciation

Deposits of Bull Lake age (Figure 5) have smooth, almost blanket-like morphology, with well-integrated drainage. Till of Bull Lake age occurs near the lower limit of glaciation, extending 2-4 km beyond the outer moraines of Pinedale age. Two stades of the Bull Lake Glaciation have been identified by Richmond (1974; 1976). However, the data in Table 4 indicate little difference between the outer (FL5 and FL6 sites) and inner (FL4) deposits of Bull Lake age. Weathering ratios, while quite distinct from that of the Pinedale deposits, show little difference between the outer (Table 3; FL5 and FL6 sites) and the inner Bull Lake (Table 3, site FL4) deposits. Weathering rinds, while thicker than on stones of Pinedale age, do not permit separation of two stades. An important aspect of the weathering rind data is that the average minimum rinds are higher on stones of Bull Lake than on Pinedale age. Weathering pit depth data permit differentiation between Pinedale and Bull Lake glaciations, but not between deposits of stadial rank identified on a morphostratigraphic basis (Richmond, 1965). Weathering pit width does not allow differentiation of deposits older than Pinedale age.

SOIL STRATIGRAPHY

Deposits in the glacial and periglacial succession can be separated on the basis of soil morphology, particle size, clay mineralogy, and soil chemistry. The state of the soil system varies with the age of the deposit and provides an important means of assessing relative age and assisting in deposit differentiation.

Soil Morphology

Deposits and soil profiles were investigated along a transect from the Titcomb Basin to Fremont Lake (Figure 1). Sixteen representative profiles are given in Appendices 1, 2 and 3 (sites are located on Figures 3, 4 and 5). Soil profiles are assigned to soil-stratigraphic units on the basis of stratigraphic position and morphological features which allow recognition in the field. Soil-stratigraphic units form, *in situ*, on deposits as pedologic processes work downward on fresh and unweathered parent materials (Cn horizons). Soil-stratigraphic units are assigned informal names coinciding with the age of the deposit on which they formed (*e.g.* use of the prefix "post" to avoid a proliferation of geologic names).

The soil properties most useful in differentiating deposits are depth of weathering, horizonation development, soil and parent material color[1], texture, structure, moist consistence, plasticity, and stickiness. For the most part

---

[1]Colors are given as moist (m) or dry (d) based on the Revised Standard Soil Color Charts (Oyama and Takehara, 1967).

soils of post-Gannett Peak age (Appendix 1) are not widespread; where present, they are very weakly developed with Cl/Cn profiles. Soil depths may reach 20 cm, and soil color is generally yellowish gray (2.5Y 5/1m) or brownish gray (10YR 6/1d) over a dark grayish yellow (2.5Y 5/2m) to light gray (10YR 7/2d) substratum. Field texture is generally bouldery pebbly coarse sandy loam. Soil structure is massive, and moist consistence is generally loose. Unconfined shear-strength (n=30), as measured by penetrometer averages 1.15 for all C horizons, and 1.75 for Cn horizons.

While Gannett Peak deposits in the Upper Titcomb Basin are generally devoid of soil cover, deposits of Audubon age contain a weakly developed Entisol with an Al/Cox/Cn horizon sequence (Appendix 1). A horizon thicknesses are 0-10 cm, while C horizons range from 40 to 60 cm in thickness. Total depth is approximately 60 cm. Soil colors in the surface horizons reflect organic matter accumulation with hues that are generally brownish-black (10YR 3/2m) to grayish yellow brown (10YR 4/2m). Cox horizons are dull yellow orange (10YR 6/3m), grading downward to unweathered materials that are grayish yellow brown (10YR 6/2m) to brownish gray (10YR 6/1m). Textures in the A horizons are usually pebbly coarse sandy loam, grading downward to bouldery pebbly coarse sandy loam or loamy sand. Surface horizons have weak granular structure, while C horizons are massive. Moist consistence in the A horizons is generally friable to very friable, while C horizons are loose. Unconfined shear-strength (n=20) is 2.0 kg/cm^2 in the A horizons; C horizons are generally too bouldery to determine shear-strength.

Post-Early Neoglacial soils (Appendix 1) have an O horizon component ranging from 1 to 2 cm in thickness. A horizons commonly have thicknesses of 10 cm, similar to post-Audubon soils, but with important color and structural differences resulting from increases in organic matter and clay content. B horizons are approximately 20 cm thick, while oxidized C horizons reach 50 cm. The total depth reaches 80 cm. Color in the O horizon is brownish black (10YR 2/2m). A horizon color is generally brownish black (10YR 2/3m), B horizons brown (10YR 4/4m) to dull yellowish brown (10YR 5/4m), while C horizons have mottled olive brown (2.5Y 4/3m) to light yellow (2.5Y 7/3d) hues. Cn horizons are grayish olive (5Y 5/3m). Textures are pebbly sandy loam in the A horizons, pebbly coarse sandy loam in the B horizons, pebbly loamy sand and pebbly coarse sandy loam in the C horizons. Structures are granular in the A horizons, weak blocky in the B horizons, and massive in the C horizons. Moist consistence is friable in the A and B horizons, and loose in the C horizons. Unconfined shear-strengths (n=50) are 2.55 kg/cm^2 in the A horizon, 2.98 kg/cm^2 in the B horizon, 2.85 kg/cm^2 in the C horizon, and 4.45 kg/cm^2 in the Cn horizon.

Post-Pinedale soils (Appendix 2) have variable horizon sequences depending on location in the alpine, krummholz (timberline ecotone), subalpine, montane, and/or upper

Sonoran vegetation associations. In the alpine and subalpine zones the post-Pinedale soil is an Alfisol with an 01/A11/A12/B2irh/IICox/IICn profile. Total depth is approximately 100 cm; A horizons are 10-15 cm, B horizons are 18 to 38 cm thick, and C horizons are approximately 50 cm. Soil colors in the surface 0 horizons range from black (10YR 2/1m) to brownish black (10YR 2/2m). A horizons are brownish black (10YR 2/3m, 3/2m), dark brown (10YR 2/3m, 3/3m); lower A12 horizons tend to be lighter in color by one or two chroma as a result of leaching. B horizons are generally dull yellowish brown (10YR 5/4m, 4/3m), dark brown (10YR 3/3m), and dull yellow orange (10YR 7/2m). The Cox horizons are dull yellow orange (10YR 6/3m, 6/4m), dull yellowish brown (10YR 5/4m) and brown (10YR 4/6m). Cn horizons have yellowish brown (2.5Y 5/3m) and olive brown (2.5Y 4/3m) colors.

Textures in the soils are sandy loam or loamy sand throughout; B horizons are generally defined on the basis of color, the increase in clay content being insufficient for a textural B horizon. C horizons generally have a higher percentage of cobbles and boulders, and a coarse sandy loam or loamy sand texture. Structure is better developed in post-Pinedale soils than in younger soils of Neoglacial age. Generally granular structure prevails in the A horizons; B horizons tend to have weak blocky structure, and C horizons are massive.

Moist consistence ranges from friable and very firm in the A horizons, and friable and very friable in the B horizons, to loose and very friable in the C horizons. Unconfined shear-strength in the A horizons (n=50) is 1.85 kg/cm^2, 2.47 kg/cm^2 (n=80) in the B horizons, 1.8 kg/cm^2 (n=40) in the C horizons, and 1.7 kg/cm^2 (n=50) in the parent material.

Under sagebrush in nearby intermontane basins, post-Pinedale soils (Appendix 3) have A/B2t/IIC/IICn profiles, ~80 cm total depth, where A horizons are 10-23 cm, B horizons 15 to 26 cm, C horizons 38 to 50 cm thick. A horizons are generally dull yellowish brown (10YR 4/3m) and brown (10YR 4/4m), B horizons range from light gray (2.5Y 7/1m) to dull yellowish brown (10YR 5/3m), and grayish yellow brown (10YR 6/2m). C horizons are dull yellow (2.5Y 6/3m) or dull yellow orange (10YR 6/3m), and Cn horizons are light gray (10YR 8/1m), grayish yellow (2.5Y 6/2m) or grayish yellow brown (10YR 6/2m). Textures vary from sandy loam to loam throughout with a tendency for a higher percentage of boulders and cobbles to appear in the subsurfaces of the profiles. Structures are generally friable in the A horizons, weak to moderate blocky in the B horizons, and massive to blocky in the C horizons. A horizons tend to have friable moist consistence, while B horizons range from friable to firm, and C horizons very friable to firm. Shear-strengths in the A horizons (n=30) are 2.5 kg/cm^2, B horizons (n=30) are 4.0 kg/cm^2, C horizons (n=50) are 3.65 kg/cm^2, and 3 kg/cm^2 (n=50) in the parent material (Cn).

Post-Bull Lake soils (Appendix 3) are more highly weathered and indurated than younger soils. A horizons are 5 to 15 cm, B horizons are 25 to 45 cm, and C horizons are 44 to 84 cm thick; total depth of the weathered zone ranges from 114 to 135 cm. In the montane forest zone these profiles generally reach 150+ cm depth. Color in the A horizons is commonly dark brown (10YR 3/4m) or dull yellowish brown (10YR 4/3m), B horizons are brown (7.5YR 4/3m), dull yellowish brown (10YR 5/3m), grayish yellow brown (10YR 5/2m), and dull yellow orange (10YR 6/3m). C horizons are dull yellow orange (10YR 6/3m), and grayish olive (5Y 6/2m), while Cn horizons are dull yellow (2.5Y 6/3m) and dull yellow orange (10YR 6/3m). Dry colors give somewhat better differentiation of soil and parent material. Textures vary from pebbly sandy loam to pebbly fine sandy loam in the A horizons, and pebbly sandy clay loam and loam in the B horizons, to pebbly loam and pebbly sandy clay loam in C horizons; Cn horizons range from pebbly coarse sandy loam to pebbly sandy clay loam. A horizons have well-developed granular structure, B horizons typically have well-developed blocky structure; C horizons are blocky to massive as are the Cn horizons. Moist consistence in the A horizons is friable to firm, firm to very firm in the B horizons, very friable to firm in C horizons, and firm to friable in the Cn horizons. Shear-strength in the A horizons (n=30) is 2.3 kg/cm^2, B horizons (n=90) are 4.4 kg/cm^2, C horizons (n=50) are 4.5+, and Cn horizons (n=50) are 4.5+.

Particle Size

Deposits of Gannett Peak age consist predominantly of an open-network of boulders set in a matrix of finer cobbles and pebbles. Soils are discontinuous and sparsely distributed on these moraine systems restricted to those areas where fines have collected. The <2 mm fractions are dominated by sand with smaller amounts of silt, and still lower amounts of clay (Table 5). The amount of clay is either similar to or less than that found in older soils of Audubon and Early Neoglacial age. Similarly, in post-Audubon profiles pebbles tend to group in large and medium grade-sizes (Table 5). The coarse grade-sizes are similar to younger post-Gannett Peak soils, while the granulometric composition of the <2 mm fractions indicates slightly higher amounts of clay without any apparent leaching effect.

Granulometric data for the Early Neoglacial soils indicate a slight shift to the right (Table 5) for coarse size material (>2 mm). The <2 mm fractions have similar ratios of sand-silt-clay with a tendency for higher clay and silt in the A horizons. Illuviation is not a widespread process in these late-Holocene profiles as it is in the Southern Rocky Mountains (Mahaney, 1974, 1976).

Particle-size data for post-Pinedale soils in the Wind River alpine and upper timberline zones are given in Table 6. Pebbles tend to concentrate in the medium to small grade-sizes, providing a significant overall texture break

Table 5. Particle-size distribution[a] for the soil horizons in Appendix 1.

Sample[b]	Depth (cm)	>2 mm Grade Sizes					<2 mm Grade Sizes		
		Very Large 64–32 mm (-6/-5ø)	Large 32–16 mm (-5/-4ø)	Medium 16–8 mm (-4/-3ø)	Small 8–4 mm (-3/-2ø)	Granule 4–2 mm (-2/-1ø)	Sand % (<2mm–63μ)	Silt % (63–4μ)	Clay % (<4μ)
TB15-C1	0-20	11.6	17.3	10.2	8.1	7.8	71.3	24.9	3.8
TB15-Cn	20+	16.9	20.1	7.5	7.0	6.7	69.6	24.6	5.8
TB5-A1	0-10	6.4	8.5	6.2	6.1	9.4	64.6	28.6	6.8
TB5-Cox	10-51	6.6	20.3	11.4	6.7	6.7	70.9	26.1	3.0
TB5-Cn	51+	8.3	22.5	16.0	11.6	9.1	77.4	20.6	2.0
TB6-A1	0-10	4.3	17.2	7.6	5.1	10.8	85.2	11.8	3.0
TB6-Cox	10-71	8.1	15.5	15.3	11.9	11.9	81.9	15.3	2.8
TB6-Cn	71+	11.8	15.7	14.9	9.6	9.1	82.2	14.8	3.0
TB14-A1	0-9	4.0	6.0	6.5	8.4	15.0	67.9	26.1	6.0
TB14-B2ir	9-18	2.7	14.8	5.8	6.9	9.0	72.7	23.3	4.0
TB14-B3	18-30	--	20.2	7.9	7.5	7.9	70.5	24.0	5.5
TB14-Cox	30-79	6.0	15.3	12.2	11.5	12.0	79.8	17.4	2.8
TB14-Cn	79+	--	11.9	10.5	9.2	9.3	70.9	24.1	5.0

[a]The data are given in weight-percent of dry mineral matter; coarse particle sizes (64 mm- 63μ) determined by dry sieving; fine particle sizes (63 - 1.95μ) determined by hydrometer; -- nil.

[b]Soil pit locations (e.g. TB5) are on Figure 3.

Table 6. Particle-size distribution[a] for the soil horizons in Appendix 2.

Sample[b]	Depth (cm)	>2 mm Grade Sizes					<2 mm Grade Sizes		
		Very Large 64-32 mm (-6/-5ø)	Large 32-16 mm (-5/-4ø)	Medium 16-8 mm (-4/-3ø)	Small 8-4 mm (-3/-2ø)	Granule 4-2 mm (-2/-1ø)	Sand % (2mm-63μ)	Silt % (63 - 4μ)	Clay % (<4μ)
TB1-A1	0-13	—	—	0.3	1.9	9.8	82.0	14.0	4.0
TB1-B1	13-24	—	—	0.3	0.8	2.5	65.2	31.6	3.2
TB1-B21r	24-53	—	—	—	0.9	3.3	68.4	28.8	2.8
TB1-B3	53-69	—	—	—	2.4	4.0	57.4	39.7	2.9
TB1-Cox	69-114	—	—	1.0	3.6	6.5	73.0	25.0	2.0
TB1-Cn	114+	—	—	2.6	1.8	2.6	87.6	10.4	2.0
TB2-A11	0-6	—	—	0.1	1.2	5.7	75.3	19.7	5.0
TB2-A12	6-10	—	8.7	1.3	1.5	3.5	58.4	29.6	12.0
TB2-B1	10-18	—	10.1	7.4	4.6	4.7	66.4	28.8	4.8
TB2-B21rh	18-28	4.1	6.3	5.6	3.5	4.3	66.3	24.7	9.0
TB2-Cox	28-86	—	11.6	9.3	10.6	10.2	79.0	20.0	1.0
TB2-Cn	86+	—	4.4	4.8	9.2	12.3	95.2	2.8	2.0
TB3-A11	0-8	—	2.0	1.9	2.7	11.7	79.2	18.0	2.8
TB3-A12	8-14	—	0.6	1.9	7.6	16.8	84.2	10.8	5.0
TB3-B1	14-25	—	2.2	3.6	4.7	10.8	84.5	11.5	4.0
TB3-B21r	25-46	—	2.0	5.3	7.2	10.9	82.4	13.6	4.0
TB3-Cox	46-109	—	0.8	3.0	3.8	7.0	81.6	16.4	2.0
TB3-C1n	109-142	—	1.0	2.1	3.5	5.9	79.2	17.8	3.0
TB3-C2n	142+	—	3.6	2.9	3.8	6.7	79.5	17.5	3.0
TB4-A11	0-10	10.6	0.6	0.8	1.7	2.4	62.5	32.7	4.8
TB4-A12	10-20	—	—	1.5	1.4	3.1	55.5	37.7	6.8
TB4-B21r	20-38	—	2.4	4.6	4.8	8.6	69.3	26.7	4.0
TB4-Cox	38-127	—	0.8	2.9	4.0	6.1	58.0	29.0	13.0
TB4-Cn	127+	6.6	2.4	2.5	6.3	9.5	66.7	28.3	5.0
TB8-A11	0-6	11.4	30.0	36.0	37.9	41.0	60.6	28.4	11.0
TB8-A12	6-13	25.8	32.6	36.6	39.2	42.5	63.9	26.1	10.0
TB8-B21r	13-36	25.1	38.2	45.6	51.8	58.3	74.8	19.2	6.0
TB8-C1ox	36-53	18.7	33.3	45.7	55.0	62.7	81.2	15.8	3.0
TB8-C2ox	53-89	—	6.4	11.3	18.3	27.0	80.3	15.7	4.0
TB8-Cn	89+	21.4	34.6	40.3	46.3	51.4	74.5	19.7	5.8
TB9-A1	0-8	3.5	6.5	8.0	10.2	13.9	82.0	12.0	6.0
TB9-B21r	8-15	—	4.9	7.7	6.2	9.0	90.5	5.5	4.0
TB9-Cox	15-51	—	1.1	1.0	2.6	10.1	83.6	12.4	4.0
TB9-Ab	51+	—	1.9	3.8	4.2	5.3	58.9	25.6	15.5

[a]The data are given in weight-percent of dry mineral matter; coarse particle sizes (64 mm- 63μ) determined by dry sieving; fine particle sizes (63 - 1.95μ) determined by hydrometer; — nil.

[b]Soil pit locations (e.g. TB8) are on Figure 4.

when compared with the younger Neoglacial soils. In the <2 mm fractions, sand is either similar to or higher than younger Neoglacial soils. On the whole, silt tends to dominate in the A and B horizons suggesting the possibility of eolian deposition. The tendency for clay to increase slightly with depth, and the presence of clay skins on peds in B horizons, suggests a low leaching effect.

Post-Pinedale soils in adjacent intermontane basins (Table 7) are dominated by medium to small pebbles and granules. Sand dominates in the <2 mm fractions, silt is highest in the A and B horizons, and is usually higher than in alpine soils of similar age. Clay is higher in the B horizons as a result of illuviation. Post-Bull Lake soils have fewer pebbles and granules. In the <2 mm fractions sand is similar to younger post-Pinedale soils, while silt is noticeably less, and clay is generally higher. Unlike post-Pinedale soils, silt is not concentrated in the A + B horizons of the post-Bull Lake profiles. When compared with percentages in the Cn horizons, silt in the A horizons of post-Bull Lake soils is generally 3 to 4 percent higher, a situation which may result from original variations of texture in the parent material as well as from loess input.

Table 7. Particle-size distribution[a] for the soil horizons in Appendix 3.

Sample[b]	Depth (cm)	>2 mm Grade Sizes					<2 mm Grade Sizes		
		Very Large 64-32 mm (-6/-5ø)	Large 32-16 mm (-5/-4ø)	Medium 16-8 mm (-4/-3ø)	Small 8-4 mm (-3/-2ø)	Granule 4-2 mm (-2/-1ø)	Sand % (2mm-63μ)	Silt % (63 - 4μ)	Clay % (<4μ)
FL1-A1	0-23	—	5.5	1.5	1.5	2.7	54.9	29.6	15.5
FL1-B2t	23-46	—	0.3	0.1	1.2	2.4	30.5	46.0	23.5
FL1-C1	46-56	—	1.9	1.5	1.5	3.1	46.9	34.1	19.0
FL1-C2	56-91	—	0.6	0.6	1.5	3.5	39.3	36.7	24.0
FL1-Cn	91+	—	0.3	1.1	1.7	2.9	49.8	33.7	16.5
FL2-A1	0-10	—	3.3	5.3	3.2	3.9	69.5	19.0	11.5
FL2-B2t	10-36	17.4	4.9	1.9	3.2	5.6	61.5	25.0	13.5
FL2-C1	36-58	—	4.1	2.3	4.9	7.8	64.1	25.4	10.5
FL2-C2	58-86	—	4.6	4.1	6.9	10.8	74.7	17.8	7.5
FL2-Cn	86+	—	2.2	1.8	3.8	6.8	69.8	22.2	8.0
FL3-A1	0-13	7.9	13.4	3.8	3.1	4.3	55.5	33.0	11.5
FL3-B2t	13-28	—	8.1	8.4	8.4	10.1	61.2	25.8	13.0
FL3-C	28-66	—	7.2	6.4	4.1	7.6	68.5	23.0	8.5
FL3-Cn	66+	11.3	11.9	6.4	5.9	7.2	62.9	27.1	10.0
FL4-A1	0-13	—	0.8	1.3	3.6	7.8	58.7	21.3	20.0
FL4-B21t	13-23	—	0.8	0.7	2.4	6.5	52.7	18.8	28.5
FL4-B22t	23-38	—	1.7	1.6	3.5	4.9	56.0	16.5	27.5
FL4-C1	38-74	—	1.1	2.3	3.5	5.1	55.2	17.8	27.0
FL4-C2	74-114	—	—	1.9	2.9	4.2	54.2	21.8	24.0
FL4-Cn	114+	—	2.0	1.9	2.7	3.8	59.0	19.5	21.5
FL5-A1	0-15	6.6	8.3	7.5	5.7	8.0	56.9	23.1	20.0
FL5-B21t	15-28	15.7	2.6	1.9	2.4	3.7	52.5	22.5	25.0
FL5-B22t	28-51	—	0.6	0.9	2.9	5.9	49.4	24.6	26.0
F15-B23t	51-81	—	0.5	2.4	3.4	5.2	50.4	23.6	26.0
FL5-C	81-125	—	3.2	3.8	5.1	6.7	59.5	19.0	21.5
FL5-Cn	125+	—	2.0	1.2	2.4	4.6	50.5	24.5	25.0
FL6-A1	0-5	—	0.7	2.1	6.0	15.5	72.2	19.8	8.0
FL6-B21t	5-15	—	2.9	4.3	7.6	14.6	52.3	25.7	22.0
FL6-B22t	15-24	—	—	1.0	3.9	10.1	50.8	24.7	24.5
FL6-B23t	24-33	—	1.3	2.9	6.0	10.5	50.2	27.8	22.0
FL6-B24t	33-51	—	—	1.0	3.8	9.0	50.4	28.6	21.0
FL6-C1	51-89	3.5	4.7	5.7	8.1	12.0	41.6	34.4	24.0
FL6-C2	89-135	3.1	4.2	6.4	9.6	15.0	53.7	30.3	16.0
FL6-Cn	135+	—	4.1	6.9	11.6	17.12	73.4	16.6	10.0

[a]The data are given in weight-percent of dry mineral matter; coarse particle sizes (64 mm - 63μ) determined by dry sieving; fine particle sizes (63 - 1.95μ) determined by hydrometer; — nil.

[b]Soil pit locations (e.g. FL1) are on Figure 5.

Analysis of particle size curves in surface (A) and subsurface (B and C) horizons indicates that there is a general fining sequence in Holocene and late-Pleistocene soils. Mean Ø increases in an orderly fashion as follows: 2.1 Ø in post-Audubon surface horizons, 2.93 Ø in post-Early Neoglacial horizons, 3.06 Ø in post-Pinedale alpine soils, 3.76 Ø in post-Pinedale basin profiles, and 4.03 Ø in post-Bull Lake horizons. In subsurface horizons the sequence is 2.31 Ø in post-Audubon, 2.70 Ø in post-Early Neoglacial, 2.77 Ø in post-Pinedale (alpine), 4.23 Ø post-Pinedale (basin), and 5.35 Ø in post-Bull Lake (basin). The data suggest that the frequency distribution center of gravity shifts to higher phi values with time, indicating a tendency for finer material to occur with increasing age.

Clay Mineralogy

The clay minerals of several profiles were analyzed by X-ray diffraction (Tables 8, 9 and 10). The variation in clay minerals between different soil profiles indicates their usefulness in differentiating soil-stratigraphic units. Clay minerals present in post-Gannett Peak and post-Audubon soils begin to show a pattern of dispersion in post-Early Neoglacial soils. Analysis of relative mineral abundance shows that kaolinite, chlorite, and illite dominate as secondary minerals, while quartz dominates as a primary mineral. In post-Early Neoglacial soils mixed-layer illite-montmorillonite, vermiculite and montmorillonite dominate along with smaller amounts of kaolinite, illite, and chlorite. The percentage of quartz is reduced somewhat in Early Neoglacial soils with no appreciable change in feldspar. The data indicate that illite decreases with time, weathering to form other 2:1 and 2:1:1 clay minerals such as illite-montmorillonite, montmorillonite, vermiculite, and chlorite. Furthermore the concentration of 2:1 and 2:1:1 clays in the B horizons indicates at least a moderate amount of weathering and leaching which assists in redistribution of relative amounts of Si^{+4} and Al^{+3}. The overall reduction in quartz in the post-Early Neoglacial soils tends to substantiate this hypothesis.

The data for post-Pinedale soils in the alpine and upper timberline zone indicate wide variation in the kinds of secondary and primary minerals present. In the 1:1 clay mineral suite kaolinite dominates over halloysite which occurs only infrequently. In the 2:1 clay mineral suite vermiculite is more common, especially in the lower sections of profiles. In contrast with the Early Neoglacial soils, montmorillonite is relatively unimportant, a factor attributed to a longer leaching period.

Post-Pinedale soils in the intermontane basins have trace to small amounts of 1:1 clay minerals, small to moderate amount of 2:1 clays, and small amounts of 2:1:1 clays. Quartz is generally present in abundance, while feldspar is in small to moderate amounts. Post-Bull Lake soils differ from younger post-Pinedale soils in that kaolinite is replaced by halloysite in trace to small amounts, and 2:1 clays such as vermiculite and montmorillonite are much more prevalent, especially in the B horizons.

Soil Chemistry

Soil pH drops from 5.8 in late-Neoglacial soils to 5.1 in the surface horizons of post-Early Neoglacial soils indicating that it takes 3000 years to achieve steady-state. This is a departure from the situation in the Front Range, Colorado, where pH in Neoglacial soils approaches steady-state in post-Audubon soils within 1000 years (Mahaney, 1974, 1976). Extractable cations in Neoglacial soils show the effects of base recycling by plants where alkaline earths and alkali metals are higher in surface horizons. In effect, the overall low amounts of extractable cations

Table 8. Mineralogy of the $< 2\mu$ size for the soil horizons in Appendix 1.

Sample[b]	Depth (cm)	Mineralogy[a]								
		Halloysite	Kaolinite	Illite	Mixed-layer Illite-Montmorillonite	Chlorite	Vermiculite	Montmorillonite	Quartz	Feldspar
TB15-C1	0-20	--	x	x	tr	x	?	tr	xx	x
TB15-Cn	20+	--	tr	xx	tr	x	--	tr	xx	x
TB5-A1	0-10	--	x	x	tr	tr	x	tr	xx	tr
TB5-Cox	10-51	--	--	tr	tr	x	?	--	xxx	x
TB5-Cn	51+	--	tr	tr	tr	tr	--	--	xxx	x
TB6-A1	0-10	--	tr	tr	--	tr	--	--	xxx	x
TB6-Cox	10-71	--	tr	tr	--	tr	--	--	xxx	x
TB6-Cn	71+	--	--	tr	--	tr	--	--	xxx	x
TB14-A1	0-9	--	tr	tr	tr	--	--	--	xx	x
TB14-B2ir	9-18	--	x	tr	xx	x	xx	xx	xx	x
TB14-B3	18-30	--	tr	x	xx	x	xx	xx	xx	x
TB14-Cox	30-79	--	tr	x	x	tr	x	x	xx	x
TB14-Cn	79+	--	tr	x	x	tr	x	x	xxx	x

[a]Mineral abundance is based on peak height: -- no detection; minor amount (tr); small amount (x); medium amount (xx); abundant (xxx).

[b]Soil pits (*e.g.* TB15) are located on Figure 3.

Table 9. Mineralogy of the <2µ size for the soil horizons in Appendix 2.

Sample[b]	Depth (cm)	Mineralogy[a]								
		Halloysite	Kaolinite	Illite	Mixed-layer Illite-Montmorillonite	Chlorite	Vermiculite	Montmorillonite	Quartz	Feldspar
TB1-A1	0-13	—	tr	x	tr	—	tr	—	xx	tr
TB1-B1	13-24	—	—	—	—	tr	—	—	xx	x
TB1-B21r	24-53	—	—	—	—	—	—	—	xxx	x
TB1-B2?	53-69	—	tr	tr	—	—	—	—	xxx	x
TB1-Cox	69-114	—	—	—	—	—	—	—	xxx	x
TB1-Cn	114+	—	—	—	—	—	—	—	xxx	x
TB2-A11	0-6	—	tr	x	x	tr	x	tr	xx	x
TB2-A12	6-10	—	x	x	xx	—	xx	—	xx	x
TB2-B1	10-18	—	x	tr	tr	—	x	—	xx	x
TB2-B21rh	18-28	—	x	tr	tr	tr	xx	—	xx	x
TB2-Cox	28-86	—	tr	—	—	—	—	—	xxx	x
TB2-Cn	86+	—	—	—	—	—	—	—	xxx	x
TB3-A11	0-8	—	—	tr	—	—	tr	—	xx	x
TB3-A12	8-14	—	tr	tr	tr	tr	tr	—	xxx	x
TB3-B1	14-25	—	tr	tr	—	tr	tr	—	xxx	x
TB3-B21r	25-46	—	tr	tr	tr	tr	x	—	xxx	x
TB3-Cox	46-109	—	x	tr	tr	tr	x	—	xx	x
TB3-C1n	109-142	—	tr	tr	tr	tr	tr	—	xx	x
TB3-C2n	142+	—	tr	tr	tr	tr	x	—	xxx	x
TB4-A11	0-10	—	tr	tr	tr	—	x	—	xxx	x
TB4-A12	10-20	—	x	tr	tr	x	xx	—	xx	x
TB4-B21r	20-38	—	x	tr	x	x	xx	—	xxx	x
TB4-Cox	38-127	—	x	tr	tr	tr	xxx	tr	xx	—
TB4-Cn	127+	—	x	tr	x	xx	—	—	xx	xx
TB8-A11	0-6	—	tr	tr	tr	tr	—	—	xx	tr
TB8-A12	6-13	—	tr	tr	tr	—	tr	—	xx	tr
TB8-B21r	13-36	—	tr	tr	tr	tr	x	tr	xxx	x
TB8-C1ox	36-53	—	tr	—	?	—	tr	—	xxx	x
TB8-C2ox	53-89	—	—	tr	tr	tr	x	—	xxx	x
TB8-Cn	89+	tr	—	tr	tr	tr	tr	—	xxx	x
TB9-A1	0-8	—	tr	—	—	—	tr	—	xxx	x
TB9-B21r	8-15	—	tr	—	tr	tr	x	—	xxx	x
TB9-Cox	15-51	—	—	tr	tr	tr	x	—	xxx	x
TB9-Ab	51+	—	x	tr	tr	x	xx	—	xx	tr

[a]Mineral abundance is based on peak height: — no detection; minor amount (tr); small amount (x); medium amount (xx); abundant (xxx).

[b]Soil pits (e.g. TB1) are on Figure 4.

parallels the low amount of clay in Neoglacial soils (Table 7). Total cation exchange capacity ranges from 4 in post-Gannett Peak soils to between 8 and 12 in the surface horizons of post-Audubon soils, and 15.6 in post-Early Neoglacial soils. Moderate to low leaching effects leave a relatively high percentage of bases and salts in most profiles. Organic matter which is initially ~0.5 increases to between 6 and 13 percent in older Neoglacial soils. The data indicate that while C/N ratios build up rather rapidly, Fe_2O_3 is generally $<1.0\%$ in most profiles.

Post-Pinedale soils in the alpine and upper timberline zone have soil reactions that are generally more acidic with depth indicating that at least moderate amounts of H^+ ions are translocated downward from the surface. The reaction is generally medium acid in the surface and strongly acid at depth. The effects of base recycling by plants is

Table 10. Mineralogy of the <2μ size for the soil horizons in Appendix 3.

Sample[b]	Depth (cm)	Mineralogy[a]								
		Halloysite	Kaolinite	Illite	Mixed-layer Illite-Montmorillonite	Chlorite	Vermiculite	Montmorillonite	Quartz	Feldspar
FL1-A1	0–23	—	tr	x	—	?	—	—	xx	x
FL1-B2t	23–46	—	x	xx	—	x	x	xx	xxx	xx
FL1-C1	46–56	—	x	xx	—	x	x	x	xx	xx
FL1-C2	56–91	—	—	x	tr	x	x	x	x	x
FL1-Cn	91+	—	x	xx	—	x	x	x	xx	x
FL2-A1	0–10	—	tr	tr	—	—	—	—	xxx	tr
FL2-B2t	10–36	—	x	xxx	tr	—	—	x	xxx	x
FL2-C1	36–58	—	x	xxx	—	—	tr	xxx	xxx	x
FL2-C2	58–86	—	tr	tr	—	—	tr	x	xxx	x
FL2-Cn	86+	—	tr	tr	—	—	—	x	xxx	x
FL3-A1	0–13	—	tr	x	x	—	—	—	xxx	tr
FL3-B2t	13–28	—	tr	xx	—	—	tr	tr	xxx	x
FL3-C	28–66	—	tr	x	—	tr	—	tr	xxx	x
FL3-Cn	66+	—	—	tr	xx	tr	—	—	xx	x
FL4-A1	0–13	—	—	?	—	—	—	—	tr	—
FL4-B21t	13–23	x	x	xx	—	—	—	xxx	xx	x
FL4-B22t	23–38	x	x	x	—	—	—	xxx	xx	xx
FL4-C1	38–74	—	—	tr	—	—	—	xxx	xx	tr
FL4-C2	74–114	tr	tr	x	—	—	xxx	xx	x	xx
FL4-Cn	114+	tr	x	x	—	—	—	xxx	xxx	tr
FL5-A1	0–15	—	—	xx	—	—	—	xx	xxx	tr
FL5-B21t	15–28	—	—	xxx	—	—	xx	x	xxx	x
FL5-B22t	28–71	—	—	xxx	—	—	xx	xxx	xxx	xx
FL5-B23t	71–122	tr	—	tr	—	—	x	xxx	xx	xx
FL5-C	122–178	—	—	xx	—	—	x	xxx	x	tr
FL5-Cn	178+	—	—	tr	—	—	x	—	x	x
FL6-A1	0–5	—	tr	x	—	—	—	x	xxx	tr
FL6-B21t	5–15	—	tr	x	tr	—	tr	xx	xx	xx
FL6-B22t	15–24	—	—	xxx	—	—	xxx	xxx	xx	xx
FL6-B23t	24–33	—	—	—	—	—	xx	xxx	xx	xx
FL6-B24t	33–51	x	tr	tr	—	—	x	xxx	xx	xx
FL6-C1	51–89	tr	—	x	—	—	x	xxx	x	xx
FL6-C2	89–135	x	tr	xx	—	—	xx	xxx	xx	xx
FL6-Cn	135+	—	tr	xx	—	—	xx	xxx	xxx	xx

[a]Mineral abundance is based on peak height: — no detection; minor amount (tr); small amount (x); medium amount (xx); abundant (xxx).

[b]Soil pit locations (e.g. FL1) are on Figure 5.

evident to somewhat greater depth than is the case in younger Neoglacial soils (Table 12). Total cation exchange capacity in post-Pinedale soils ranges from 11 to 27 (meq/100 g) indicating greater amounts of clay and humus. The tendency for base saturation to decrease in the B horizon is probably the result of low to moderate leaching power. Organic matter, organic carbon and nitrogen are similar to younger Neoglacial soils indicating that steady-state is achieved by Early-Neoglacial time. Iron oxide development is generally higher than in Neoglacial soils, and compares favorably with data for soils of similar age in the alpine zone of the Front Range, Colorado (Mahaney and Fahey, 1976).

Post-Pinedale soils in intermontane basins (Table 13) have a surface pH that is slightly acid to neutral and subsurface pH's that are neutral to very strongly alkaline. This trend is followed by decreasing H^+ ion concentration

Table 11. Selected chemical properties[a] of the <2 mm fraction of the material in the soil horizons of Appendix 1.

Sample[b]	Depth cm	Extractable Cations (meq/100g)						CEC meq/100g	Base Saturation %	Salts mmhos/cm 25°C	Organic matter	Organic carbon	N %	Fe_2O_3 %	Carbon/nitrogen
		pH	Exch. H^+	Na^+	K^+	Ca^{+2}	Mg^{+2}								
TB15-C1	0–20	5.7	2.4	0.1	0.1	1.3	0.3	4.1	41	3.8	0.3	0.17	0.008	...	21
TB15-Cn	20+	5.8	2.6	0.1	0.1	1.4	0.4	4.5	42	1.6	0.2	0.12	0.006	...	20
TB5-A1	0–10	5.3	11.9	.09	.19	2.4	.46	8.6	37	.32	6.0	3.48	.18	1.05	19
TB5-Cox	10–51	5.5	3.7	.09	.04	.4	.10	1.95	32	.15	1.32	.77	.02	.40	39
TB5-Cn	51+	5.9	1.9	.09	.05	.7	.16	1.52	66	.19	.57	.33	.01	.28	33
TB6-A1	0–10	5.8	7.5	.07	.40	6.1	.89	12.30	61	.22	13.17	7.64	.30	.71	25
TB6-Cox	10–71	5.6	2.4	.07	.08	1.0	.25	1.85	76	.17	.10	.06	.02	.37	3
TB6-Cn	71+	5.6	1.1	.07	.05	.6	.16	1.66	53	.15	.94	.55	.004	.23	138
TB14-A1	0–9	5.1	7.7	0.5	0.3	3.5	1.6	15.6	50	5.2	5.9	3.42	0.320	...	11
TB14-B21r	9–18	5.0	5.9	0.8	0.1	1.4	0.7	10.4	43	2.9	1.7	0.99	0.108	...	9
TB14-B3	18–30	5.0	3.9	0.5	0.1	0.9	0.4	7.3	46	1.6	1.0	0.58	0.059	...	10
TB14-Cox	30–79	4.9	3.5	0.8	0.1	0.6	0.3	6.4	45	1.3	0.4	0.23	0.015	...	15
TB14-Cn	79+	5.3	5.9	0.1	0.1	2.9	0.8	9.8	40	0.5	0.2	0.12	0.008	...	15

[a] ... not analyzed.

[b] Soil pits (e.g. TB15) are on Figure 3.

with depth, as well as an increase in basic cations, especially Ca^{+2}, with depth. The CEC ranges from 7 to 9 (meq/100 g) in surface horizons, while base saturation increases with depth. The build-up of salts with depth, as in profile FL1 indicates relatively low leaching effects. Organic matter ranges from 3 to 4 percent paralleled by carbon/nitrogen ratios ~12. $CaCO_3$ is highest in the subsurface horizons, and less than 1.7 percent.

Table 12. Selected chemical properties[a] of the <2 mm fraction of the material in the soil horizons of Appendix 2.

Sample[b]	Depth cm	pH	Extractable Cations (meq/100g) Exch. H^+	Na^+	K^+	Ca^{+2}	Mg^{+2}	CEC meq/100g	Base Saturation %	Salts mmhos/cm 25°C	Organic matter %	Organic carbon %	N %	Fe_2O_3 %	Carbon/nitrogen
TB1-A1	0-13	5.8	0.3	.23	.55	9.7	2.0	14.96	83	.85	11.1	6.44	.33	.82	20
TB1-B1	13-24	5.4	0.5	.10	.01	1.3	.20	2.85	56	.19	1.27	.75	.04	.30	19
TB1-B21r	24-53	4.8	0.3	.09	.01	.3	.10	1.47	34	.14	.66	.38	.01	.19	38
TB1-B3	53-69	5.0	9.7	.09	.01	.3	.03	1.0	43	.14	.57	.33	.01	.16	33
TB1-Cox	69-114	5.1	4.2	.10	.005	.3	.03	.81	54	.13	.57	.33	.004	.12	83
TB1-Cn	114+	5.0	7.7	.10	.01	.3	.03	.57	77	.10	.28	.16	.003	.07	53
TB2-A11	0-6	5.2	17.6	.17	.50	5.5	1.1	13.92	52	.62	9.82	5.70	.35	1.68	16
TB2-A12	6-10	5.0	16.1	.13	.09	2.5	.46	12.26	26	.23	4.63	2.69	.16	1.68	17
TB2-B1	10-18	5.1	9.8	.13	.02	.7	.10	5.89	16	.22	1.98	1.15	.07	.83	16
TB2-B21rh	18-28	4.9	8.5	.13	.01	.4	.10	5.84	11	.15	2.27	1.32	.06	.85	22
TB2-Cox	28-86	5.2	6.2	.07	.02	.3	.03	1.52	28	.12	.52	.30	.01	.31	30
TB2-Cn	86+	4.8	2.3	.07	.01	.3	.03	.57	72	.10	.85	.49	.004	.10	123
TB3-A11	0-8	6.4	6.6	.09	.55	11.2	2.7	16.25	89	.42	9.24	5.36	.36	.58	15
TB3-A12	8-14	5.7	5.8	.10	.14	5.0	.92	7.08	87	.15	5.04	2.92	.14	.50	21
TB3-B1	14-25	5.9	4.3	.10	.11	2.3	.43	3.42	86	.05	1.96	1.14	.06	.40	19
TB3-B21r	25-46	5.6	2.9	.13	.15	1.0	.30	2.42	65	.06	1.23	.71	.04	.31	18
TB3-Cox	46-109	5.1	2.9	.17	.15	.40	.10	2.09	39	.05	.78	.45	.02	.29	23
TB3-C1n	109-142	5.5	1.4	.17	.14	.60	.16	1.90	56	.08	.67	.39	.02	.23	20
TB3-C2n	142+	5.2	1.6	.09	.09	.40	.10	1.57	43	.06	.45	.26	.01	.20	26
TB4-A11	0-10	4.7	30.0	.15	.70	8.8	1.8	27.41	42	.74	20.33	11.79	.65	1.27	18
TB4-A12	10-20	4.5	31.0	.21	.37	3.0	.46	21.14	19	.40	11.42	6.62	.40	2.04	17
TB4-B21r	20-38	4.8	8.3	.15	.08	.6	.10	3.94	24	.08	1.40	.81	.04	.64	20
TB4-Cox	38-127	5.2	13.1	.17	.09	.6	.10	7.08	14	.01	1.79	1.04	.06	1.02	17
TB4-Cn	127+	5.8	6.9	.10	.07	.4	.03	1.85	32	.01	.50	.29	.01	.41	29
TB8-A11	0-6	5.0	10.1	0.1	0.7	5.4	0.9	17.2	41	2.3	8.0	4.6	0.391	...	12
TB8-A12	6-13	4.9	9.9	0.1	0.5	4.5	0.9	15.9	38	2.7	6.3	3.7	0.315	...	12
TB8-B21r	13-36	4.6	4.4	0.1	0.1	0.5	1.6	6.7	34	1.4	1.9	1.10	0.088	...	13
TB8-C1ox	36-53	4.6	4.0	<0.1	<0.1	0.5	0.7	5.2	27	0.7	1.0	0.58	0.055	...	11
TB8-C2ox	53-89	4.8	2.2	<0.1	<0.1	0.4	0.4	3.0	33	0.4	0.3	0.17	0.014	...	12
TB8-Cn	89+	4.7	1.8	<0.1	<0.1	0.3	0.2	2.3	30	0.4	0.2	0.12	0.010	...	12
TB9-A1	0-8	4.3	20.6	0.1	0.6	4.7	1.1	27.1	24	2.2	8.7	5.05	0.569	...	9
TB9-B1r	8-15	4.5	3.1	0.1	0.1	0.3	0.1	3.7	16	2.7	0.9	0.52	0.051	...	10
TB9-Cox	15-51	5.2	2.8	0.1	0.1	0.3	0.1	3.4	18	0.6	0.8	0.46	0.037	...	12
TB9-Ab	51+	5.1	8.6	0.1	0.1	1.8	0.2	10.8	20	0.9	3.2	1.86	0.148	...	13

[a] ... not analyzed.

[b] Soil pits (e.g. TB1) are on Figure 4.

Post-Bull Lake soils in intermontane basins have surface and subsurface pH's that are neutral to mildly alkaline. Exchangeable H^+ ions are generally lower than in post-Pinedale soils. In the basic cation group, only Ca^{+2} shows an appreciable increase when compared with younger post-Pinedale soils. In surface horizons, CEC is generally twice as high (15.0 meq/100 g), and in subsurface horizons, 3 times as high as in younger profiles. Base saturation is constant at 100 percent indicating lower leaching than in post-Pinedale soils. Salts are higher, while organic matter and organic carbon are similar to younger post-Pinedale soils. Fe_2O_3 and carbon/nitrogen ratios are similar to younger soils, while $CaCO_3$ is noticeably higher (7.2 percent), but not high enough to qualify as calcic horizons (Soil Survey Staff, 1960).

Table 13. Selected chemical properties[a] of the <2 mm fraction of the material in the soil horizons of Appendix 3.

Sample[b]	Depth cm	pH	Extractable Cations (meq/100g) Exch. H^+	Na^+	K^+	Ca^{+2}	Mg^{+2}	CEC meq/100g	Base Saturation %	Salts mmhos/cm 25°C	Organic matter %	Organic carbon %	N %	Fe_2O_3 %	Carbon/nitrogen	$CaCO_3$ %
FL1-A1	0-23	6.4	5.1	.17	.70	5.9	1.4	9.41	87	.3	3.1	1.8	.15	.83	12	.03
FL1-B2t	23-46	7.0	1.8	.83	.19	4.0	2.7	6.75	100	1.0	.62	.36	.03	.54	12	.02
FL1-C1	46-56	8.2	1.6	2.1	.18	43.0	3.3	4.32	100	9.0	.57	.33	.02	.40	17	.40
FL1-C2	56-91	9.6	0.1	4.7	.30	37.0	6.4	4.56	100	8.98	.29	.17	.01	.40	17	1.72
FL1-Cn	91+	9.1	1.2	3.3	.22	6.4	2.1	3.18	100	8.2	tr	tr	.002	.34	<5	.18
FL2-A1	0-10	6.5	3.5	.09	.65	4.8	.69	7.17	87	.32	3.01	1.75	.10	.59	17	.05
FL2-B2t	10-36	6.4	1.5	.09	.36	4.2	.89	5.32	100	.10	1.05	.61	.04	.56	15	.03
FL2-C1	36-58	6.8	2.0	.07	.18	3.4	.82	4.13	100	.10	.52	.30	.02	.37	15	.02
FL2-C2	58-86	6.6	1.5	.09	.11	2.7	.69	3.04	100	.09	.24	.14	.01	.30	14	.02
FL2-Cn	86+	6.9	1.4	.07	.08	2.5	.69	2.99	100	.10	.43	.25	.005	.26	50	.03
FL3-A1	0-13	7.0	4.8	.10	.81	7.2	1.5	9.6	100	.37	4.01	2.33	.17	.79	14	.02
FL3-B2t	13-28	7.1	2.1	.37	.52	3.2	.69	4.23	100	.25	.86	.50	.03	.41	17	.02
FL3-C	28-66	7.0	1.5	.23	.24	2.5	.59	3.18	100	.20	.43	.25	.02	.25	13	.02
FL3-Cn	66+	8.9	0.1	.26	.08	34.0	1.4	2.28	100	.26	1.14	.66	.01	.23	66	1.30
FL4-A1	0-13	7.5	3.0	.09	.79	10.6	3.6	15.11	100	.38	3.39	1.97	.13	.66	15	0.3
FL4-B21t	13-23	7.3	3.0	.13	.29	17.3	6.3	18.91	100	.22	1.43	.83	.07	.61	12	0.3
FL4-B22t	23-38	8.3	2.5	.15	.19	27.0	4.7	17.05	100	.32	1.24	.72	.06	.86	12	.39
FL4-C1	38-74	8.8	0.3	.39	.18	37.0	5.6	11.50	100	.40	1.48	.86	.02	.30	43	7.2
FL4-C2	74-114	9.1	0.7	1.58	.19	34.0	5.9	13.78	100	1.60	.19	.11	.01	.52	11	1.4
FL4-Cn	114+	8.9	1.3	1.49	.19	23.0	5.0	12.11	100	1.85	.43	.25	.01	.44	25	.57
FL5-A1	0-15	7.2	3.2	.09	.93	9.0	2.5	11.45	100	.40	2.34	1.36	.12	.53	11	.02
FL5-B21t	15-28	7.1	2.4	.07	.86	9.7	3.4	12.54	100	.30	1.53	.89	.06	.35	15	.02
FL5-B22t	28-71	7.0	2.0	.09	.58	9.3	3.4	12.78	100	.20	.52	.30	.02	.32	15	.02
FL5-C1	71-122	7.0	2.1	.13	.36	9.3	3.4	12.30	100	.20	.48	.28	.02	.29	14	.02
FL5-C2	122-178	6.7	2.4	.19	.16	9.5	3.4	10.26	100	.19	.43	.25	.01	.29	25	.02
FL5-Cn	178+	8.7	1.5	.17	.18	27.0	4.2	11.26	100	.30	.48	.28	.01	.30	28	.71
FL6-A1	0-5	6.3	0.2	0.2	1.0	10.2	4.4	16.0	99	10.6	2.9	1.68	0.127	...	13	0.7
FL6-B21t	5-15	6.3	0.3	0.1	0.3	10.4	4.0	15.1	98	2.3	0.9	0.52	0.054	...	10	0.9
FL6-B22t	15-24	6.5	0.2	0.1	0.2	10.5	4.6	15.6	99	2.3	0.6	0.35	0.041	...	9	0.9
FL6-B23t	24-33	6.8	<0.1	0.2	0.2	9.6	4.2	14.2	100	2.3	0.5	0.29	0.029	...	10	0.9
FL6-B24t	33-51	7.0	<0.1	0.2	0.2	8.3	4.0	12.4	100	2.4	0.5	0.29	0.029	...	10	0.6
FL6-C1	51-89	8.0	<0.1	0.6	0.2	—	—	11.6	100	4.2	0.5	0.29	0.025	...	12	8.3
FL6-C2	89-135	8.4	<0.1	2.1	0.4	—	—	9.2	100	27.9	0.2	0.12	0.007	...	17	2.8
FL6-Cn	135+	8.1	<0.1	1.1	0.3	6.3	1.8	9.0	100	43	0.1	0.06	0.005	...	12	0.9

[a]— no detection; ... not analyzed.

[b]Soil pits (e.g. FL1) are on Figure 5.

CONCLUSIONS

Soil profiles in the Titcomb Basin and Fremont Lake areas are representative of similar soil-stratigraphic horizons in other areas of the Rocky Mountains. The relict soils discussed in this paper have progressed through stages of development, occurring as separate segments of a time continuum.

Five major soil-stratigraphic units are common throughout the Wind River Range. Changes in horizon sequence, depth of weathering, color, field texture, structure, moist consistence, shear-strength, water retention, particle size, clay-mineral assemblage, pH, extractable cations, organic constituents, and free iron oxide are significant in profile differentiation. Soil profiles have great utility in deposit differentiation depending on elapsed time since deposition. The data indicate that soil properties approach steady-state with different velocities producing a situation in which some soil properties are useful only over short time spans, while others are longer-lived. Organic parameters, for instance, approach steady-state within a few thousand years, while depth of weathering, percent clay, clay-mineral assemblages, and certain soil-chemical parameters require greater time to achieve equilibrium.

ACKNOWLEDGMENTS

I thank E.B. Evenson, Lehigh University; and B.D. Fahey, Guelph University, for critical reviews of this paper. N. Stokes assisted in the field in 1972. Students in my mountain geomorphology course, from 1973-1975, assisted with the field work, and are too numerous to name in full. L.J. Gowland helped with the laboratory analyses, and the soil chemistry was completed in the soil testing laboratories of Colorado State University and Oregon State University. D. Halvorson identified some of the mineral suites. Research was supported by grants from York University.

REFERENCES CITED

Atwood, W.W., 1940, The Physiographic Provinces of North America: N.Y., Blaisdell, 536 p.

Baker, F.S., 1944, Mountain climates of the Western U.S.: Ecol. Monogr., v. 14, p. 223-254.

Barry, R.G., and Chorley, R.J., 1970, Atmosphere, Weather, and Climate: N.Y., Holt, Rinehart and Winston, 320 p.

Bartlett, R.A., 1962, Great Surveys of the American West: Norman, Okla., University of Oklahoma Press, 408 p.

Birkeland, P.W., 1973, Use of relative age-dating methods in a stratigraphic study of rock glacier deposits: Mt. Sopris, Colorado, Arctic and Alpine Research, v. 5, no. 40, p. 401-416.

_______________, 1974, Pedology, Weathering and Geomorphological Research: N.Y., Oxford Univ. Press, 285 p.

Birkeland, P.W., and Miller, C.D., 1974, Probable pre-Neoglacial age of the Type Temple Lake Moraine, Wyoming: Discussion and Additional relative age data: Arctic and Alpine Research, v. 6, no. 3, p. 301-306.

Blackwelder, E., 1915, Post-Cretaceous History of the Mountains of Central Western Wyoming: Jour. Geology, v. 23, nos. 2, 3, and 4, p. 97-117, 193-217, and 307-340.

Bower, C.A., and Wilcox, L.V., 1965, Soluble salts, *in* Black, C.A., ed., Methods of Soil Analysis, Part 2: Madison, Wisc., Amer. Soc. Agron., p. 933-951.

Bouyoucos, G.J., 1962, Hydrometer method improved for making particle-size analyses of soils: Agron. Jour., v. 54, p. 464-465.

Bradley, W.H., 1964, Geology of the Green River Formation and associated Eocene rocks in southwestern Wyoming and adjacent parts of Colorado and Utah, U.S. Geol. Survey Prof. Pap. 496-A: Washington, D.C., U.S. Gov't Printing Office, 86 p.

Branson, E.B., and Branson, C.C., 1941, Geology of the Wind River Mountains, Wyoming: Amer. Assoc. Pet. Geol. Bull., v. 25, p. 120-151.

Bremner, J.R., 1965, Total nitrogen, *in* Black, C.A., ed., Methods of Soil Analysis, Part 2: Madison, Wisc., Amer. Soc. Agron., p. 1149-1176.

Currey, D.R., 1974, Probable pre-Neoglacial age of the type Temple Lake Moraine, Wyoming: Arctic and Alpine Research, v. 6, no. 3, p. 293-300.

Day, P., 1965, Particle fractionation and particle size analysis, *in* Black, C.A., ed., Methods of Soil Analysis: Madison, Wisc., Amer. Soc. Agron., p. 545-567.

DeVoto, B., 1964, Across the Wide Missouri: Cambridge, Mass., Riverside Press, 451 p.

Fenneman, N.M., 1931, Physiography of Western U.S.: N.Y., McGraw-Hill, 534 p.

Folk, R.L., 1968, Petrology of Sedimentary Rocks: Austin, Tex., Hamphill Press, 170 p.

Griggs, R.F., 1938, Timberlines in the Northern Rocky Mountains: Ecology, v. 19, p. 548-564.

Hayden, F.V.,. 1863, On the Geology and Natural History of the Upper Missouri being the substance of a report to

Lieut. G.K. Warren, T.E.U.S.A.: Trans. Amer. Phil. Soc., New Series, v. XII, p. 1-231.

Irving, W., 1836, Astoria or anecdotes of an enterprise beyond the Rocky Mountains, Rodd, E.W., ed.: Norman, Okla., University of Oklahoma Press, 556 p.

King, C., 1870, Copy Book of Letters: Washington, D.C., National Archives.

Mahaney, W.C., 1974, Soil stratigraphy and genesis of Neoglacial deposits in the Arapaho and Henderson Cirques, Central Colorado Front Range, *in* Mahaney, W.C., ed., Quaternary Environments, Proceedings of a Symposium: Toronto, Geographical Monographs, no. 5, p. 197-240.

______________, 1975, Soils of post-Audubon age, Teton Glacier area, Wyoming: Arctic and Alpine Research, v. 7, no. 2, p. 141-154.

Mahaney, W.C., and Fahey, B.D., 1976, Quaternary soil stratigraphy of the Front Range, Colorado, *in* Mahaney, W.C., ed., Quaternary Stratigraphy of North America: Stroudsburg, Penna., Dowden, Hutchinson and Ross, p. 319-352.

Mehra, O.P., and Jackson, M.L., 1960, Iron oxide removal from soils and clays by a dithionite citrate system buffered with sodium bicarbonate, *in* Swineford, A., ed., Natl. Conf. on Clays and Clay Minerals, 1958: London, Pergamon, p. 317-327.

Miller, C.D., and Birkeland, P.W., 1974, Probable pre-Neoglacial age of the type Temple Lake Moraine, Wyoming: Discussion and additional relative age data: Arctic and Alpine Research, v. 6, no. 3, p. 301-306.

Moss, J.H., 1951, Glaciation in the Wind River Mountains and its relation to Early Man in the Eden Valley, *in* Eiseley, L., ed., Early Man In the Eden Valley: Phil., Penna., Univ. of Pennsylvania Mus. Monogr. 6, p. 9-94.

__________, 1951, Late glacial advances in the Southern Wind River Mountains, Wyoming: Amer. Jour. Sci., v. 249, p. 865-883.

Olsen, S.R., and Dean, L.A., 1965, Phosphorus, *in* Black, C.A., ed., Methods of Soil Analysis, Part 2: Madison, Wisc., Amer. Soc. Agron., p. 1035-1048.

Outcalt, S.I., and MacPhail, D., 1965, A survey of Neoglaciation in the Front Range of Colorado, University of Colorado Stud. Ser. Earth Science, no. 4: Boulder, Univ. of Colorado, 124 p.

Oyama, M., and Takehara, H., 1967, Revised Standard Soil Color Charts, Japan Research Council for Agriculture, Forestry and Fisheries.

Peech, M.L., Alexander, T., Dean, L.A., and Reed, J.F., 1947, Methods of soil analyses for soil fertility investigations: Washington, D.C., U.S.D.A. Circ. 757, p. 25.

Reed, R.M., 1971, Aspen forests of the Wind River Mountains, Wyoming: Amer. Midland Naturalist, v. 86, no. 2, p. 327-343.

__________, 1976, Coniferous forest habitat types of the Wind River Mountains, Wyoming: Amer. Midland Naturalist, v. 95, no. 1, p. 159-173.

Richmond, G.M., 1965, Type moraines of the Pinedale Glaciation, *in* Schultz, C.B., and Smith, H.T.U., eds., Guidebook for field Conference E: Lincoln, Nebr., Nebr. Acad. Sci. for VII INQUA Congr., p. 34-36.

______________, 1974, Geologic map of the Fremont Lake South quadrangle, Sublette County, Wyoming: U.S. Geol. Survey, Geol. Quad. Map GQ-1138.

______________, 1976, Pleistocene stratigraphy and chronology in the mountains of Western Wyoming, *in* Mahaney, W.C., ed., Quaternary Stratigraphy of North America: Stroudsburg, Penna., Dowden, Hutchinson and Ross, p. 353-379.

Russell, O., 1965, Journal of a Trapper, 1834-1843, *in* Haines, A.L., ed.,: Lincoln, Nebr., Univ. of Nebraska Press, 191 p.

Schollenberger, C.J., and Simon, R.H., 1945, Determination of exchange capacity and exchangeable bases in soils-ammonium acetate method: Soil Sci., v. 59, p. 13-24.

Sharp, R.P., 1969, Semiquantitative differentiation of glacial moraines near Convict Lake, Sierra Nevada, California, Jour. Geol., v. 77, p. 68-91.

Soil Survey Staff, 1951, Soil Survey Manual, U.S. Dept. Agric. Handbook 18: Washington, D.C., U.S. Gov't Printing Office, 503 p.

_________________, 1960, Soil Classification: 7th Approximation: Washington, D.C., U.S. Gov't Printing Office, 265 p.

U.S. Dept. Commerce, 1965, Climatography of the U.S., No. 86-42, Climatic Summary of the U.S., Supplement for 1951-1960, Wyoming: Washington, D.C., U.S. Gov't Printing Office, 77 p.

Walkley, A., and Black, I.A., 1934, An examination of the Degtjareff method for determining soil organic matter, and a proposed modification of the chromic acid titration method: Soil Sci., v. 34, p. 29-38.

Whittig, L.D., 1965, X-ray diffraction techniques for mineral identification and mineralogical composition, *in* Black, C.A., ed., Methods of Soil Analysis: Madison, Wisc., Amer. Soc. Agron., p. 671-696.

Worl, R.G., 1968, Taconite in the Wind River Mountains, Sublette County, Wyoming, Preliminary Report 10: Laramie, Wyoming, Geol. Survey Wyoming, 15 p.

Appendix 1. Soil profiles[a] in the Upper Titcomb Basin.[b]

Profile: TB15

Location: end moraine of Gannett Peak Age (Fig. 3)

Elevation: 3,475 m

Parent material: granite, dioritic gneiss, quartz-dioritic gneiss, granodioritic gneiss and quartz-monzonitic gneiss

Topography: northeast-facing aspect

Vegetation: vascular plant cover nil; lichen cover <1%; lichen species include *C. elegans*, *L. thomsonii*, and *R. geographicum*

Soil horizons	Depth below surface (cm)	Description
C1	0-23	Yellowish gray (2.5Y 5/1m) and brownish gray (10YR 6/1d) pebbly coarse sandy loam; cobbles and boulders; massive structure; very friable to loose moist consistence; nonplastic and slightly sticky.
Cn	23+	Dark grayish yellow (2.5Y 5/2m) and light gray (10YR 7/1d) pebbly coarse sandy loam; cobbles and boulders; massive structure; loose to very friable moist consistence; nonplastic and nonsticky.

Profile: TB5

Location: end moraine of Audubon age (Fig. 3)

Elevation: 3,410 m

Parent material: granite, granodioritic gneiss, gabbro-dioritic gneiss, and porphyroblastic garnet-rich gneiss

Topography: end moraine crest

Vegetation: vascular plant cover variable, 20-40 percent; lichen cover 20-40 percent; lichen species include *C. elegans*, *L. aspicilia*, *L. thomsonii*, *L. atrobrunnea*, and *R. geographicum*

Soil horizons	Depth below surface (cm)	Description
A1	0-10	Brownish black (10YR 3/2m) and dull yellowish brown (10YR 5/3d) pebbly sandy loam; weak granular structure; friable moist consistence; plastic and sticky. Roots reach 4 mm diameter and penetrate into the Cox horizon.
Cox	10-51	Dull yellow orange (10YR 6/3m) and dull yellow orange (10YR 7/3d) pebbly coarse sandy loam; cobbles and boulders; massive structure; friable moist consistence; plastic and sticky.
Cn	51+	Grayish yellow brown (10YR 6/2m) and grayish yellow (2.5Y 7/2d) pebbly coarse loamy sand; cobbles and boulders; massive structure; loose to very friable moist consistence; nonplastic and nonsticky.

Profile: TB6

Location: end moraine of Audubon Age (Fig. 3)

Elevation: 3,400 m

Parent material: granite, granodioritic gneiss and gabbro-dioritic gneiss

Topography: end moraine crest

Vegetation: vascular plant cover variable, 25–40 percent; lichen cover 25–35 percent; lichen species include *C. elegans*, *L. aspicilia*, *L. thomsonii*, *L. atrobrunnea*, *R. geographicum*, and *R. oreina*

Soil horizons	Depth below surface (cm)	Description
A1	0–10	Grayish yellow brown (10YR 4/2m, d); pebbly coarse loamy sand; weak granular structure; very friable moist consistence; slightly plastic, and slightly sticky.
Cox	10–71	Dull yellow orange (10YR 6/3m) and dull yellow orange (10YR 7/3d) pebbly coarse sandy loam; cobbles and boulders; massive structure; loose moist consistence; non-plastic and nonsticky.
Cn	71+	Brownish gray (10YR 6/1m) and light gray (5Y 7/2d) pebbly coarse loamy sand; cobbles and boulders; massive structure; loose moist consistence; nonplastic and nonsticky.

Profile: TB14

Location: ground moraine of Early Neoglacial Age (Fig. 3)

Elevation: 3,405 m

Parent material: granite, quartz-dioritic gneiss, granodioritic gneiss, quartz-monzonitic gneiss.

Topography: ground moraine crest

Vegetation: vascular plant cover ~ 50 percent; lichen cover 75 percent; lichen species include *A. chlorophana*, *C. elegans*, *L. aspicilia*, *L. thomsonii*, *L. atrobrunnea*, *R. geographicum*, and *R. oreina*

Soil horizons	Depth below surface (cm)	Description
O1	0–1	Brownish black (10YR 2/2m).
A1	0–9	Brownish black (10YR 2/3m) and dark brown (10YR 3/3d), pebbly sandy loam; granular structure; friable moist consistence; plastic and slightly sticky; roots reach 7 mm in diameter and penetrate through the B2ir and B3 horizons; horizon boundaries regular and distinct
B2ir	9–18	Brown (10YR 4/4m) and dull yellow orange (10YR 6/4d) pebbly coarse sandy loam; cobbles and boulders; weak blocky structure; friable moist consistence; plastic and slightly sticky.
B3	18–30	Dull yellowish brown (10YR 5/4m) and dull yellow orange (10YR 6/3d) pebbly coarse sandy loam; cobbles and boulders; weak blocky structure; friable moist consistence; plastic and slightly sticky.
Cox	30–79	Olive brown (2.5Y 4/3m) and light yellow (2.5Y 7/3d) pebbly loamy sand; cobbles and boulders; massive structure; loose moist consistence; nonplastic and nonsticky.
Cn	79+	Grayish olive (5Y 5/3m) and grayish yellow (2.5Y 7/2d) pebbly coarse sandy loam; cobbles and boulders; massive structure; loose moist consistence; nonplastic and nonsticky.

[a] Terms and horizon nomenclature follow the Soil Survey Staff (1951, 1960).

[b] Soil pits (e.g. TB15) are on Figure 3.

Appendix 2. Soil profiles[a] in the Lower Titcomb Basin[b].

Profile: TB1

Age: post-Pinedale

Location: lower Titcomb Basin (Fig. 4)

Elevation: 3,245 m

Parent material: granite, dioritic gneiss, granodioritic gneiss and quartz-monzonitic gneiss

Topography: inner end moraine crest

Vegetation: vascular plant cover 75 percent; lichen cover 75 percent; lichen species includes *A. chlorophana*, *C. elegans*, *L. aspicilia*, *L. Thomsonii*, *L. atrobrunnea*, and *R. geographicum*

Soil horizons	Depth below surface (cm)	Description
01	0–1	Black (10YR 2/1m).
A1	0–13	Brownish black (10YR 2/3m) and brownish black (10YR 3/2d) pebbly loamy sand; granular structure; friable moist consistence; slightly plastic and slightly sticky; lower horizon boundary undulatory and irregular.
B1	13–24	Mottled dull yellow orange (10YR 6/3m, d) coarse sandy loam; weak blocky structure; friable moist consistence; slightly plastic and slightly sticky; upper and lower horizon boundaries undulatory and irregular.
B21r	24–53	Mottled dull yellowish brown (10YR 5/4m) and dull yellow orange (10YR 7/3d) pebbly sandy loam; weak blocky structure; friable to very friable moist consistence; slightly plastic and slightly sticky.
B3	53–69	Mottled grayish yellow brown (10YR 6/2m) and dull yellow orange (10YR 7/2d) pebbly sandy loam; massive structure; very friable moist consistence; slightly plastic and slightly sticky.
IICox	69–114	Dull yellow orange (10YR 6/3m) and dull yellow orange (10YR 7/2d) pebbly loamy sand, cobbles and boulders; massive structure; very friable moist consistence; nonplastic and nonsticky.
IICn	114+	Grayish yellow (2.5Y 7/2m) and light gray (5Y 7/2d) pebbly sand; cobbles and boulders; massive structure; very friable moist consistence; nonplastic and nonsticky.

Profile: TB2

Age: post-Pinedale

Location: lower Titcomb Basin (Fig. 4)

Elevation: 3,230 m

Parent material: granite, quartz-dioritic gneiss and granodioritic gneiss

Topography: outwash fan crest

Vegetation: vascular plant cover 75 percent; lichen cover 60 percent; lichen species similar to site TB1

Soil horizons	Depth below surface (cm)	Description
01	0–3	Brownish black (10YR 2/2m).
A11	0–6	Dark brown (10YR 2/3m) and dull yellowish brown (10YR 4/3d) fine sandy loam; granular structure; friable moist consistence; slightly plastic and slightly sticky.
A12	6–10	Brownish black (10YR 3/2m) and dull yellowish brown (10YR 5/3d) fine sandy loam; granular structure; friable moist consistence; slightly plastic and slightly sticky.
B1	10–18	Dark brown (10YR 3/3m) and dull yellow orange (10YR 6/4d) coarse sandy loam; weak blocky structure; friable moist consistence; slightly plastic and slightly sticky.
B21rh	18–28	Brown (7.5YR 4/4m) and dull yellow orange (10YR 6/3d) pebbly sandy loam, cobbles; weak blocky structure; friable moist consistence; slightly plastic and slightly sticky; roots up to 8 mm penetrate to base of B2 horizon.
IICox	28–86	Bright yellowish brown (10YR 6/8m) and dull yellow orange (10YR 6/4d) pebbly loamy sand, cobbles and boulders; massive structure; loose to very friable moist consistence; nonplastic and nonsticky.
IICn	86+	Dull yellow orange (10YR 6/3m), pale yellow (2.5Y 8/3d) and dull yellow (2.5Y 6/3d) pebbly coarse sand; cobbles and boulders; massive structure; loose moist consistence; nonplastic and nonsticky.

Profile: TB3

Age: post-Pinedale

Location: lower Titcomb Basin (Fig. 4)

Elevation: 3,235 m

Parent material: granite, dioritic gneiss, granodioritic gneiss, and quartz-monzonitic gneiss

Topography: outer end moraine crest

Vegetation: vascular plant cover 75 percent; lichen cover 75 percent; lichen species similar to site TB1

Soil horizons	Depth below surface (cm)	Description
O1	0–2	Black (10YR 1/1m)
A11	0–8	Brownish black (10YR 3/2m, d) pebbly loamy sand, cobbles; granular structure; friable moist consistence; slightly plastic and slightly sticky.
A12	8–14	Dull yellowish brown (10YR 4/3m) and grayish yellow brown (10YR 5/2d) pebbly loamy sand, cobbles; granular structure; friable moist consistence; nonplastic and nonsticky.
B1	14–25	Dull yellowish brown (10YR 5/4m) and dull yellowish brown (10YR 5/3d) pebbly loamy sand, cobbles and boulders; weak blocky structure; friable moist consistence; nonplastic and nonsticky.
B21r	25–46	Dull yellow orange (10YR 6/3m, d) pebbly loamy sand, cobbles and boulders; weak blocky structure; friable moist consistence; slightly plastic and slightly sticky; roots up to 8 mm penetrate to base of B horizon.
Cox	46–109	Dull yellowish brown (10YR 5/4m) and dull yellow orange (10YR 6/3d) pebbly coarse loamy sand, cobbles and pebbles; massive structure; loose to very friable moist consistence; nonplastic and nonsticky.
C1n	109–142	Dull yellowish brown (10YR 5/4m) and dull yellow orange (10YR 7/3d) pebbly coarse loamy sand, cobbles and boulders; massive structure; very friable moist consistence; nonplastic and nonsticky.
C2n	142+	Dull yellow orange (10YR 6/4m) and light yellow (2.5Y 7/3d) pebbly coarse loamy sand, cobbles and boulders; massive structure; loose to very friable moist consistence; nonplastic and nonsticky.

Appendix 2 Continued. Soil profiles in the Lower Titcomb Basin

Profile: TB4

Age: post-Pinedale

Location: lower Titcomb Basin (Fig. 4)

Elevation: 3,240 m

Parent material: dioritic gneiss, granodioritic gneiss, and quartz-monzonitic gneiss

Topography: stone-banked lobe crest

Vegetation: krummholz of procumbent willow (*Salix* spp.?) and subalpine fir (*Abies lasiocarpa*); vascular plant cover 75 percent; lichen cover 80 percent; lichen species similar to site TB1

Soil horizons	Depth below surface (cm)	Description
01	0-3	Black (10YR 1/1m).
A11	0-10	Brownish black (10YR 2/2m) and brownish black (10YR 3/2d) pebbly sandy loam, cobbles and boulders; granular structure; friable moist consistence; slightly plastic and slightly sticky.
A12	10-20	Brownish black (10YR 3/2m) and dark brown (10YR 3/3d) fine sandy loam, cobbles and boulders; granular structure; friable moist consistence; plastic and sticky.
IIB2ir	20-38	Brown (7.5YR 4/4m) and dull yellow orange (10YR 6/4d) pebbly coarse sandy loam; cobbles and boulders; weak blocky structure; very friable moist consistence; slightly plastic and slightly sticky.
IICox	38-127	Dark brown (10YR 3/4m) and dull yellow orange (10YR 6/3d) pebbly fine sandy loam; cobbles and boulders; weak blocky structure; very friable moist consistence; slightly plastic and slightly sticky.
IICn	127+	Yellowish brown (2.5Y 5/3m) and light yellow (2.5Y 7/3d) pebbly coarse sandy loam; cobbles and boulders; massive structure; loose to very friable moist consistence; nonplastic and nonsticky.

Profile: TB8

Age: post-Pinedale

Location: Island Lake (Fig. 4)

Elevation: 3,200 m

Parent material: granite, quartz-diorite gneiss, granodioritic gneiss and amphibolite schist

Topography: ground moraine crest

Vegetation: scattered krummholz of subalpine fir (*Abies lasiocarpa*), spruce *Picea engelmanni*), whitebark pine (*Pinus albicaulis*) and limber pine (*P. flexilis*); vascular plant cover 60 percent; lichen cover 75 percent; lichen species similar to site TB1

Soil horizons	Depth below surface (cm)	Description
01	0-3	Black (10YR 2/1m).
A11	0-6	Brownish black (10YR 2/3m) and dull yellowish brown (10YR 4/3d) pebbly fine sandy loam; cobbles granular structure; friable moist consistence; plastic and nonsticky.
A12	6-13	Dark brown (10YR 3/3m) and dull yellowish brown (10YR 4/3d) pebbly sandy loam; cobbles; weak granular structure; firm moist consistence; plastic and slightly sticky.
IIB2ir	13-36	Brown (7.5YR 4/6m) and dull yellow orange (10YR 6/4d) pebbly coarse sandy loam; cobbles and boulders; weak blocky structure; very friable moist consistence; plastic and slightly sticky.
IIC1ox	36-53	Brown (10YR 4/6m) and dull yellowish brown (10YR 5/4d) pebbly loamy sand; cobbles and boulders; massive structure; loose moist consistence; nonplastic and nonsticky.
IIC2ox	53-89	Dull yellowish brown (10YR 4/3m) and dull yellow orange (10YR 6/3d) pebbly loamy sand; cobbles and boulders; massive structure; loose moist consistence; nonplastic and nonsticky.
IICn	89+	Olive brown (2.5Y 4/3m) and grayish yellow (2.5Y 7/2d) pebbly coarse sandy loam; cobbles and boulders; massive structure; loose moist consistence; nonplastic and nonsticky.

Profile: TB9

Age: buried soil, post-Pinedale

Location: lower Titcomb Basin (Fig. 4)

Elevation: 3,230 m

Parent material: granite, quartz-dioritic gneiss, granodioritic gneiss, and quartz-syenitic gneiss

Topography: narrow beach plain

Vegetation: kobresia meadow including *Caltha* spp.?; vascular plant cover 75 percent; lichen cover 50 percent; lichen species similar to site TB1

Soil horizons	Depth below surface (cm)	Description
O1	0-3	Black (10YR 2/1m).
A1	0-8	Very dark brown (7.5YR 2/3m) and brownish black (10YR 3/2d) pebbly loamy sand; granular structure; friable moist consistence; plastic and nonsticky.
B2ir	8-15	Dark brown (7.5YR 3/4m) and brown (10YR 4/4d) pebbly coarse sand; massive structure; loose moist consistence; slightly plastic and nonsticky.
Cox	15-51	Brown (10YR 4/6m) and dull yellow orange (10YR 6/3d) pebbly loamy sand; massive structure; loose moist consistence; slightly plastic and nonsticky.
Ab	51+	Dark brown (10YR 3/3m) and dull yellow orange (10YR 6/3d) pebbly fine sandy loam; laminated; very firm moist consistence; plastic and sticky.

[a]Terms and horizon nomenclature follow the Soil Survey Staff (1951, 1960).

[b]Soil pits (e.g. TB1) are on Figure 4.

Appendix 3. Soil profiles[a] in the Fremont Lake area.[b]

Profile: FL1

Age: post-Pinedale

Location: Fremont Lake (Fig. 5)

Elevation: 2,267 m

Parent material: granite, quartz-dioritic gneiss, granodioritic gneiss, and quartz-monzonitic gneiss

Topography: inner end moraine crest

Vegetation: sagebrush *(Artemisia)* and grasses.

Soil horizons	Depth below surface (cm)	Description
A1	0–23	Dull yellowish brown (10YR 4/3m) and dull yellowish brown (10YR 5/3d) pebbly fine sandy loam; granular structure; friable moist consistence; slightly plastic and slightly sticky; lower horizon boundary undulatory and irregular.
B2t	23–46	Light gray (2.5Y 7/1m) and dull yellow orange (10YR 7/2d) loam; cobbles; blocky structure; firm moist consistence; plastic and sticky.
IIC1	46–56	Dull yellow orange (10YR 6/3m) and grayish yellow brown (10YR 6/2d) pebbly loam; cobbles and boulders; blocky structure; friable moist consistence; slightly plastic and slightly sticky; calcium carbonate coatings on few large stones.
IIC2	56–91	Dull yellow (2.5Y 6/3m) and dull yellow orange (10YR 7/2d) pebbly loam; cobbles and boulders; blocky structure; firm moist consistence; plastic and sticky; calcium carbonate coatings on large stones.
IICn	91+	Grayish yellow (2.5Y 6/2m) and light gray (5Y 7/1d) pebbly loam; cobbles and boulders; massive structure; firm moist consistence; plastic and sticky.

Profile: FL2

Age: post-Pinedale

Location: Fremont Lake (Fig. 5)

Elevation: 2,275 m

Parent material: granite, quartz-dioritic gneiss, and granodioritic gneiss

Topography: intermediate moraine crest between inner Pinedale (FL1) and outer Pinedale (FL3) moraines

Vegetation: sagebrush *(Artemisia)* and grasses

Soil horizons	Depth below surface (cm)	Description
A1	0–10	Dull yellowish brown (10YR 4/3m) and grayish yellow brown (10YR 5/2d) pebbly sandy loam; cobbles; weak granular structure; friable moist consistence; slightly plastic and slightly sticky.
B2t	10–36	Dull yellowish brown (10YR 5/3m) and dull yellow orange (10YR 6/3d) sandy loam; cobbles; weak blocky structure; friable to firm moist consistence; plastic and sticky.
C1	36–58	Dull yellow (2.5Y 6/3m) and dull yellow orange (10YR 6/3d) pebbly sandy loam; cobbles and boulders; weak blocky structure; friable moist consistence; slightly plastic and slightly sticky.
C2	58–86	Dull yellow (2.5Y 6/3m) and dull yellow orange (10YR 7/2d) pebbly coarse sandy loam; cobbles and boulders; very weak blocky to massive structure; very friable moist consistence; slightly plastic and slightly sticky.
Cn	86+	Grayish yellow brown (10YR 6/2m) and light gray (5Y 7/2d) pebbly sandy loam; cobbles and boulders; massive structure; friable moist consistence; slightly plastic and slightly sticky.

Profile: FL3

Age: post-Pinedale

Location: Fremont Lake (Fig. 5)

Elevation: 2,313 m

Parent material: granite, quartz-dioritic gneiss, granodioritic gneiss, gabbro-dioritic gneiss, and quartz-monzonitic gneiss

Topography: outermost Pinedale moraine crest

Vegetation: sagebrush *(Artemisia)* and grasses.

Soil horizons	Depth below surface (cm)	Description
A1	0-13	Brown (10YR 4/4m) and dull yellowish brown (10YR 5/3d) pebbly fine sandy loam; cobbles; granular structure; friable to firm moist consistence; slightly plastic and slightly sticky.
IIB2t	13-28	Grayish yellow brown (10YR 6/2m) and dull yellow orange (10YR 6/3d) pebbly sandy loam; cobbles and boulders; blocky structure; firm to friable moist consistence; plastic and sticky.
IIC	28-66	Dull yellow orange (10YR 6/3m) and dull yellow orange (10YR 7/3d) pebbly sandy loam; cobbles and boulders; weak blocky structure; friable moist consistence; nonplastic and nonsticky.
IICn	66+	Light gray (10YR 8/1m) and light gray (5Y 7/1d) pebbly sandy loam; cobbles and boulders; massive structure; firm moist consistence; plastic and sticky.

Profile: FL4

Age: post-Bull Lake

Location: Fremont Lake (Fig. 5)

Elevation: 2,270 m

Parent material: granite, quartz-diorite gneiss, granodioritic gneiss, granitic gneiss, and limonite

Topography: inner Bull Lake moraine crest

Vegetation: sagebrush *(Artemisia)* and grasses

Soil horizons	Depth below surface (cm)	Description
A1	0-13	Dull yellowish brown (10YR 4/3m) and grayish yellow brown (10YR 5/2d) pebbly fine sandy loam; granular structure; friable moist consistence; slightly plastic and slightly sticky.
B21t	13-23	Brown (7.5YR 4/4m) and dull yellow orange (10YR 6/3d) pebbly sandy clay loam; weak blocky structure; firm to very firm moist consistence; plastic and sticky.
B22t	23-38	Dull yellowish brown (10YR 5/4m) and dull yellow orange (10YR 7/3d) pebbly sandy clay loam; cobbles; blocky structure; firm moist consistence; plastic and sticky; roots up to 15 mm penetrate into Clox horizon.
Clox	38-74	Dull yellow orange (10YR 7/2m) and light gray (10YR 8/2d) pebbly sandy clay loam; cobbles and boulders; blocky and prismatic structure; firm moist consistence; plastic and sticky.
C2	74-114	Dull yellow (2.5Y 6/3m) and grayish yellow (2.5Y 7/2d) pebbly sandy clay loam; cobbles and boulders; weak blocky structure; firm moist consistence; plastic and sticky.
Cn	114+	Dull yellow (2.5Y 6/3m) and light gray (5Y 7/2d) pebbly sandy clay loam; cobbles and boulders; massive structure; firm moist consistence; plastic and sticky.

Appendix 3 Continued. Soil profiles in the Fremont Lake area.

Profile: FL5

Age: post-Bull Lake

Location: Fremont Lake (Fig. 5)

Elevation: 2,240 m

Parent material: granite, quartz-dioritic gneiss, and granodioritic gneiss

Topography: outer Bull Lake moraine, slope 8°

Vegetation: sagebrush *(Artemisia)* and grasses

Soil horizons	Depth below surface (cm)	Description
A1	0-15	Dull yellowish brown (10YR 4/3m) and grayish yellow brown (10YR 5/2d) pebbly sandy loam; cobbles; granular structure; firm moist consistence; slightly plastic and slightly sticky.
B21t	15-28	Grayish yellow brown (10YR 5/2m) and grayish yellow brown (10YR 6/2d) pebbly sandy clay loam; cobbles; blocky structure; firm moist consistence; plastic and sticky.
B22t	28-51	Dull yellow orange (10YR 6/3m, d) pebbly sandy clay loam; cobbles and boulders; blocky structure; firm moist consistence; plastic and sticky.
B23t	51-81	Dull yellow orange (10YR 6/4m) and grayish yellow (2.5Y 7/2d) pebbly sandy clay loam; cobbles and boulders; blocky structure; firm moist consistence; plastic and sticky.
Cox	81-125	Dull yellow orange (10YR 6/3m) and light yellow (2.5Y 7/3d) pebbly sandy clay loam; cobbles and boulders; massive structure; firm moist consistence; plastic and sticky.
Cn	125+	Dull yellow orange (10YR 6/3m) and grayish yellow (2.5Y 7/2d) pebbly sandy clay loam; cobbles and boulders; massive structure; firm moist consistence; plastic and sticky.

Profile: FL6

Age: post-Bull Lake

Location: Fremont Lake (Fig. 5)

Elevation: 2,249 m

Parent material: granite, quartz-dioritic gneiss, and granodioritic gneiss

Topography: outer Bull Lake moraine

Vegetation: sagebrush *(Artemisia)* and grasses

Soil horizons	Depth below surface (cm)	Description
A1	0-5	Dark brown (10YR 3/4m) and dull yellowish brown (10YR 5/3d) pebbly sandy loam; cobbles; granular structure; firm to friable moist consistence; slightly plastic and sticky.
B21t	5-15	Brown (7.5YR 4/3m) and dull yellow orange (10YR 6/4d) pebbly sandy clay loam; cobbles; blocky structure; firm to very firm moist consistence; very plastic and very sticky.
B22t	15-24	Dull yellowish brown (10YR 5/4m) and dull yellow orange (10YR 6/3d) pebbly sandy clay loam; cobbles and boulders; blocky structure; very firm moist consistence; very plastic and very sticky.
B23t	24-33	Dull yellowish brown (10YR 5/3m) and dull yellow orange (10YR 7/3d) pebbly sandy clay loam; cobbles and boulders; blocky structure; firm moist consistence; very plastic and very sticky.
B24t	33-51	Yellowish brown (2.5Y 5/3m) and dull yellow orange (10YR 7/2d) pebbly loam; cobbles and boulders; blocky structure; firm moist consistence; very plastic and very sticky.
C1	51-89	Grayish olive (5Y 6/2m) and light gray (2.5Y 7/1d) pebbly loam; cobbles and boulders; weak blocky structure; very friable moist consistence; very plastic and very sticky.
C2	89-135	Grayish olive (5Y 5/2m) and light gray (5Y 7/2d) pebbly sandy loam; cobbles and boulders; weak blocky structure; firm moist consistence; very plastic and very sticky.
Cn	135+	Dull yellow (2.5Y 6/3m) and grayish yellow (2.5Y 7/2d) pebbly coarse sandy loam; cobbles and boulders; massive structure; friable moist consistence; slightly plastic and nonsticky.

[a] Terms and horizon nomenclature follow the Soil Survey Staff (1951, 1960).

[b] Soil pits (e.g. FL1) are on Figure 5.

The Glacial Geology of the Copper Basin, Idaho: A Morphologic, Pedologic, and Proventologic Approach

William C. Wigley, Thomas A. Pasquini, Edward B. Evenson

ABSTRACT

The present surficial geology of the Copper Basin, Idaho, is largely the product of three late-Quaternary glaciations. A fourth, older glaciation may account for a single exposure of an elevated bedded gravel, but the evidence is inconclusive. The deposits of the three youngest glaciations are tentatively correlated with the Bull Lake, Pinedale, and Neoglacial glaciations of the Wind River Mountains, Wyoming. Arcuate terminal moraines sourcing well preserved terrace sets indicate that: the Bull Lake Glaciation probably consisted of only a single stade, the Pinedale Glaciation consisted of four stades, and the Neoglacial consisted of a numerous, but as yet undetermined number of intervals.

Differentiation of complex lobal affinities and cross cutting relationships between the thirteen valley glaciers which flowed into the Copper Basin are delineated from the analysis of morainic morphology, stratigraphic and down valley position. Quantitative support for these relationships and delineation of additional relationships are obtained from detailed provenance investigation of pebble and boulder lithologies and heavy mineral suites. Five source areas are defined on the basis of these parameters. Each source area is uniquely defined by characteristic indicator lithologies and heavy mineral suites.

Although clearly differentiable from analysis of morphologic and stratigraphic criteria, the till of the Pinedale and Bull Lake Age moraines cannot be differentiated on the basis of pedologic development. Further, the quantitative measurements of weathering features on surface erratics are of limited use in the distinction of deposits on a local level, and meaningless in the development of basin wide correlations.

Strong clay mineral alteration is evidenced within the soils developed in the basin, but shows no consistent relationship to the apparent age of the soil. Illite dominates surface horizons, forming from the mechanical breakdown of free soil micas, and is apparently altering to montmorillonite at depth in a poorly leached cation rich environment.

INTRODUCTION

The objectives of this investigation are twofold. First, to locate, map and assign ages to all glacial deposits in the Copper Basin utilizing detailed investigations of morphologic character, stratigraphic associations and provenance. Second, to analyze a number of parameters as tools for quantitative differentiation of those deposits. This second phase focuses on the alteration of surface erratics and formation of soils.

Previous Work

The quantitative aspects of glacial geology and pedology have received much attention in recent years. Numerous workers (Porter, 1969, 1975; Carroll, 1974; Sharp, 1969) have advocated the use of surface boulder features as criteria for age distinction. Soils have been used extensively (Richmond, 1962a, 1964a; Birkeland, 1973; Scott, 1977) as an aid to the subdivision of local glacial successions, to provide information on the length of interglacial intervals and to facilitate local and regional correlations throughout the Western United States (Mears, 1974; Scott, 1977). Quantitative examination of pedologic properties of soils (texture and clay mineral suites) has also been successfully utilized (Holmes and Moss, 1955; Birkeland and Janda, 1971; Stewart and Mickelson, 1976; Mahaney, 1974 and this volume) to differentiate glacial advances.

The Copper Basin area first received geologic attention in connection with the various mining industries present in the Basin and surrounding areas. The main interest was centered around investigations of the replacement mineralization associated with the Tertiary intrusions, (Farwell and Full, 1944; Ross, 1937; Sweeney, 1957; Umbleby, *et al.*, 1930) and Paleozoic sedimentary facies changes (Paull, *et al.*, 1972; Erwin, 1972; Volkmann, 1972; and Rothwell, 1973). There have been no specific studies of the extensive glacial deposits of the basin although several (Sweeney, 1957; Nelson and Ross, 1969; Erwin, 1972; and Rothwell, 1973) investigations include mention of glacial activity. Nelson and Ross (1969) mapped and described two tills which were inferred to be of Pinedale and Bull Lake age, however these deposits were mapped only in a generalized manner (scale 1:62,500).

Location and Setting

The Copper Basin is a small (400 km^2) intermontane

basin located in South-Central Idaho (Figure 1). The basin separates the White Knob and Pioneer Mountains of Custer County, and lies approximately equidistant (32 km) from the metropoli of Arco and Ketchum, Idaho, on the southeast and northwest respectively. The maximum relief in the basin is approximately 1300 m (4250 ft). The lowest point, 2300 m (7540 ft), is along the east fork of the Big Lost River at the distal end of the basin, with elevations rising to a maximum of 3600 m (11,800 ft) along the northwest rim of the basin. Sixteen major canyons encircle the basin, forming the headwaters of the east fork of Big Lost River (Figure 2).

The geomorphic origin of the basin is not positively known. Tertiary block faulting is the general consensus of previous workers (Sweeney, 1957; Nelson and Ross, 1969). Possibly this faulting was the result of isostatic adjustment due to the depression of the adjacent Snake River Lava Plains that began in Oligocene time.

The bedrock of the Copper Basin is extremely diverse. General lithologies include the predominantly clastic Copper Basin Group of late Paleozoic age, three acidic Tertiary plutons, and the predominantly andesitic Challis Volcanics of Tertiary age. The distribution of the major lithologic units in the Copper Basin will be described and discussed in a subsequent section (provenance) of this paper.

The climate is typical of an alpine semi-arid intramontane basin. Rainfall is 40-50 cm (16-20 in) per year, averaging 46 cm. Maximum precipitation occurs during the winter months (Dec.-Jan.) averaging 5.3 - 5.6 cm (2.1 - 2.2 in). During the summer months (Aug.-Sept.) rainfall averages only 0.5 cm (0.2 in) per month. The mean annual air temperature is 6.4°C (43.5°F) with a July average of 20°C (68° F) and January average of -7.4°C (18.7°F). Vegetation is that of a Sagebrush Steppe Community (Knoll and Dort, 1973) composed primarily of *Artemisia tridentata* (big sage), *Festuca idahoensis* (Idaho fescue), *Agropyron spicatum* (bluebunch wheatgrass) and scattered communities of *Poa* (sandburg bluegrass) and *Phlox* (longleaf and Hood's phlox).

SURFICIAL GLACIAL GEOLOGY

The location and age of all glacial deposits in the basin, mapped on the basis of morphologic and stratigraphic association and provenance are shown in generalized form on Figure 3. The deposits are assigned to the pre-Bull Lake, Bull Lake, Pinedale, and Neoglacial glaciations of the Rocky Mountain model (Mears, 1974) but as this correlation is indirect, these assignments should be regarded as extremely tentative. For a detailed map and discussion of the glacial features, see Wigley (1976, Plate 1 and text).

Pre-Bull Lake Deposits

A small slump scar in the mouth of Broad Canyon exposes

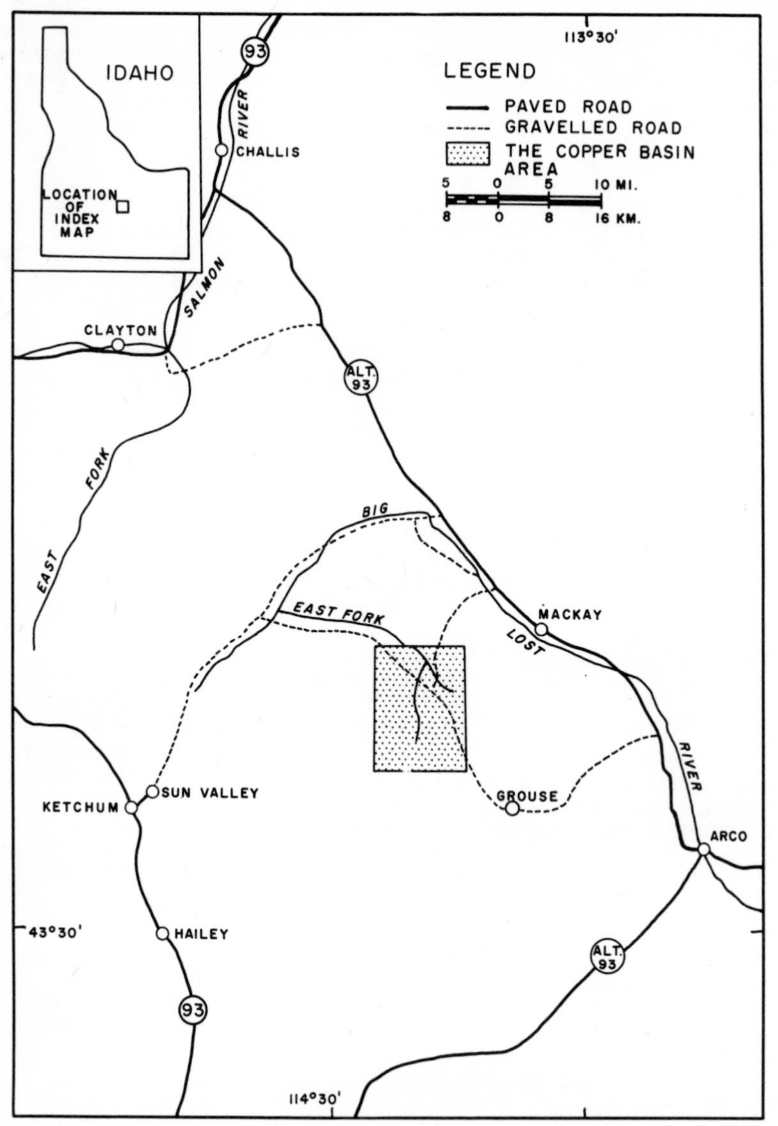

Figure 1 Location map of the Copper Basin, Idaho. Stippled area marks approximate boundaries of the basin drainage as shown in Figure 2.

a high gravel which is assigned a Pre-Bull Lake age. The gravel is 250 m above the present Broad Canyon drainage, and consists of large (10-15 cm) well-rounded, bedded cobbles, composed primarily of fresh Challis Volcanics. A distinctive red (2.5YR 5/6) color is well developed and seems to be primarily a product of staining of the sand matrix. Highly altered boulders are also exposed at the

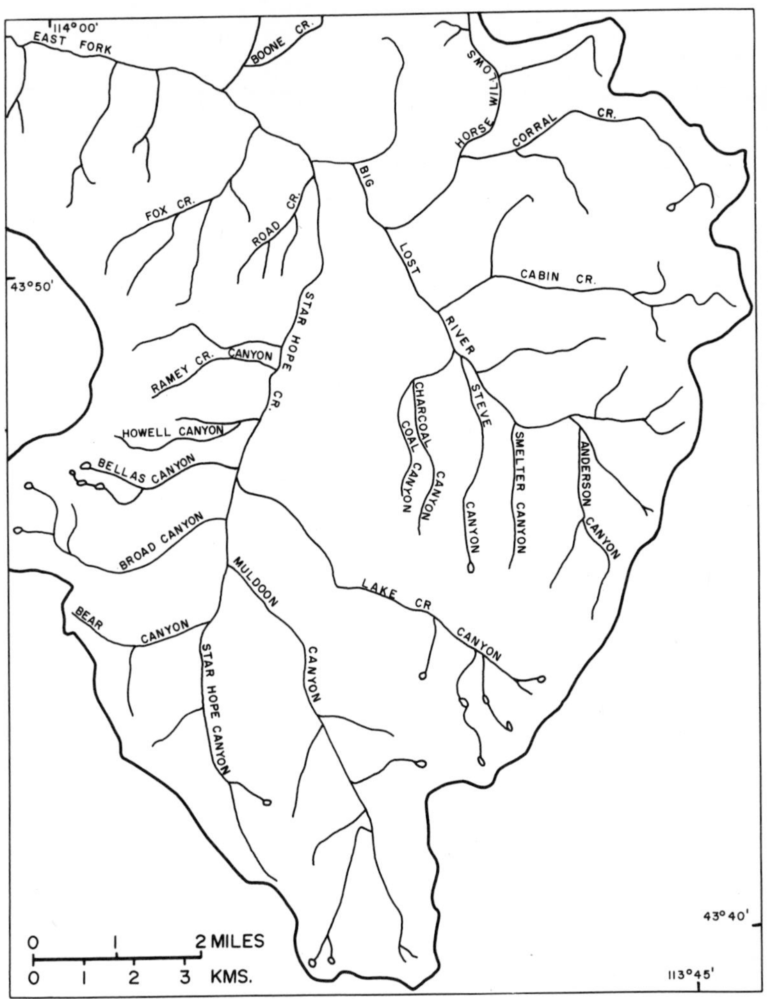

Figure 2 Major drainages of the Copper Basin, Idaho. Area roughly corresponds to the stippled area in Figure 1.

top of the slump scar, however these do not appear to be stratigraphically or lithologically related to the gravels. The altered boulders are granitic and are assumed to be part of a Bull Lake till sheet veneering the gravels.

The topographic position of these gravels suggests two possible origins; fault control or deposition prior to formation of the present drainage system (Knoll and Dort, 1973). Evidence to support any structural control is lacking, and at this time, the latter hypothesis is favored but not demonstrated. Several workers (Richmond, 1973; Dort, 1962; Knoll and Dort, 1973) have assigned similar

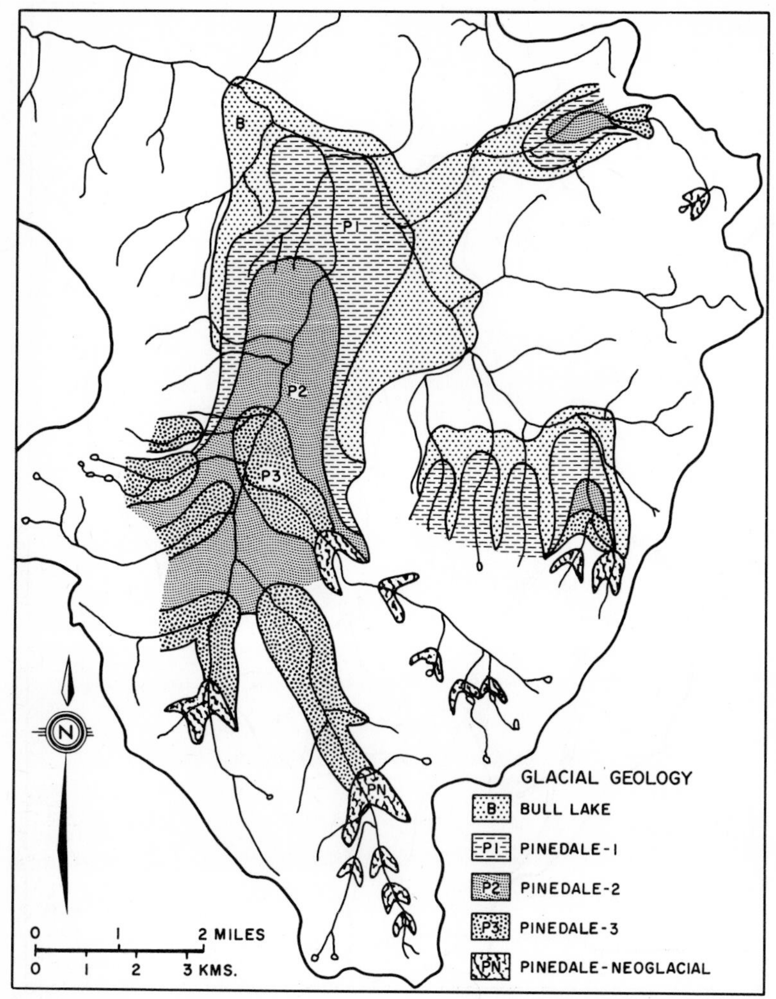

Figure 3 Generalized map of glacial deposits in the Copper Basin, Idaho. For detailed map of features, see Wigley, 1976.

elevated outwash remnants to the pre-Bull Lake, and it seems possible that this is the case in the Copper Basin.

Further evidence to support a pre-Bull Lake Glaciation in the basin was not located. Drainage divides and elevated flats were thoroughly investigated. If pre-Bull Lake deposits were ever present at any locality in the basin other than the gravels previously mentioned, they have evidently been removed or covered by subsequent activity.

Bull Lake Deposits

Till of the Bull Lake Glaciation is the oldest well

preserved glacial deposit that clearly postdates the erosion of the deep valleys ringing the basin. Generally, Bull Lake till extends 0.5 to 1.0 km farther down valley than the Pinedale terminal moraines (Figure 4). High elevation remnants of Bull Lake lateral moraines are also present at various locations up-valley from Pinedale terminal moraine positions where they occur at elevations higher than that covered by Pinedale lateral moraines (Figure 3).

Bull Lake moraines are distinguished from younger deposits on the basis of morphology and down valley position. Moraines are large and bulky and well dissected by tributary streams. Moraine surfaces are irregular but mature in appearance. Surface depressions are generally filled or breached by drainage, with drainage moderately well developed on slopes. Surface boulder frequencies are low to moderate, characterized by scattered large (2-4 m) boulders with relatively few intermediate (0.5-2 m) sizes. Boulder surfaces are well pitted, removing all evidence of ice scour features.

Figure 4 Pinedale and Bull Lake moraine and outwash complexes at the mouth of Anderson Canyon (Figure 2). Note subdued topography on Bull Lake deposits and contrasting "fresh" topography on Pinedale deposits.

Pinedale Deposits

Pinedale age moraines comprise 80% of the glacial deposits preserved in the basin. Pinedale deposits are easily distinguished on the basis of their surface morphology. The amount of surface relief is considerable (± 10 m). Moraines are steep and sharp crested. Surfaces are rugged and extremely kettled. Depressions are not breached, and many contain water. Drainage is non-existent or poorly developed, and strongly controlled by the trend of lateral moraines (Figure 5). Surface boulder frequencies are very high, encompassing a complete spectrum of sizes. Some surface boulders appear as highly weathered as their Bull Lake equivalents, however subsurface weathering is less severe with ice scour features being preserved on protected surfaces.

Four sets of Pinedale moraines associated with well developed terrace sets constitute the main evidence that Pinedale Glaciation in the basin consisted of four distinct stades, designated as Pinedale I, II, III, and IV (Figure 3). However, the Pinedale IV stade may incorporate deposits more correctly identified as belonging either to the Pinedale III or the pre-Altithermal Temple Lake Glaciations. Evidence is lacking to positively identify any moraine as a Temple Lake equivalent.[1] Because the deposits are clearly younger than those mapped as Pinedale III, the use of the "Pinedale-Neoglacial" designation is adopted. Arcuate trends of minor elevated ridges behind major Pinedale moraines (Figure 5) suggest minor readvances or controlled disintegration associated with downwashing.

Neoglacial Deposits

Neoglacial activity in the Copper Basin is evidenced by numerous rock glaciers, protalus ramparts, and extensive talus deposits. Cross-cutting and overriding relationships suggest multiple ages for these deposits. However, these relationships were not documented in detail and all deposits are grouped for mapping purposes as undifferentiated rock glaciers or talus accumulations. Climatic and insolation effects were critical in the formation of Neoglacial ice masses. Not all cirques occupied by the Pinedale ice exhibit Neoglacial features, and all Neoglacial deposits are restricted to high level, well protected locations.

METHODS

Provenance

At each sample location (N=84), a pit was excavated and 100 pebbles (5 to 8 centimeters) were collected, split and identified. From the base of each pit, 100 grams of till matrix was collected for heavy mineral analysis. Also, at each sample site, 25 surface boulders from a small radius

---

[1]See Mahaney (1975, and this volume) for a discussion of the term Temple Lake.

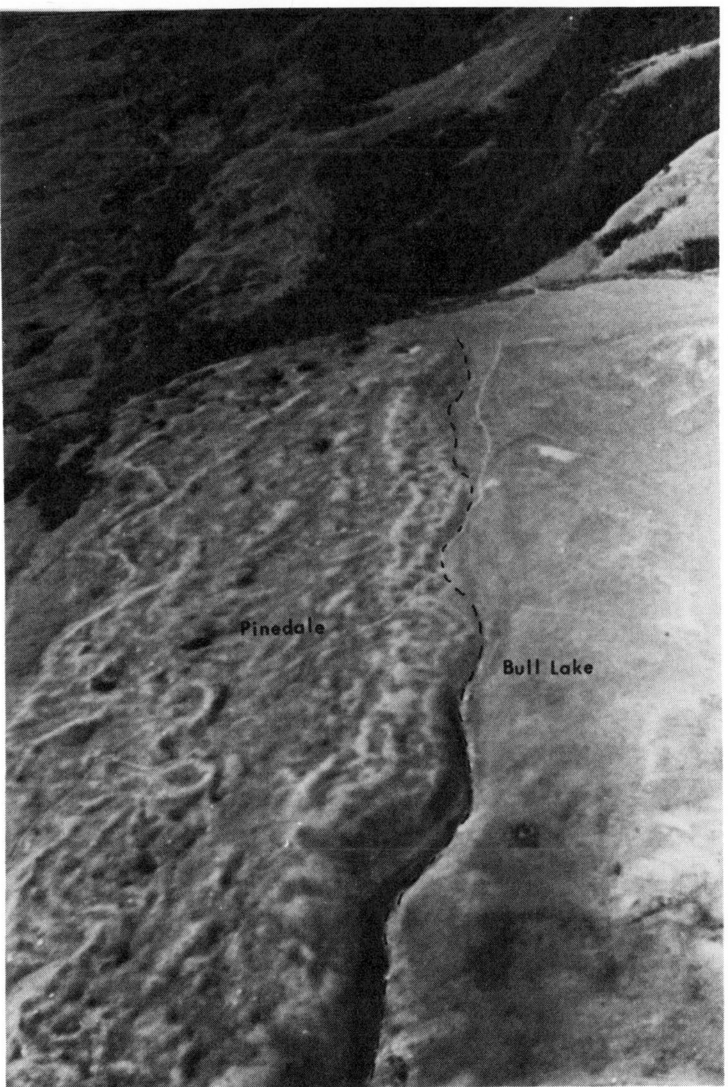

Figure 5 Morphology of Pinedale and Bull Lake moraines near Ramey Creek (Figure 2). Ice disintegration features are fresh on Pinedale moraine and absent on Bull Lake.

around the pit were lithologically classified. Additional boulder counts were taken at significant locations (N=161).

Samples taken from the base of selected sample pits (N=50) were sieved and the 3 phi size fraction was selected for standard heavy mineral analysis. Two hundred and fifty grains per slide were identified.

All field and laboratory data was analyzed by computerized statistical techniques. Boulder and pebble count data were analyzed by a Q-mode cluster analysis. Heavy mineral data was first examined by R-mode factor analysis. Heavy minerals indicated as most significant from the factor analysis were then contoured. Normalized factor loading values for each heavy mineral sample site were grouped into populations by the Q-mode cluster analysis.

Pedology

Soil samples were collected from 21 dug pits. 200 gram samples were taken at the textural "A" horizons, at the textural and color "A"/"B" contact and at 15 cm intervals throughout the "B" horizons. All samples (N=121) were pretreated for the removal of organic material and carbonate utilizing the methods of Jackson (1956). After sieve and pipette size analyses [Ingram, 1971 (sieve); Galehouse, 1971 (pipette)] the less than two micron fraction was retained for X-ray analyses. Clay sized sediments were freed of amorphous iron and manganese by the sodium dithionite-sodium citrate extraction technique of Mehra and Jackson (1960). Final dispersion and clean-up of the clays in a sodium saturated environment was conducted according to the methods of Heath (1968).

Oriented, glass mounted slides of the ($<2\mu$) fraction were prepared by the filter-membrane peel technique of Drever (1973). All slides were glycolated prior to X-ray analysis according to the methods of Brunton (1955). The samples were X-rayed from 2° to 14° 2θ, on a standard Norelco wide angle X-ray diffractometer using high intensity nickel filtered copper Kα radiation. Relative percentages of the various clay mineral species present were determined by measuring peak areas under appropriate peaks with a polar planimeter.

Illite was identified by a sharp symmetrical peak, between 9.4Å and 9.8Å (Figure 6), that was not effected by glycolation or heating. Illite is used in a broad sense to include all "mica" groups, as no attempts were made to distinguish polymorphs.

Chlorite displayed a sharp peak between 13.0 and 13.4Å in both a glycolated and heated state. Chlorite was probably mixed with montmorillonite (Biscaye, 1964) to form a mixed peak around 11.3Å in the untreated state.

Montmorillonite showed one or more broad peaks between 13.4Å and 15.2Å. These peaks expanded to 15.5Å upon glycolation and collapsed to 11.2Å upon heating to 550° for 2 hours (Figure 6).

A mixed kaolinite-chlorite group diffracted at 7Å in both an untreated and glycolated state with kaolinite disappearing upon heating.

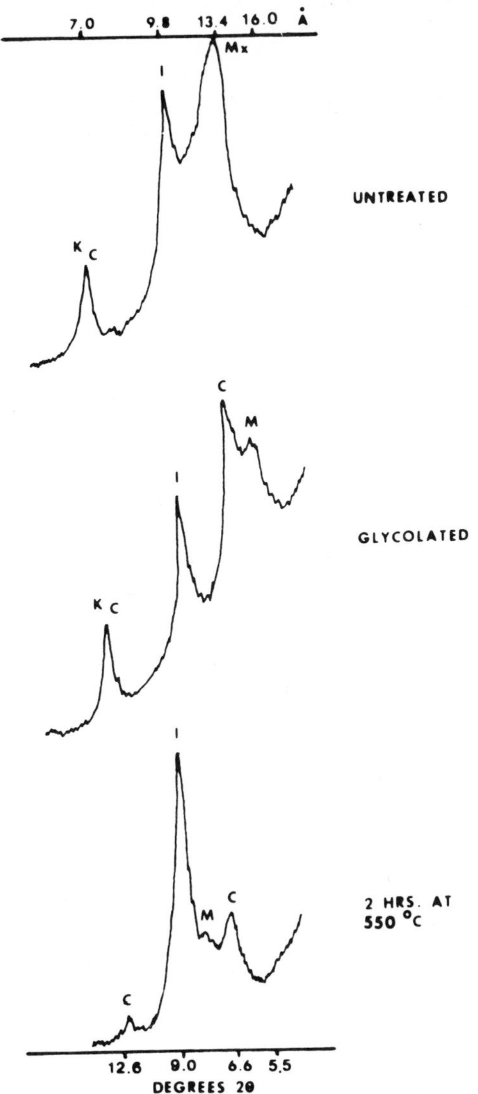

Figure 6 Examples of Diagnostic, Na Saturated X-ray diffractograms. C, Chlorite; I, Illite; K/C, Kaolinite and Chlorite; M, Montmorillonite; Mx, Mixed.

Statistical Treatment of Soil Data

A search for quantitative parameters useful in the differentiation of till units was based on statistical analyses of data generated from field and laboratory results.

121 samples were described by eight variables:

1) percentage of 5.5Å clay (montmorillonite)
2) percentage of 6.7Å clay (chlorite)
3) percentage of 9.0Å clay (illite)
4) percentage of 12.5Å clay (kaolinite/chlorite)
5) normalized depth of sample[a]
6) percentage of sand
7) percentage of silt
8) percentage of clay

[a]The total depth of soil development was extremely variable across the basin. Soils of similar ages were not necessarily developed to the same depth. Correlations, whether stratigraphic or statistical, would have been erroneous if effected by absolute (abs.) depth, so a normalized (norm.) depth term defined as:

$$\text{depth (norm)} = \frac{\text{depth (abs)}}{\text{thickness of B horizon}} \tag{1}$$

was substituted for each absolute sample depth to eliminate this problem.

All data were analyzed by R-mode factor analysis. Essentially, this technique searches the data space for those combinations of variables which minimize the error variance throughout the entire sample population. Each combination of variables (factor) accounts for a certain percentage of the total sample variance and the program generates additional factors until the collective variances exceeds a designated value (80%) or the variance increment accounted for is less than 5% of the total variance. The analysis assigns normalized factor coefficients or loadings (0 to ± 1.0) to the individual variables comprising each factor, and in doing so numerically illustrates correlations between variables. In addition to a variable correlation matrix, the R-mode factor analysis also generated normalized factor measures for each sample. These factor measures replace the original data matrix, and are analyzed by Q-mode cluster analysis using a distance function as a classifying criteria to ascertain whether the data defined groups that were areally or stratigraphically related. The sequence of programs was modified from Parks (1970).

It should be noted that the use of data from closed number systems (relative clay mineral abundance, and size fraction percentages) in a factor analysis may yield spurious results in the correlation matrix (Chayes, 1960). The degree to which these closed number systems affect the results is probably tempered by the use of two unrelated systems. All closed number effects were probably removed by drawing final conclusions only from the actual data, and not the factor or cluster analysis. That is, these statistical treatments were not used as end-products, rather they revealed the significant variable and sample site combinations which were then further studied by referring back to the actual data.

RESULTS

Provenance

The source areas for all glacial deposits in the basin are clearly delineated by analysis of boulder and pebble lithologies and heavy mineral suites in the till. Figure 7 shows the location of 84 boulder counts, 84 pebble counts and 50 identified heavy mineral suites.

Using R-mode factor analysis and Q-mode cluster analysis, five major source areas are defined (Figure 8; b-f). Each source area is characterized by distinctive indicator lithologies and heavy mineral suites as shown by arrows. Contour diagrams of individual heavy minerals show similar trends (8e-f). The percentage of diopside plus epidote (Figure 8e) shows a strong source from Lake Creek Canyon. The diopside is derived from contact metamorphism of an isolated outcrop of limestone (Figure 8a) and the epidote is sourced from the Challis Volcanics. Monazite (Figure 8f) shows a strong source from the western canyons: Broad, Bellas, Howell, and Ramey Canyons. The monazite is sourced from the pink granite outcropping in this area (Figure 8a).

Many minor ice flow patterns that likely would have been overlooked in a standard morphologic mapping program were determined or confirmed by our provenance investigations. These include: up canyon ice flow (Figure 8b), cross-cutting relationships, topographic diversion, and watershed breaching by ice (Figure 8b). These flow peculiarities will not be discussed here (for details see Pasquini, 1976) but have been incorporated in the construction of the glacial map of the basin (Figure 3). Information gained from provenance investigations in this study demonstrates that it is an invaluable tool required for a complete understanding of the ice flow history of a complex area.

Quantitative Distinction of Glacial Deposits

Boulder Weathering. Although widely applied (Porter, 1969, 1975; Carroll, 1974) as criteria for age distinction, surface boulder features are of limited use for assigning relative age to drifts in the Copper Basin. The textural character of the diverse bedrock lithologies (Figure 8a) sourcing the surface boulders at differing locations in the basin were such that the use of three quantifiable weathering indicators:

1) percent of pitting on exposed boulder surfaces,
2) maximum depth of pitting on exposed boulder surfaces and,
3) maximum depth of weathering rinds developed on exposed boulders,

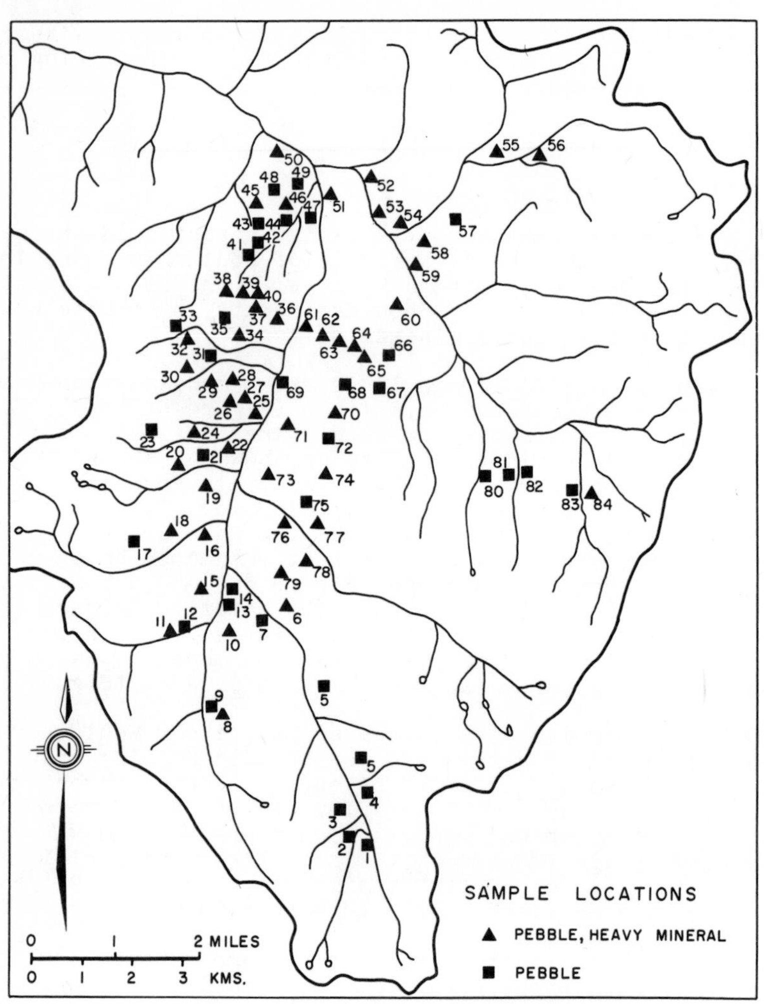

Figure 7 Location of pebble and heavy mineral sample sites. Boulder lithology analysis was performed at all sites shown.

varied so widely, strictly as a function of local lithology, as to render relative age assignments based on these parameters meaningless. The igneous lithologies of the basin, while extremely useful as indicators of source area (Figure 8b), exfoliated so rapidly that age designations based on small scale surface irregularities probably reflected age in hundreds, rather than thousands, of years.

Soil Development. The Copper Basin is an excellent locality for the investigation of pedogenic development.

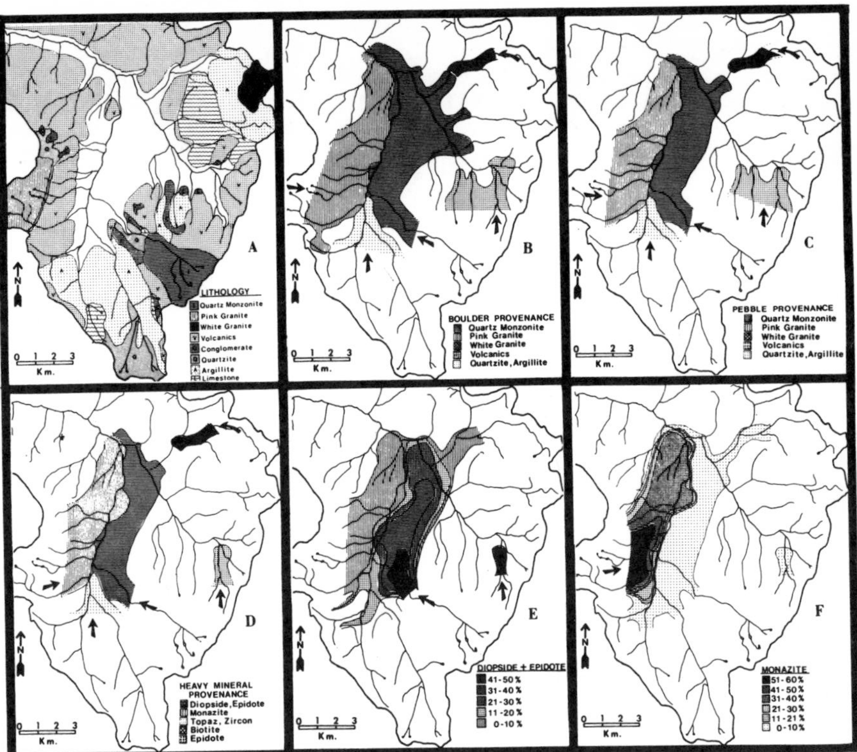

Figure 8 Generalized bedrock geology (A); Boulder Provenance (B); Pebble Provenance (C); Heavy Mineral Provenance (D); Contour Diagrams for Diopside and Epidote (E) and Monazite (F). Note clear indication of source area (arrows) for each component.

A number of the major soil forming factors can be considered areally constant. All of the sites selected for pedogenic study are encased between the 2190 and 2320 m (7200 and 7600 ft) contours and the vegetative cover (Sagebrush Steppe Community; *Artemisia tridentata*, *Poa* and *Festuca idahoensis*), rainfall (40.6 - 50.8 cm, ave. 45.7 cm) and temperature are essentially uniform. Parent material (till) lithology is also very uniform. The till lithology differences that are so clearly demonstrated by our provenance investigations are a function of subtle differences (grain size, texture) in the igneous rocks that cropout around the basin. The bulk of the till matrix is composed of a relatively uniform admixture of materials obtained from the more easily erodable clastic sequence which occurs throughout the basin. Unweathered till from various parts of the basin is virtually identical in composition and granulometric properties.

If the above conditions are indeed uniform and if soil

sites are "slope matched" and carefully investigated to avoid colluviation and/or erosion, we can then assume that any variation in pedogenic development is a function of: 1) time elapsed since deposition of the till and, 2) past climatic variation--the same variables that are responsible for the observed, and previously described, differences in morphology on deposits mapped as Pinedale and Bull Lake.

For detailed pedologic investigation, a total of 19 soil pits were dug, described and sampled (four on deposits of assumed Bull Lake age, 12 on deposits mapped as Pinedale and 3 on deposits of post-Pinedale age). Age assignments (BL=Bull Lake; P=Pinedale) were made on the basis of morphologic criteria discussed previously.

Field observations and laboratory investigations combined with preliminary statistical analyses indicated that, in order to gain any meaningful results from the statistical interpretations of soil profile data, profiles had to be subdivided into equivalent categories. Sites characterized by poorly developed profiles: BLE-1, PE-1, PE-4, PE-6, PE-7, PS-1, PW-4, PW-5, MC-1, MC-2, and MC-3; were strongly developed: BLE-2, BLW-1, BLW-2, PE-2, PE-5, PW-1, PW-2, and PW-3 (see Wigley, 1976; plate 1 for locations). Data from the poorly developed profiles was withdrawn from further consideration because meaningful trends which might have developed in these profiles were evidently so altered by external influences as to render their distinction impossible. The poorly developed profiles were not restricted to young deposits, and were in fact located on moraines of all ages, suggesting that these soils were not true relict soils, but that they had developed under the influence of localized erosional or colluvial effects. All further discussion of soils, in this study, is based upon the eight moderately developed profiles, and not the eleven judged to be altered.

All soils analyzed are calciorthids (Cca horizon present) or camborthids, where the calcic horizon is not developed at the base of the solum (see Appendix 1). Profiles are weakly to moderately developed, characterized by ochric epipedons, cambic "B" horizons, and in most cases a strongly effervescent Stage I (Birkeland, 1974) calcium carbonate accumulation at the base of the solum. In general, these soils are indicative of a poorly leached, closed chemical system (Buol, *et al.*, 1973) typified by high base saturation levels in the cambic horizons.

Soil Texture. Texturally, the soils of the Copper Basin are loams at the surface which grade into sandy loams with depth (Appendix 1). Statistical treatments indicate that textural variations show no stratigraphically (assigned age) or areally recognizable trends. Factors 2 and 3 indicate that sand, silt, and clay percentages vary independently of depth. Cluster analyses of the factor values grouped individual samples in a geologically meaningless manner. That is, statistical treatment of textural variations suggests that there are no significant variations within the soils sampled from the standpoint of either age (as assigned on morphologic criteria) or depth. Inspection of several typical profiles supports this conclusion. Replicate sample analyses further revealed that minor variations within individual samples (on the order of 2-3%) were not reproduceable and also not significant. A single profile (BLW-2, Appendix 2) is the lone exception and will be discussed later in this section. All other profiles show no significant textural variation with either depth or age.

Texture, defined as sand, silt and clay percentages, is a rather insensitive indicator of minor textural variations. To verify this hypothesis and search for more subtle textural trends, four profiles (BLE-2B, BLW-2, PE-5 and PW-3) were subjected to a complete (0 to 11Ø interval) granulometric analysis. The results (Appendix 3) are similar to those of the gross sand, silt, clay ratios. Again, except for the BLW-2 profile, there is no significant textural variation with either depth or age.

Soil Clays. All profiles in the basin are characterized by cambic "B" horizons, defined on the basis of clay mineral alteration and color rather than textural (clay percentage) accumulation. Clay mineral species are altering strongly, however, compositional variations are closely related to depth and not to age. That is, the strong clay mineral alterations evident within any given profile are not consistant with other profiles of the same age. Therefore, the clay mineral alterations seen in the Copper Basin do not appear to be useful as a means of differentiating ages of glacial deposits. In short, clay mineral alterations are not useful (on a statistical or intuitive level) as indicators of relative age within this study area.

Clay Mineral Alteration. Although clay mineral alteration cannot be used to assign relative age, the individual clay mineral verses depth trends (Figures 9-11) do reveal interesting aspects of the soil forming environment on the basin floor. Illite, chlorite, and montmorillonite dominate the clay mineral species of the basin. Illite dominates surface horizons (Figure 9) and shows a strong tendency (4 = 0.68) to decrease rapidly with depth. The low strength of the chlorite trend (Figure 10, r= 0.29; r= 0.23 is the minimum value for significance; Snedecor and Cochran, 1967) suggests that chlorite is not playing a strong role in the clay mineral alteration. Kaolinite, while not analyzed directly, is inversely related to the chlorite content (see Appendix 2) and therefore decreases with depth. However, kaolinite is never present in quantity in any sample, and makes a negligible contribution to the clay fraction regardless of age or depth.

Textural variations indicate that there is no significant change in the total clay content accompanying these clay mineral alterations. Clearly then, the change in clay species cannot be related to illuvial effects or airborne contributions and must therefore be occurring from *in situ*

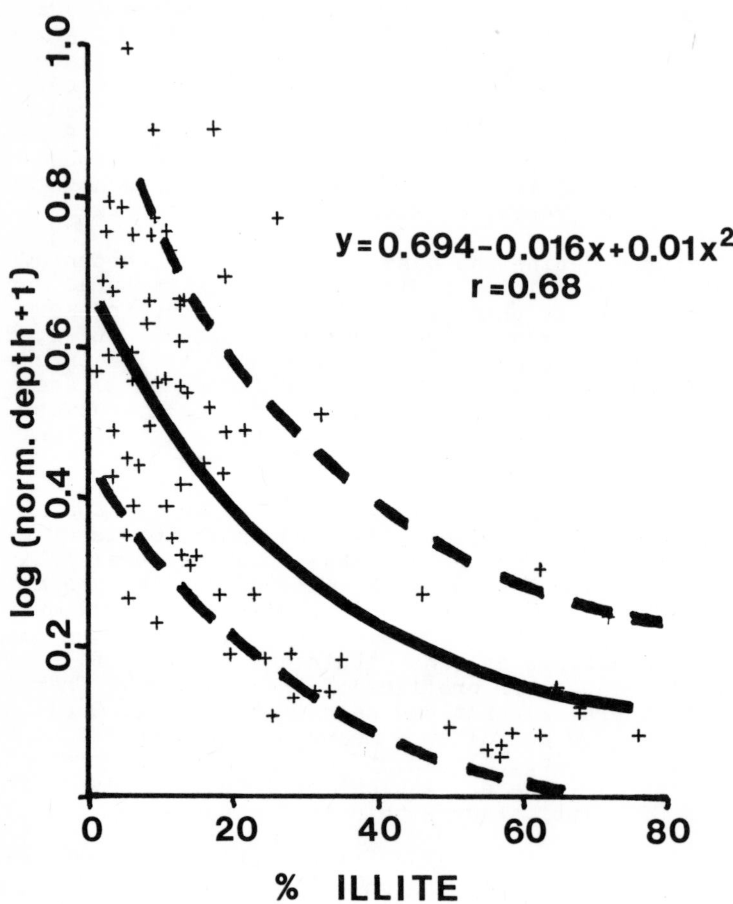

Figure 9 Percent illite vs. normalized depth for soil profiles. Plot reveals strong ($r = 0.68$) basin wide trend to decrease in illite with depth, a trend that is apparently unrelated to the age of the sample.

alteration. Illite, appearing in high concentrations in surface horizons, is probably the product of intense mechanical breakdown of free soil micas weathered from the granitic lithologies present in the parent till. In the poorly leached high base saturation environment of the cambic B horizons, this illite is apparently chemically altering to montmorillonite. Chlorite, not closely related to depth, is fairly uniform with depth, and probably primary in origin. If illite is in fact altering to montmorillonite, then the change is from a physically larger clay species to a smaller one; a change that should be reflected in the fine clay fraction. The detailed size analyses (Appendix 2) do not

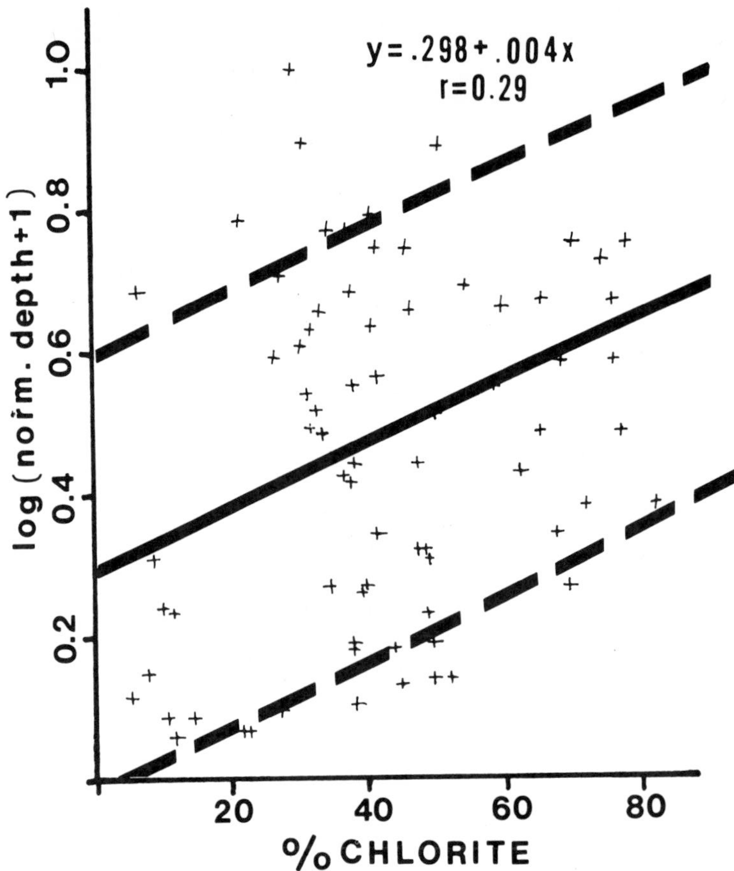

Figure 10 Percent chlorite vs. normalized depth for soil profiles. Plot indicates a weak (r = 0.29) increase in chlorite with depth. This trend is probably due to chlorite-poor surface samples with chlorite being essentially uniform with depth regardless of the age of the soil.

support this contention, except for the BLW-2 profile.

At this point in the investigation of the glacial geology of the Copper Basin, a multitude of questions have been answered, but a few are only partially resolved. In almost all cases, unanswered questions are directly related to the BLW-2 profile. In the light of the twenty other profiles analyzed, this BLW-2 profile is clearly unusual. This single profile best exhibits those characteristics commonly associated with age; *i.e.*, well developed soil structure, deep red color, compactness, and marked clay accumulation with depth. Further, the clay mineral suite is dominated by montmorillonite (88% in deepest sample) with 28% of the total clay fraction (32% of total size fractions) concentrated in the <11Ø fraction. Conclusions based on a single profile are not warranted, but the situation is enigmatic. Whether this profile represents a pre-Bull Lake soil or the only completely developed well preserved Bull Lake soil in the basin is not known.

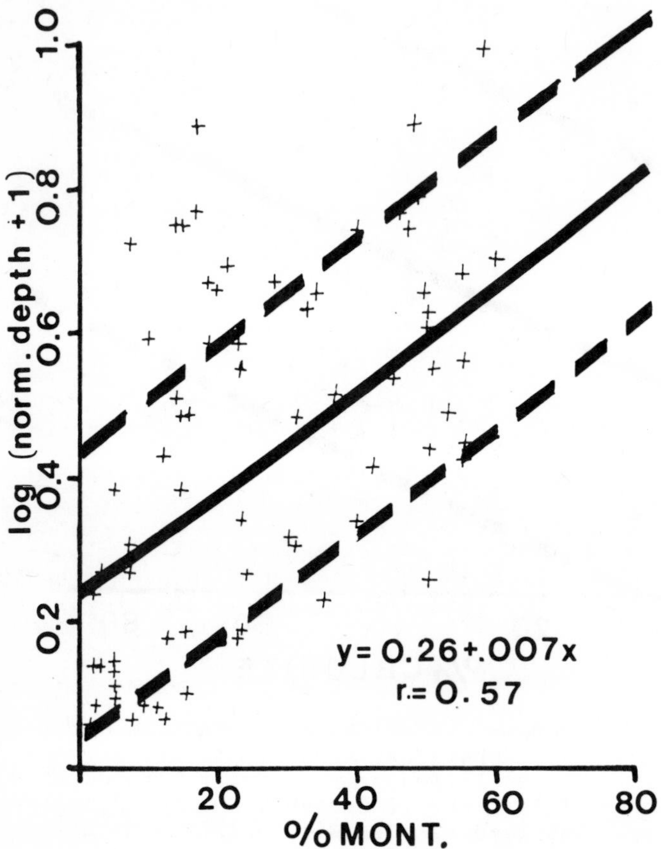

Figure 11 Percent montmorillonite vs. normalized depth for soil profiles. Plot reveals a relatively strong (r = 0.57) increase in montmorillonite content with depth throughout the basin, regardless of the age of the sample.

An alternate explanation for the apparent textural and clay mineral uniformity of Bull Lake and Pinedale soils is the possibility that all soils in the basin are at "terminal grade" for the climatic conditions under which they developed. That is, the soil forming processes of the basin are such that the uniform degree of soil development observed in both the Pinedale and Bull Lake profiles is at or very near the maximum that could be expected under the basin's soil forming environment regardless of the amount of time since deposition of the parent material. Without further data, this hypothesis cannot be completely dismissed. However, numerous authors (Madole, 1969; Mahaney, 1973a,b; this volume; and Mahaney and Fahey, 1976) have reported well developed profiles from environments similar to the basin suggesting that attainment of terminal grade is not occurring in those areas. Alternately, erosion of soils in the basin may be severe enough to be limiting pedogenic development to the uniform degree observed.

CONCLUSIONS

The present topography of the Copper Basin, Idaho, is largely the product of three late-Quaternary glaciations. A single exposure of an elevated bedded gravel may support an additional "Pre-Canyon" glaciation, however the evidence is not conclusive. The deposits of the three recognized glaciations are tentatively correlated with classical Rocky Mountain Glacial Model (Mears, 1974) as "Bull Lake" (oldest), "Pinedale" (middle), and "Neoglacial" (youngest) Glaciations.

Poorly preserved outwash remnants suggest that there may have been two Bull Lake stades, however morainic evidence to support more than one Bull Lake interval is lacking. Four arcuate terminal moraines sourcing well preserved terrace sets indicate that the Pinedale Glaciation consisted of 4 distinct stades. The deposits of the youngest stade (Pinedale IV) may include deposits more correctly correlated with either the Pinedale III stade, or a younger pre-Altithermal interval. Crosscutting and overriding relationships in the Neoglacial deposits of the basin suggest more than one interval of cooling for this glaciation, however this is not well documented.

Detailed analyses of Bull Lake and Pinedale Tills support the following conclusions:

(1) Differentiation of complex lobal affinities and crosscutting relationships between the thirteen valley glaciers which developed during the major glacial intervals can be clearly delineated from the combined analysis of:

a) morphologic character,
b) moraine position, and
c) detailed provenance investigations.

(2) Surface boulder weathering characteristics are of limited use in assessment of relative age on a local level, and of no use in the development of basin wide age correlations.

(3) The relict soils developed on till within the basin show no significant textural variations with either assigned age or depth.

(4) Strong clay mineral alterations are occurring throughout the basin, but show no consistent relationship to age. Clay minerals are developing from *in situ* alteration, and not translocation from the surface. Illite dominates surface horizons, probably forming from the physical breakdown of free soil micas. Montmorillonite is forming at depth from the chemical alteration of illite in a poorly leached cation rich environment. Chlorite is fairly uniform with depth and is not playing a major role in the clay mineral alteration.

(5) Based on the localities sampled, Pinedale and Bull Lake tills cannot be differentiated on the basis of pedologic development.

REFERENCES CITED

Birkeland, P.W., 1973, Use of relative age-dating methods in a stratigraphic study of rock glacier deposits, Mount Sopris, Colorado: Arctic Alpine Research, v. 5, p. 401-416.

_______________, 1974, Pedology, Weathering, and Geomorphological Research: New York, Oxford University Press, 285 p.

Birkeland, P.W., and Janda, R.S., 1971, Clay Mineralogy of Soils developed from Quaternary deposits of the eastern Sierra Nevada, Calif.: Geol. Soc. America Bull., v. 82, p. 2495-2514.

Biscaye, P.E., 1964, Distinction between Kaolinite and Chlorite in recent sediments by X-ray diffraction: Am. Mineral., v. 49, p. 1281-1289.

Brunton, G., 1955, Vapor pressure glycolation of oriented clay minerals: Am. Mineralogist, v. 40, p. 124-126.

Buol, S.W., Hole, F.E., and McCracken, R.J., 1973, Soil genesis and classification: Ames, Iowa State University Press, 360 p.

Carroll, T., 1974, Relative age dating techniques and a late Quaternary chronology, Arikaree cirque, Colo.: Geology, v. 2, p. 321-325.

Chayes, F., 1960, On correlation between variables of Constant Sum.: Jour. of Geophysical Research, v. 65, no. 12, p. 4185-4193.

Dort, Wakefield, 1962, Multiple glaciation of southern Lemhi Mountains, Idaho-preliminary reconnaissance report: Idaho State Coll. Mus. Jour., v. 5, p. 2-17.

Drever, J.I., 1973, The preparation of oriented clay mineral specimens for X-ray diffraction analysis by a filter-membrane peel technique: American Min., v. 58, p. 553-554.

Erwin, C.R., 1972, Geology of Mackay 2SE, 2NE, and part of the 1NW quadrangles, Custer County, Idaho, (M.S. Thesis): Milwaukee, University of Wisconsin, 210 p.

Farwell, F.W., and Full, B.P., 1944, Geology of the Empire Mine near Mackay, Idaho: U.S. Geol. Survey, Open-file Report, 184 p.

Galehouse, J.S., 1971, Sedimentation analysis, *in* Carver, R.E., ed., Procedures in Sedimentary Petrology: New York, Wiley-Interscience, p. 69-94.

Heath, G.R., 1968, Mineralogy of Cenozoic deep-sea sediments from the equatorial Pacific Ocean: Unpublished (Ph.D. Dissert.): San Diego, University of California, 168 p.

Holmes, G.W., and Moss, J.H., 1955, Pleistocene geology of the southewestern Wind River Mountains, Wyoming: Geol. Soc. America Bull., v. 66, p. 629-654.

Ingram, R.L., 1971, Sieve Analysis, *in* Carver, R.E., ed., Procedures in Sedimentary Petrology: New York, Wiley-Interscience, p. 49-67.

Jackson, M.L., 1956, Soil Chemical Analysis - Advanced Course: Madison, Wisconsin, University of Wisconsin, Dept. Soils, 991 p.

Knoll, K.M., and Dort, W., Jr., 1973, Detailed alpine glacial chronology from central Lemhi Range, Idaho: Geol. Soc. America, Abstract with Programs, v. 5(6), p. 489.

Madole, R.F., 1969, Pinedale and Bull Lake Glaciation in upper St. Vrain Drainage Basin, Boulder County, Colorado: Arctic Alpine Research, v. 1, p. 279-287.

Mahaney, W.C., 1973a, Neoglacial chronology in the Fourth of July Cirque, central Colorado Front Range: Geol. Soc. America Bull., v. 84, p. 161-170.

____________, 1973b, Neoglacial chronology in the Fourth of July Cirque, central Colorado Front Range: Reply: Geol. Soc. America Bull., v. 84, p. 3767-3772.

____________, 1974, Soil Stratigraphy and Genesis of Neoglacial deposits in the Arapaho and Henderson Cirques, central Colorado Front Range, *in* Mahaney, W.C., ed., Quaternary Environments, Proceedings of a Symposium, Geographical Monographs, no. 5: Toronto, York Univ. Ser. Geog., p. 197-240.

Mahaney, W.C. and Fahey, B.D., 1976, Quaternary soil stratigraphy of the Front Range, Colorado, *in* Mahaney, W.C., ed., The Quaternary Stratigraphy of North America: Stroudsburg, Pa., Dowden, Hutchinson and Ross, p. 319-352.

Mears, B., 1974, The Evolution of the Rocky Mountain Glacial Model, *in* Coates, D.R., ed., Glacial Geomorphology: Binghamton, N.Y., Publications in Geomorphology, p. 11-40.

Mehra, O.P., and Jackson, M.L., 1960, Iron oxide removal from soils and clays by a dithionite-citrate system buffered with sodium bicarbonate, *in* Swineford, A., ed., Clays and clay minerals, Proc. 7th Natl. Conf.: London, Pergamon, 369 p.

Nelson, W.H., and Ross, C.P., 1969, Geology of the Mackay 30-minute quadrangle: United States Dept. of the Interior, Geol. Survey, Open-file Report, 215 p.

Parks, J.M., 1970, Fortran IV Program for Q-mode Cluster Analysis on distance function with printed dendogram: Computer contribution 46, State Geol. Surv.: Lawrence, Univ. of Kansas, 31 p.

Pasquini, T.A., 1976, Provenance investigation of the Glacial Geology of the Copper Basin, Custer Co., Idaho (M.S. Thesis): Bethlehem, Pa., Lehigh University, 79 p.

Paull, R.A., Wolbrink, M.A., Volkmann, R.G., and Grover, R.L., 1972, Stratigraphy of Copper Basin Group, Pioneer Mountains, south-central Idaho: American Association Petroleum Geologists Bulletin, v. 56, p. 1370-1401.

Porter, S.C., 1969, Relative dating of Alpine drift sheets using weathering rinds: Geol. Soc. America, Abstracts for 1968, Spec. Paper 121, p. 545.

____________, 1975, Weathering rinds as a relative age criterion: Application to subdivision of glacial deposits in the Cascade Range: Geology, v. 3, no. 3, p. 101-104.

Richmond, G.M., 1962a, Three pre-Bull Lake Tills in the Wind River Mountains, Wyoming: U.S. Geol. Survey Prof. Paper 450-D, 4 p.

____________, 1964a, Glaciation of Little Cottonwood and Bells Canyons, Wasatch Mountains, Utah: U.S. Geol. Survey Prof. Paper 454-D, 41 p.

____________, 1973, Geologic map of the Fremont Lake South Quadrangle, Sublette County, Wyoming: Geol. Quad. Map, GQ-1138.

Ross, C.P., 1937, Geology and Ore Deposits of the Bayhorse Region, Custer County, Idaho: U.S. Geol. Survey Bull. 877, 161 p.

Rothwell, A., 1973, Geology of Standhope Peak, and parts of Mackay 2SW & 2NW Quadrangles, Custer Co., Idaho (M.S. Thesis): Milwaukee, Univ. of Wisconsin, 74 p.

Scott, W.E., 1977, Quaternary glaciation and volcanism, Metolius River Area, Oregon: Geol. Soc. America Bull., v. 88, no. 1, p. 113-124.

Sharp, R.P., 1969, Semiquantitative differentiation of glacial moraines near Convict Lake, Sierra Nevada, Calif.: Jour. Geology, v. 77, p. 68-91.

Snedecor, G.W., and Cochran, W.G., 1967, Statistical Methods: Ames, Iowa State University Press, 6th ed., 593 p.

Soil Survey Staff, 1951, Soil Survey Manual: Washington, D.C., U.S. Gov't Printing Office, 503 p.

________________, 1960, Soil Classification, a comprehensive system (7th approximation): U.S. Dept. Agri., 265 p.

Stewart, M.T., and Mickelson, D.M., 1976, Clay Mineralogy and Relative Ages of Tills in North-Central Wisconsin: Jour. Sed. Pet., v. 46, no. 1, p. 200-205.

Sweeney, G.T., 1957, Geology of Copper Basin, Custer County, Idaho (M.S. Thesis): Moscow, University of Idaho, 66 p.

Umpleby, J.B., Westgate, L.G., and Ross, C.P., 1930, Geology and ore deposits of the Wood River Region, Idaho: U. S. Geol. Survey Bull. 814, 250 p.

Volkmann, R.G., 1972, Geology of the Mackay 3NE quadrangle, Custer, Blaine, and Butte Counties, Idaho (M.S. Thesis): Milwaukee, University of Wisconsin, 86 p.

Wigley, W.C., 1976, The Glacial Geology of the Copper Basin, Custer County, Idaho: A Morphologic and Pedogenic Approach (M.S. Thesis): Bethlehem, Pa., Lehigh University, 106 p.

APPENDIX 1

This appendix contains detailed descriptions of all soil profiles sampled. Observations are an integration of field logging and laboratory analyses (textural). Classification is according to the 7th Approximation (Soil Survey Staff, 1960). Descriptions utilize the nomenclature developed by the U.S.D.A., Soil Survey Staff (1951).

Profile: BLE-1

Age: Bull Lake

Location: Corral Creek: up valley advance from Star Hope Creek lobe: 2395 m (7860 ft)

Parent Material: Till, mixed lithologies with limestone locally predominate.

Classification: Calciorthid

Soil Horizons	Depth below surface (cm)	Description
01	0-1	
A1	0-9	Dark grayish brown (10YR 4/2) fine clay loam; massive; sticky when moist.
B2(?)	9-40	Grayish brown (10YR 5/2) loam; massive; loose, very gravelly composition. Angular and rounded cobbles. Poorly developed profile, could be colluvial deposit.

Profile: BLE-2

Age: Bull Lake

Location: Potholes, north of Copper Basin Flats; 2440 m (8000 ft)

Parent Material: Till, mixed lithologies, Lake Creek intrusive

Classification: Camborthid

Soil Horizons	Depth below surface (cm)	Description
01	0-3	
A2	0-14	Dark grayish brown (10YR 4/2) fine sandy loam; massive, slightly sticky when moist.
B1	14-52	Yellowish brown (10YR 5/4 sandy loam. Poorly developed crumb peds (<50 mm) are present. Matrix is quite sandy, and cobbles are not highly weathered. Gradation to B2 horizon is diffuse.

Cca	40	
Cox	40-64	Stage I accumulation, violent reaction. $CaCO_3$ percentage increases with depth to Stage II. Light brownish gray (2.5Y 6/2) sandy loam: massive and loose.
Cn(?)	64-95+	Stage II $CaCO_3$. Light gray (2.5Y 7/2) coarse sandy loam. Possibly highly calcareous unweathered till. Massive, very loose and stony.

Sampling [No. - depth (cm)]: #1-6; #2-15; #3-30; #4-45; #5-60; #6-75; #7-90.

B2	52-143+	Light yellowish brown (10YR 5/4) loam. Large (50-150 mm) subangular porous peds are moderately well developed. Size may be due in part to moisture content. Obvious increase in weathering effects from above, coarse grain clasts are completely grusified. No $CaCO_3$ encountered at any depth.

Sampling [No. - depth (cm)]: #1-8; #2-14; #3-29; #4-44; #5-59; #6-74; #7-89; #8-104.

(Replicate samples taken from opposite side of pit).

Profile: BLW-1

Age: Bull Lake

Location: Ramey Creek Flats near head of Fox Creek: 2515 m (8260 ft)

Parent Material: Till, mixed lithologies, Broad Canyon intrusive

Classification: Camborthid

Soil Horizons	Depth below surface (cm)	Description
01	0-3	
A1	0-12	Brown (10YR 4/3) very fine sandy loam; massive and slightly sticky when moist.
B2	12-125	Brown (7.5YR 5/4) to reddish brown (5YR 5/4) sandy loam. Moderate structure (5-200 mm) subangular porous peds are developed at depth; however, structure is dominated by indistinct boulder ghosts (granitic). Soil is moderately compact, but very friable and nonsticky.
Cox	125+	Diffuse boundary with above, no Cca encountered however, material is lighter (10YR 6/4). Coarse loamy sand, massive and loose at depth.

Sampling [No. - depth (cm)]: #1-7; #2-16; #3-31; #4-46; #5-61; #6-76; #7-91; #8-106; #9-121.

Profile: BLW-2

Age: Bull Lake

Location: Top of Challis Knobs near Road Creek; 2470 m (8100 ft)

Parent Material: Till, mixed lithologies, Broad Canyon intrusive

Classification: Calciorthid

Soil Horizons	Depth below surface (cm)	Description
01	0-1	
A1	0-8	Light brownish gray (10YR 6/2), fine sandy loam; massive and slightly sticky when wet.
B2t	8-53	Reddish brown (5YR 5/3) sandy clay loam. Soil is extremely compact and probably over consolidated. Well developed prismatic peds (4-5 cm) are present. Soil is not very stony, and granite cobbles present are not completely grusified, however, red color is well developed, and soil is noticeably clayey.
Cca	53	Stage I accumulation, strong reaction, possibly this $CaCO_3$ layer is secondary in this highly indurated profile, with the rest of the B horizon extending down to a much deeper level.

Sampling [No. - depth (cm)]: #1-8; #2-20;

Profile: PE-1

Age: Pinedale II

Location: Potholes near Willow Springs; 2320 m (7620 ft)

Parent Material: Till, mixed lithologies

Classification: Calciorthid

Soil Horizons	Depth below surface (cm)	Description
01	0-3	
A1	0-8	Dark grayish brown (10YR 4/2) loam; slightly sticky when wet.
B2	8-41	Brown (10YR 4/3) loamy sand, massive structure, friable when moist. Horizon boundaries are clear.
Cca	41	Stage I carbonate accumulation. Violent reaction with HCl.
Cox	41-56^{+}	Light brownish gray (2.5Y 6/2) extremely cobbly (well-rounded) sand. Till, but could be reworked outwash deposits.

Sampling [No. - depth (cm)]: #1-5; #2-16; #3-29: #4-44.

Profile: PE-2

Age: Pinedale I

Location: Pinedale I terminal position near mouth of Basin; 2320 m (7620 ft)

Parent Material: Till, mixed lithologies

Classification: Calciorthid

Soil Horizons	Depth below surface (cm)	Description
01	0-2	
A1	0-13	Dark grayish brown (10YR 4/2) silty loam.
B2	13-85	Light grayish brown (2.5Y 6/2) fine sandy loam. Weakly developed crumb (1-2 mm) peds present. Soil is fairly compact but friable and nonsticky when moist. Soil is quite stony, but noticeably fresher than older sites; however, the difference may be due to a lack of granitic cobbles rather than age.
Cca	85	Stage I accumulation; violet reaction with HCl.
Cox	85-110^{+}	Light olive brown (2.5Y 5/4) sandy loam. Loose, massive nonsticky.

Sampling [No. - depth (cm)]: #1-8; #2-20; #3-35; #4-50; #5-65; #6-80; #7-95.

Profile: PE-4

Age: Pinedale I

Location: Outer Pinedale potholes near the east fork of the Big Lost River; 2395 m (7860 ft)

Parent Material: Till, mixed lithologies, Lake Creek intrusive

Classification: Camborthid

Soil Horizons	Depth below surface (cm)	Description
01	0-1	
A1	0-18	Dark brown (10YR 3/3) fine sandy loam; massive and nonsticky when moist.
Cox (?)	18-58	Light brownish gray (2.5Y 6/2) sandy loam. Poorly developed peds (2-3 mm) are apparent, but close inspection (lack of clay skins) suggests that they are developed more as a function of moisture content and compaction than clay accumulation. Soil is generally quite stony and nonsticky. Horizons are diffuse. No $CaCO_3$ is present at any depth.
D	58^+	Light gray (2.5Y 7/2) sandy loam. Massive, nonsticky and very stony. Textural differentiation with Cox is indistinct. Separation is based on a lack of weathered products in evidence in the lower horizon.

Sampling [No. - depth (cm)]: #1-10; #2-19; #3-29; #4-45; #5-107.

Profile: PE-5

Age: Pinedale I

Location: Top of potholes, near Lake Creek Canyon; 2570 m (8430 ft)

Parent Material: Till, mixed lithologies, Lake Creek intrusive

Classification: Calciorthid

Soil Horizons	Depth below surface (cm)	Description
01	0-3	
A1	0-15	Grayish brown (10YR 5/2) fine sandy loam; massive; nonsticky.
B2	15-71	Brown (10YR 5/3) sandy loam. Poorly developed structure. Small (1-3 mm) porous peds apparent at depth, but general profile is massive. Moderately to very stony; relatively small amounts of sand matrix. Cobbles are fresh or slightly weathered, and generally fresher than in other Pinedale I soils.
Cca	71	Stage I accumulation, violent reaction.
Cox	$71\text{-}95^+$	Pale brown (10YR 6/3) sandy loam; massive; loose; very stony.

Sampling [No. - depth (cm)]: #1-9; #2-20; #3-35; #4-50; #5-65; #6-80.

Profile: PE-6

Age: Pinedale II

Location: Below main potholes, near mouth of Lake Creek Canyon; 2500 m (8200 ft)

Parent Material: Till, mixed lithologies, Lake Creek intrusive

Classification: Camborthid

Soil Horizons	Depth below surface (cm)	Description
01	0-1	
A1	0-7	Dark grayish brown (10YR 4/2) loam; massive, slightly sticky when moist.
B2 (?)	7-45	Light brownish gray (2.5Y 6/2) sandy loam. Structure is very poor if developed at all. Minor development of porous crumbs (1-2 mm) at depth. Soil is very stony, evidence of cobble weathering is lacking. Profile could be A-Cox.
Cca	45	Stage I accumulation, violent reaction.

Sampling [No. - depth (cm)]: #2-13; #3-28; #4-43.

Profile: PE-7

Age: Pinedale I

Location: Star Hope Lobe at mouth of Corral Creek; 2365 m (7770 ft)

Parent Material: Till, mixed lithologies.

Classification: Calciorthid

Soil Horizons	Depth below surface (cm)	Description
01	0-2	
A1	0-15	Dark grayish brown (10YR 4/2) loam; massive; slightly sticky when wet.
B2 (?)	15-68	Light olive brown (10YR 5/4) sandy loam. Massive, basically unweathered, and very loose. Could be B1 or Cox horizon.
Cca	68	Stage I accumulation, violent reaction.

Sampling [No. - depth (cm)]: #2-17; #3-32; #4-47; #5-62.

Profile: PS-1

Age: Pinedale I

Location: Post moraine, mouth of basin; 2280 m (7480 ft)

Parent Material: Till, mixed lithologies

Classification: Calciorthid

Soil Horizons	Depth below surface (cm)	Description
01	0-2	
A1	0-10	Grayish brown (10YR 5/2) loam; massive; slightly sticky when moist.
B2	10-42	Yellowish brown (10YR 5/4) sandy loam. Very stony. Structure is generally massive although poorly developed crumb (<3 mm) seem to develop with depth. Horizons are clear.
Cca	42	Stage I accumulation, strong reaction with HCl.
Cox	42-57+	Light yellowish brown (2.5Y 6/4). Massive and very cobbly loamy sand. Well rounded cobbles probably is reworked outwash.

Sampling [No. - depth (cm)]: #1-5; #2-10; #3-15; #4-33; #5-48.

Profile: PW-1

Age: Pinedale I

Location: Outer Pinedale, near Ramey Creek Flats; 2500 m (8210 ft)

Parent Material: Till, mixed lithologies, Broad Canyon intrusive

Classification: Calciorthid

Soil Horizons	Depth below surface (cm)	Description
01	0-3	
A1	0-16	Dark grayish brown (10 YR 4/2) very fine sandy loam; massive; slightly sticky when wet.
B2	16-98	Light brown (7.5YR 6/4) sandy loam. Moderate structure developed; subangular blocky (50-150 mm) porous peds are present. Soil is moderately compact, and slightly sticky when moist. Horizons are distinct; however, entire Bt is remarkably uniform and pebble free.
Cca	98	Stage I, violent reaction.
Cox	98-163	Light yellowish brown (10YR 6/4) coarse loam. Loose, massive, unweathered granites encountered at bottom.

Sampling [No. - depth (cm)]: #1-6; #2-16; #3-31; #4-46; #5-61; #6-76.

Profile: PW-2

Age: Pinedale II

Location: Terminal position at head of Road Creek; 2370 m (7780 ft)

Parent Material: Till, mixed lithologies, Broad Canyon intrusive

Classification: Calciorthid

Soil Horizons	Depth below surface (cm)	Description
01	3	
A1	0-20	Grayish brown (10YR 5/2) fine grained loam; massive; slightly sticky when wet; thicker than usual.
B2	20-117	Light brown (7.5YR 6/4) sandy loam. Weakly developed porous peds (2-20 mm) developing with depth. Structure is slightly compact, and slightly sticky when moist. Noticeable absence of coarse granular texture found in older soils.
Cca	117	Stage I, violent reaction with HCl.
Cox	117+	Differentiation is based solely on Cca horizon. Any textural gradation from Bt is diffuse.

Sampling [No. - depth (cm)]: #1-10; #2-20; #3-25; #4-40; #5-55; #6-70; #7-85; #8-100; #9-115.

Profile: PW-3

Age: Pinedale I

Location: Outer Pinedale above Road Creek; 2410 m (7920 ft)

Parent Material: Till, mixed lithologies, Broad Canyon intrusive

Classification: Calciorthid

Soil Horizons	Depth below surface (cm)	Description
01	0-2	
A1	0-14	Grayish brown (10YR 5/2) fine sandy loam; massive; slightly sticky when moist.
B1	14-55	Light yellow brown (10YR 6/4) crumbly poorly developed structure; nonsticky; coarser matrix than Bt. Horizons are diffuse.
B2	55-112	Brownish yellow (10YR 6/6) sandy loam. Structure is moderately compact; subangular blocky (50-75 mm) peds are present. Matrix is more clayey than above, and slightly sticky when moist.
Cca	112	Stage I; violent reaction with HCl.

Sampling [No. - depth (cm)]: #1-8; #2-19; #3-34; #4-49; #5-64; #6-79; #7-94; #8-109.

Profile: PW-4

Age: Pinedale III

Location: Lake Creek Lobe, near confluence of Howell and Star Hope Creeks; 2355 m (7730 ft)

Parent Material: Till, mixed lithologies, Lake Creek intrusive

Classification: Calciorthid

Soil Horizons	Depth below surface (cm)	Description
01	0-2	
A1	0-9	Grayish brown (10YR 5/2) fine loam; massive; slightly sticky when moist.
B2 (?)	9-62	Light brownish gray (2.5Y 6/2) sandy loam. Massive, nonsticky, poorly developed structure (1-2 mm) may be present at depth, but seems to be a function of moisture content. Very cobbly soil, no evidence of weathering at any depth.
Cca	62	Stage I, strong reaction to HCl.

Sampling [No. - depth (cm)]: #1-5; #2-16; #3-31; #4-46; #5-61.

Profile: PW-5

Age: Pinedale II (?)

Location: Mouth of Howell Canyon near Loop Road; 2350 m (7720 ft)

Parent Material: Till; primarily granitic and volcanic but mixed lithologies are present, Broad Canyon intrusive

Classification: Camborthid

Soil Horizons	Depth below surface (cm)	Description
01	0-3	
A1	0-12	Dark grayish brown (10YR 4/2) sandy loam; massive; slightly sticky when moist.
B2	12-84+	Light brownish gray (2.5Y 6/2) loam; matrix is silty. Soil is fairly compact with structure; (50 mm) subangular peds poorly developed. Transitions are gradual, with no internal horizons developed. No $CaCO_3$ was encountered at any depth.

Sampling [No. - depth (cm)]: #1-5; #2-20; #3-35; #4-50; #5-65; #6-80

<u>Profile</u>: MC-1

<u>Age</u>: Pinedale IV

<u>Location</u>: Muldoon Canyon; 2620 m (8600 ft)

<u>Parent Material</u>: Till, volcanic and clastic sources.

<u>Classification</u>: Cryochrepts (?)[a]

Soil Horizons	Depth below surface (cm)	Description
01	0-2	
A1	0-11	Dark grayish brown (10YR 4/2) loam; massive but very stony.
Cox	11-28	Light brownish gray (2.5Y 6/2) sandy loam. Massive, extremely cobbly with little matrix present; no evidence of weathering on cobble surfaces. Transitions are gradual.
D	28^+	Olive yellow (2.5Y 6/6). No $CaCO_3$ present, very stony and loose.

<u>Sampling</u> [No. - depth (cm)]: #1-5: #2-20.

[a]-*probable classification is due to inferred, rather than factual, soil temperature data.*

<u>Profile</u>: MC-2

<u>Age</u>: Pinedale IV

<u>Location</u>: Muldoon Canyon; 2610 m (8560 ft)

<u>Parent Material</u>: Till, clastic sources

<u>Classification</u>: Cryochrepts (?)

Soil Horizons	Depth below surface (cm)	Description
01	0-2	
A1	0-9	Brown (10YR 4/3) silty loam; massive but very stony with little matrix.
Cox	9-24	Light brownish gray (2.5Y 6/2) sandy loam; massive but very strong with little matrix; unweathered; transitions are gradual.
D	24^+	Olive yellow (2.5Y 6/6). No $CaCO_3$ present, very stony and loose.

<u>Sampling</u> [No. - depth (cm)]: #1-5; #2-20.

Profile: MC-3

Age: Pinedale IV

Location: Muldoon Canyon at the Springs; 2520 m (8260 ft)

Parent Material: Till; volcanic and clastic sources.

Classification: Cryochrepts (?)

Soil Horizons	Depth below surface (cm)	Description
01	0-3	
A1	0-14	Grayish brown (10YR 5/2) loam; massive; noticeably silty matrix.
Cox	14-34	Light yellowish brown (10YR 6/4) sandy loam. Massive, but more matrix is present than MC-1 or MC-2; transitions are clear.
D	34^+	Olive yellow (2.5Y 6/6) loamy sand. Massive, loose, and very stony.

Sampling [No.- depth (cm)]: #1-5; #2-20; #3-40.

APPENDIX 2

Laboratory Results

This appendix contains the tabulated results of the laboratory analyses of the 121 soil samples collected in the field. Investigations considered the clay mineralogies and texture of those samples. The abbreviations used in the column headings translate as:

(M)
5.5Å - Percentage of Montmorillonite with diagnostic glycolated diffraction spacing

(C)
6.75Å - Percentage of Chlorite with diagnostic glycolated diffraction spacing

(I)
9.0Å - Percentage of Illite with diagnostic glycolated diffraction spacing

(K/CH)
12.5Å - Percentage of grouped Kaolinite/Chlorite with diagnostic glycolated diffraction spacing

Depth
(abs) - Absolute as defined in
(norm)- Normalized Methods section

% SD - Percentage of sand

% ST - Percentage of silt

% CY - Percentage of clay

Sample No.	(M)° 5.5A	(C)° 6.75A	(I)° 9.0A	(K/Ch)° 12.5A	Depth (abs)	Depth (norm)	% Sand	% Silt	% Clay
BLE 1/1	18.1	6.5	54.8	14.2	6	.16	31.1	41.9	27.0
1/2	39.4	41.5	13.7	5.4	15	.39	38.3	37.9	23.8
1/3	48.9	46.6	3.4	1.2	30	.79	55.1	23.7	21.2
1/4	57.3	38.9	1.5	2.3	45	1.18	54.4	27.8	17.8
1/5	28.1	51.1	12.5	8.2	60	1.58	54.0	30.0	16.8
1/6	56.8	36.4	2.1	4.7	75	1.99	57.0	28.1	14.9
1/7	71.7	20.4	3.6	4.3	90	2.37	61.1	23.6	15.3
BLE 2/1A	9.7	11.1	61.1	18.1	8	.09	56.6	27.4	16.0
2/2A	2.1	52.0	32.5	13.4	14	.15	52.9	37.1	10.0
2/3A	3.2	39.8	45.6	11.3	29	.31	47.8	38.8	13.4
2/4A	5.4	82.4	6.2	6.0	44	.47	54.8	33.3	11.8
2/5A	15.0	65.4	18.4	1.1	59	.63	50.8	36.5	12.7
2/6A	23.5	68.5	4.9	3.1	74	.80	49.6	38.0	12.4
2/7A	19.1	76.2	3.2	1.5	89	.96	61.1	29.6	9.3
2/8A	14.3	70.2	11.2	4.2	104	1.12	53.3	34.9	11.8
BLE 2/1B	2.7	10.8	75.6	10.9	8	.09	65.5	16.2	18.3
2/2B	3.2	49.8	31.2	15.8	14	.15	40.6	34.7	12.3
2/3B	7.5	69.3	17.7	5.5	29	.31	48.6	37.4	14.0
2/4B	14.7	72.1	10.7	2.5	44	.47	51.0	33.8	13.8
2/5B	16.1	77.4	3.9	2.6	59	.63	54.8	36.8	8.4
2/6B	19.0	76.2	2.8	2.0	74	.80	60.1	30.2	9.7
2/7B	23.5	68.5	4.9	3.1	89	.96	48.1	38.6	13.3
2/8B	15.3	78.0	2.7	4.1	104	1.12	50.6	37.1	12.3
BLW 1/1	1.6	12.0	56.1	30.3	7	.06	50.1	29.2	20.7
1/2	5.2	45.0	27.9	6.4	16	.14	61.6	21.8	16.6
1/3	2.3	10.1	71.4	16.2	31	.27	64.4	21.5	9.9
1/4	23.7	67.8	5.1	3.4	46	.41	73.4	18.8	7.8
1/5	12.5	62.5	18.8	6.3	61	.54	66.2	22.9	10.9
1/6	14.3	50.0	32.1	3.6	76	.67	72.9	22.6	4.4
1/7	10.3	26.9	52.5	10.7	91	.81	69.3	22.2	8.5

Sample No.	(M)° 5.5A	(C)° 6.75A	(I)° 9.0A	(K/Ch)° 12.5A	Depth (abs)	Depth (norm)	% Sand	% Silt	% Clay
BLW 1/8	20.1	59.9	13.3	6.7	106	.94	68.5	23.2	8.3
1/9	7.8	74.4	11.5	6.4	121	1.07	73.0	19.8	7.2
BLW 2/1	15.8	38.6	25.4	20.2	5	.11	54.6	28.6	16.9
2/2	24.3	34.5	23.0	18.3	14	.31	50.5	20.6	28.9
2/3	53.7	32.0	8.6	5.7	29	.64	49.2	18.8	31.9
2/4	88.3	6.5	3.2	1.9	44	.98	50.5	22.9	26.6
PE 1/1	21.8	34.5	29.1	14.5	5	.16	48.1	37.5	14.5
1/2	31.2	49.1	13.9	5.8	14	.44	83.0	15.0	2.0
1/3	30.6	34.1	17.6	17.6	29	.91	77.4	14.3	8.3
1/4	23.6	30.3	28.1	18.0	44	1.38	85.6	8.4	6.0
PE 2/1	5.4	5.4	67.1	22.1	8	.12	35.6	47.9	16.5
2/2	50.7	39.4	5.6	4.2	20	.30	46.3	48.1	5.6
2/3	55.7	36.7	3.1	4.4	35	.53	61.8	23.7	14.5
2/4	55.7	41.5	1.2	1.5	50	.76	53.4	29.4	17.2
2/5	55.6	37.9	2.2	4.3	65	.98	52.8	31.2	16.0
2/6	49.5	40.7	3.1	6.7	80	1.21	53.3	32.0	14.7
2/7	48.9	31.0	9.5	10.6	95	1.44	52.7	32.1	15.2
PE 4/1	9.6	15.4	63.5	11.5	10	.50	52.9	36.2	10.9
4/2	7.0	45.8	35.2	12.0	19	.95	62.5	23.1	14.5
4/3	11.4	55.4	22.9	10.3	29	1.45	56.7	36.1	7.2
4/4	20.0	42.3	20.7	16.9	45	2.25	64.8	29.7	5.5
4/5	12.1	57.1	17.9	12.9	107	5.35	55.8	34.1	10.1
PE 5/1	5.1	7.7	64.1	23.1	9	.16	52.2	34.3	13.6
5/2	7.4	8.8	61.8	22.0	20	.36	55.6	36.4	8.0
5/3	31.9	33.6	21.8	12.6	35	.63	52.4	32.5	15.1
5/4	33.2	41.1	16.4	9.2	50	.89	55.1	31.8	13.1
5/5	17.2	34.5	26.2	22.1	65	1.16	67.9	22.1	10.1
5/6	17.4	50.7	17.4	14.5	80	1.43	61.3	27.0	11.7
PE 6/2	11.4	33.1	31.4	24.0	13	.38	58.0	34.2	7.9
6/3	18.1	30.5	26.0	25.4	28	.82	60.7	27.6	11.7

Sample No.		(M) 5.5Å	(C) 6.75Å	(I) 9.0Å	(K/Ch) 12.5Å	Depth (abs)	Depth (norm)	% Sand	% Silt	% Clay
PE	6/4	9.8	32.8	23.0	35.2	43	1.26	60.1	28.6	11.3
PE	7/2	7.0	8.8	68.4	15.8	17	.29	43.0	46.0	11.0
	7/3	31.9	27.5	26.1	14.5	32	.55	44.5	45.3	10.2
	7/4	25.9	24.1	39.7	10.3	47	.81	59.1	35.9	5.0
	7/5	22.8	47.0	18.8	11.4	62	1.07	65.2	29.4	5.4
PS	1/1	34.5	24.1	22.4	19.0	5	.12	49.5	34.2	16.3
	1/2	43.2	45.2	5.2	6.5	9	.21	73.3	21.9	4.8
	1/3	48.9	45.1	2.7	3.3	18	.43	73.5	16.6	10.0
	1/4	57.6	33.2	4.9	4.2	33	.79	69.8	20.4	9.9
	1/5	64.9	25.3	5.4	4.4	48	1.14	80.3	11.8	7.9
PW	1/1A	12.6	22.9	56.3	8.0	6	.07	63.2	24.3	10.5
	1/2A	23.2	43.9	24.4	8.5	16	.20	59.3	30.9	9.8
	1/3A	30.7	47.2	14.7	7.4	31	.38	58.2	28.9	12.9
	1/4A	51.0	38.0	7.0	4.0	46	.56	52.2	34.5	13.4
A	1/5A	51.6	37.9	6.3	4.2	61	.74	50.0	38.0	12.0
	1/6A	50.2	33.4	8.3	8.1	76	.93	53.2	35.1	11.8
	1/7A	48.0	41.5	6.2	4.4	91	1.11	61.8	28.8	9.4
PW	1/1B	7.8	22.1	54.5	15.6	6	.07	64.2	22.6	13.2
	1/2B	13.0	38.0	35.0	14.0	16	.20	62.2	25.5	12.2
	1/3B	30.6	48.5	12.7	8.2	31	.38	58.5	27.8	13.8
	1/4B	29.4	47.3	16.0	7.3	46	.56	59.8	29.0	11.2
	1/5B	23.7	58.7	9.5	8.1	61	.74	55.0	31.9	13.1
	1/6B	34.6	46.5	12.6	6.3	76	.93	55.2	32.0	12.8
	1/7B	40.5	46.0	8.8	4.7	91	1.11	62.0	28.6	9.4
PW	1/Till	58.8	29.3	5.9	6.0	140	1.71	53.2	31.9	14.9
PW	2/1	5.4	27.3	49.1	18.2	10	.10	44.9	32.1	22.9
	2/2	15.7	38.0	28.1	18.2	20	.21	46.4	32.6	21.0
	2/3	35.5	48.8	9.5	6.2	25	.26	49.4	26.5	24.1
	2/4	40.3	41.4	11.7	6.6	40	.41	52.5	25.7	21.9
	2/5	56.0	34.8	5.4	3.8	55	.57	52.0	25.0	23.0

Sample No.		(M) 5.5Å	(C) 6.75Å	(I) 9.0Å	(K/Ch) 12.5Å	Depth (abs)	Depth (norm)	% Sand	% Silt	% Clay
PW	2/6	45.7	31.4	12.9	10.0	70	.72	53.9	26.0	20.1
	2/7	50.8	32.2	8.0	9.0	85	.88	51.9	28.0	20.1
	2/8	60.6	27.7	4.7	6.9	100	1.03	50.4	29.4	20.2
	2/9	66.4	21.6	4.7	7.3	115	1.19	50.9	29.4	19.6
PW	3/1	11.6	14.5	58.0	15.9	8	.09	51.8	36.2	12.0
	3/2	23.8	49.6	19.6	7.0	19	.20	64.8	25.1	10.1
	3/3	31.3	49.2	14.1	5.5	34	.36	48.9	43.9	7.1
	3/4	42.7	37.6	12.8	6.8	49	.52	68.6	19.7	11.7
	3/5	37.3	32.5	16.9	13.3	64	.68	65.7	25.0	9.3
	3/6	50.6	30.6	12.7	6.1	79	.84	66.9	21.8	11.4
	3/7	48.9	35.6	10.1	5.4	94	1.00	69.1	20.4	10.6
	3/8	46.7	37.3	9.3	6.7	109	1.16	66.7	22.1	11.2
PW	4/1	6.4	9.7	54.9	29.0	5	.10	37.1	44.9	18.1
	4/2	9.8	68.2	9.4	12.6	16	.33	52.7	31.6	15.7
	4/3	17.7	53.8	16.1	12.3	31	.63	63.1	24.6	12.3
	4/4	14.6	48.5	20.1	16.8	46	.94	56.2	29.2	14.6
	4/5	19.0	43.3	17.5	20.2	61	1.24	57.3	29.6	13.1
PW	5/1	1.9	17.6	58.8	21.6	5	.07	43.2	38.9	18.0
	5/2	6.1	52.5	21.2	20.2	20	.27	46.1	38.3	15.6
	5/3	9.8	52.4	21.0	16.8	35	.47	49.0	35.0	16.0
	5/4	4.6	55.1	24.6	15.8	50	.68	44.9	33.4	21.7
	5/5	14.8	46.4	27.8	11.0	65	.88	41.1	33.6	25.3
	5/6	27.5	43.3	19.4	9.7	80	1.08	42.1	31.4	26.5
	5/7	1.4	51.7	30.4	16.4	95	1.28	41.6	33.7	24.7
MC	1/1	3.8	28.8	32.7	34.6	5	.21	45.4	41.9	12.7
	1/2	7.3	30.5	26.8	35.4	20	.83	70.0	25.3	4.7
MC	2/1	3.7	25.9	37.0	33.3	5	.24	32.5	51.7	15.8
	2/2	17.6	42.9	16.5	23.1	20	.95	63.0	25.0	12.0
MC	3/1	7.8	16.9	46.1	29.2	5	.19	40.0	41.0	19.0
	3/2	7.6	25.2	40.3	25.9	20	.77	55.7	39.3	5.0
	3/3	7.7	19.2	30.8	42.3	40	1.54	82.0	15.0	2.9

APPENDIX 3

Detailed Size Analysis

This appendix contains the tabulated results of the detailed size analysis (0Ø through 11Ø, at 1Ø intervals) of four selected profiles. Not all "soil" profiles were subjected to this detailed analysis, as the intent was to test whether the gross, sand, silt, clay percentages given in Appendix II were insensitive to minor textural variations.

% OF TOTAL SAMPLE WEIGHT

Sample No.	0ϕ	1ϕ	2ϕ	3ϕ	4ϕ	5ϕ	6ϕ	7ϕ	8ϕ	9ϕ	10ϕ	11ϕ	<11ϕ
BLE 2/1B	12.9	11.3	8.3	7.2	7.1	15.0	11.4	3.7	3.7	2.8	5.5	1.5	9.5
2/2B	8.1	9.9	9.4	9.1	9.0	14.2	13.5	2.9	4.3	4.3	2.9	2.3	9.9
2/3B	6.5	9.3	10.2	11.5	9.6	12.4	13.4	3.7	3.7	4.4	3.1	2.0	10.0
2/4B	8.9	10.1	11.8	11.1	10.2	9.2	10.8	4.7	4.0	3.3	3.5	2.4	9.9
2/5B	6.6	8.8	9.8	10.8	10.3	12.4	12.4	4.9	3.7	4.8	2.8	3.5	8.7
2/6B	12.2	14.5	11.7	10.1	9.2	10.7	10.7	3.3	3.0	3.5	3.8	2.3	5.9
2/7B	5.5	7.8	9.8	11.3	11.5	12.0	14.3	4.5	3.7	6.3	2.1	3.5	8.2
2/8B	7.1	9.8	11.2	11.3	9.9	11.9	12.1	3.4	2.2	6.6	4.4	2.4	7.7
BLW 2/1	6.8	10.1	13.0	12.6	9.2	7.7	7.7	4.5	3.2	5.0	3.5	3.7	13.0
2/2	6.0	10.1	12.2	11.9	8.6	7.7	5.6	2.4	2.4	2.8	2.6	2.6	24.1
2/3	6.2	9.8	12.2	11.8	8.6	5.7	4.4	4.1	0.8	2.3	3.9	2.6	27.7
2/4	8.7	12.2	13.1	11.0	7.2	5.2	7.7	3.0	3.2	4.5	1.7	1.7	20.8
PW 3/1	14.3	15.1	11.1	8.0	6.0	8.9	10.7	4.8	2.0	4.8	2.1	2.7	9.5
3/2	14.9	17.7	14.5	10.2	6.8	6.9	5.6	3.3	1.3	3.3	2.3	2.8	10.4
3/3	14.0	17.0	15.2	11.4	8.0	6.4	6.0	2.6	2.4	2.3	2.6	2.3	9.8
3/4	18.5	20.6	15.4	9.4	5.9	5.1	4.5	2.2	2.8	3.8	1.8	1.6	7.7
3/5	17.5	19.2	13.9	9.0	6.6	6.5	8.0	2.0	3.2	2.2	2.2	2.7	7.0
3/6	17.4	18.8	14.6	9.0	5.8	7.4	6.9	2.0	3.1	3.2	3.0	2.2	6.6
3/7	16.3	18.4	14.3	9.8	7.6	10.0	5.1	2.7	3.1	2.7	1.2	2.7	5.9
3/8	14.8	18.1	15.8	10.9	7.9	8.5	5.7	2.0	2.2	3.0	1.7	1.7	7.2
PE 5/1	10.1	13.2	11.9	9.9	8.3	10.6	8.7	4.8	3.4	5.3	2.6	2.1	9.0
5/2	8.8	10.5	11.3	10.8	9.1	11.3	10.5	4.5	4.1	4.5	2.3	2.1	10.2
5/3	8.3	11.0	11.5	10.1	10.0	11.7	10.3	3.0	4.8	4.2	3.0	2.2	9.9
5/4	10.2	13.1	12.7	11.1	9.1	10.2	7.4	4.1	4.7	3.8	1.9	1.9	9.3
5/5	14.6	16.5	15.6	12.7	9.6	5.3	6.7	3.3	3.6	3.6	1.0	1.4	6.7
5/6	10.1	13.5	13.8	13.1	11.5	10.1	7.3	3.8	3.2	3.2	1.6	1.4	7.5

Climatic Trends During the Quaternary in Central Yukon Based Upon Pedological and Geomorphological Evidence

N. W. Rutter, A. E. Foscolos, O. L. Hughes

ABSTRACT

Soils and paleosols were investigated from pre-Reid (early Pleistocene), Reid (Illinoian or early Wisconsinan) and McConnell (Classical Wisconsinan) surfaces in central Yukon. The paleosol on the pre-Reid surface suggests that it was subjected to two distinct climates, an initial one which was warm and subhumid with grassland-shrub vegetation and later a more temperate and humid climate characterized by the development of a Luvisol with a red, textural B horizon, in places over 190 cm thick. Subsequently, the climate became colder, resulting in the Reid glaciation. Thermal contraction cracks developed in the pre-Reid deposits beyond the limit of Reid glaciation and filled with eolian sand, plus minor silt and clay to form sand wedges. During the subsequent Reid-McConnell interglacial, a cool-subhumid climate prevailed as evidenced by the Brunisolic characteristics of a paleosol on deposits of Reid age. A cold period followed which climaxed with the advent of the McConnell glaciation. Sand wedges also formed in the deposits of the Reid glaciation; the wedges are shallower and narrower than those on the pre-Reid surface, suggesting a shorter cold period. During retreatal stages of the McConnell glaciation, a thin blanket of loess was deposited over McConnell, Reid and pre-Reid surfaces, covering the soils on the Reid and pre-Reid surfaces during postglacial (Holocene) time. Finally Brunisolic soils developed on the loess blanket, and locally, where the loess is very thin or lacking, on deposits of McConnell age.

INTRODUCTION

Four major advances of the Cordilleran ice sheet in central Yukon were first recognized by Bostock (1948). Although Bostock referred deposits at certain localities to the respective advances, the chronology is based primarily on geomorphic evidence, mainly moraines and ice-marginal features and their degree of preservation, rather than on stratigraphic evidence. By inference, there are three intervening interglacial intervals. Interglacial beds are known from a number of sections along major rivers of the region, but these are not firmly correlated with Bostock's morphostratigraphic chronology, and the flora and fauna of the beds and their paleoclimatic implications remain to be studied.

In reconnaissance mapping it was found that in going from the youngest to the oldest of the outwash gravel deposits of the respective advances, there is a marked increase in the depth of soil development, the thickness and redness of the B horizon, and the degree of weathering of rock clasts within the soil profile. Character of the soils proved to be a useful adjunct to geomorphic criteria in distinguishing between deposits of the respective advances. These observations led to the present study, in which paleopedology is used to expand the rather meagre understanding of changing climates of central Yukon during the Quaternary.

Paleopedology, which is the study of ancient soils, has been used to assess changes in vegetation during the Quaternary period (Bailey *et al.*, 1964; Jenny, 1941, 1958; Jungerius, 1969; Ruhe, 1969; Ruhe and Cady, 1969; Pettapiece, 1969; and Smith *et al.*, 1950) and to evaluate paleoclimates (Janda and Croft, 1967; and Birkeland and Janda, 1971). Birkeland (1969) has used clay mineralogy to decipher paleoclimates of the Sierra Nevada - Great Basin region of the southwestern United States. The present study thus follows already well-established principles of paleopedology.

REGIONAL PLEISTOCENE GEOLOGY AND CHRONOLOGY

The Pleistocene chronology of central and southern Yukon can be considered in terms of the following four regions (Figure 1). (1) The central and southern parts of Yukon Plateau, together with the flanks of the Coast Mountains to the southwest, and the flanks of the Selwyn Mountains to the east and northeast. This region was subjected to repeated glaciations by Cordilleran ice sheets that extended from the Interior Plateau region of British Columbia; (2) The east flank of St. Elias Mountains, together with Shakwak Trench and Ruby Range on the southwest fringe of Yukon Plateau. Large valley glaciers advanced repeatedly from St. Elias Mountains, and coalesced as a piedmont glacier in Shakwak Trench. Tongues of the piedmont glacier pushed into Ruby Range and into Wellesley basin north of the Ruby Range. Valley glaciers originating in the Ruby Range were in part coalescent with the piedmont glacier, and the piedmont glacier at times coalesced with the main Cordilleran ice sheet; (3) Ogilvie Mountains, which supported independent valley glaciers that moved southwestward toward and locally into Tintina Trench, and northward into the headwaters of the Peel River; (4) The unglaciated northwestern portion of Yukon Plateau (essentially the Klondike Plateau as defined by Bostock, 1964).

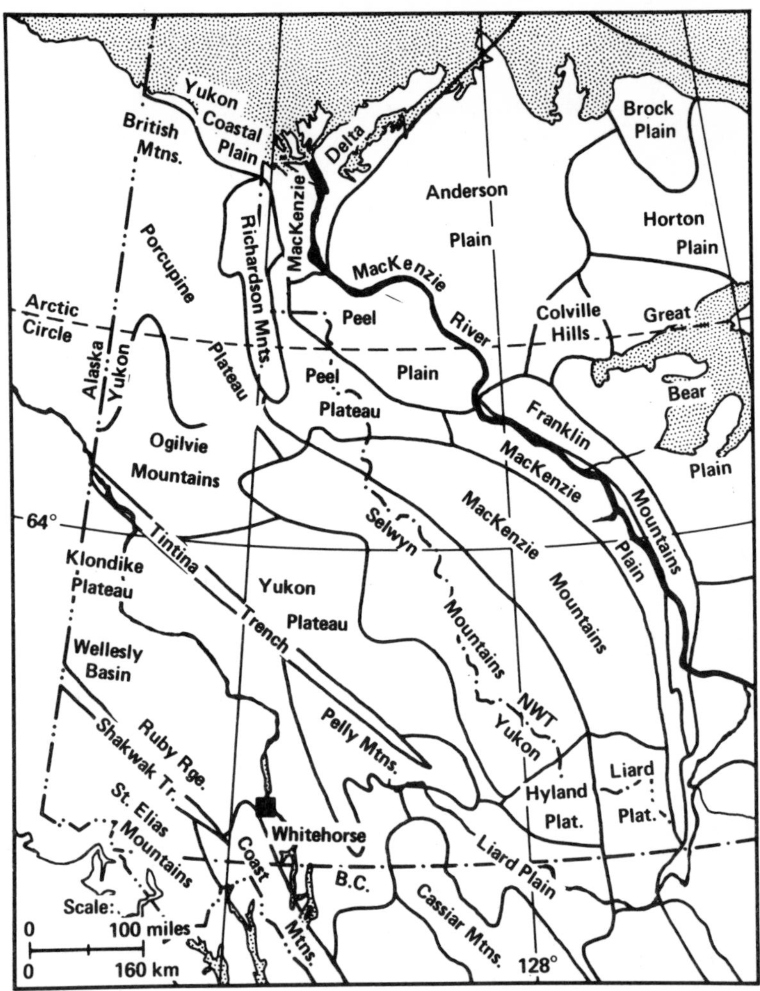

Figure 1 Physiographic map of Yukon Territory and Western District of MacKenzie (after Geol. Surv. Canada Map 13-1964, 1964, By H.S. Bostock).

Soils described in this report are all from localities within the region covered by the main Cordilleran ice sheet, except for Locality I in Tintina Trench (Figures 1 and 2) where the Quaternary deposits are the product of alpine glaciers emanating from North and South Klondike valleys. These glaciers probably coalesced to form a piedmont glacier in Tintina Trench, which may in turn have coalesced with a sub-lobe of the Cordilleran ice sheet that pushed northwestward along Tintina Trench beyond Stewart River.

Bostock (1966) found evidence for four advances of the

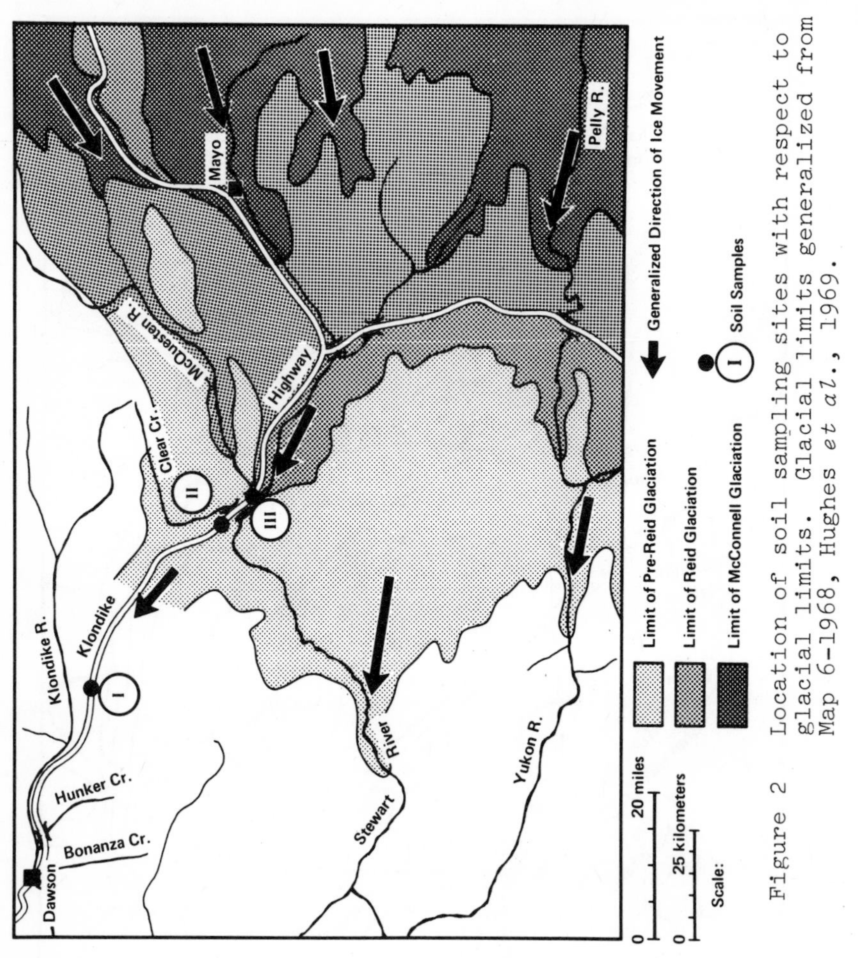

Figure 2 Location of soil sampling sites with respect to glacial limits. Glacial limits generalized from Map 6-1968, Hughes *et al.*, 1969.

Cordilleran ice sheet, which he named (from youngest to oldest), McConnell, Reid, Klaza and Nansen advances. The digitate outline of the Cordilleran ice sheet margin at the maximum of the McConnell advance can be traced almost continuously from fresh moraines and other ice-marginal features (Hughes *et al.*, 1969). Where the ice sheet formed sublobes in major valleys, as at the type locality in Stewart Valley, broad outwash terraces extend several tens of kilometers down valley from the limit of the advance. Extensive ice-contact glaciofluvial deposits and retreatal outwash are found throughout the Yukon south and east of the McConnell limit.

Although moraines and other ice-marginal features of the earlier, more extensive Reid advance are subdued relative to those of the McConnell advance, they can likewise be traced almost continuously and outline the digitate margins of the former ice sheet. Outwash terraces can also be traced several tens of kilometers along major valleys beyond the Reid limit, and extensive ice-contact glaciofluvial deposits and retreatal outwash of Reid age are found between the Reid and McConnell limits.

Evidence of yet older more extensive glaciations is found in the form of glacial erratics, outwash gravel and local till occurrences, as far as 50 miles beyond the Reid limit. Moraines and other ice-marginal features have been almost wholly obliterated by weathering and erosion.

At least three advances of valley glaciers are recorded in Ogilvie Mountains (Vernon and Hughes, 1966; Ricker, 1968; Hughes *et al.*, 1972). The youngest relatively fresh moraines, judged to be correlative with moraines of McConnell age, indicate a restricted advance of alpine glaciers in tributary valleys. In an earlier advance, glaciers of these tributary valleys merged to form trunk glaciers that occupied the larger valleys of Ogilvie Mountains. The trunk glaciers of North Klondike and Chandindu Valleys extended to Tintina Trench. Their limits are marked by moraines and other ice-marginal features that are judged, on the basis of geomorphic comparison, to be of Reid age. Glacial erratics and patches of very subdued moraine are found beyond moraines of presumed Reid age in the valleys of north-flowing rivers, indicating one or more pre-Reid glaciations. Beyond the presumed Reid limit of ice flowing from North Klondike Valley, Tintina Trench contains a sequence of glaciolacustrine, glacial and glaciofluvial sediments of pre-Reid age, named "Flat Creek beds" by McConnell (1905, p. 24B). The former extent of alpine ice origin in Tintina Trench during pre-Reid glaciation(s) has not been determined. Near its juntions with North Klondike River (Figure 2), South Klondike River is deeply incised into the Flat Creek beds. The beds are now exposed only in intermittent slumped cuts along an abandoned road that ascends the south wall of South Klondike Valley, and in shallow road cuts along the Klondike Highway. The lower beds are mainly glaciolacustrine silt and clay, and the upper beds gravel with silty gravel beds that may be ablation till. Lack of a continuous section through the beds precludes any firm interpretation of whatever part of pre-Reid history is represented by the beds. Deposits of alpine glaciations correlative with Nansen and Klaza advances, respectively, of the Cordilleran ice sheet may well occur within the thick and poorly known sequence.

Northwestward along Klondike River to its juntions with Yukon River, a high terrace of Klondike gravels (McConnell, 1905), consisting of quartzite, diabase and syenite derived from Ogilvie Mountains, appears to form a continuation of the upper surface of the Flat Creek beds. High terraces underlain by quartzose White Channel gravel along Hunker, Bonanza and other north-flowing tributaries of Klondike River, merge with the Klondike gravel terrace (Figure 2). McConnell (1905, p. 33B) considered the White Channel gravels to be at least as old as Pliocene. More recently, Hughes *et al.* (1972) have suggested that the uppermost part of the

White Channel gravels, together with the Klondike gravels, are of early Pleistocene age. For the purpose of this study, the apparently correlative upper part of the Flat Creek beds, on which the paleosol of Locality 1 is developed, is assigned a pre-Reid age.

Flat Creek type deposits appear to underlie the gently undulant floor of Tintina Trench southeastward almost to Stewart River, where the deposits are the product of pre-Reid advances of the main Cordilleran ice sheet. At Belleview Point, on the west side of Clear Creek about 8 km north of its juntion with Stewart River, till regarded as belonging to the Klaza advance (Bostock, 1966, p. 8) overlies brown gravel. The gravel, part of the "brown drift" of Bostock, contains foreign pebbles derived from Ogilvie Mountains. Bostock (1966, p. 9) considered the brown drift to be part of the Flat Creek beds of McConnell, and correlated it with the Nansen drift.

Locality II (Figure 2) is on an outwash terrace of Reid age whereas Locality III is on an outwash terrace of McConnell age. Age assignments of these localities are considered to be highly secure because of their relationship to mapped glacial limits or to outwash terraces beyond the glacial limits.

RADIOCARBON DATING

Radiocarbon dates from the area are inadequate to provide a firm chronology. In a section on the west bank of Stewart River, 12 km above the mouth of McQuesten River, till of Reid age overlies gravel that may be retreatal outwash of the Klaza glaciation, and is overlain by outwash of Reid age. A broad channel cut into the Reid outwash is filled by organic silt and sand with woody and peaty lenses. Wood from near the base of the fill has been dated as greater than 42,900 years BP (GSC-524, Lowdon and Blake, 1968). Wood from beneath till of McConnell age in a section on the north bank of Stewart River 2.4 km downstream from Mayo has been dated as more than 46,580 years old BP (GSC-331, Dyke *et al.*, 1966). The dates suggest that the Reid glaciation ended, and the Reid-McConnell interglacial began more (possibly considerably more) than 46,580 years ago BP, but do not fix the date of onset of the McConnell glaciation. There are no dates from above McConnell drift. The correlative last advance of alpine glaciers in Ogilvie Mountains culminated sometime before 13,740 years ago BP (Hughes *et al.*, 1972, p. 36).

EOLIAN DEPOSITS

Thin loess deposits occur widely on flat or gently sloping surfaces in the glaciated parts of central Yukon, commonly extending as a discrete, recognizable layer to 300 m or more above the floors of major valleys. Loess is similarly disposed in the unglaciated area adjacent to Yukon River and major tributaries such as Pelly, Stewart, Klondike and White Rivers that carried glacial meltwater beyond ice frontal positions during successive glaciations. Loess thickness is typically 30 to 60 cm, locally 90 to 120 cm on surfaces of Reid age and older. On till surfaces of McConnell age, the thickness is typically 30 cm or less, and on outwash of the same age, 15 cm or less. However, in both the glaciated and unglaciated areas, there are locally thicker deposits of organic silt that appear to be reworked loess concentrated from adjacent slopes. Such organic silt is common on the valley floors and low terraces of gold-producing creeks of the Klondike district, where silt may be 3 m or more thick, and is commonly overlain by several meters of woody, silty peat. The silt and peat are collectively termed "muck" by placer miners of the district. In the glaciated area, wedge-shaped accumulations of reworked loess, up to 3 m or more thick, are common along the toes of the slopes.

Loess of the region is believed to have been derived during successive glaciations from the broad outwash plains that extended down major valleys beyond the ice fronts. Outwash plains of McConnell age retain a braided channel pattern characteristic of glacial streams such as those of the upper reaches of present-day Kaskawalsh, Slims, Donjek and White Rivers, which head in major glaciers of the St. Elias Range. One might expect that field studies of the oldest glacial deposits and adjacent unglaciated surfaces would reveal four-fold loess sequences related to the four regional glacial advances and containing paleosols developed during intervening nonglacial intervals. However, although organic reworked loessal silts in lower Hunker Creek in Klondike district have a radiocarbon age greater than 35,000 years BP [I(GSC)-181] and may be of Reid age or older, loess sequences with recognizable paleosols have not been found. At each of the localities reported herein, glacial or glaciofluvial deposits ranging in age from probable early Pleistocene (Nansen) to classical Wisconsin (McConnell), are covered by a single thin loess layer. The soils developed in loess at the various localities are all essentially similar and are judged to have developed in post-McConnell time, and hence the loess at the various localities is of McConnell age. Where then are the loess deposits of earlier glaciations?

The following three phenomena common to the pre-McConnell localities are considered as highly significant towards providing an explanation: (1) The loess lies directly on truncated paleosols; (2) Ventifacts are common at the interface between the loess and the truncated paleosol; (3) Glacial or glaciofluvial deposits containing the paleosols are cut by sand wedges. These indicate severe climatic conditions in the immediate periglacial area during successive glaciations.

The wedges consist mainly of clean well-sorted sand with traces of silt and clay and contrast sharply in texture and color to the paleosols in which they are developed. They range in width from 10 to 30 cm at the top, terminating 1 to 1.5 m below the top of the paleosol. The wedges are therefore quite unlike fossil ice-wedges, which typically are filled with material sloughed from the sides of the former ice-wedge, or material from an overlying stratum. They resemble sand wedges forming actively under polar desert conditions at McMurdo Sound area, Antarctica (Péwé, 1959, 1962, 1974; Berg and Black, 1966).

Sand wedges that form when seasonal thermal contraction cracks become filled with wind-blown sand are closely akin in origin to the much commoner ice-wedges of arctic regions. The wedges can grow by annual increments of a few millimeters to attain widths at the top of the wedge of a meter (Péwé, 1974, p. 41). Whether sand wedges or ice-wedges develop in an area where thermal contraction cracking takes place seasonally depends mainly on the presence or absence of surface water prior to closing of the cracks during the warm season. Sand wedge formation is favored by low winter snowfall, but perhaps more particularly by high evaporation potential prior to the main warm season, so that the snow is removed mainly by sublimation rather than melting. For the cracks to be filled with windblown sand, there must be a sand source, with vegetation sparse or lacking. In Antarctica and Greenland, the small quantity of sand required is derived from coarse glacial debris and coarse fan debris, respectively. It is of major significance to our interpretation of sand wedges in soils of central Yukon that actively-forming sand wedges have been reported only from the immediate periphery of continental-size glaciers.[1]

PEDOLOGICAL INVESTIGATIONS

Morphological Characteristics and Soil Classification

A sequence of three moderately well to very well-drained soils located on outwash gravel of pre-Reid, Reid and McConnell age was sampled. The site of each soil is presented in Figure 2. Each soil profile is derived from a paleosol and surface neosol, which has been developed on loess of McConnell age. The paleosols were developed on drifts ranging from pre-Reid to McConnell age. In order to use the morphological characteristics of each soil as an aid to elucidate the Quaternary history of the area an effort was made to classify the neosols separately from the paleosol, whenever that was possible. The morphological characteristics and classification of each soil are presented in Appendix 1.

Laboratory Methods for Chemical Analysis and Mineral Determination

---

[1]An extensive summary of the origin of various types of frost fissure fillings and criteria for their identification have been provided by Romanovskij (1973).

Soil samples were collected from each soil horizon, dried in the laboratory and passed through a 2 mm sieve. Subsequently, portions of the soil were subjected to the following analyses:

(1) Detailed mechanical analysis following the A.S.T.M. method (1964), pH measurements as described by Peech (1965a, b) and total carbon and carbonates following the method of Foscolos and Barefoot (1970a). The results of these analyses are reported in the description of soils.

(2) Percent quartz in the very coarse, coarse, medium and fine sand fractions was determined following Jackson's (1965) method in order to delineate the depth of the aeolian deposits in each soil (Barshad, 1955). Quartz and feldspars were also determined in the clay fraction in order to quantize mica in each horizon of the paleosols from the K_2O content of the total elemental analysis of the <2 μm fraction.

(3) Elemental analysis of the <2mm and <2 μm fraction was conducted using the $LiCO_3$:H_3BO_3 fusion method (Foscolos and Barefoot, 1970b).

(4) Determination of oxalate extractable Fe and Al to aid in differentiating the soils (McKeague and Day, 1966).

(5) Mineral identification on the whole rock and the <2 μm fraction by X-ray diffraction using CoK α radiation, iron filter, setting of 40Kv-20ma, scanning speed of 1°/min/2 cm, time constant 2 and range of 2° to 40°2θ. The clay fraction, which was collected after sedimentation, was saturated with Ca^{2+} and K^+. X-ray diffractograms of Ca^{2+} saturated specimens were obtained using (a) an inert gas atmosphere (N_2) to induce 0% relative humidity (R.H.); (b) 50% R.H.; (c) glycerol solvation; and (d) heat treatment at 550°C for two and one half hours. The K-saturated specimens were heated to 100°C for one hour and then analyzed under (a) an inert gas atmosphere (N_2) to induce 0% R.H.; and (b) 50% R.H. Additional treatment on the clay size fraction was undertaken in order to differentiate between chlorite and kaolinite by X-rays. Warm dilute HCl was used to destroy chlorite thus enabling the identification of kaolinite in the sample. Finally, the identification of discrete and mixed layered silicates was carried out following the criteria described in Brown (1961), Harward *et al.* (1968), and Kodama and Brydon (1968).

(6) Differential thermal analysis was performed on Ca-saturated samples in order to semiquantize the ratio of 1:1+2:2 clay minerals against 2:1 layer silicates on the basis of crystal lattice water as discussed by Barshad (1965). Forty mg of sample was weighed and run from ambient temperature to 1000°C at a rate of 10° C/min. using a R.L. Stone T.G.S.-5b unit for the thermal analysis.

Results of Chemical Analysis

The elemental analysis of the whole soil and the percent quartz in four sand fractions of the studied sites are presented in Tables 1 to 3. These analyses indicate that the delineation of the paleosols on the basis of chemistry coincide with field observations. Elemental analysis was also used to study chemical trends within the paleosols. The data in Table 4 indicates that the SiO_2/R_2O_3 ($R_2O_3 = Al_2O_3 + Fe_2O_3$) molar ratio varies considerably among the different paleosols. In the pre-Reid paleosol of site I, the molar ratio of SiO_2/R_2O_3 is 0.5 in the IIB horizon and increases to 1.0 in the IIC horizon. In the Reid paleosol the IIB horizon has a molar ratio of SiO_2/R_2O_3 of 0.8 which increases to 1.0 in the IICk horizon. The same trend is obtained in the McConnell paleosol.

From elemental analysis of the <2 µm fraction, after the removal of quartz and feldspars, the K_2O/SiO_2 ratio was used to quantize the clay minerals and to assist in the interpretation of X-ray patterns. The data are presented in Tables 5 to 12. The results in Table 11 indicate that in the pre-Reid paleosol of site I, the molar ratio of K_2O/SiO_2 ranges from 1.7×10^{-2} in the paleo B horizon to 2×10^{-2} in the paleo C horizon. In the Reid paleosol the molar ratio of K_2O/SiO_2 ranges from 4.2×10^{-2} in the paleo B horizon to 5.0 in the paleo Ck horizon. A similar trend is observed in the McConnell paleosol.

An effort was also made in this study to evaluate the percent mica, expandables, and kaolinite and chlorite on the basis of chemical analysis of the <2 µm fraction, after the removal of quartz and feldspar (Table 12). The objective was to assess the relation, if any, between the type and concentration of different layer silicates with paleoclimate and/or age of paleosols. Thus by assuming an average K_2P contact of 8.5% for mica (Barshad, 1965) the concentration of mica was obtained. Also, by assuming an average value of 4.65% crystal lattice water for the 2:1 layer silicates at 400°C and an average value of 13.7% crystal lattice water for kaolinite and chlorite at 400°C (Barshad, 1965) the concentration of expandables was obtained.[1]

---

[1]Example: Assuming that mica, expandables, kaolinite and chlorite minerals have been identified from the X-ray diffractograms of the <2 µm fraction, and assuming that the elemental analysis yields a crystal lattice water value on a 400°C basis of 8.1% then if x is assigned to all 2:1 layer silicates and y to kaolinite and chlorite we have:

$$0.046tx + .1370y = 0.081$$
$$x + y = 1.000$$
$$0.0465\ (1-y) + 0.1375y = 0.081$$
$$0.0465 - 0.0465y + 0.1375y = 0.081$$
$$0.091y = 0.0345$$
$$6 = 0.0345/0.091 - .379 \text{ or } 38\%$$

kaolinite chlorite

2:1 layer silicates 100 - 38 62%

If mica has been valued on the basis of K_2O content at 21% then, 41% is the concentration of expandables.

Table 1. Site I, Elemental Analysis, Percent Quartz in Four Sand Fractions (2.00-1.00 mm, 1.00-0.50 mm, 0.50-0.25 mm and 0.250-0.125 mm) and Oxalate Extractable Iron and Aluminum of a Soil Profile Developed on Pre-Reid Outwash Gravel

Horizon	Depth in cm	Total Qtz.	Elemental Analysis in Percent											Percent Oxalate Extractable		
			SiO_2	Al_2O_3	TiO_2	MnO	Fe_2O_3	Na_2O	K_2O	MgO	CaO	L.O.I.	Total	Fe	Al	(Al+Fe)
Ahe	0.0-13.5	5.25	79.06	8.69	0.66	0.03	3.68	2.00	0.71	1.01	0.88	3.02	99.82	0.48	0.30	0.78
Bm1	13.5-18.5	3.85	76.02	9.88	0.65	0.03	4.08	2.00	0.71	1.04	0.91	3.17	98.49	0.29	0.38	0.67
II Bt1	18.5-46.0	26.02	83.12	7.08	0.33	0.03	3.57	0.71	0.57	0.43	0.25	3.30	99.39	0.20	0.28	0.48
II Bt2	46.0-117	31.26	84.15	6.57	0.31	0.03	3.43	0.57	0.61	0.42	0.26	3.11	99.46	0.23	0.16	0.39
II BC	117-190	25.24	87.04	5.30	0.17	0.02	2.57	0.55	0.60	0.38	0.27	2.37	99.27	0.15	0.12	0.27
II C	190+	20.08	89.23	4.01	0.17	0.02	1.97	0.61	0.58	0.28	0.20	1.88	98.95	0.11	0.10	0.21
		3.85	78.53	9.83	0.51	0.02	3.85	2.05	0.79	0.93	0.84	2.67	100.02	0.33	0.22	0.52
sand wedge																

Table 2. Site II, Elemental Analysis, Percent Quartz in Four Sand Fractions (2.00-1.00 mm, 1.00-0.50 mm, 0.50-0.25 mm and 0.250-0.125 mm) and Oxalate Extractable Iron and Aluminum of a Soil Profile Developed in Reid Outwash Gravel

Horizon	Depth in cm	Total Qtz.	Elemental Analysis in Percent											Percent Oxalate Extractable		
			SiO_2	Al_2O_3	TiO_2	MnO	Fe_2O_3	Na_2O	K_2O	MgO	CaO	L.O.I.	Total	Fe	Al	(Al+Fe)
Bm1	0.0-14.5	8.24	80.17	7.62	0.48	0.07	3.68	1.53	0.61	0.79	0.91	2.98	99.84	0.74	0.28	1.02
Bm2	14.5-30.0	8.69	80.97	7.09	0.50	0.06	4.34	1.35	0.74	0.86	0.78	2.50	99.19	0.98	0.28	1.26
II Bm1	30.0-49.0	38.46	83.96	5.14	0.49	0.06	3.71	1.48	0.65	0.71	0.68	2.16	99.04	0.23	0.15	0.38
II Bm2	49.0-68.0	33.94	86.52	5.01	0.27	0.03	2.85	1.04	0.60	0.66	0.59	1.81	99.38	0.28	0.19	0.47
II BC	68.0-93.0	32.20	86.78	4.31	0.33	0.02	2.71	1.94	0.58	0.50	0.24	1.80	99.21	0.39	0.32	0.71
II Ck	93+	37.36	87.04	4.20	0.28	0.05	2.69	1.97	0.57	0.58	0.45	1.47	99.30	0.20	0.12	0.32

Table 3. Site III, Elemental Analysis, Percent Quartz in Four Sand Fractions (2.00-1.00 mm, 1.00-0.50 mm, 0.50-0.25 mm and 0.250-0.125 mm) and Oxalate Extractable Iron and Aluminum of a Soil Profile Developed in McConnell Outwash Gravel

Horizon	Depth in cm	Total Qtz.	Elemental Analysis in Percent											Percent Oxalate Extractable		
			SiO_2	Al_2O_3	TiO_2	MnC	Fe_2O_3	Na_2O	K_2O	MgO	CaO	L.O.I.	Total	Fe	Al	(Al+Fe)
Bm	0.0-12.5	20.50	84.26	6.19	0.50	0.03	3.43	1.04	0.59	0.68	0.49	2.68	99.89	0.17	0.23	0.40
Bm	12.5-21.0	40.60	88.03	4.16	0.27	0.02	3.15	0.84	0.54	0.51	0.32	1.72	99.50	0.17	0.07	0.24
C	21.0-42.0	40.63	89.17	3.98	0.15	0.02	2.28	1.03	0.54	0.45	0.25	1.10	98.97	0.19	0.09	0.28
C	42+	43.33	89.32	3.68	0.17	0.03	1.85	0.89	0.62	0.50	0.95	1.37	99.38	0.18	0.05	0.23

Table 4. Ratio of SiO_2/R_2O_3 of the Horizon to SiO_2/R_2O_3 of Parent Material in the <2mm, Fraction of the Paleosols Developed on Gravel Surfaces of Pre-Reid, Reid and McConnell Age.

Parent Material	Age of Paleosols					
	Horizon	Pre-Reid	Horizon	Reid	Horizon	McConnell
		site I				
		$\frac{SiO_2/R_2O_3 \text{ of horizon}}{SiO_2/R_2O_3 \text{ of p.m.}}$		$\frac{SiO_2/R_2O_3 \text{ of horizon}}{SiO_2/R_2O_3 \text{ of p.m.}}$		$\frac{SiO_2/R_2O_3 \text{ of horizon}}{SiO_2/R_2O_3 \text{ of p.m.}}$
Gravel	II Bt1	0.5	II Bm1	0.8		
	II Bt2	0.6	II Bm2	0.8	II Bm	0.8
	II BC	0.7	II BC	1.0	II C	0.9
	II C	1.0	II Ck	1.0	III C	1.0

Table 5. Site I, Elemental Analysis of the Ca-Saturated <2 μm Fraction of a Soil Profile Developed on Pre-Reid Outwash Gravel

Horizon	Depth in cm	Elemental Analysis											Loss on Ignition		
		SiO_2	Al_2O_3	Fe_2O_3	TiO_2	P_2O_5	MnO	MgO	CaO	Na_2O	K_2O	BaO	H_2O^+	105°-350°C	350°-1000°C	Total
Ahe	0.0-13.5	43.03	19.43	11.45	1.30	0.35	0.05	2.03	1.41	0.35	1.10	0.45	5.00	4.50	5.90	98.35
Bm1	13.5-18.5	44.86	20.15	11.77	1.70	0.28	0.06	2.05	1.70	0.35	1.10	0.45	4.30	4.80	5.50	99.07
II Bt1	18.5-46.0	44.85	20.63	12.62	1.50	0.14	0.04	1.32	0.26	1.57	1.70	0.22	2.60	4.60	6.90	98.95
II Bt2	46.0-117	44.82	19.92	12.01	1.20	0.16	0.06	1.31	0.31	1.54	1.20	0.22	3.20	5.60	7.20	98.75
II BC	117-190	46.01	19.81	12.04	1.20	0.16	0.06	1.33	0.25	1.59	1.40	0.22	3.40	4.80	6.80	99.07
II C	190+	46.07	19.89	12.33	1.00	0.18	0.06	1.42	0.25	1.67	1.40	0.33	3.60	4.60	7.40	100.20
sand wedge		44.01	18.92	11.72	1.30	0.20	0.05	1.91	1.60	0.19	1.70	0.11	5.90	3.40	7.50	98.51

Table 6. Site II, Elemental Analysis of the <2μm Ca-Saturated Fraction of a Soil Profile Developed on Reid Outwash Gravel

Horizon	Depth in cm	Elemental Analysis											Loss on Ignition		
		SiO_2	Al_2O_3	Fe_2O_3	TiO_2	P_2O_5	MnO	MgO	CaO	Na_2O	K_2O	BaO	H_2O^+	105°-350°C	350°-100°C	Total
Bm1	0.0-14.5	45.61	19.02	11.74	0.84	0.06	0.12	2.22	2.03	0.24	1.30	0.67	3.40	5.10	5.90	98.25
Bm2	14.5-30.0	44.83	19.43	14.32	0.84	0.04	0.11	2.23	1.95	0.22	1.70	0.78	2.20	5.30	5.60	99.55
II Bm1	30.0-49.0	42.22	19.47	16.93	0.84	0.11	0.22	2.35	1.52	0.59	3.10	1.00	2.70	4.00	4.00	99.05
II Bm2	49.0-68.0	45.24	20.68	11.25	0.84	0.18	0.07	3.16	1.34	0.27	2.80	0.56	2.10	3.70	6.20	98.39
II BC	68.0-93.0	45.83	20.32	11.47	0.84	0.18	0.08	3.12	1.21	0.22	3.00	0.56	4.00	3.00	5.10	98.93
II Ck	93.0+	43.85	20.41	12.90	0.67	0.20	0.15	2.74	1.28	0.24	3.20	0.56	3.30	4.00	5.80	99.30

Table 7. Site III. Elemental Analysis of <2µm Ca-Saturated Fraction of a Soil Profile Developed on McConnell Outwash Gravel

Horizon	Depth in cm	Elemental Analysis												Loss on Ignition		
		SiO_2	Al_2O_3	Fe_2O_3	TiO_2	P_2O_5	MnO	MgO	CaO	Na_2O	K_2O	BaO	H_2O^+	105°-350°C	350°-1000°C	Total
Bm	0.0-12.5	48.22	13.41	8.67	1.20	0.20	2.20	2.11	5.40	1.00	1.20	3.23	1.50	3.40	7.50	99.24
II Bm	12.5-21.0	60.03	13.43	8.92	0.50	0.72	0.06	1.97	3.30	0.67	2.20	1.12	1.00	2.90	3.10	99.92
II C	21.0-42.0	45.86	18.92	12.31	0.50	0.72	0.50	2.13	1.50	0.43	2.80	1.23	0.60	4.50	6.30	98.30
III C	42.0+	46.23	18.96	12.34	0.33	0.69	0.52	1.94	1.70	0.38	2.80	1.45	0.80	5.70	4.60	98.44

Table 8. Site I, Elemental Analysis of the <2µm Fraction at 400°C After the Removal of Quartz and Feldspars of a Soil Profile Developed on Pre-Reid Drift

Horizon	Depth in cm	Elemental Analysis											
		SiO_2	Al_2O_3	Fe_2O_3	TiO_2	P_2O_5	MnO	MgO	CaO	Na_2O	K_2O	BaO	H_2O^+	Total
Ahe	0.0-13.5	45.08	24.22	14.58	1.72	0.45	0.06	2.61	1.81	0.28	1.10	0.59	7.50	100.00
Bm	13.5-18.5	44.73	24.79	14.43	2.07	0.35	0.08	2.53	2.20	0.25	1.00	0.57	7.00	100.00
II Bt1	18.5-46.0	45.63	23.67	14.76	1.84	0.16	0.05	1.54	0.33	1.86	1.80	0.26	8.10	100.00
II Bt2	46.0-117	48.21	22.95	13.74	1.46	0.19	0.08	1.56	0.40	1.77	1.30	0.24	8.10	100.00
II BC	117.0-190	48.50	22.96	13.83	1.37	0.22	0.08	1.35	0.40	1.84	1.50	0.15	7.80	100.00
II C	190+	48.93	22.64	13.77	1.33	0.18	0.07	1.52	0.28	1.80	1.50	0.18	7.80	100.00
sand wedge		49.34	22.02	13.60	1.10	0.20	0.07	1.61	0.27	1.80	1.40	0.39	8.20	100.00

Table 9. Site II, Elemental Analysis of the <2µm Fraction at 400°C After the Removal of Quartz and Feldspars of a Soil Profile Developed on Reid Outwash Gravel

Horizon	Depth in cm	Elemental Analysis												
		SiO_2	Al_2O_3	Fe_2O_3	TiO_2	P_2O_5	MnO	MgO	CaO	Na_2O	K_2O	BaO	H_2O^+	Total
Bm1	0.0-14.5	46.86	22.46	14.49	1.01	0.07	0.15	2.73	2.50	0.26	1.40	0.77	7.30	100.00
Bm2	14.5-30.0	43.12	22.17	17.14	1.06	0.05	0.13	2.65	2.30	0.23	3.00	0.85	6.70	100.00
II Bm1	30.0-49.0	44.52	23.98	13.17	1.03	0.19	0.18	3.46	2.10	0.33	2.90	0.84	7.30	100.00
II Bm2	49.0-68.0	47.29	23.54	11.70	0.98	0.21	0.08	3.62	1.50	0.20	3.00	0.58	7.30	100.00
II BC	68.0-93.0	47.83	22.40	12.76	1.04	0.20	0.09	3.51	1.40	0.08	3.20	0.59	6.90	100.00
II Ck	93.0+	44.52	23.37	15.20	0.79	0.24	0.18	3.25	1.40	0.10	3.50	0.65	6.80	100.00

Table 10. Site III, Elemental Analysis of the <2 µm Fraction at 400°C After the Removal of Quartz and Feldspar of a Soil Profile Developed on McConnell Outwash Gravel

Horizon	Depth in cm	Elemental Analysis												
		SiO_2	Al_2O_3	Fe_2O_3	TiO_2	P_2O_5	MnO	MgO	CaO	Na_2O	K_2O	BaO	H_2O^+	Total
Bm	0.0-12.5	49.84	15.76	14.07	1.43	0.23	1.59	2.56	2.66	1.12	1.20	2.74	6.80	100.00
II Bm	12.5-21.0	48.63	17.07	15.64	0.76	1.09	0.09	2.66	2.40	0.42	3.10	1.44	6.70	100.00
II C	21.0-42.0	40.87	23.53	15.51	0.90	1.46	0.24	2.63	2.20	0.49	3.20	1.27	7.70	100.00
III C	42.0+	42.78	22.75	14.90	0.61	0.88	0.61	2.62	1.80	0.47	3.40	1.48	7.70	100.00

Table 11. Ratio of K_2O/SiO_2 in the <2 µm Fraction at 400°C After the Removal of Quartz and Feldspars of Paleosols Developed on Gravel Surfaces of Pre-Reid, Reid and McConnell Age.

Parent Material	Age of Paleosols					
	Horizon	Pre-Reid	Horizon	Reid	Horizon	McConnell
		site I				
		K_2O/SiO_2 x 10^{-2}		K_2O/SiO_2 x 10^{-2} / x 10^{-2}		K_2O/SiO_2 x 10^{-2}
Gravel	II Bt1	2.5	II Bm1	4.2		
	II Bt2	1.7	II Bm2	4.3	II Bm	4.1
	II BC	2.0	II BC	4.3	II C	5.0
	II C	2.0	II Ck	5.0	III C	5.1

Table 12. Percent Mica, Kaolinite+Chlorite and Expandable Clay Minerals in the <2µm Fraction at 400°C After the Removal of Quartz and Feldspar of Paleosols Developed on Gravel of Pre-Reid, Reid and McConnell Age

Parent Material	Age of Paleosols											
	Horizon	Pre-Reid			Horizon	Reid			Horizon	McConnell		
		site I										
		Mica	Expand	Kaol + Chl		Mica	Expand	Kaol + Chl		Mica	Expand	Kaol + Chl
Gravel	II Bt	21	41	38	II Bm	34	37	29				
	II Bt	15	47	38	II Bm	35	36	29	II Bm	36	41	23
	II BC	18	47	35	II BC	38	37	25	II C	38	29	33
	II C	18	47	35	II Ck	41	35	24	III C	41	25	34

The concentration of layer silicates which include also the end members of the mixed layers, indicate that the mica concentration is low in the older paleosols and high in the younger paleosols. Specifically, the mica content ranges from 15% to 21% in the pre-Reid paleosol and from 35% to 41% in the Reid paleosol. The range of mica concentration of the paleosol developed on McConnell outwash gravel is similar to that of the Reid paleosol.

Results of Clay Mineral Determination

X-ray patterns of the <2 µm specimen, of the soil and the paleosol horizons are presented in Figures 3 to 12.

Based upon the clay minerals, the soils of this study can be placed into two categories. The first encompasses the Reid and McConnell paleosols, whereas the second includes the pre-Reid paleosol. Soils of the first category contain kaolinite, illite, chlorite, vermiculite and chloritic intergrades resulting from aluminum or iron interlayering of depotassified mica and vermiculite. The soil of the second category contains kaolinite, illite and montmorillonite-kaolinite mixed layers.

Kaolinite has been confirmed from the d_{001} peak at 7.Å after the removal of chlorite by warm dilute HCl. The specimen has been subjected to X-rays under an inert gas atmosphere in order to avoid any contribution to the d_{001} spacing of kaolinite, from other concomitant clay minerals which are not destroyed by HCl, *e.g.* H-vermiculite (Brown, 1961). Figure 10 shows the diffraction patterns of an HCl treated specimen of <2µm soil clays under different X-ray conditions. Additional evidence for the presence of kaolinite was also obtained from the presence of the 3.57Å and 3.54Å double peaks representing the d_{002} and d_{004} planes of kaolinite and chlorite in the untreated <2µ clay samples, *e.g.*, Figure 3. This distinction is somewhat difficult when the sample is subjected to CuKα radiation and the use of Debye-Scherrer camera or when the diffraction patterns are recorded at a speed chart of 1°/cm/min. However, the separation of the doublet is possible when CoKα is used and the diffraction patterns are recorded at a chart speed of 2 cm/min and a scan speed of 1°2θ/min. The presence of kaolinite was also confirmed by the disappearance of the 3.57Å peak upon heating the sample to 550°C for two and one half hours. Thus, the presence of kaolinite has been verified in all diffractograms (Figures 3 to 12).

Chlorite has been confirmed by the presence of d_{001} spacing at 14Å after heating the sample to 550°C for two and one half hours. Thus chlorite has been ascertained in the loess deposits as well as in the paleosols of Reid and McConnell age (Figures 3, 7 to 12).

Mica has been identified by the presence of d_{001} spacing at 10Å in all Ca-saturated specimens which have been subjected to 50% R.H. or glycerol solvation (Figures 3 to 12).

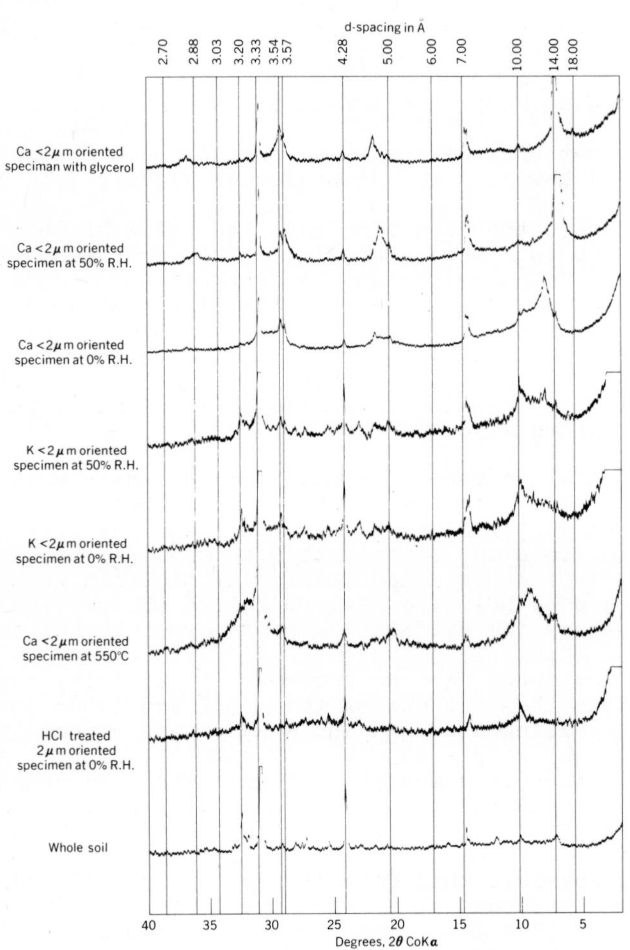

Ahe horizon

Figure 3 X-ray diffraction patterns of unoriented soils and < 2μm oriented fraction under different treatments from a soil developed on a pre-Reid outwash gravel surface, site I.

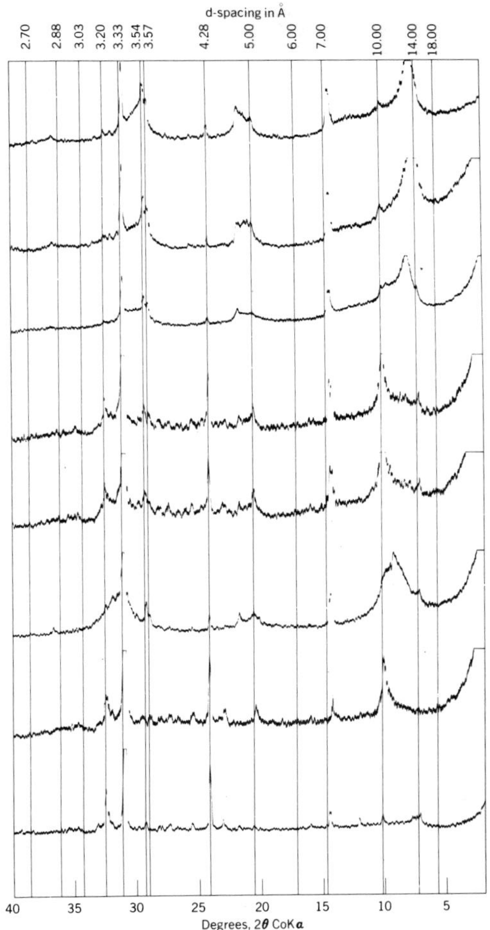
d-spacing in Å
2.70
2.88
3.03
3.20
3.33
3.54
3.57
4.28
5.00
6.00
7.00
10.00
14.00
18.00
40
35
30
25
20
15
10
5
Degrees, 2θ CoKα

Bm horizon

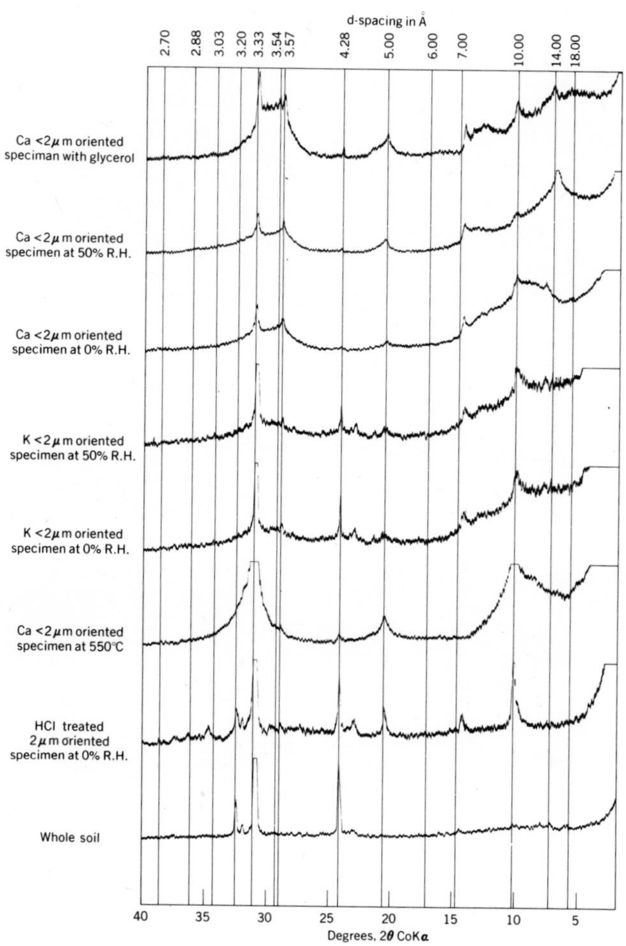

II Btl horizon

Figure 4 X-ray diffraction patterns of unoriented soils and <2μm oriented fraction under different treatments from a soil developed on a pre-Reid outwash gravel surface, site I.

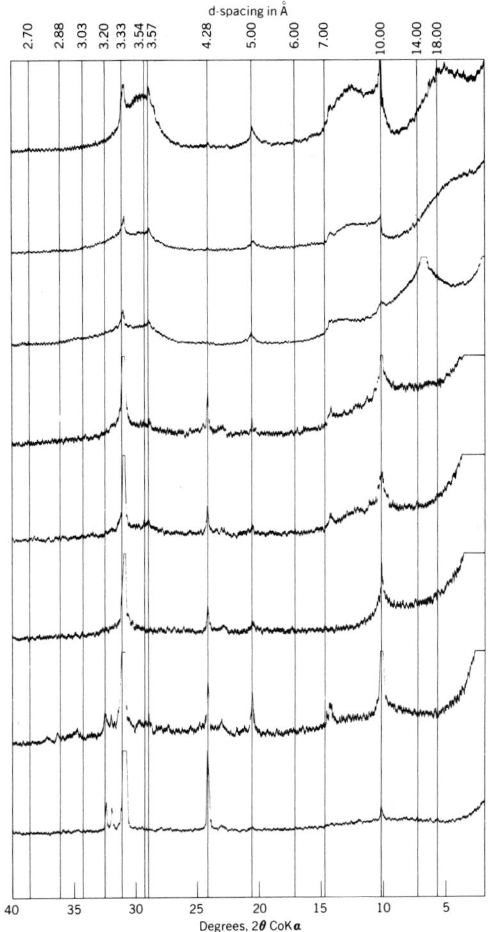

II Bt2 horizon

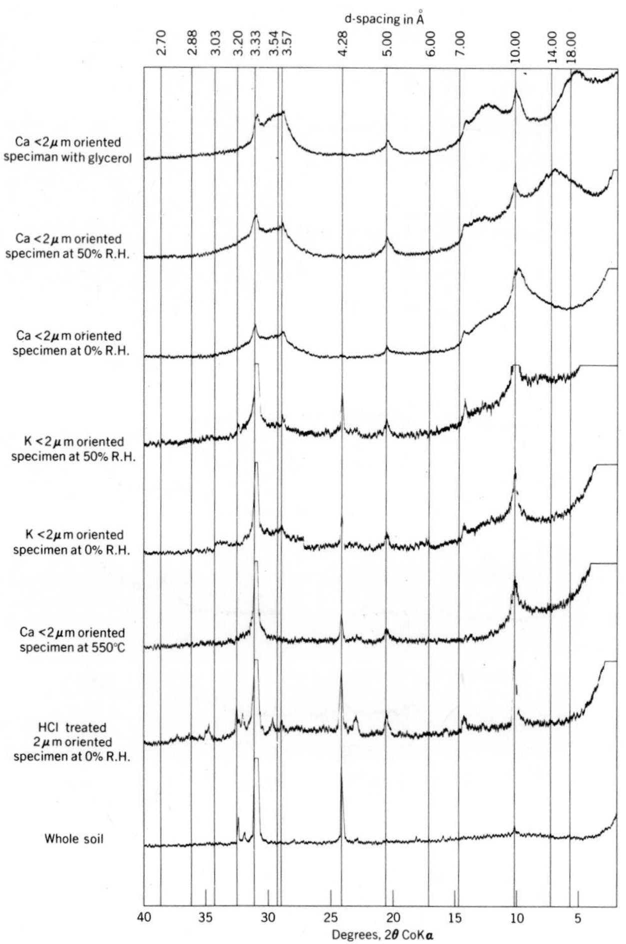

II BC horizon

Figure 5 X-ray diffraction patterns of unoriented soils and < 2μm oriented fraction under different treatments from a soil developed on a pre-Reid outwash gravel surface, site I.

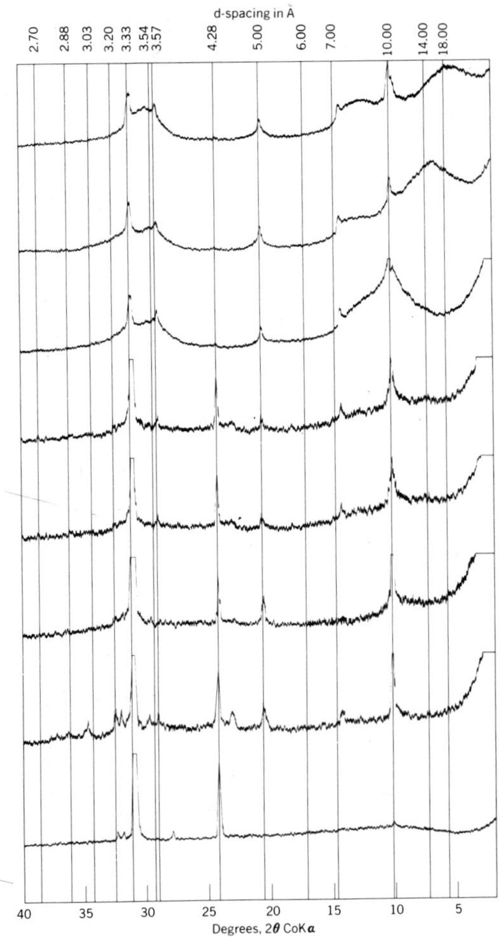
d-spacing in Å
2.70
2.88
3.03
3.20
3.33
3.54
3.57
4.28
5.00
6.00
7.00
10.00
14.00
18.00
40
35
30
25
20
15
10
5
Degrees, 2θ CoKα

II C horizon

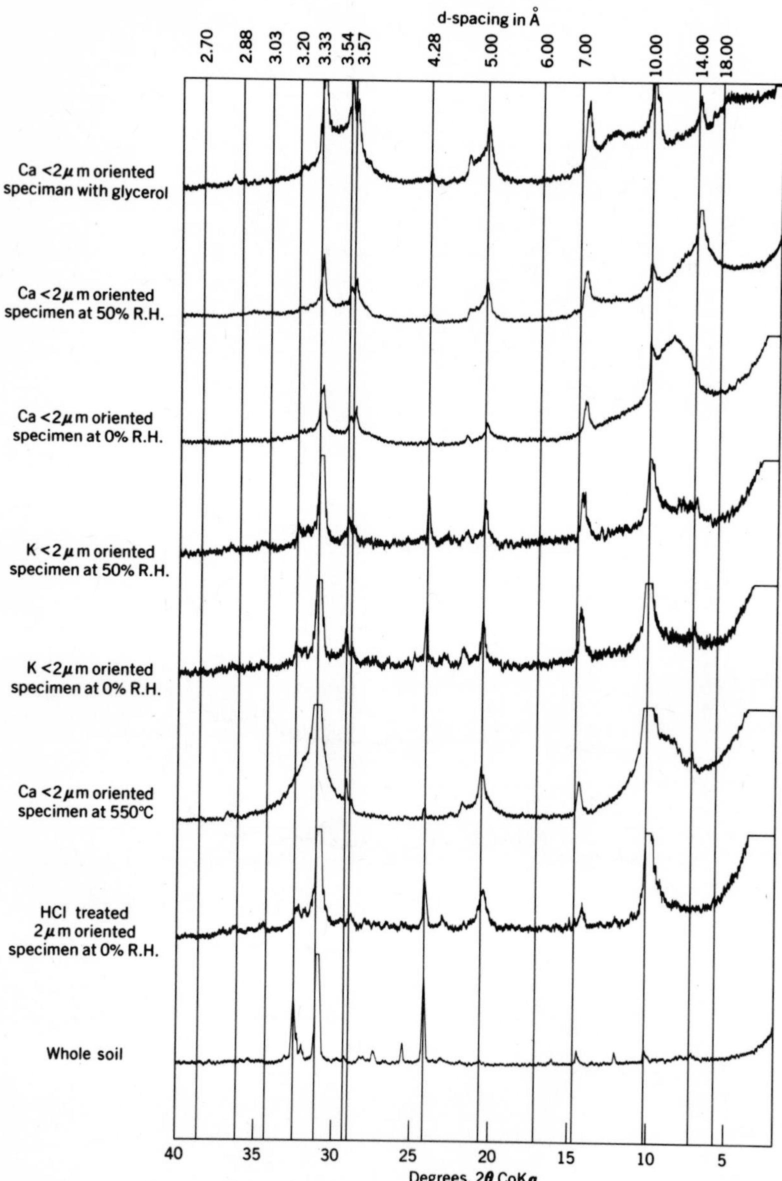

Figure 6 X-ray diffraction patterns of unoriented soils and < 2μ m oriented fraction under different treatments from a sand wedge located in a soil developed on pre-Reid outwash gravel surface, site I.

However, the intensity of 10Å peak varies with the depth and location of the paleosol. This trend is also verified from the K_2O content presented in Tables 7 to 9. Thus depotassification of mica seems to be related to the degree of weathering of the paleosol.

Vermiculite in presence of illite, d_{001} 10Å, has been identified on the basis of the following mineralogical and chemical criteria. X-ray diffractograms obtained at 0% R.H., 50% R.H. and glycerol solvation indicate a simultaneous decrease of 10Å peak intensity and increase in the 14Å peak intensity on the same ca-saturated <2μ m specimen. At 0% R.H. the d_{001} spacing at 10Å and 14Å has a given peak height. At 50% R.H. the d_{001} spacing at 10Å decreases whereas, the d_{001} spacing of 14Å increases, indicating that an expandable mineral has shifted its d_{001} spacing from 10Å to 14Å. Upon glycerol solvation the d_{001} spacing of 10Å and 14Å remains the same. This indicates that further expansion did not take place either from 10Å to 14Å, or to higher spacing than 14Å. This behavior is typical of vermiculites since their maximum expansion is achieved with relatively low R.H. values, with little additional change for samples equilibrated at higher R.H. values, or by glycerolation (Harward *et at.*, 1968; Kodama and Brydon, 1968). Additional evidence for the presence of vermiculite was obtained by comparing the K-saturated <2μ m specimens at different R.H. values with the Ca-saturated <2μ m specimen at the same R.H. values. For the K-saturated specimens, the d_{001} peak intensity at 10Å and 14Å was constant regardless of the value of the relative humidity. However, by comparing the diffractograms of the Ca-saturated specimens vs. the K-saturated ones at 50% R.H., it was observed that in the former the peak intensity at 10Å decreases and the peak intensity of 14Å increases whereas with the latter the opposite is observed. This behavior is attributed to the presence of vermiculite in the specimen (Harward *et al.*, 1968).

The presence of vermiculite seems to be related to depotassification process of mica because wherever the mica increases in the soil profile, vermiculite seems to disappear from the X-ray patterns.

The presence of chloritic intergrades has been confirmed by the X-ray patterns (Appendix: Figures 3, 7, 8, 9 (II BC horizon), 10, 11). These intergrades seem to result by partial interlayering of hydroxy aluminum of hydroxy iron material in depotassified mica and vermiculite. The occurrence of partially interlayered phyllosilicates is suggested by the resistance to complete collapse to 10Å upon heating to 550°C. The possibility of having an expandable component in the chloritic intergrades is speculated from the behavior of the Ca-saturated <2μ m specimen when subjected to different relative humidities and glycerol solvation treatments. For example, in Figure 11 the chloritic intergrades in the Ca-saturated specimens occupy the region between 10Å and 14Å. At 0% R.H. the chloritic intergrades

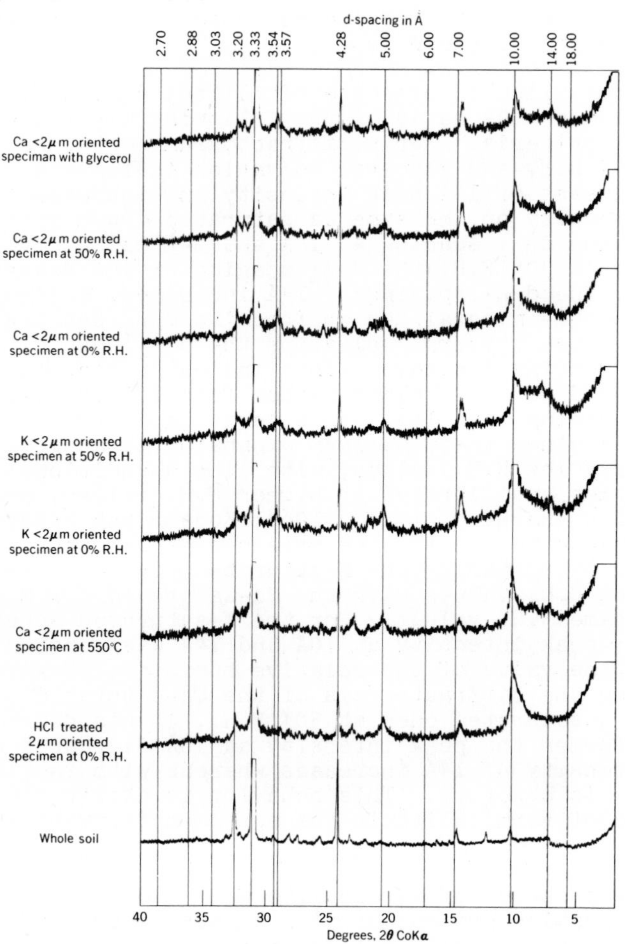

Bm1 horizon

Figure 7 X-ray diffraction patterns of unoriented soils and < 2μm oriented fraction under different treatments from a soil developed on Reid outwash gravel surface, site II.

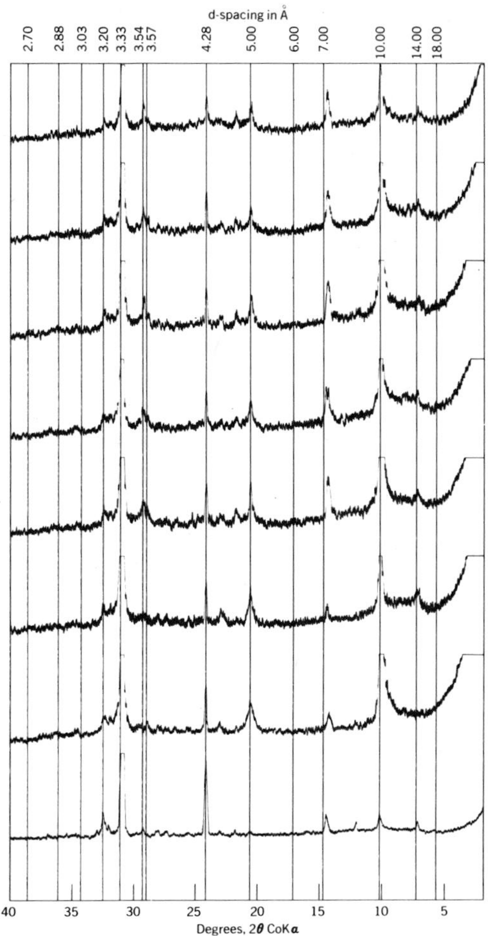
d-spacing in Å
2.70
2.88
3.03
3.20
3.33
3.54
3.57
4.28
5.00
6.00
7.00
10.00
14.00
18.00
40
35
30
25
20
15
10
5
Degrees, 2θ CoKα

Bm2 horizon

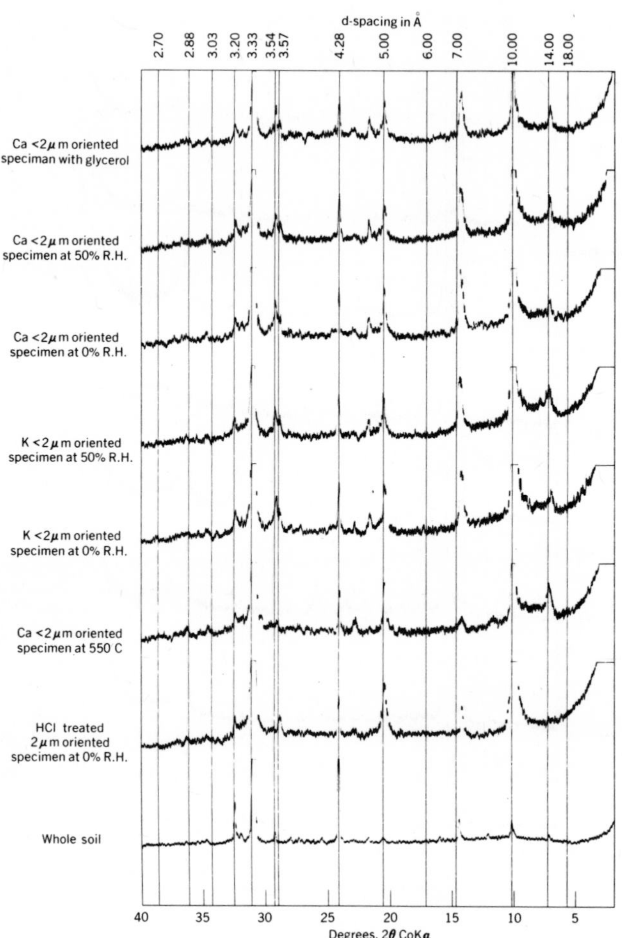

II Bml horizon

Figure 8 X-ray diffraction patterns of unoriented soils and < 2μm oriented fraction under different treatments from a soil developed on Reid outwash gravel surface, site II.

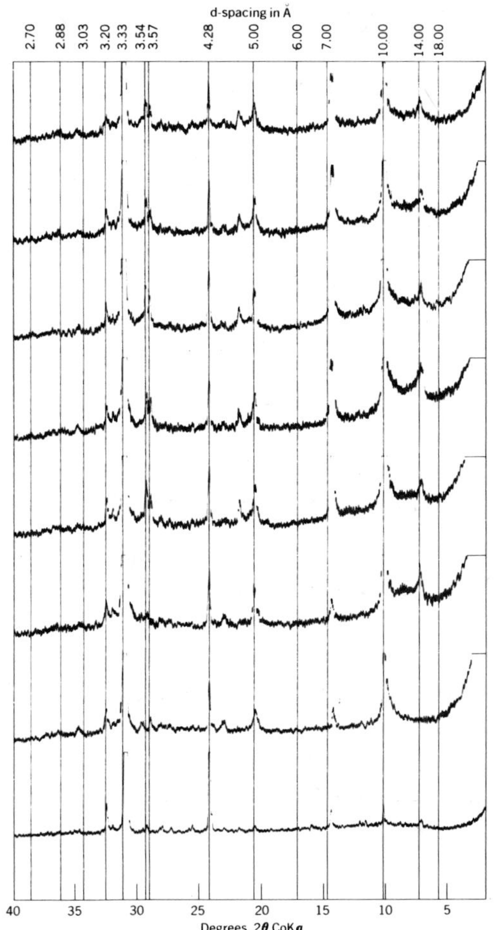

II Bm2 horizon

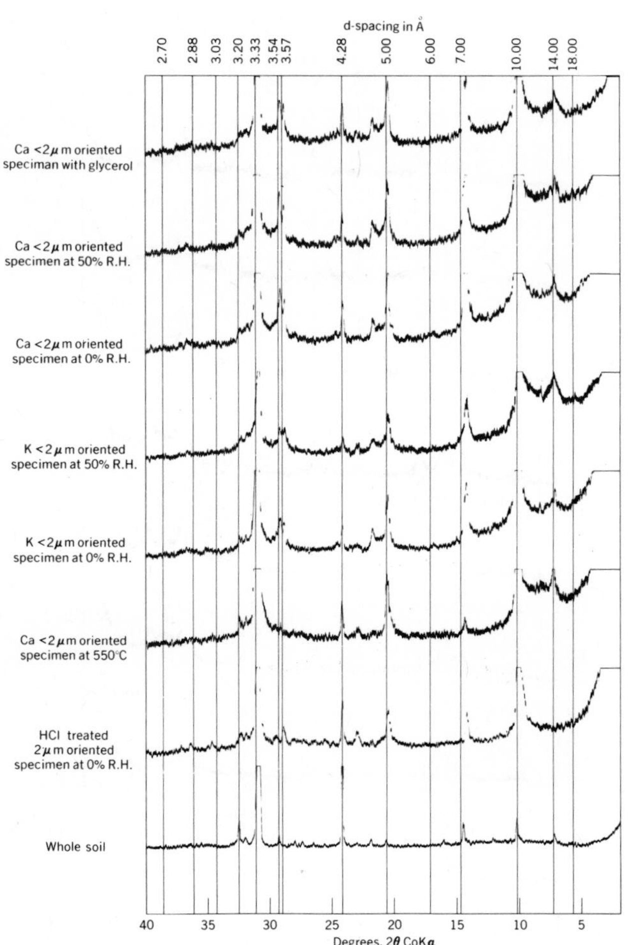

II BC horizon

Figure 9 X-ray diffraction patterns of unoriented soils and <2μm oriented fraction under different treatments from a soil developed on Reid outwash gravel surface, site II.

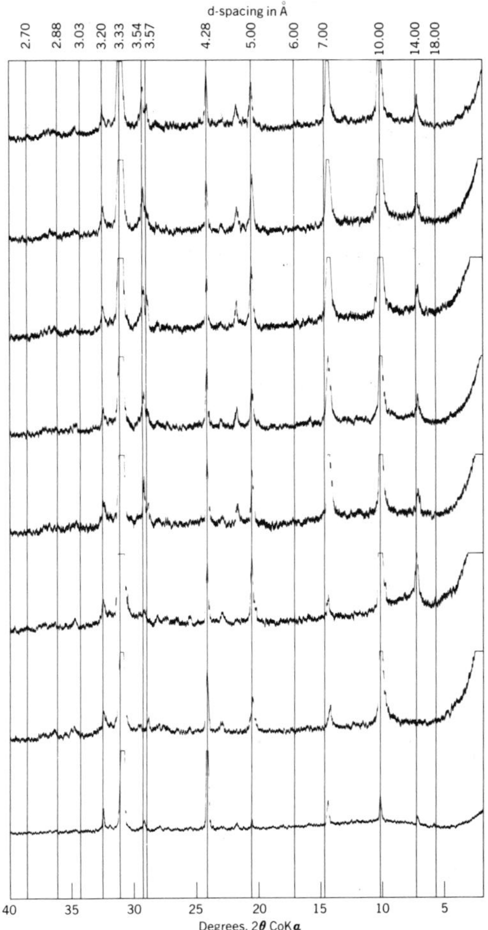

II Ck horizon

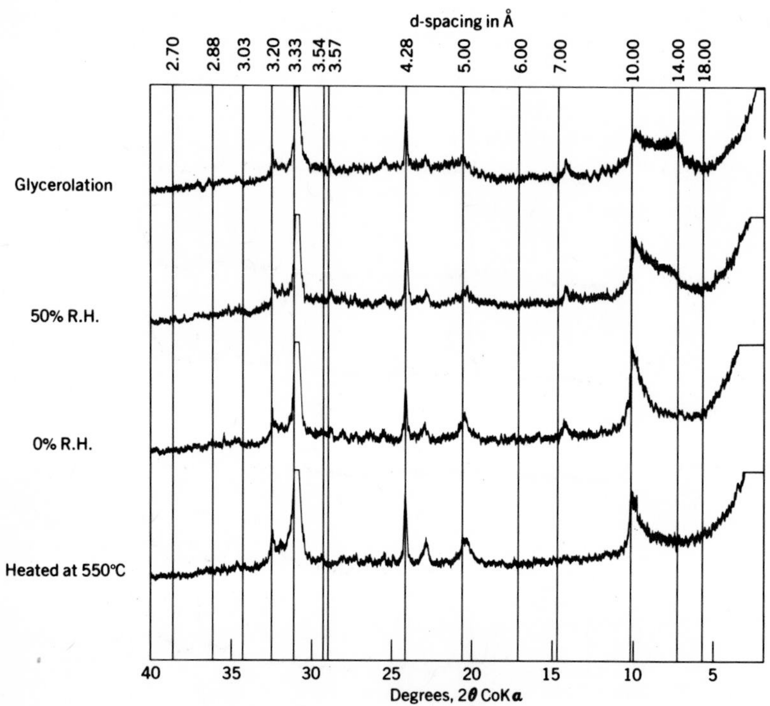

Figure 10 X-ray diffractograms of < 2μm oriented fraction from the Bml horizon after the removal of chlorite by HCl treatment. X-ray patterns of the same specimen under different conditions (a) glycerolation, (b) 50% R.H., (c) 0% R.H., (d) heat treatment at 550°C for 2 1/2 hours.

tails off towards 14Å. At 50% R.H. the chloritic intergrades form a continuous plateau between 10Å and 14Å. Finally, upon glycerol solvation the intergrades taper off towards the 10Å spacing. This continuous change of the plateau shape with different relative humidities and glycerol treatment implies that there is a mineral in the chloritic intergrades which expands to a certain point by absorbing water molecules and expands even further upon glycerol solvation thus inducing changes in the shape of the X-ray patterns in the region between 10Å and 14Å. Though this behavior can be attributed mainly to vermiculite because the expansion is limited between 10Å and 14Å, it is probable that some montmorillonite also may be present. If vermiculite was the only expandable component of the intergrades then the expansion or the change of the X-ray pattern between 10Å and 14Å should have been limited between 0% R.H. and 50% R.H. because vermiculite reaches its maximum expansion at 50% R.H. As a result, upon glycerol solvation of the <2μm specimen, the shape of the X-ray pattern between 10Å and 14Å should have stayed the same as it is under 50% R.H. However, upon glycerolation the X-ray pattern changes, tailing off towards 10Å. This suggests that very small amounts of montmorillonite might also be present in the chloritic intergrades.

The overall reaction for the formation of intergrades can be summarized as follows:

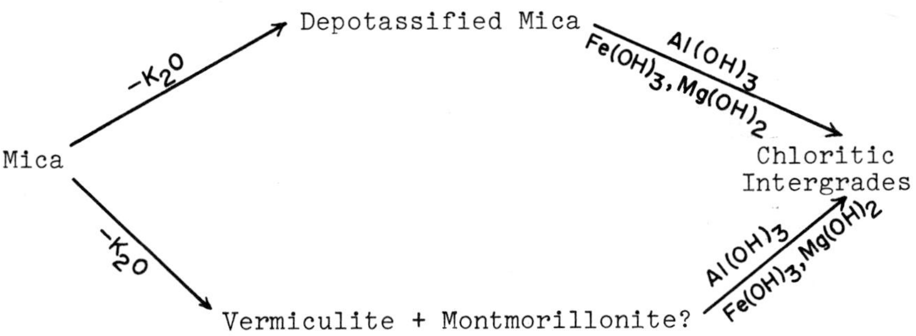

The presence of excess aluminum and iron from the formation of chloritic intergrades can be verified from the chemical analysis (Tables 7 to 9). These analyses show high concentration of Al_2O_3 and Fe_2O_3 in the studied paleosol.

In the pre-Reid paleosol illite, and kaolinite have been identified as previously described. Chlorite is absent and vermiculite also seems to be absent from the X-ray patterns of the <2μm specimen. Chloritic intergrades are present, though not as conspicuous as in the Reid and McConnell paleosols. For example, in Figure 3, the Ca-saturated <2μm specimen of the IIBt2 horizon treated at 550°C for two and one half hours indicates resistance to complete collapse to 10Å for the 2:1 layer silicates. The latter is indicative of chloritic intergrades. For deeper horizons (IIBt1 and IIBc, Figures 4 and 5) the presence of chloritic intergrades becomes less conspicuous and finally for the IIC horizon (Figure 4), the chloritic intergrades disappear completely. Again the explanation for the presence of chloritic intergrades in the pre-Reid paleosol lies probably in the depotassification process of mica and the absorption of hydroxy aluminum and hydroxy iron complexes.

Evidence for the presence of interstratified kaolinite-montmorillonite is presented in Figures 4, 5 and 6. In Figure 5, the Ca-saturated <2μm specimen indicates that upon increasing the R.H. from 0% to 50% and upon glycerolation there is a progressive movement of a poorly defined

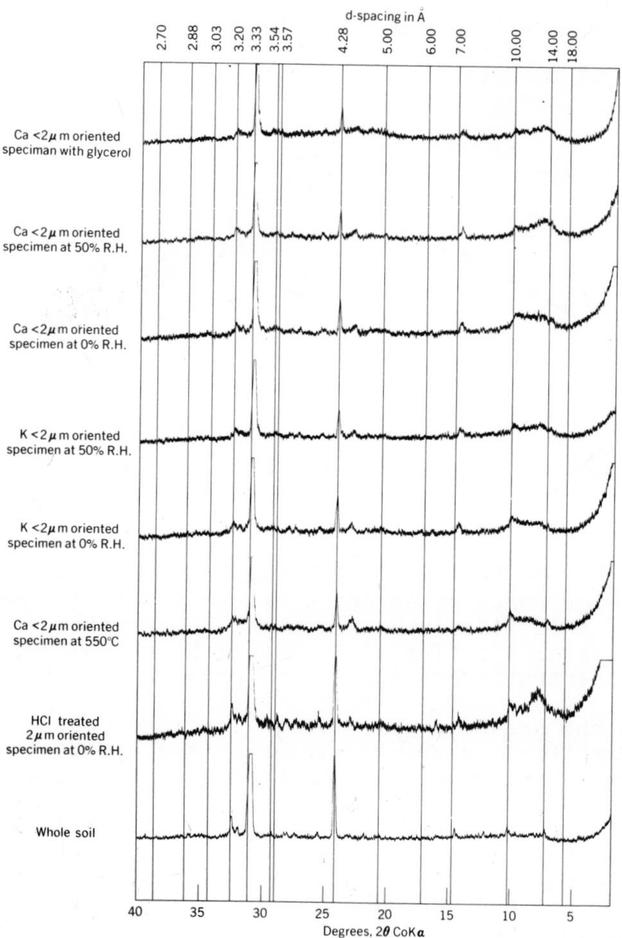

Bm

Figure 11 X-ray diffraction patterns of unoriented soils and <2μm oriented specimen under different treatments of a soil developed on McConnell outwash gravel, site III.

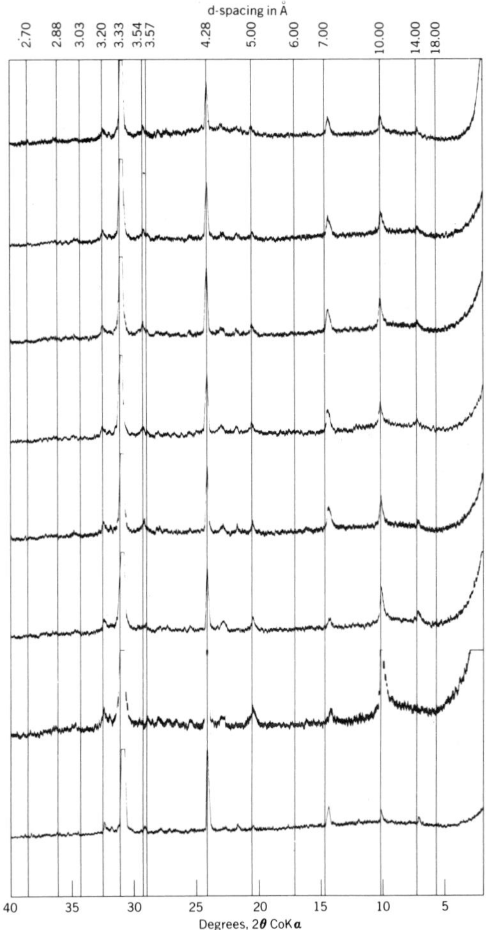
d-spacing in Å
2.70
2.88
3.03
3.20
3.33
3.54
3.57
4.28
5.00
6.00
7.00
10.00
14.00
18.00
40
35
30
25
20
15
10
5
Degrees, 2θ CoKα

II Bm

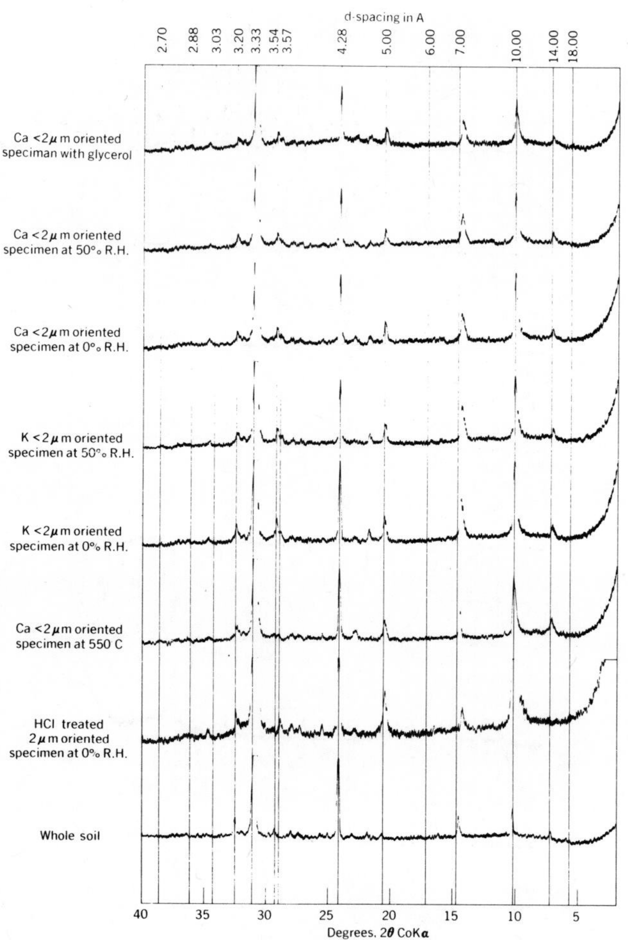

II C

Figure 12 X-ray diffraction patterns of unoriented soils and < 2μm oriented specimen under different treatments of a soil developed on McConnell outwash gravel, site III.

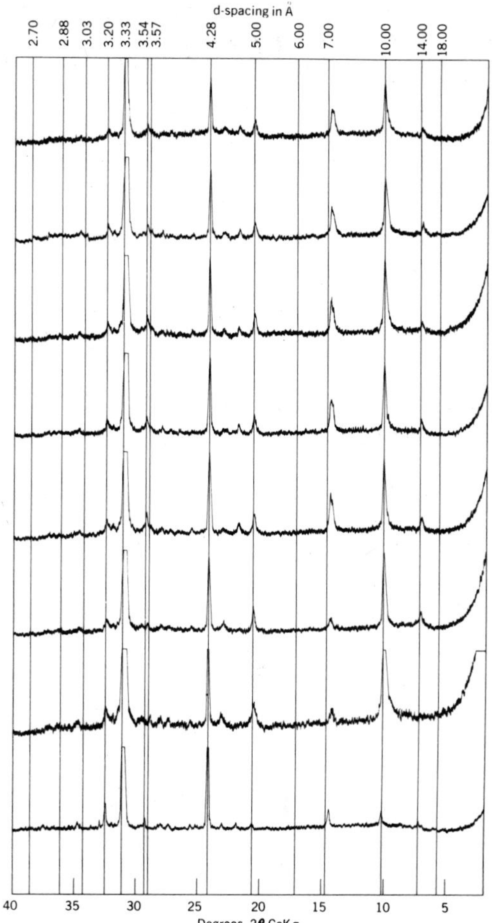
d-spacing in Å
2.70
2.88
3.03
3.20
3.33
3.54
3.57
4.28
5.00
6.00
7.00
10.00
14.00
18.00
40
35
30
25
20
15
10
5
Degrees, 2θ CoKα

III C

peak from 10Å to 15Å to 18Å. At 18Å the peak has a leading tail towards the direction of lower angles. This behavior of the interstratified clays indicates the presence of a montmorillonitic component in the mixed layer. Another line of evidence for the presence of a montmorillonitic component in the mixed layers is obtained by comparing the X-ray diffractograms of K-saturated $<2\mu m$ specimen at 0% R.H. and at 50% R.H. In Figure 4, the IIBC horizon of site I shows that the X-ray pattern of the K-saturated specimen at 50% R.H. differs from the X-ray pattern of the very same specimen at 0% R.H. At 0% R.H. all 2:1 layer silicates upon K^+ saturation collapse, to yield a well-defined 10Å peak. At 50%, the very same specimen exhibits a plateau from 11Å to almost 19Å. This differential behavior is attributed to a montmorillonitic component in the interstratified clays. Evidence for the presence of the kaolinitic component in the interstratified clays is obtained from the Ca-saturated specimen at 0% R.H., 50% R.H., glycerol solvation and heat treatment at 550°C. In Figure 4, the Ca-saturated $<2\mu m$ specimen of the IIBt2 horizon exhibits a plateau between 7Å and 10Å. Upon increasing the R.H. to 50% the plateau becomes more humpy and upon glycerol solvation there is a hump with a maximum at 8.1Å. This hump, as well as kaolinite, disappears upon heating to 550°C for two and one half hours.

The Interstratified mixture of kaolinite-montmorillonite is identified in the pre-Reid paleohorizons of site I as well as in the clay fraction of the sand wedge of site I (Figure 5).

DISCUSSION

Geomorphic and Stratigraphic Evidences of Climatic Fluctuations

Climatic fluctuations of the region during the Quaternarey, as inferred from stratigraphic and geomorphic evidence, can be summarized as follows:

(1) Pre-Reid glaciation(s). The parent material of the paleosol at Locality I was deposited during a glaciation which preceded the Reid glaciation. However, there is no firm geomorphic, stratigraphic or pedologic evidence by which the parent material can be assigned to either Klaza or Nansen glaciation, or correlative advances of alpine glaciers in Ogilvie Mountains.

(2) Pre-Reid interglacial. Geomorphic evidence, such as major deepening of Klondike River valley prior to the Reid glaciation, and virtual elimination of characteristic glacial landforms on pre-Reid deposits, demonstrate that a major interglacial period preceded the Reid glaciation. No deposits of the interglacial have been found in the area, so that soil development on pre-Reid deposits herein provides the first evidence of climatic conditions.

(3) Reid glaciation. During the Reid glaciation, much of the Yukon Plateau was again covered by a Cordilleran

ice sheet. Glacial till was deposited over much of the region, and outwash was deposited beyond the ice front in major valleys. Similarly, alpine glaciation was renewed in Ogilvie Mountains, with a trunk glacier extending down North Klondike almost to the junction of North and South Klondike Rivers. There, terminal moraine and other ice-marginal features are inset in an inner valley below the pre-Reid surface. Restricted areas of pre-Reid deposits, with deep soils developed during a pre-Reid interglacial, remained icefree but were subjected to intense periglacial climate, significantly colder and drier than the present. Periglacial climatic conditions are inferred from the widespread occurrence of sand wedges that transect the paleosol developed on pre-Reid deposits, and of ventifacts at the upper surface of the paleosol. The sand is accompanied by minor amounts of the same pedogenetic clays as those found in adjacent paleosol, indicating local derivation of the sand filling. Deflation of the Ae horizon, transfer of the sand to the developing wedges, and development of ventifacts indicate that there could have been at most only sparse steppe vegetation.

During retreatal stages of the Reid, with increasing vegetation cover, loess may have been deposited widely over the region but if so, it was removed by wind erosion during the subsequent McConnell glaciation.

(4) Reid-McConnell interglacial. Nonglacial deposits are known from beneath till of the McConnell advance and from above Reid glacial deposits that lie beyond the McConnell limit and are being studied currently. To date, however, soil development on Reid deposits (see Appendix 1) provides the only evidence of climatic conditions during the Reid-McConnell interglacial.

(5) McConnell glaciation. During the McConnell glaciation much of Yukon Plateau was again covered by the Cordilleran ice sheet. Areas of Reid drift beyond the ice margin were subjected to a cold dry climate, with resultant development of sand wedges, deflation of the A horizon of the Brunisol soil and development of ventifacts on quartz and quartzite pebbles. Sand wedges that transect the paleosol developed on parent materials of Reid age are narrower than those in pre-Reid drift, suggesting either a less intense or briefer periglacial climate (or both) during the McConnell glaciation than during the Reid.

Sand wedges of McConnell age were not identified in soils developed on pre-Reid parent material, although our interpretation of climatic history suggests that they should be present. They may have been locally, lacking or were not distinguished from older wedges.

Deposition of McConnell loess on pre-Reid, Reid and newly exposed McConnell drift probably took place in waning stages of the McConnell glaciation when, with climatic amelioration, vegetation was sufficient for entrapment of loess.

There are no palynological or other paleobotanical data from the study area from which climate during the McConnell glaciation can be inferred. However, in a pollen sequence from southwest Yukon (Rampton, 1971) the lowermost zone (Zone 1), dated about 31,000 years BP or possibly somewhat older, indicates either fell-field (rock desert) or sedge-moss tundra vegetation. The fell-field interpretation, which best fits our interpretation, would indicate a cold dry climate comparable to that of the high arctic (Rampton, 1971. p. 970 and Figure 7). The remainder of the sequence shows a warming, a reversal to colder and drier, and then general warming to the present.

(6) Post-McConnell. As with the McConnell glaciation, there are no palynologic data from the study area from which post-McConnell climate can be inferred. Rampton's palynologic data do not indicate a hypsithermal warming; however, in the same region, radiocarbon dates on logs indicate that the tree line, and hence summer temperature, fluctuated above present levels between 6000 and 1200 years BP (Rampton, 1971, p. 959). Soils on McConnell drift were therefore developed under climatic conditions the same as or somewhat warmer than those of the present. The same conclusion has been reached from the study of shelled invertebrates in the Richardson Mountains area, north of the studied area (Delorme *et al.*, 1975).

INTERGLACIAL CLIMATES AS INFERRED FROM PEDOLOGIC EVIDENCE

The geomorphic and stratigraphic evidence summarized above establishes an alternation of glacial intervals in which parent materials were deposited, and interglacial intervals (including the present Holocene) in which paleosols and the neosol on the youngest (McConnell) deposits were developed. However, except for the present interglacial interval, there is no evidence except what can be derived from the paleosols themselves, as to the duration of the interglacial soil-forming intervals, or the climate prevailing during these intervals. Were the climates of the interglacials, which gave rise to strikingly different soils, markedly different to that of the present, or is the difference in soil development attributable to progressively longer interglacial intervals as we go back in time? Examination of moisture and temperature requirements for formation of the older soils indicates that differences in climate rather than duration of the interglacials, was the dominant factor.

The appendix indicates that the soil developed on site I (Figure 2) is a truncated Luvisol. Also the appendix shows that soils developed on Reid and McConnell drifts as well as on loess deposits are Brunisols. However, the Brunisols developed on McConnell drift, site III, and on loess deposits are markedly thinner in thickness, than those developed on Reid drift, site II, Figure 2.

Evidence for the difference in climate between the

pre-Reid and the Reid-McConnell interglacial periods, is obtained from the depths at which pedogenetic clays are encountered in the pre-Reid and Reid paleosols. Montmorillonite-kaolinite mixed layers are identified beyond 190^{+} cm depth in the pre-Reid paleosol of site I, whereas vermiculite and chloritic intergrades are found in Reid paleosol of site II to a depth of 93 cm. Since water must be considered for both the depth of weathering and for either the clay translocation or the generation of clays, then it is possible to calculate the amount of water needed to reach these depths, using the equation which describes the water movement in unsaturated material. This equation which is presented by Birkeland (1974) is as follows:

$$d = \frac{Pw}{100} \times P_b \times D \qquad (1)$$

where d = is the amount of water in cm
Pw = is the difference between field capacity (F.C.) and permanent wilting point (P.W.P). In this case it represents the water holding capacity per horizon expressed in cm^3 H_2O/g soil
Pb = is the bulk density in g/cm^3
D = is the depth of water penetration in cm.

On the basis of studies by Kelly *et al.* (1946), who have published solid phase suction-water retention curves for a number of California soils, it can be assumed that a sandy clayey loam soil has a Pw value around 15% and a loamy sand or a sandy loam soil has a Pw value around 10%. Assuming a bulk density of 1.5 g/cm^3 then the amount of water needed for the pre-Reid paleosol at 190^{+} cm depth is:

$$d = \frac{15}{100} \times 1.5 \times 117^{3} + \frac{10}{100} \times 1.5 \times (190-117) = 36.4 \text{ cm or } 364 \text{ mm.}$$

For the Reid paleosol the obtained amount of water to reach the 93 cm depth of weathering is:

$$d = \frac{10}{100} \times 1.5 \times 93 = 14 \text{ cm or } 140 \text{ mm.} \qquad (2)$$

These calculations assume that water (or rainfall) has been precipitated at once and that evapotranspiration and water runoff is zero. What the calculations indicate is that the climate during the pre-Reid interglacial period was much more humid or that the water regime for the development of

---

[3] Up to 117 cm depth the texture is sandy clayey loam.

pre-Reid soils was of a higher magnitude than that responsible for the development of Reid soils.

From the type of pedogenetic clay minerals encountered in the paleosols another hint concerning the climatic differences between pre-Reid-Reid and Reid-McConnell interglacial periods is obtained. Montmorillonite in soils developed on acid or basic igneous parent material is found under low precipitation, M.A.P. of less than 100 cm and M.A.T. of 12° to 18°C (Barshad, 1966) or under higher M.A.P., and higher M.A.T., (Sherman, 1952). To transform montmorillonite to kaolinite as it is encountered in pre-Reid paleosols a high degree of leaching should have been taking place.

Thus, the type of clay minerals in the pre-Reid soil in conjunction with the red color of the textural B horizon suggests that initially the climate in the pre-Reid era should have been warm and dry to induce the formation of montmorillonite, and thereafter it became more humid.

Then was the climate which existed in the Reid-McConnell interglacial period similar to today's climate? The answer seems again to be no. On the basis of climatological data (Thomas, 1953), today's climate in Central Yukon is characterized by a M.A.T. of -7° to -10°C, and M.A.P. (snow + rain) or 25 cm to 37.5 cm. Furthermore, soils are frozen for 7 to 8 months thus being pedogenetically inactive for this period of time. Considering therefore the rainfall of 12.7 cm which precipitates in spring, summer and fall in the area, and dividing this precipitation to three equal portions, of 4.25 cm for the three seasons, then, the maximum depth of water penetration on the basis of the previously described equation of water movement in unsaturated media assuming no evapotranspiration, a Pw or 15% and Pb of 1.5, is D =

$$\frac{d \times 100}{Pw \times Pb} \quad 19 \text{ cm.}$$

The depth of water penetration of 19 cm is inadequate to explain the pedogenetic formation of clays at a depth of 93 cm found in the pre-Reid paleosol on Reid outwash of site II. Therefore, though the soil developed postglacially on McConnell deposits and those developed during the Reid-McConnell interval have the same type of clay minerals and both are classified as Brunisols we must consider that the moisture regime during the Reid-McConnell period was much higher than today's. The suggested transition is from subhumid climate during the Reid-McConnell interglacial period to subarctic-semiarid of today.

From the preceding discussions it is apparent that today's climate can neither explain the genesis of the thick, reddish brown, textural B horizons, nor the type of clay minerals encountered in the pre-Reid paleosols. If time is the missing element to convert the McConnell Brunisol to a pre-Reid Luvisol then we should have seen some similarities between Reid, which has the same type of clay minerals as the McConnell paleosol, and pre-Reid paleosol. Birkeland

(1974), summing up the work of many scientists, states that clay minerals formed in soils from parent materials low in clay content, as is the case in this study, probably form mineral assemblages stable in that environment and therefore, variation in the minerals with age is not found and not expected. As a result, it seems that dissimilar climates existed between the interglacial periods. Work by Zoltai and Tarnocai (1974), in the plains east of the investigated area and by Pettapiece and Zoltai (1974) in the subarctic of northwestern Canada indicates a more humid climate with somewhat higher temperatures during the postglacial period than what is found today. The same conclusion has been reached from the study of shelled invertebrates (Delorme *et al.*, 1975). The latter have been used as paleoclimatic indicators in the southern Richardson Mountains, which are north of the investigated area.

QUATERNARY CLIMATES OF CENTRAL YUKON

Combining the geological studies with soil morphology, clay mineralogy and clay chemistry of the paleosols and neosols along with evidence from other research studies the following sequence of Quaternary climates is suggested for the investigated area.

The pre-Reid paleosol surface was subjected initially to a climate which was warm and subhumid with grassland-shrub vegetation, and favorable for the generation of montmorillonite (Strakhov, 1967). Former grassland environment in Klondike district, adjacent to the study area, during some part of the Quaternary, is well documented from fossil remains of large grazing herbivores (Harington and Clulow, 1973). Age of the fossil assemblages relative to the chronology of the present study area has not been established, and it is not our intent to suggest a correlation with the pre-Reid interglacial. The fossil evidence does however lend credence to the interpretation, from pedologic evidence, of a former grassland environment in a region now occupied by boreal forest. Subsequently more temperate and humid climate appeared inducing the degradation of montmorillonite to kaolinite through the intermediate step of mixed-layer montmorillonite-kaolinite. In addition, this type of climate was responsible for the development of the Luvisol with the red, very thick, textural paleo B horizon to a depth of 190^+ cm. Later, the climate became colder with the onset of the Reid glaciation. The very cold climate induced thermal contraction cracks in soils, which were filled progressively with sand plus minor silt and clay to form sand wedges. The sand was derived locally by deflation of the Ae horizon of the previously formed luvisol. During the subsequent Reid-McConnell interglacial period it seems that a cool subhumid climate prevailed. This climate was responsible for the development of the Brunisolic characteristics encountered in the paleosols developed on deposits of Reid age. A cold period ensued resulting in the McConnell glaciation. This period must have been severe but shorter than the Reid as deduced from the presence of narrower and shallower sand wedges in the Reid paleosols. In the waning stages of the McConnell glaciation, loessal silt derived from extensive outwash trains was deposited as a thick blanket over the truncated paleosols of the pre-Reid and Reid surfaces and over newly uncovered McConnell surfaces. The transition from deflation and truncation of the earlier formed paleosols to loess deposition indicates significant increase in vegetative cover that stabilized the soil surface and served to trap the loess.

Finally, the postglacial climate has imparted to McConnell deposits and the overlying loess the characteristics of a Brunisol soil. Today's subarctic-semiarid climate is different from that occurring during the retreat of McConnell ice and shortly after. Evidence presented by Zoltai and Tarnocai (1974), Pettapiece and Zoltai (1974), and Delorme *et al.* (1974) in areas adjacent to the investigated area suggest that the climate was somewhat warmer than what it is today and more humid, probably a subarctic-subhumid type of climate.

ACKNOWLEDGMENTS

The authors wish to thank Drs. M. Dudas, Department of Soils, University of Alberta, W.W. Pettapiece, Canada Department of Agriculture, University of Alberta, R.O. Van Everdingen, Inland Waters Branch, Environment Canada, R.W. Klassen and G.E. Reinson of the Terrain Sciences and Atlantic Geoscience Center, respectively, Geological Survey of Canada, for constructively criticizing and reviewing the manuscript. Appreciation is also expressed to Miss L. Bevington and Messrs. A.G. Vilonyay, A.G. Heinrich and R.R. Barefoot of the Institute of Sedimentary and Petroleum Geology, Geological Survey of Canada, for their technical assistance.

REFERENCES CITED

American Society for Testing and Materials, 1964, Grain size analysis of soils in the 1964 book of A.S.T.M. Standards, 4th Ed.: Philadelphia, Am. Soc. Test. Mater., p. 95-106.

Bailey, L.W., Odell, R.T., and Boggers, W.R., 1964, Properties of selected soils developed near the forest-prairie border in east-central Illinois: Soil Sci. Soc. America Proc., v. 28, p. 256-263.

Barshad, I., 1955, Soil development, *in* Bear, E.E., ed., Chemistry of the Soil: New York, Reinhold Publishing Corp., p. 1-52.

___________, 1965, Thermal analysis techniques for mineral identification and mineralogical composition, *in* Black, C.A., ed., Methods of Soil Analysis, part 1: Madison, Amer. Soc. Agron., p. 699-742.

___________, 1966, The effect of a variation in precipitation on the nature of clay mineral formation in soils from acid and basic igneous rocks: Jerusalem, Israel, Intl. Clay Conf. Proc., p. 157-173.

Berg, T.E., and Black, R.F., 1966, Preliminary measurements of growth of nonsorted polygons, Victoria Land, Antarctica: Antarctic Soils and Soil Forming Processes, Am. Geophys. U.: Washington, D.C., Antarctic Research Series, v. 8, p. 61-108.

Bostock, H.S., 1948, Physiographic subdivisions of the Canadian Cordillera north of the fifty-fifth parallel, in Physiography of the Canadian Cordillera, with special reference to the area north of the fifty-fifth parallel, Mem. 247: Geol. Survey Can., Map 922A.

_____________, 1964, Provisional physiographic map of Canada: Geol. Survey Can., Map. 13-1964.

_____________, 1966, Report of Activities, May to October, 1964, Jenness, S.E., ed.: Ottawa, Geol. Survey Can., 210 p.

Birkeland, P.W., 1969, Quaternary paleoclimate implication of soil clay mineral distribution in a Sierra Nevada - Great Basin transect: Jour. Geology, v. 77, p. 289-302.

_______________, 1974, Pedology, weathering and geomorphological research: New York, Oxford University Press.

Birkeland, P.W., and Janda, R.J., 1971, Clay mineralogy of soils developed from Quaternary deposits of the eastern Sierra Nevada, California: Geol. Soc. Amer. Bull., v. 80, p. 2495-2514.

Brown, G., 1961, X-ray diffraction and crystal structure of clay minerals, 2nd Ed.: London, Mineralogical Soc.

Delorme, L.D., Zoltai, S.C., and Callas, L.L., 1975, Shelled invertebrates as indicators of paleoclimate in N.W. Canada during the late glacial period: Kyoto, Japan, Internatl. Conf. on Global Scale Paleolimnology and Paleoclimate.

Dyke, W., Lowdon, J.A., Fyles, J.G., and Blake, W. Jr., 1966, Radiocarbon Dates V: Geol. Survey Can., Paper 66-48.

Foscolos, A.E., and Barefoot, R.R., 1970a, A rapid determination of total, organic and inorganic carbon in shales and carbonates: Geol. Survey Can., Paper 70-11.

_______________, 1970b, A buffering and standard addition technique as an aid in the comprehensive analysis of silicates by atomic absorption spectroscopy: Geol. Survey Can., Paper 70-16.

Harington, C.R., and Clulow, F.V., 1973, Pleistocene mammals from Gold Run Creek, Yukon Territory: Can. Jour. Earth Sci., v. 10, no. 5, p. 697-759.

Harward, M.E., Corstea, D.D., and Sayegh, A.H., 1968, Properties of vermiculites and smectites: Expansion and collapse: Clays and Clay Minerals, v. 16, p. 437-447.

Hughes, O.L, Campbell, R.B., Muller, J.E., and Wheeler, J.O., 1969, Glacial map of Yukon Territory, south of 65 degrees north latitude, *in* Geol. Survey Can., Paper 68-34: Geol. Survey Can., Map 6-1968.

Hughes, O.L., Rampton, V.N., and Rutter, N.W., 1972, Quaternary geology and geomorphology, southern and central Yukon: XXIV Internatl. Congr., Montreal 1972, Field Excursion A-11.

Jackson, M.L., 1965, Free oxides, hydroxides, and amorphous aluminosilicates, *in* Black C.A., ed., Methods of Soil Analysis, Pt. 1: Madison, Amer. Soc. Agron., p. 578-603.

Janda, R.J., and Croft, M.G., 1967, Stratigraphic significance of a sequence of noncalcic brown soils formed on Quaternary alluvium in the north-eastern San Joaquin Valley, California, *in* Morrison, R.B., and Wright H.E. Jr., eds., Quaternary Soils: Internatl. Assoc. Quaternare Research, VII Cong., Proc., v. 9, p. 151-196.

Jenny, H., 1941, Factors of soil formation: New York, McGraw-Hill.

__________, 1958, Role of plant factor in the pedogenic functions: Jour. Ecology, v. 39, p. 5-16.

Jungerius, P.D., 1969, Soil evidence of postglacial tree line fluctuations in the Cypress Hills area, Alberta, Canada: Arctic and Alpine Research, v. 1, p. 235-245.

Kelly, O.J., Hunter, A.S., Haise, H.R., and Hobbs, C.H., 1946, A comparison of methods of measuring soil moisture under field condition: Jour. Amer. Soc. Agron., v. 38, p. 759-784.

Kodama, H., and Brydon, J.E., 1968, A study of clay minerals in podzol soils in New Brunswick, eastern Canada: Clay Minerals Bull., v. 7, p. 295-309.

Lowdon, J.A., and Blake, W. Jr., 1968, Radiocarbon dates VII: Geol. Survey Can., Paper 68-2, Pt. B.

McConnell, R.G., 1905, Report on the Klondike gold fields: Geol. Survey Can., Annual Rept. (misc.), v. XIV (1901), Pt. B, p. 1-71. (Also separate Publ. No. 884).

McKeague, J.A., and Day, J.H., 1966, Dithionite and oxalate-extractable Fe and Al as aids in differentiating various classes of soils: Can. Jour. Soil Sci., v. 46, p. 13-22.

Peech, M., 1965a, Exchange acidity, *in* Black C.A., ed., Methods of Soil Analysis, Pt. 2: Madison, Amer. Soc. Agron., p. 905-913.

_________, 1965b, Hydrogen-ion activity, *in* Black, C.A., ed., Methods of Soil Analysis, Pt. 2: Madison, Amer. Soc. Agron., p. 914-926.

Pettapiece, W.W., 1969, The forest grassland transition, *in* Pawluk, S., ed., Pedology and Quaternary Research: Edmonton, University of Alberta, p. 103-113.

Pettapeice, W.W., and Zoltai, S.C., 1974, Soil environments in the western Canadian subarctic, *in* Mahaney, W.C., ed., Quaternary Environments: Proc. of a Symposium, Geographical Monographs: Toronto, York Univ. Ser. Geog., p. 279-292.

Péwé, T.L., 1959, Sand-wedge polygon (Tesselations) in the McMurdo Sound Region, Antarctica: A progress report: Amer. Jour. Sci., v. 257, p. 545-552.

_________, 1962, Age of moraines in Victoria Land, Antarctica: Jour. Glaciology, v. 4, p. 93-100.

_________, 1974, Geomorphic processes in polar deserts, *in* Smiley, T.L., and Zumberge, J.H., eds., Polar Deserts and Modern Man: Tuscon, University of Arizona Press.

Rampton, V., 1971, Late Quaternary vegetational and climatic history of the Snag-Klutlan area, southwest Yukon Territory, Canada: Geol. Soc. Amer. Bull., v. 82, p. 959-978.

Ricker, K.E., 1968, Quaternary geology in the southern Ogilvie Ranges, Yukon Territory (M.Sc. Thesis): Dept. of Geology, University of British Columbia.

Romanovskij, N.N., 1973, Regularities in formation of frost-fissures and development of frost-fissure polygons: Internatl. Symp. of the Commission on Periglacial Morphology of the I.G.U. and Subcommission of Paleogeographical Maps and Atlases of the INQUA, Proc., Yakutsk-Moscow, 1969, Biuletyn Peryglacjalny, No. 23.

Ruhe, R.V., 1969, Soils, paleosols and environment in Pleistocene and Recent environments of the central Great Plains, *in* Dort, Jr., W., and Jones, Jr., J.K., eds.: Lawrence, University Press of Kansas, p. 37-52.

Ruhe, R.V., and Cady, J.C., 1969, The relation of Pleistocene geology and soils between Bentley and Adair in southwestern Iowa: U.S. Dept. Agric. Tech. Bull., v. 1349, p. 1-92.

Sherman, G.D., 1952, The genesis and morphology of the alumina-rich laterite clays, *in* Problems of Clay and Laterite Genesis: New York, Am. Inst. Mining Metal. Eng., p. 154-161.

Smith, G.D., Alloway, W.H., and Riecken, F.F., 1950, Prairie soils of the upper Mississippi Valley: Advances in Agron., v. 2, p. 157-205.

Strakhov, N.M., 1967, Principles of lithogenesis, Vol. 1: Oliver and Boyd Ltd., Edinburgh, 245 p.

Thomas, M.K., 1953, Climatological atlas of Canada: Publ. of the Division of Building Research N.R.C. and Meteorological Division, Dept. of Transport, Canada, N.R.C. No. 3151, D.B.R. No. 41.

Vernon, P., and Hughes, O.L., 1966, Surificial geology, Dawson, Larsen Creek and Nash Creek map-areas, Yukon Territory: Geol. Survey Can. Bull., v. 136.

Zoltai, S.C., and Tarnocai, C., 1974, Perennially frozen peatlands in the western Arctic and subarctic of Canada: Can. Jour. Earth Sci., v. 12, p. 28-43.

APPENDIX 1

DESCRIPTION OF SOIL PROFILES

Soil Profile in Site I

This site is located on a well to moderately well drained pre-Reid outwash surface with a slope of ≃2°. A sand wedge extends from 13 cm to 117 cm depth. The vegetation consists predominantly of poplar with some white spruce and with a ground cover of clump grass and some Labrador tea. The neosol is classified as a Brunisol (Eutric) and the paleosol, which has been developed in the pre-Reid drift, as a Luvisol.

HORIZON	DEPTH (cm)	DESCRIPTION
H	3.5-0	Very dark greyish brown (10YR 3/2, d)[1]; very dark brown (10YR 2/2, m); extremely decomposed, humified organic matter; plentiful fine roots, clear wavy boundary; 3-5 cm thick; pH = 5.4.
Ahe	0-13	Brownish yellow (10YR 6/6, d); dark yellowish brown (10YR 4/4, m); silt loam; compacted; single grain; very friable; few fine roots; gradual, smooth boundary; 12-15 cm thick; pH = 4.9; O.M. = 0.7%.
BM	13 -18	Yellowish brown (10YR 5/4, d); brown (7.5YR 4/4, m); silt loam; weak, fine

[1]Soil colors are given as moist (m), and dry (d).

HORIZON	DEPTH (cm)	DESCRIPTION
		granular structure; friable, very few fine roots; very few pebbles; abrupt, clear boundary; 5-7 cm thick; pH = 5.0; O.M. = 0.5%
II Bt1	18 -40	Reddish brown (5YR 5/5, d); reddish brown (5YR 4/5, m); sandy clay loam; medium, granular structure, friable; clear, irregular boundary 15-25 cm thick; no roots; plentiful shattered and weathered stones; pH = 4.9; O.M. = 0.3%.
IIBt2	40 -117	Brown (7.5YR 5/2, d); dark brown (7.5YR 4/4, m); sandy clay loam; single grain; abundant medium size pebbles with thin weathering rinds; pH = 5.2; O.M. = 0.2%.
II BC	117 -190	Yellowish brown (10YR 5/6, d); dark yellowish brown (10YR 4/4, m); sandy loam, single grained abundant medium size pebbles with weathering rinds; pH = 6.1; O.M. = 0.2%.
II C	190^{+}	Yellowish brown (10YR 5/6, d); dark yellowish brown (10YR 4/4, m); loamy sand; single grain; abundant, medium and large gravels with some weathering rinds; pH = 6.1; O.M. = 0.1%.

Soil Profile in Site II

This site is located on a flat lying well drained, gravelly outwash terrace of Reid age. The vegetation consists of spruce and poplar with a ground cover of clump grass. A sandy wedge extends from 22.5 cm to 93 cm depth. The neosol as well as the paleosol are classified as Brunisols (Eutric).

HORIZON	DEPTH (cm)	DESCRIPTION
L-H	2.5-0	Very dark brown (10YR 2/2, d); partially decomposed organic matter; fibrous; predominately moss; abundant fine and medium roots; abrupt, smooth boundary; 2.3-3.0 cm thick; pH = 6.3.
Bm1	0-14.5	Reddish brown (5YR 5/4, d); reddish brown (5YR 4/4, m); silt loam; single grained, friable; few, fine and very fine roots; gradual boundary; smooth, 13-16 cm thick; pH - 5.9; O.M. = 1.7%.

HORIZON	DEPTH (cm)	DESCRIPTION
Bm2	14.5-30	Dark yellowish brown (10YR 4/4, d); dark yellowish brown (10YR 3/4, m); loam, single grained; friable; few, fine and very fine roots; gradual boundary; smooth, 7-9 cm thick; pH = 6.0; O.M. = 1.0%.
II Bm1	30-49	Yellowish brown (10YR 5/4, d); brown (10YR 4/3, m); sandy loam; fine, weak crumb; very friable; very few fine roots; no pores; no stones; smooth abrupt boundaries, except in presence of small sand wedges; 6-10 cm thick; pH = 6.4; O.M. = 0.7%.
II Bm2	49-68	Yellowish brown (10YR 5/4, d); brown (10YR 4/3, m); very gravelly sandy loam; single grained, friable; no roots, no pores; abrupt smooth boundary; 6-10 cm thick; pH = 6.4; O.M. = 0.4%.
II BC	68-93	Light olive-brown (2.5YR 5/4, d); olive-brown (2.5Y 4/4, m); very gravelly sandy loam, single grained, friable; no roots; no pores; diffuse boundary; abundant gravel; pH = 6.9; O.M. = traces.
II Ck	93^{+}	Light olive-brown (2.5Y 5/4, dry); olive-brown (2.5Y 4/4, m); very gravelly sandy loam, single grain; friable; no roots; no pores; abundant gravel; pH = 7.1; moderately calcareous, no O.M.

Soil Profile in Site III

The site is located on a well drained, flat lying McConnell outwash surface. The natural vegetation consists of spruce and poplar. Both neosol and paleosol are classified as Dystric Brunisols.

HORIZON	DEPTH (cm)	DESCRIPTION
F-H	2.0-0.0	Dark brown (10YR 3/3, d); very dark brown (10YR 2/2, m); partially decomposed organic matter; fibrous; abundant medium and fine roots; clear, wavy boundary, 3-5 cm thick; pH = 5.8.
Bm	0-12.5	Yellowish brown (10YR 5/4, d); dark yellowish brown (10YR 4/4, m); loamy sand; single grained to amorphous (m); friable; few fine roots; gradual boundaries; smooth, 11-14 cm thick; pH = 4.9; O.M. = 2.2%.

HORIZON	DEPTH (cm)	DESCRIPTION
II Bm	12.5-21	Yellowish brown (10YR 5/4, d); dark yellowish brown (10YR 4/4, m); loamy; sandy; single grained; friable; very few fine roots; abrupt, clear boundary; 5-7 cm thick; pH = 5.4; O.M. = 0.9%.
II C	21-42	Dark greyish brown (10YR 4/2, d); very dark greyish brown (10YR 3/2, m); very stony loamy sand; single grained; clear smooth boundary; 18-24 cm thick; very stony; pH - 6.2; O.M. = 0.5%.
III C	42^+	Dark greyish brown (10YR 4/2, d); very dark greyish brown (10YR 3/2, m); very stony loamy sand; single grained; pH = 6.2; O.M. = traces.

Soil Morphogenesis

Quantification and Pedological Processes

L. J. Evans

ABSTRACT

Various pedological processes are often invoked to explain soil morphology without first establishing the exact nature of the soil parent material. The identification of lithologic discontinuities in parent materials is an essential prerequisite in differentiating soil properties that are pedological in origin from those that are geological. Quantitative estimates of the extent and nature of any pedological change during soil development can best be made by assuming that certain minerals remain unaltered and immobile during pedogenesis. The quantification of pedological processes through the use of these minerals, *i.e.* internal standards, has proved valuable in a number of pedological studies.

Luvisolic soils (Udalfs) in the Central Plain of Ireland have developed from a calcareous glacial till derived from Carboniferous sandstones and shales. Decalcification of surface horizons and the subsequent movement of clay-sized material are the major pedological processes occurring. Variations in the nature and degree of profile development are related to relief. Quantification of clay movement using acid residue as an internal standard has shown that this movement is greatest in Luvisolic Gleysols (Aqualfs) and that movement in paleosols showing profile development is comparable to that in Orthic Gray Luvisols (Hapludalfs) Clay mineral analyses and constructed balance sheets for K_2O suggest that variations in clay content in Bt (argillic) horizons are associated with the movement of illite.

In mid-Wales many Brunisolic soils (Dystrochrepts) developed on non-calcareous mudstones and shales and situated at the lower end of scree slopes show yellow-brown to brown colors due to the presence of large amounts of amorphous Fe-oxides. Similar Podzolic soils (Aquods) situated farther up valley sides not only display these brownish colors in their Bf (spodic) horizons but also have eluvial horizons containing large amounts of silt plus clay. Using acid residue as an internal standard it was found that although substantial amounts of Fe and Al were released during pedogenesis from highly weathered horizons the mobilised Fe, in contrast to mobilised Al, was almost completely redeposited in illuvial horizons. Balance sheets constructed for Fe suggest that the high contents of Fe-oxides in Brunisolic soils is due to the lateral movement of Fe down slope. Quantitative assessments of changes during pedogenesis indicate that the mobilisation of Fe, Al and Mg, the textural degradation of the parent rock and the weathering of chlorite are concurrent.

INTRODUCTION

The morphology of soils is an expression of the variation of soil properties with depth and various pedological processes are often invoked to explain these variations. Theories in pedogenesis abound with postulations of transformations and redistributions of soil constituents, although very little work has been done to corroborate many of these theories. In fact there is a paucity of data available in the literature on the actual amount and nature of change that has occurred during pedogenesis. Many published papers in pedology, in fact, invoke changes that are either untenable, unjustified or even at variance with the data presented.

Relative changes of soil constituents can only be assessed using an internal standard that is itself inert and immobile during pedogenesis. Quantitative pedology is concerned with the construction of soil constituent balance sheets using internal standards. These balance sheets indicate the extent of pedological change during soil development and, although rarely used yet in Quaternary soil research, will prove invaluable for use in soil stratigraphic studies.

Assessment of Lithologic Uniformity

A major problem in quantitative pedology is in the identification of the parent material. For soils developed directly on bedrock in temperate regions, few difficulties are encountered, the unweathered rock being considered the parent material. However for deeply weathered soils in tropical regions and for soils developed on glacial materials in temperate regions the problem is more difficult. The success of quantitative methods in elucidating soil forming processes depends to a large extent on the identification of the parent material at the start of soil formation.

An important characteristic of the soil parent material for quantitative pedological studies is that its inherent variability should be no greater than the variability likely to be brought about in the soil profile by pedological processes. The identification of lithologic discontinuities is therefore an essential prerequisite in differentiating soil properties that are pedological from those that are geological. Variations in the lithology of soil parent materials may arise for a number of reasons. These may include glacial or periglacial mixing of different geological materials, as well as additions of aeolian, colluvial and alluvial material during soil development.

Because of the great variety and variability of soil parent materials there are various criteria for establishing their uniformity. Barshad (1964) lists a number of commonly used criteria, of which the most useful are:

1. particle-size distribution of resistant minerals. *e.g.* percentage of total rutile in each size fraction.

2. particle-size distribution of non-clay fraction *e.g.* $\frac{\text{fine sand}}{\text{silt}}$ or $\frac{\text{coarse sand}}{\text{fine sand}}$

3. ratio of the contents of two resistant minerals in any one size fraction. *e.g.* quartz/zircon in silt fraction.

Other somewhat more subjective criteria such as the presence of unexpected minerals or a non-pedological distribution of soil constituents may also be used. Recently the use of phosphorus for identifying paleosols has been suggested (Runge *et al.*, 1974).

The nature of the parent material itself will dictate to a large extent the most useful method of assessing its uniformity. The particle size distribution of quartz, ZrO_2 and TiO_2 for instance did not distinguish between two visibly different tills in Canada (Sudom and St. Arnaud, 1971) probably because of the presence of inclusions which render quartz and zircon susceptible to physical breakdown (Raeside, 1959). Particle-size distribution of the non-clay fraction has proved a useful parameter for identifying lithologic discontinuities in some soils (Smith and Wilding, 1972; Raad and Protz, 1971) but is of limited value if particulate disaggregation of sand and silt sized particles occurs during pedogenesis (Evans and Adams, 1975a).

Probably the most common method of assessing the uniformity of soil parent materials is to compare the ratio of the amounts of two resistant minerals in one or more size fractions (Roonwall and Bhumbla, 1969; Evans and Adams, 1975a). When a chemical element is present exclusively in the resistant mineral, such as may be the case of zirconium in zircon, titanium in rutile or anatase, or boron in tourmaline, the resistant mineral may be assessed by chemical rather than mineral analysis (Khangarot *et al.*, 1971; Ritchie *et al.*, 1974). General agreement however is lacking as to what degree of variation in ratios constitutes a definite discontinuity in the parent material. Variations of 28% and 81% from the mean for ZrO_2: TiO_2 ratios have been reported for uniform soils developed from sandstones and siltstones, and cherty limestone respectively (Chapman and Horn, 1968). Drees and Wilding (1973) from a study of elemental variation within a sampling unit believe that significant differences between any two horizons must exceed random variation by the value of t for any given probability level.

Estimates of uniformity should be based on as many lines of evidence as possible. Often the presence of discontinuities can be recognised clearly in the field because of sudden changes in such properties as color, structure, or texture. Where variations in these properties are not well defined the mineralogical and particle size distribution of the material must be studied in detail and the data analysed using a number of the suggested criteria to establish parent material uniformity.

Use and Assessment of Internal Standards

Determination of the extent and manner in which the soil parent material is altered during pedogenesis, and assessments of the consequential redistribution of soil constituents either in solution or in colloidal form can most easily be made by assuming that certain minerals remain unaltered and static during soil development. Assessments of change can thus be made relative to the unaltered mineral.

The problem is to choose the most stable and immobile constituent so that the least possible error will be made in calculations in soil formation. However, no mineral is completely stable under all conditions, and stability and immobility depend on a large number of factors including grain size, specific surface, cleavage, coefficient of expansion, inclusion and cracks in crystal and solubility under the specific environment prevailing. This problem is unresolved at present except to the extent that the range of proposed constituents is limited.

Much attention has focused on the heavy minerals zircon, tourmaline, rutile and garnet, proposed originally by Marshall (1940), as well as quartz, albite and orthoclase (Barshad, 1964). The use of chemical references has also been suggested; TiO_2 for the podzol zone and SiO_2 for the chernozem zone (Joffe, 1936), and Al_2O_3 for arid soils (Muir, 1951; Roonwall and Bhumbla, 1969). Calculations of losses and gains of soil constituents have also been made by assuming there is no change in volume during pedogenesis (Wild, 1961). Such an assumption is only likely to be valid for a few selected soils. The use of coarse sand fractions, presumed to contain only resistant minerals, has also been used (Reid and Matthews, 1954; Wang and Arnold, 1973).

In spite of the fact that no internal standard is completely resistant to weathering the majority of workers have used only one internal standard to assess pedological change. Zircon for instance is generally regarded as one of the most resistant of the heavy minerals and has been widely used as an index mineral (Khan, 1959; Green, 1966; Smeck and Runge, 1971) although it has been shown to weather appreciably in a number of environments (van der Marel, 1949; Carroll, 1953; Kimura and Swindale, 1967; Gilkes *et al.*, 1973). Similarly quartz has been used extensively as an index mineral (Bourne and Whiteside, 1962; Schlichting and Blume, 1961; Redmong and Whiteside, 1967) but is obviously unstable under lateritic conditions or when subjected to thermal expansion (Raeside, 1959).

Recently the residue obtained after treatment of soil with a tri-acid mixture of sulphuric, hydrochloric and nitric acids has been suggested to be of use as an internal standard (Akhtyrtsev, 1968; Evans and Adams, 1975a). Such a treatment leaves a residue dominated by quartz and feldspars and dissolves minerals readily weathered in a pedological environment (Hardy and Follet-Smith, 1931). Claisse (1968, 1972) has shown that for clay sized material up to 5% pure quartz and 12% quartz extracted from soil may be dissolved by such a procedure.

There have been a number of studies comparing various internal standards. Sudom and St. Arnaud (1971) found that TiO_2 is generally less reliable an internal standard than quartz or ZrO_2 even though there was some evidence of Zr translocation in clay fractions. From a study of the same standards, Eaqub (1972) concluded that 20-50 µm quartz was the most reliable standard as it gave results consistent with the soil's morphology. Evans and Adams (1975a) showed that there was little to choose between ZrO_2, quartz and acid residue using subjective criteria but acid residue was chosen as the most suitable standard because (1) its measurement by gravimetric methods was both simple and precise and (2) it constituted a large proportion of the soil.

Construction of Soil Balance Sheets

Having decided on a suitable internal standard with which to assess changes in soil constituents, the predicted amount of any constituent can be calculated as follows:

$$\text{predicted content in any soil horizon} = \text{content in parent material} \times \frac{\text{content of internal standard in horizon}}{\text{content of internal standard in parent material}} \quad (1)$$

For a knowledge of the predicted and known amounts of a particular constituent its percentage gain or loss can be calculated. However, to assess the extent of redistribution of constituents within the soil profile as a whole it is necessary to be able to reconstruct the amount of material in a soil core of known dimensions. Chemical and mineralogical analysis of soils are of necessity performed on a fine earth sample (*i.e.* material less than 2mm) often after ignition to 375°C to remove organic matter. The amount of soil within a core of cross-sectional area of 1 sq cm can be calculated from a knowledge of:

1. soil bulk density
2. organic matter content of fine earth
3. depth of each horizon
4. the proportion of fine earth in the total soil of each horizon.

It is only when all these properties are known that a soil balance sheet can be constructed for each soil constituent. Summations of changes for each horizon, expressed in gm per sq cm of horizon core, allow the calculation of the gross overall change in a soil constituent during pedogenesis, *i.e.* not only its extent of redistribution but also its total loss or gain from the profile.

It is possible therefore using the methodology of quantitative pedology to not only assess the extent of release of elements from minerals during weathering but also to examine their resulting redistribution. I will illustrate the use of quantitative pedology in providing valuable information in some pedological studies by reference to two studies of my own.

QUANTITATIVE PEDOLOGICAL STUDIES

Studies in Ireland

The bedrock geology of the Central Plain of Ireland is dominated by sedimentary rocks of the Lower Carboniferous series, with sandstones and limestones predominant. These rocks are overlain by a calcareous stony glacial till containing up to 80% stones of limestone origin. The retreat of the glaciers from the region some 12,000 - 15,000 years ago left a gently undulating topography with many enclosed basins in which late-glacial inorganic and later organic lacustrine deposits began to accumulate (Carey and Hammond, 1970). With a deterioration in climate in the postglacial Atlantic period tree growth was retarded and ombrogenous bog peat formation took place.

Owing to the relatively uniform nature of the glacial drift, variations in the nature and degree of horizon development within mineral soils are associated with relief, and major pedological processes that have taken place are decalcification and subsequent eluviation of clay-sized material. The mineral soils are often completely surrounded by organic soils and exist as mineral 'islands' within the landscape. A typical sequence of soils consists of base-rich Eutric Brunisols (Eutrochrepts) showing little or no profile development occupying the knolls of small hills, with Luvisolic soils (Hapludalfs) containing horizons of clay accumulation occupying intermediate positions and finally as the soils become progressively wetter Luvisolic

Gleysols (Umbraqualfs) give way to organic soils. Such a sequence is shown in Figure 1. Sub-peat mineral soils (Paleosols) also show features which are attributable to their position on the bog floor (Hammond, 1968), horizonation being least developed at the lowest elevations.

There has been no intensive investigation of the mineralogy of the sedimentary rocks of Central Ireland or of their associated glacial tills. Clay mineralogical investigations of some Upper Carboniferous rocks in Ireland indicate that illite is ubiquitous with kaolinite or chlorite being the other clay mineral. Mineralogical investigations on a number of glacial tills at the study area of Lullymore, Co. Kildare showed that quartz and calcite are dominant with smaller amounts of illite and feldspars. The clay contents of the tills ranged from 11 to 16%, and illite, kaolinite, vermiculite and small amounts of chlorite were identified. The relative amounts of the phyllosilicates varied however with kaolinite ranging from dominant to rare.

A typical catenary sequence of soils developed on calcareous glacial till was studied at Lullymore. Three mineral soils showing a range of profile development, together with two sub-peat mineral soils (paleosols) were investigated (Figure 1). Their lithologic uniformity was assessed using both mineralogical and textural information (Table 1) and no lithologic breaks could be detected. The relatively constant ratios of coarse sand/fine sand and fine sand/silt in spite of a loss of carbonates indicated that calcite was not acting as a rock cement.

Using acid residue as an internal standard the redistribution of clay was assessed (Table 2). Results confirmed morphological indications that little or no clay movement had occurred in the Eutric Brunisol, that substantial movement had occurred in the Gray Luvisol and that the greatest movement had occurred in the Luvisolic Gleysol. The amount of clay movement in the paleosol showing marked horizonation was of the same order of magnitude as that occurring in the Gray Luvisol. The pH's throughout the Eutric Brunisol and in the paleosols not showing horizonation were always in excess of 7.50 - 7.0. Decalcification therefore is a necessary prerequisite for clay movement.

Clay mineral analyses of the various soil horizons showed that illite and pedogenetic chlorite occur in the greatest amounts in the Bt (argillic) horizons. Quantitative estimates of the amount of redistribution of K_2O (Table 2) suggest that the relative increases in clay content of Bt horizons can be accounted for by the movement of illite. Indeed illite was practically absent from surface horizons of luvisolic soils but increased in content into the Bt horizons. The relative decrease in vermiculite content and increase in pedogenetic chlorite with depth in luvisolic soils suggests that movement of Al into interlayer positions also occurs in Bt horizons.

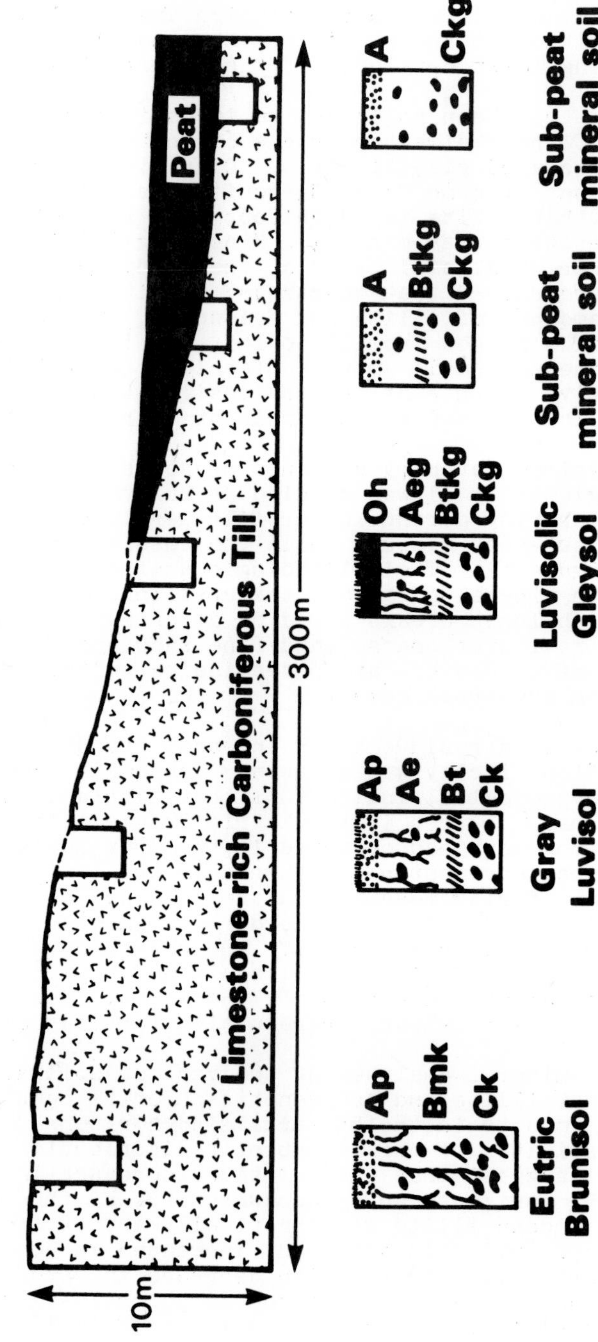

Figure 1 Slope Sequence of Soils in Central Ireland

Table 1. Criteria used to assess parent material uniformity for soil profiles in central Ireland.

	Coarse sand/ Fine Sand	Fine sand/ Silt	Quartz/ Feldspars
Eutric Brunisol			
Ap	0.42	1.82	11.2
Bmk1	0.47	2.33	11.4
Bmk2	0.66	1.84	11.9
Ck	0.50	1.82	12.6
Gray Luvisol			
Ap	0.44	1.81	10.9
Ae	0.38	2.35	10.6
Bt	0.37	1.98	9.9
Ck	0.50	2.02	10.0
Luvisolic Gleysol			
Oh	0.25	1.35	9.5
Aeg	0.40	1.41	11.2
Btkg1	0.32	1.46	11.3
Btkg2	0.43	1.36	11.2
Ckg	0.54	1.21	12.2
Sub-peat mineral soil			
Ah	0.48	2.02	9.9
Btkg	0.45	2.10	9.4
Ckg	0.53	2.60	9.6

Studies in Wales

The rocks of mid-Wales consist of relatively uniform Lower Silurian and Upper Ordovician non-calcareous mudstones and shale, interbedded with sandstones and grits. The mineralogy of the parent rocks is dominated by illite, quartz and Fe-rich chlorite, with lesser amounts of albite, orthoclase and rutile. Calculated structural formulae for chlorite indicate that it contains both a dioctahedral and a trioctahedral layer, with the latter layer containing most of the rock Fe and Mg (Evans and Adams, 1975c). The total amount of rock Fe, Mg, and Al in chlorite amounts to 95%, 80% and 35% respectively.

During the Pleistocene Epoch phases of maximum cold and glacial advance were separated by warmer interglacial periods and the soils existing at the time were largely removed or buried by glacial drift. The ice finally retreated some 12,000 years ago and the vast majority of the soils of mid-Wales have developed since that time. Many of the topographic features found in mid-Wales today are the result of

Table 2. % Changes in clay and K_2O contents relative to their contents in the parent material.

	Predicted Clay	Actual Clay	% Change	Predicted K_2O	Actual K_2O	% Change
Eutric Brunisol						
Ap	23.14	22.47	-3.0	1.22	1.12	-8.2
Bmk1	22.54	23.43	+0.5	1.33	1.08	-18.8
Bmk2	22.22	22.37	-0.7	1.20	1.01	-15.8
Ck	22.07	22.07	0.0	0.93	0.93	0.0
Gray Luvisol						
Ap	12.83	12.85	+0.2	1.09	0.86	-22.9
Ae	12.46	11.45	-8.8	1.28	0.87	-32.0
Bt	15.14	17.46	+13.3	1.03	1.27	+23.3
Ck	13.88	13.88	0.0	0.82	0.82	0.0
Luvisolic Gleysol						
Oh	15.37	14.50	-6.0	1.10	1.00	-9.1
Aeg	15.35	13.45	-14.1	1.34	0.83	-38.1
Btkg1	17.57	24.95	+29.6	1.18	1.27	+7.6
Btkg2	20.21	30.73	+34.2	1.01	1.38	+36.6
Ckg	12.54	12.54	0.0	0.73	0.73	0.0
Sub-peat mineral soil						
Ah	14.59	12.30	-18.6	1.27	0.90	-15.7
Btkg	15.02	17.03	+11.8	1.22	1.08	+4.1
Ckg	15.89	15.89	0.0	0.83	0.83	0.0

glacial and periglacial activity, although extensive signs of a vanished ice sheet are not evident. Terrace gravels exist in major valleys and solifluction terraces are found on valley sides. Glacial drift remains only on hill plateaus and valley bottoms. Exposed rock was shattered by severe frost action to form the scree slopes upon which pedogenesis has subsequently progressed.

The soil profiles studied in the valley of Cwmcadian represent a typical catenary sequence of soil development on scree slopes in mid-Wales (Figure 2), ranging from a Dystric Brunisol (Dystrochrept) through to a well developed Humo-Ferric Podzol (Placaquod). The soils at the lower end of the valley consist of freely draining Brunisols under a canopy of *Quercus* and *Betula*, and are classified as Denbigh Series (mor phase). Many of the brunisolic soils in the

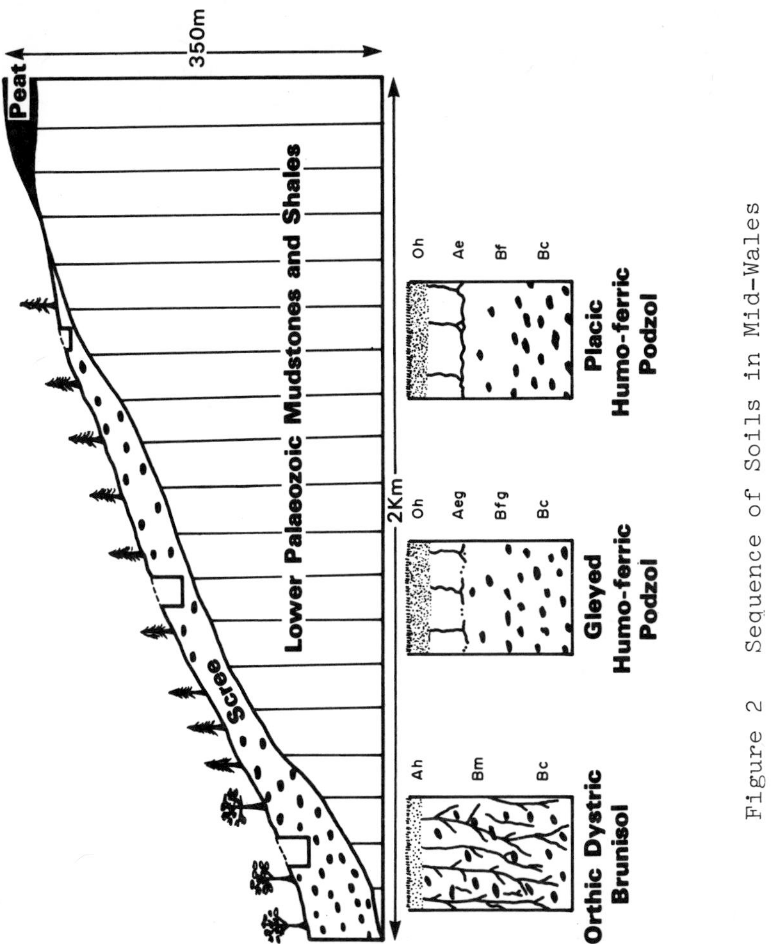

Figure 2 Sequence of Soils in Mid-Wales

valley have thin organic surface horizons and evidence of podzolization within their top few inches. These soils contain abundant stones and have developed on sharp angular scree. The soils have warm brown colors due to large amounts of amorphous Fe-oxide coatings, and these high contents of Fe have led to the suggestion that the soils are probably truncated podzols (Robinson *et al.*, 1949).

Further up on valley sides there is a sharp change to well-developed podzols with deep peat deposits found on the plateau overlooking the valley. These podzols are classified as Hiraethog Series. All the podzols studied contain charcoal layers and indicate that burning of the previous vegetation has occurred. This burning may have accentuated the podzol process or may even have initiated it. The ground flora of the podzols is principally *Deschampsia flexuosa*.

Highly organic surface horizons, generally between 10-20 cm thick, overlie grey silty clay eluvial horizons, that contain numerous weathered stones. An iron pan (placic horizon) if present, is generally found immediately above an orange brown Bf (spodic) horizon that merges into sharp angular scree. The highly weathered eluvial horizons consist almost entirely of silt plus clay whereas the textures of the illuvial horizons range from clay loams to sandy loams, with stone contents commonly in excess of 50%.

The uniformity of the profiles studied was investigated using as the criteria of uniformity the constancy of the ratios of Zr/Sr and quartz/illite in the silt fractions (Table 3). No lithologic breaks could be detected in any of the soil profiles studied using the currently accepted limits of detection. The near constant ratio of silt to clay in the fine earth fractions not only provides further evidence of uniformity within the profiles but indicates that the parent rock is breaking down to give silt plus clay in a constant ratio.

Table 3. Criteria used to assess parent material uniformity for soil profiles in mid-Wales.

	Zr/Sr in silt fraction	Quartz/Illite in silt fraction
Dystric Brunisol		
Ah	2.19	1.50
Bm1	2.09	1.69
Bm2	1.99	1.80
BC	2.09	1.82
Placic Humo-Ferric Podzol		
Oh	1.33	1.58
Ae	1.30	2.00
Bf	1.16	1.40
BC	1.24	1.59
Gleyed Humo-Ferric Podzol		
Oh	1.50	1.21
Ae	1.39	1.57
Bfg	1.28	1.38
BCg	1.77	1.05

Using acid residue as an internal standard the redistribution of the major elements in the rock was investigated. The soil parent material was assumed to be gravel-sized material separated from the profiles. Balance sheets were constructed for Si, Al, Fe and Mg (Table 4). The only element that was substantially redistributed during pedogenesis is Fe, with two of the profiles showing net gains for Fe in the total profile. These gains are probably due to lateral movement of Fe into the illuvial horizons from further up the slope. Recent work on similar soils (Adams, 1976, pers. commun.) suggests that truncation may have taken place on some soils. By calculating the net loss of each element from the profiles as a whole the order of mobility was found to be Mg > Al > Si > Fe (Evans and Adams, 1975b). However, if the order of release from crystalline minerals is considered then Fe becomes the most mobile of the elements studied (Table 5). It appeared therefore that the illuvial horizons of these soils have a tremendous capacity to retain the Fe released from the eluvial horizons. Al is not however accumulated, probably because of excess acidity in the Bf (spodic) horizons. The pH's of these horizons was generally less than 4.5. By summing the mean losses or gains of each element from each profile it is possible to calculate the following overall changes (in kg/ha) :-1.4 SiO_2, -1.5 Al_2O_3, -0.27 MgO and +0.24 Fe_2O_3. These values can be used to calculate actual amounts of elements entering drainage waters (Evans and Adams, 1975b).

Table 4. Changes in oxide content (g) in horizon cores 1 cm^2 in cross section.

	SiO_2	Al_2O_3	Fe_2O_3	MgO
Dystric Brunisol				
Ah	+0.01	-0.05	0.00	-0.011
Bm1	-0.27	-0.29	+0.01	-0.046
Bm2	-2.25	-1.94	+0.55	-0.277
BC	-0.52	-0.41	+0.35	-0.032
	-3.03	-2.69	+0.91	-0.366
Placic Humo-Ferric Podzol				
Oh	-0.38	-0.49	-0.16	-0.084
Ae	-0.87	-0.95	-0.55	-0.163
Bf	-0.16	-0.23	+0.24	-0.047
BC	-0.12	-0.06	+0.34	-0.025
	-1.53	-1.73	-0.13	-0.312
Gleyed Humo-Ferric Podzol				
Oh	-0.00	-0.27	-0.13	-0.064
Ae	-0.98	-1.15	-0.30	-0.249
Bfg	-0.01	-0.12	+0.34	-0.023
BCg	-0.03	-0.05	+0.71	-0.025
	-1.02	-1.59	+0.62	-0.361

The chief mineralogical change during pedogenesis is the weathering of chlorite, probably through the formation of vermiculite and amorphous silica (Evans, 1972). Using the amount of silt plus clay in the total mineral matter as an index of weathering (Stewart *et al.*, 1970), the total

Table 5. % Changes[a] in elemental content in crystalline minerals relative to their contents in the parent material.

	SiO_2	Al_2O_3	Fe_2O_3	MgO
Dystric Brunisol				
Ah	-4.0	-25.5	-81.7	-68.2
Bm1	-13.6	-24.2	-80.2	-67.1
Bm2	-12.1	-21.0	-73.2	-32.1
BC	-11.7	-19.2	-62.9	-18.4
Placic Humo-Ferric Podzol				
Oh	-16.0	-37.0	-98.5	-79.6
Ae	-14.4	-36.5	-95.7	-82.3
Bf	-9.8	-20.2	-82.4	-53.8
BC	-10.2	-10.3	-59.0	-40.4
Gleyed Humo-Ferric Podzol				
Oh	-13.1	-33.2	-99.6	-71.5
Ae	-9.9	-26.8	-91.9	-81.9
Bfg	-2.4	-17.4	-56.2	-30.2
BCg	-4.1	-5.3	-17.0	-21.0

[a]percent

amount of elements released on complete weathering can be calculated by plotting their losses from crystalline material against the weathering index (Evans and Adams, 1975b). The calculated losses of Fe (95%), Mg (80%) and Al (35%) on complete weathering (*i.e.*,100% silt plus clay) agree well with their calculated contents in chlorite. It can be concluded therefore that the release of silt plus clay from the rock, the losses of Fe, Mg and Al and the weathering of chlorite are all concurrent processes (Adams *et al.*, 1971), with the trioctahedral layer of chlorite probably acting as a rock cement. All of the released Fe is subsequently immobilised in sub-surface horizons although lateral movement of Fe may occur. Much of the brown to yellow brown color displayed in the soil profiles is due to this lateral movement of Fe.

CONCLUSION

It is possible by using an internal standard to quantify the actual changes in content of soil constituents that have occurred during pedogenesis. These changes may involve both a redistribution within the soil profile as well as an overall net loss from or gain into the profile. If both chemical and mineralogical information is available then assessments can be made not only on the extent of release of elements from minerals during weathering but also on their subsequent movement and possible incorporation into other minerals.

Not all soil forming materials or soils themselves are suitable for quantitative studies. Where the inherent variability of the material is great, any small changes induced by pedogenesis may be overlooked. Variations not only in the chemistry and mineralogy but also in organic matter content, stone content, bulk density and depth of horizon may make the material unsuitable for quantitative studies. The identification of the true soil parent material and the recognition of any lithologic breaks are also necessary prerequisites for such studies.

In spite of the fact that no internal standard is completely resistant to weathering the majority of workers have used only one internal standard to assess change and have not therefore had the confidence in their data that using two or more internal standards would have given them. Rarely also is the amount of predicted pedological change subjectively assessed.

Although the use of an internal standard has certain drawbacks and inherent errors due to the number of soil properties that need to be known, it is as yet the only reliable way of estimating the actual change that has occurred during pedogenesis and should allow many pedological postulations from soil data to be checked in an objective manner.

ACKNOWLEDGMENTS

I wish to thank Dr. W.A. Adams of the Soil Science Unit, University College of Wales, Aberystwyth for initiating my interest in quantitative pedology and to Mr. R. Hammond, An Foras Taluntais, Lullymore, Co. Kildare and Drs. G.W. Smillie and J.F. Collins of the Soil Science Department, University College, Dublin for help in the studies in Ireland. Finally I thank Dr. Ward Chesworth of the Department of Land Resource Science, University of Guelph for suggesting improvements to the manuscript.

REFERENCES CITED

Adams, W.A., Evans, L.J., and Abdulla, H.H., 1971, Quantitative studies on soils derived from Silurian mudstones. III. Laboratory and *in situ* weathering of chlorite: Jour. Soil Sci., v. 22, p. 158-165.

Akhtyrtsev, B.P., 1968, Removal and accumulation of oxides in soils of broadleaf forests in the central Russia Forest Steppe: Soviet Soil Sci., v. 1968, p. 1423-1434.

Barshad, I., 1964, Chemistry of soil development, *in* Bear, F.E., ed., Chemistry of the soil: London, Chapman and Hall Ltd., p. 1-71.

Bourne, W.C., and Whiteside, E.P., 1962, A study of the morphology and pedogenesis of a medial chernozem developed on loess: Soil Sci. Soc. America, Proc., v. 26, p. 484-490.

Carey, M.L., and Hammond, R.F., 1970, The soils beneath the Midland peats: Irish Forestry, v. 27, p. 23-36.

Carroll, D., 1953, Weatherability of zircon: Jour. Sed. Pet., v. 23, p. 100-116.

Chapman, S.L., and Horn, M.E., 1968, Parent material uniformity and origin of silty soils in north west Arkansas based on zirconium - titanium ratios: Soil Sci. Soc. America, Proc., v. 32, p. 265-271.

Claisse, G., 1968, Etude experimentale de l'analyse aux 3 acides: Cahiers ORSTOM, Serie Pedologie, v. 6, p. 129-149.

__________, 1972, Etude sur la solubilisation du quartz en voie d'alteration: Cahiers ORSTOM, Serie Pedologie, v. 10, p. 970-122.

Drees, L.R., and Wilding, L.P., 1973, Elemental variability within a sampling unit: Soil Sci. Soc. America Proc., v. 37, p. 82-87.

Eaqub, M., 1972, Studies on soil profile development: profile balance using mineralogical and chemical data on an illimerzed chernozem on loess: Zeit. Pflanz, Dung. Boden., v. 133, p. 102-110.

Evans, L.J., 1972, Mineralogical changes in Lower Palaeozoic mudstones during pedogenesis in mid-Wales: Welsh Soils Disc. Grp. Report, v. 13, p. 91-105.

Evans, L.J., and Adams, W.A., 1975a, Quantitative pedological studies on soils derived from Silurian mudstones. IV. Uniformity of the parent material and evaluation of internal standards: Jour. Soil Sci., v. 26, p. 319-326.

__________, 1975b, Quantitative pedological studies on soils derived from Silurian mudstones. V. Redistribution and loss of mobilised constituents: Jour. Soil Sci., v. 26, p. 327-335.

__________, 1975c, Chlorite and illite in some Lower Palaeozoic mudstones of mid-Wales: Clay Minerals, v. 10, p. 387-397.

Gilkes, R.J., Scholz, G., and Dimmock, G.M., 1973. Lateritic deep weathering of granite: Jour. Soil Sci., v. 24, p. 523-536.

Green, P., 1966, Mineralogical and weathering study of a red-brown earth formed on granodiorite: Austr. Jour. Soil Res., v. 4, p. 181-197.

Hammond, R.F., 1968, Studies in the development of a raised bog in central Ireland: Quebec, Third Internatl. Peat

Congress, p. 109-115.

Hardy, F., and Follet-Smith, R.R., 1931, Studies in tropical studies. II. Some characteristic igneous rock soil profiles in British Guiana, South America: Jour. Agric. Sci., v. 21, p. 739-761.

Joffe, J.S., 1936, Pedology: New Brunswick, Rutgers University Press, 575 pp.

Khan, D.H., 1959, Studies in translocations of chemical constituents in some red-brown soils, terra rosas and rendzinas using zirconium as a weathering index: Soil Sci., v. 88, p. 196-200.

Khangarot, A.S., Wilding, L.P., and Hall, G.F., 1971, Composition and weathering of loess mantled Wisconsin - and Illinoian-age terraces in Central Ohio: Soil Sci. Soc. America Proc., v. 35, p. 621-626.

Kimura, H.S., and Swindale, L.D., 1967, A discriminant function using zirconium and nickel for parent rocks of strongly weathered Hawaiian soils: Soil Sci., v. 104, p. 69-76.

Marel, H.W. van der, 1949, Mineralogical composition of a heath podzol profile: Soil Sci., v. 69, p. 193-207.

Marshall, C.E., 1940, A petrographic method for the study of soil formation processes: Soil Sci. Soc. America Proc., v. 5, p. 100-103.

Muir, A., 1951, Notes on the soils of Syria: Jour. Soil Sci., v. 2, p. 163-182.

Raad, A.T., and Protz, R., 1971, A new method for the identification of sediment stratification in soils of the Blue Springs Basin, Ontario: Geoderma, v. 6, p. 23-41.

Raeside, J.D., 1959, Stability of index minerals in soils with particular reference to quartz, zircon and garnet: Jour. Sed. Pet., v. 29, p. 493-502.

Redmond, C.E., and Whiteside, E.P., 1967, Some till-derived chernozem soils in eastern North Dakota. II Mineralogy, micromorphology and development: Soil Sci. Soc. America Proc., v. 31, p. 100-107.

Reid, R.F., and Matthews, B.C., 1954, The genesis of the Oneida series in southern Ontario: Can. Jour. Agric. Sci., v. 34, p. 614-623.

Ritchie, A., Wilding, L.P., Hall, G.F., and Stahnke, C.R., 1974, Genetic implications of B horizons in aqualfs of northeastern Ohio: Soil Sci. Soc. America Prof., v. 38, p. 351-358.

Robinson, G.W., Hughes, D.O., and Roberts, E., 1949, Podzolic soils of Wales: Jour. Soil Sci., v. 1, p. 50-62.

Roonwall, G.S., and Bhumbla, D.R., 1969, Contributions to the mineralogy of the sand fraction and geochemistry of the soils developed over gneissic rocks in the Kulu area (Central Himalayas, India): Geoderma, v. 2, p. 309-319.

Runge, E.C.A., Walker, T.W., and Howarth, D.T., 1974, A study of Late Pleistocene loess deposits, south Canterbury, New Zealand. I. Forms and amounts of phosphorus compared with other techniques for identifying paleosols: Quaternary Research, v. 4, p. 76-84.

Schlichting, E., and Blume, H.P., 1961, Art und Aussmass der Veranderungen des Tonmineralbestandes typischer Boden aus jungpleistozanen Geshiebemergel und ihrer Horizonte: Zeit. Pflanz. Dung. Boden., v. 95, p. 227-239.

Smeck, N., and Runge, E.C.A., 1971, Phosphorus availability and redistribution in relation to profile development in an Illinois landscape segment: Soil Sci. Soc. America Proc., v. 35, p. 952-959.

Smith, H., and Wilding, L.P., 1972, Genesis of argillic horizons in ochraqualfs derived from fine textured till deposits of northwestern Ohio and southeastern Michigan: Soil Sci. Soc. America Proc., v. 36, p. 808-815.

Stewart, V.I., Adams, W.A., and Abdulla, H.H., 1970, Quantitative pedological studies on soils derived from Silurian mudstones. I. The parent material and the significance of the weathering process: Jour. Soil Sci., v. 21, p. 242-247.

Sudom, M.D., and St. Arnaud, R.J., 1971, Use of quartz, zirconium and titanium in pedological studies: Can. Jour. Soil Sci., v. 51, p. 385-396.

Wang, C., and Arnold, R.W., 1973, Quantifying pedogenesis for soils with discontinuities: Soil Sci. Soc. America Proc., v. 37, p. 271-278.

Wild, A., 1961, Loss of zirconium from 12 soils derived from granite: Austr. Jour. Agric. Res., v. 12, p. 300-305.

Soil Development in Arctic and Subarctic Areas of Quebec and Baffin Island

T. R. Moore

ABSTRACT

Soil development has been examined at nine sites along a transect from the boreal forest to the subarctic and arctic zones. There are pronounced variations in climate, vegetation and period for soil development along the transect, but the parent materials are generally similar. Soil studies have been confined to sites relatively unaffected by frost heave activities.

Soils at the boreal forest and low subarctic sites (Sept Iles, Esker and Ross Bay Junction) show strong morphological and chemical evidence of podzolisation. Ae horizons are 5 to 15 cm thick and the profiles show translocation of iron, aluminum and organic carbon. Cemented subsoil horizons (Bfc) occur frequently in freely drained sites and are related to amphibole-rich parent materials and cementation by amorphous iron and aluminum compounds. Strong cementation requires about 5000 yr for development.

Soils at the high subarctic sites (Cambrian Lake and Fort Chimo) show less intensive podzolisation with the horizons 1 to 5 cm thick and smaller translocations of sesquioxides and organic carbon. Morphological evidence of podzolisation is weak in soils derived from iron-rich materials of the Labrador Trough (Schefferville), but chemical analyses reveal movement of large amounts of iron, aluminum and organic carbon.

North of the treeline (Payne Bay) evidence of podzolisation is restricted to this Ae horizons (1 cm thick) and minor translocation. In the arctic sites (Frobisher Bay and Clyde River) podzolisation is confined to bleached sand grains at the base of the surface organic horizon. Polar desert soils occur on ridge tops at Clyde River.

Morphological evidence of gleying is weak in most soils;

chemical analyses reveal the translocation of iron and manganese. Reducing conditions develop slowly at cold temperatures.

Clay fractions from representative soils have been analysed by X-ray diffraction. Smectite and vermiculite occur in the Ae horizon of Podzols, and the Bf horizon is dominated by chlorite, vermiculite and interstratified and amorphous minerals. Clay mineral transformations are best developed in soils of the Labrador Trough, which possess clay mineral structures inherited from the parent materials. Chlorite, illite, vermiculite and interstratified minerals occur in the arctic soils, and clay mineral transformations within profiles are weak. Analyses of laboratory solutions in equilibrium with representative horizons showed that the solutions fall into the kaolinite field of the mineral stability phase diagrams.

Reasons for the decrease in podzolisation along the transect are examined. Fulvic acid is an important portion of the organic matter in Bf horizons, suggesting its role in the sesquioxide translocation. Fulvic acid also occurs in organic surface horizons, forming 15 to 30% of the organic carbon in arctic soils and 5 to 15% in Podzols. The decrease in podzolisation in these soils may be a function of organic matter production and decomposition and the short active season for pedogenesis, rather than specifically vegetation cover. The possibility of iron and aluminum translocation in inorganic forms should also be recognised.

INTRODUCTION

Until recently, relatively little was known about soil development in subarctic and arctic areas. In North America, most of this recent work has concentrated on studies in Alaska and western Canada (Pettapiece and Zoltai, 1974; Retzer, 1965; Tedrow and Cantlon, 1958). Studies in eastern Canada have been restricted mainly to the east coast of Hudson Bay and Riviere Arnaud area (Payette, 1968, 1973; Payette and Morisset, 1974).

The study of subarctic and arctic soils is important because of their extensive spatial distribution in Canada and Alaska, and because most of the soils of temperate regions of Canada have been subject to subarctic and arctic soil forming processes in the past. The increasing utilisation of palaeosols as indicators of previous environments also demands a thorough understanding of the relationships between present day soils and environments.

This paper examines the variations in soil development along a transect from the boreal forest near Sept Iles, through the subarctic of northern Quebec and into the arctic areas of Baffin Island (Figure 1). The distribution of soils within each study area will be examined first and will be followed by a more detailed examination of the variation in pedogenic processes along the transect. Particular

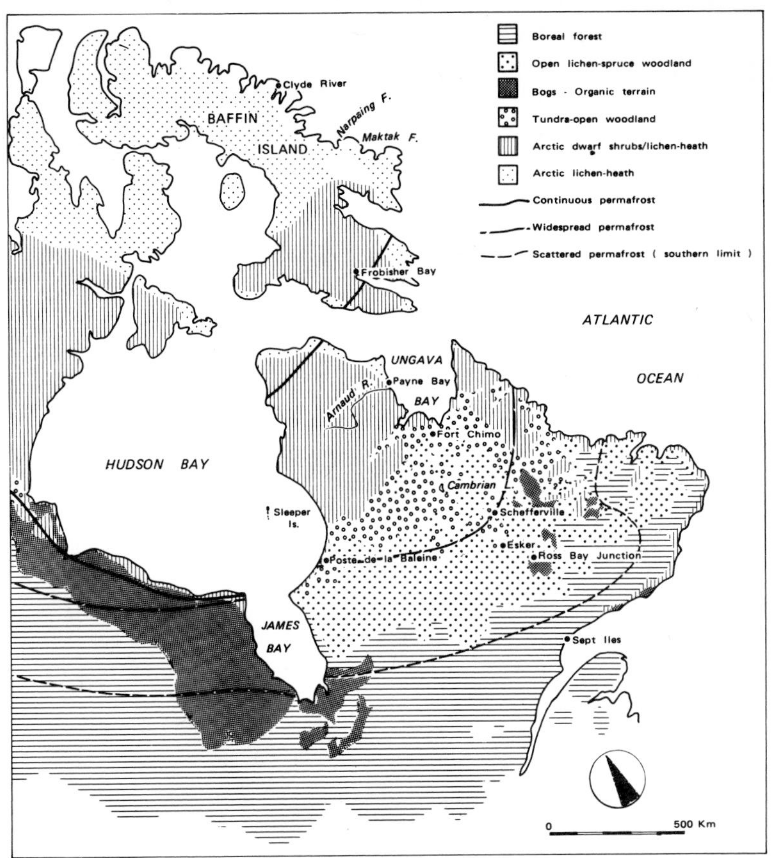

Figure 1 Location of sites discussed in this paper from Sept Iles in the northern boreal forest to Baffin Island in the Eastern Canadian Arctic.

attention will be paid to chemical translocation and clay mineral transformation, and their variations related to the major environmental factors of soil formation. Finally, the position of this type study within the context of pedology and its methodology is examined.

THE ENVIRONMENTAL SETTING

There are strong environmental gradients along the transect. Climatic variations are indicated in Table 1. Annual precipitation decreases from about 100 cm near Sept Iles to 70 cm at Schefferville, to 35 cm at Frobisher Bay and to 20 cm at Clyde River; about half the precipitation falls as snow (Canada Dept. of Transport, 1962). The prediction of the amount of water available for percolation through the soil profile is difficult because of the paucity

Table 1. Climatic data[a] for stations representative of the transect

	Stations							
	Sept Iles		Schefferville		Frobisher Bay		Clyde River	
	air temperature	precipi- tation	air temperature	precipi- tation	air temperature	precipi- tation	air temperature	precipi- tation
Month	(°C)	(cm)	(°C)	(cm)	(°C)	(cm)	(°C)	(cm)
January	-16.0	10.4	-24.4	4.1	-26.6	1.8	-26.9	1.1
February	-13.9	9.7	-21.5	4.1	-26.3	2.3	-27.4	0.7
March	- 7.9	7.9	-13.9	4.8	-20.9	2.0	-26.3	0.5
April	- 0.6	5.6	- 5.9	3.6	-14.6	2.0	-18.7	0.8
May	5.3	7.9	1.1	3.3	- 4.3	2.0	- 6.9	1.4
June	11.0	4.8	9.3	7.9	3.2	2.8	0.9	1.1
July	15.1	11.4	12.8	8.9	7.6	4.1	4.6	2.3
August	14.3	8.4	10.8	8.9	6.9	5.1	4.0	3.0
September	9.5	9.4	5.7	7.9	1.9	4.3	-0.1	3.9
October	3.7	7.9	-0.7	7.4	-4.4	2.8	-6.6	3.3
November	-3.1	9.9	-9.0	6.4	-10.9	2.8	-16.9	1.8
December	-11.1	10.4	-18.0	3.0	-19.7	2.5	-24.2	0.8
Year	0.6	103.7	-4.5	70.3	-9.0	34.5	-12.2	20.6

[a]Mean monthly temperature and precipitation.

and unreliability of climatic data. Annual potential evapotranspiration rates vary from about 45 cm at Sept Iles to 25 cm at Schefferville and lower values at Frobisher Bay and Clyde River (Canada Dept. Energy, Mines and Resources, 1974). Hare and Hay (1971), using annual precipitation and runoff data, estimate annual apparent evapotranspiration to vary from about 30 cm at Sept Iles to 10 cm at Schefferville and lower values at Frobisher Bay and Clyde River. The lichen mat keeps the soil close to field capacity through much of the summer, and reduces evapotranspiration (Rouse and Kershaw, 1971). Much of the spring snowmelt probably passes across the frozen soil surface, and thus plays little role in pedogenic processes. The above speculations suggest that water for soil percolation varies from 20 to 30 cm at Sept Iles to 5 to 10 cm on Baffin Island, though much variation will be caused by topographic position and vegetation cover. The growing season decreases from about 140 days at Sept Iles to 40 days or less at Frobisher Bay and Clyde River (Canada Dept. Energy, Mines and Resources, 1974).

Permafrost occurs along most of the transect (Figure 1). Whilst the overall pattern can be related to mean air temperatures, there are many local variations caused by topography, geology, hydrology and vegetation cover.

Most of the transect is underlain by igneous and metamorphic rocks of Archean and Proterozoic Age, such as granites, gneisses, anorthosite and syenite. The Labrador Trough stretches south from the west coast of Ungava Bay down through Cambrian Lake and Schefferville to Wabush. It contains sedimentary and metamorphic rocks of Proterozoic Age, such as quartzite, shale, slate, dolomite and iron formations. Other sedimentary rocks occur on the western coast of Baffin Island, and around the coasts (Douglas, 1970).

Nearly all of the transect has been affected by the Wisconsin glaciation, which removed most of the unconsolidated materials. The period since deglaciation varies from about 13,000 years at Sept Iles to about 6,000 years at Schefferville, and then increases again to 7,000 to 8,000 years at Fort Chimo, Frobisher Bay and Clyde River (Bryson *et al.*, 1969). Marine transgressions occurred along the coastal areas, and postglacial emergence varies from up to 200 m on the coast of Hudson Bay and Ungava Bay to about 50 m around the other coasts (Bird, 1972). Fluvial, fluvioglacial and lacustrine deposits occur frequently, as well as glacial till.

The vegetation pattern along the transect is shown in Figure 1. In the south, the boreal forest occurs on freely drained sites, dominated by black and white spruce (*Picea mariana* and *P. glauca*) and balsam fir *(Abies balsamifera)* with tamarack *(Larix laricina)* in poorly drained sites. Further north, open spruce-lichen woodlands occur on freely drained sites, and grade into closed spruce-moss forests in the wetter sites. Lichen-heath tundra occurs on exposed ridges in the subarctic. The treeline runs close to the east coast of Hudson Bay, through Fort Chimo and down the east coast of Labrador. North of the treeline, mixtures of dwarf shrubs, lichens, grasses and mosses dominate, and bare patches of soil and cryoturbation features become more important in Baffin Island. Mosses, sedges and grasses again occupy the poorly drained sites. More detailed descriptions of the major plant associations can be found in Hare (1959) and Rousseau (1968). The major vegetation associations moved northwards as the ice retreated; thus, the southernmost soils have been occupied by arctic and subarctic as well as boreal vegetation. In the Parc des Laurentides area, Richard (1975) has found that the tundra vegetation occupied the area for about 3500 years, was displaced by open spruce-lichen woodlands for about 3000 years and then was finally replaced by spruce-fir forest about 5000 years BP. A similar progression of vegetation associations may be expected to have occurred in other areas in northern Quebec. Fir is an important component of the environment, particularly in the subarctic, where open spruce-lichen woodlands are very susceptible to burning. Although causing short-term changes in organic matter and nutrient distribution, these fires probably have little longterm effect on the soil properties (Viereck, 1973).

METHODS

A major problem in the study of soils along such a diverse transect is to ensure an adequate sampling of the major types of soils, as indicated by variations in the environmental soil forming factors. This is particularly difficult where background information is weak and access difficult. Because of these restrictions, soils were studied and sampled at only 9 locations: Sept Iles, Ross Bay Junction and Esker along the QNSL Railway, Schefferville, Cambrian Lake, Fort Chimo, Payne Bay, Frobisher Bay and Clyde River (Figure 1). Soils at these locations probably show most of the variations in properties and genesis that one could expect along the transect.

Within each location, soil distribution was examined by transects; 5 to 50 km^2 were covered in each location. Representative soil profiles were described and sampled. Horizon samples were air dried, ground and passed through a 2 mm sieve and the fine earth fraction analysed by the following methods:

pH in 1:1 s/w/ soil:water and soil:0.01M $CaCl_2$ suspensions (Peech, 1965); organic carbon by dichromate digestion (Tinsley, 1950); dithionite-extractable iron and manganese (Fe_d and Mn_d) by the dithionite-citrate-bicarbonate method (Mehra and Jackson, 1960); oxalate-extractable iron and aluminum (Fe_o and Al_o) by the acidified ammonium oxalate method (McKeague, 1967); pyrophosphate-extractable iron, aluminum and carbon (Fe_p, Al_p and Cp) by the 0.1M sodium pyrophosphate method (McKeague, 1967); humic acid and fulvic acid fractions were separated by the sodium pyrophosphate-sodium hydroxide extraction of Kononova (1966); clay mineral assemblages were determined by X-ray diffraction (Whittig, 1965). The soils have been classified within the Canadian Classification System (Canada Soil Survey Committee, 1974).

RESULTS

From the large number of soil profiles described and analysed, 17 have been selected as representative of the major types of soils along the transect. Their morphological properties are described in Appendix 1 and analytical data presented in Appendix 2. The major properties of the soils will be discussed in the context of the arbitrary zones in which they occur: the boreal forest, the low subarctic, the Labrador Trough, the high subarctic, the low arctic and the middle arctic. These zones divide the transect in terms of the major variation in climate and vegetation.

The Boreal Forest

Dense spruce-fir stands are developed on coarse fluvio-glacial and beach deposits along the Quebec North Shore, around Sept Iles (Figure 1). Organic surface horizons are generally thick (5 to 15 cm) and the subsoils show strong morphological evidence of podzolisation: bleached Ae horizons 5 to 15 cm thick overlying reddish brown Bf horizons. The soils are strongly acid in the upper horizons, and evidence of leaching (as indicated by decreases in pH and base saturation) extends down to 70 to 80 cm in freely drained sites. Analyses of extractable iron, aluminum and organic carbon show translocation of these elements from the A to B horizons, though absolute quantities are small probably because of the coarse nature of the parent material (Soil 1, Appendix 1). Many of the soils contain a high proportion of iron-bearing minerals such as hornblende, ilmenite and magnetite. Weathering of these minerals has led to the formation of strongly cemented, reddish brown subsoil horizons (Bfc). The cementation appears to be due to amorphous iron and aluminum compounds; the organic matter content of these horizons is low, suggesting that the translocation of iron and aluminum in inorganic forms may be important, as the soils are very acid. Examination of coastal soil chronosequences suggests that the cementation takes 5000 to 6000 years to develop. Richard (1975) reports a period of 4000 years necessary for cementation to develop in subsoils in the Parc des Laurentides region, based on palynological evidence. More detailed discussion of the cemented horizons can be found in Moore (in press).

Gleysols and organic soils occur in the poorly drained depressions, and there is strong subsoil mottling. Orthic and Ortstein Humo-Ferric Podzols and Degraded Dystric Brunisols dominate on the freely drained sites.

The Low Subarctic

Soils of this zone were examined by Ross Bay Junction

and Esker (Figure 1); open spruce-lichen woodlands occur on freely drained sites, with dense spruce-moss forests and muskeg in poorly drained sites. Morphological and chemical evidence of leaching and podzolisation is still strong in these soils, but the surface organic and Ae and Bf horizons are thinner than at Sept Iles and the soils are not so acid in the surface horizons (Soils 2 and 3, Appendix 1). Soils are classified as Degraded Dystric Brunisols or Ortstein or Orthic Humo-Ferric Podzols. Cemented Bfc horizons occur less frequently on freely drained sites than at Sept Iles, though the soils still show strong evidence of sesquioxide and organic matter translocation. These soils have been exposed for about 6000 to 7000 years, which is close to the time required for cementation to develop strongly in the soils near Sept Iles. Cemented horizons also occur in seasonally waterlogged small depressions (Soil 3, Appendix 1) but the cemented horizon shows more organic matter and manganese than its freely drained equivalents, reflecting the importance of waterlogging.

The Labrador Trough

Parent material variations in the Labrador Trough are sufficient to distinguish these soils from those developed on the adjacent Canadian Shield (Figure 1). The iron formations and sedimentary rocks such as shales influence the parent material, which, in the Schefferville area, usually consists of glacial till of a strong reddish brown color, rich in iron and fine fractions (commonly 10 to 20% clay and 20 to 40% silt). The reddish brown color of the parent materials and the iron minerals present subdue the morphological representation of the soil forming processes. On freely drained sites beneath open spruce-lichen woodland, Ae horizons vary from a few bleached sand grains to 5 cm in thickness. The underlying Bm or Bf horizons are difficult to differentiate in the field because of the reddish brown color of the parent materials. Chemical analyses, however, reveal the translocation of large amounts of extractable iron and smaller amounts of aluminum and organic matter (Soils 4 and 5; Appendix 1).

Because of strong chemical translocation, many of the soils readily fall into the Podzol order in soil classification schemes in which criteria of translocation of sesquioxides and organic matter is important. By contrast, the soils derived from adjacent igneous and metamorphic rocks often fail to meet the chemical criteria, yet show stronger morphological evidence of podzolisation.

The heavy texture of the parent materials and the topography at Schefferville (Figure 1) combine to create seasonally waterlogged depressions. Soils in these depressions are Gleyed Humo-Ferric Podzols and Gleysols and often show evidence of both gleying and podzolisation: mottling and iron, aluminum, manganese and organic matter translocation. Gleysols occur in sites which are waterlogged for most of the summer, but mottling is restricted to the subsoil horizons directly beneath the organic surface horizons, again reflecting the subduing role the parent materials play in the development of soil morphology (Soil 6; Appendix 1). Chemical analyses again show the translocation of large amounts of extractable iron and manganese. The influence on soil formation of haematite-containing reddish brown parent materials has been observed by McKeague and Cann (1969) in the Atlantic Provinces.

Regosols, Brunisols and cryoturbed soils occur on the ridgetops where the vegetation cover is sparse and consists of lichen and heath. More detailed discussion of soil formation in the Schefferville area can be found in Nicholson and Moore (in press).

The High Subarctic

In the Cambrian Lake and Fort Chimo locations (Figure 1) the woodlands become less dense, forests occupy a smaller proportion of the landscape, and tundra vegetation becomes more extensive. Cryoturbation features become more common and permafrost occurs at depths of 0.5 to 1 m beneath the closed canopy forests. Leaching and podzolisation are still important, as shown by the acidic surface horizons and Ae horizons 1 to 4 cm thick beneath the open spruce-lichen woodlands. However, morphological and chemical evidence of Bf horizons is weaker than at Esker or Ross Bay Junction, and analytical data show only minor translocation of iron, aluminum and organic matter (Soils 7 and 8: Appendix 1). Most of the soils are Orthic or Degraded Dystric Brunisols.

Variations in parent material affect the intensity of podzolisation. At Cambrian Lake, soils developed on quartzite deposits from the Labrador Trough show thick Ae horizons (2-8 cm) and translocations of large amounts of iron, aluminum and organic matter (Soil 9), sufficient for inclusion into the Podzolic order in the Canadian classification scheme.

Gleysols occur in poorly drained depressions (Soil 10), but gleying is often weakly expressed and permafrost is close to the surface. More detailed discussion of soils in the Cambrian Lake area can be found in Moore (1974); similar soils and pedogenic processes are found in soils of the Fort Chimo area.

The Low Arctic

North of the treeline, in the Payne Bay and Riviere Arnaud areas (Figure 1), Regosols and cryoturbed soils occur commonly, with Brunisols developed on the more stable sites and beneath the denser lichen-heath tundra vegetation. Ae horizons are thin and discontinuous and analyses of soils 11 and 12 (Appendix 1) and those quoted by Payette (1968) show that there has been little release of sesquioxides by weathering, or translocation in the profile. Degraded and Orthic Dystric Brunisols dominate, with Gleysols again occurring in poorly drained sites with weak evidence of gleying.

The Middle Arctic

At Frobisher Bay and Clyde River (Figure 1), the vegetation cover becomes sparser, with lichen-heath tundra occupying the more sheltered sites, grasses and mosses in the poorly drained depressions and only a thin cover of lichens on the thinnest soils and exposed sites. Cryoturbation is strong, especially where the soils are wet, as on flood plains. Evidence of leaching is weak, with only mildly acid surface horizons (pH 4.5 to 5.5), and leaching depths of only 20 to 30 cm (Soils 13, 14 and 16; Appendix 1). Subsoil sesquioxide and organic matter contents are low and evidence of podzolisation is restricted to a few bleached sand grains at the base of the surface horizons. Cryic Dystric Brunisols dominate where the soil cover is thick.

In the Clyde River area, exposed ridges of beach material are sparsely vegetated and the equivalent of Polar Desert Soils occur (Tedrow, 1968). In these soils (Soil 17; Appendix 1) there is minimal accumulation of organic matter, neutral pH values, weak soil colors and low release of sesquioxides by weathering. They fall into the Orthic and Cryic Regosol subgroups of the Canadian classification scheme.

Cryic Gleysols and Cryic Humic Gleysols occur in poorly drained sites, where permafrost is close (20 to 40 cm) to the surface. Morphological evidence of gleying is weak, but a Rego Gleysol (Soil 15; Appendix 1) developed on a floodplain at Frobisher Bay shows that much iron and manganese can be translocated within the profile, despite the short active season and cold soil temperatures.

THE PEDOGENIC PROCESSES

The preceding section has described the distribution and properties of the major types of soils recognised along the transect, in terms of the arbitrary zones established. Within the soil profiles, several important pedogenic processes have been identified and their magnitude assessed through morphology and chemical analyses. In this section, the pedogenic processes will be examined in more detail, and the causes for the variation in their magnitude discussed.

Leaching

Leaching involves the removal from the soil of material in solution, and, in this context, also includes the replacement of cations on the exchange complex by hydrogen or aluminum, and the subsequent lowering of pH. In the soils of the boreal forest near Sept Iles, pH and base saturation data for freely drained soils suggest that the leaching process has reached depths of about 70 to 80 cm, and it decreases to about 50 cm at Esker and Ross Bay Junction. In the more northerly soils, at Cambrian Lake and Payne Bay, leaching depths rarely exceed 30 to 40 cm and are only 20 to 30 cm at Frobisher Bay and 10 to 20 cm at Clyde River. The decrease in pH also changes, with very acid conditions (pH 3.5 to 4.0) in the southern soils, and only mildly acid conditions (pH 5.0 to 6.0) in the northern soils.

The northwards decrease in leaching depth and intensity can be related to the decrease in effective precipitation from about 30 cm near Sept Iles to 10 cm or less at Clyde River and Frobisher Bay. The primary productivity of the ecosystems is also important, because this involves root respiration and organic matter which decomposes on the soil surface to organic and carbonic acids, which are important in the leaching process. Annual primary productivity varies from 0.4 to 2.0 kg/m^2 in the boreal forest to 0.2 to 1.0 kg/m^2 in open spruce-lichen woodlands and to 0.1 to 0.4 kg/m^2 in lichen-heath tundra (Rodin and Bazilevec, 1966).

Gleying

Gleying involves the formation of reducing conditions, usually when the soil is saturated, and under these conditions, iron and manganese are reduced and mobilised, being either translocated into mottles within the soil body, or removed from the soil profile. Waterlogged conditions exist in many soils along the transect, but morphological evidence of gleying becomes weaker in the northern soils. However, even Frobisher Bay and Clyde River soils (*eg.* Soil 15) show evidence of iron and manganese mobilisation. The short unfrozen season and the cold soil temperatures impede the development of reducing conditions, which are essentially microbiologically induced. However, McKeague (1965) has shown that reducing conditions can be developed at low temperatures ($<5°C$) and Gersper and Challinor (1975) noted an increase in the prominence of mottling in tundra soils near Barrow, Alaska over a 6-year period after the soils had been compacted by vehicles and the waterlogging had become more extensive.

Clay Minerals

The formation and transformation of clay minerals are important aspects of pedogenesis. Clay mineral assemblages in soils are a function of their inheritance from the parent material, their addition to the soil as dust, alluvium or colluvium and the pedologic processes operating in the soil, particularly the composition of the soil solution. It has been shown that the activity of the cations and anions in the soil solution are affected by and control the stability of the aluminosilicate and clay minerals (Kittrick, 1969). As soil formation proceeds, the soil solution will change and affect the types of clay mineral decomposition and formation processes.

In most of the soils examined, the clay (<2µ) content is low, generally 5 to 10%, and in many cases is dominated by primary minerals such as quartz, feldspars and ferro-magnesians. Thus, clay mineral formation in these soils, derived from coarse igneous and metamorphic rocks is low. The exceptions are the soils of the Schefferville area derived from sedimentary rocks, which frequently contain 10 to 20% clay, much of it as well defined clay minerals.

The clay mineral assemblage from representative horizons of 13 transect soils has been examined by X-ray diffraction, and the results are presented in Table 2. In addition to the clay minerals, the clay fractions also contain quartz, feldspars and amphiboles.

Soils which show strong morphological and chemical evidence of podzolisation (Soils 1, 2, 3 and 9; Appendix 1) exhibit the formation of smectite, Al interlayered vermiculite, and in some cases Mg interlayered vermiculite in the Ae horizon. In the Bf horizon, Al interlayered vermiculite, chlorite and amorphous and interstratified minerals are common. In the C horizons only Al interlayered vermiculite, and interstratified minerals occur regularly. The same trends occur in the podzolised soils at Schefferville (Soil 5) which possesses chlorite, illite and kaolinite in the parent material. Similar, but weaker, transformations occur in the Fera Gleysol at Schefferville (Soil 6). The Schefferville soils possess strongly crystalline clay minerals and no amorphous, interstratified or vermiculite minerals. This reflects the occurrence of a well defined clay mineral assemblage in the parent material, or compared to the soils derived from coarse textured materials of igneous and metamorphic rocks.

In the less strongly podzolised soils of the subarctic (Soils 7 and 8) smectite formation does not occur in the Ae or Aej horizons and chlorite and Al interlayered vermiculite are common, along with interstratified minerals. In the arctic soils at Frobisher Bay (Soils 13 and 15) chlorite, Al interlayered vermiculite and interstratified minerals dominate, but small amounts of smectite also occur. At Clyde River (Soils 16 and 17), illite, interstratified and Mg interlayered vermiculite dominate, with small amounts of chlorite and smectite.

The clay mineral assemblages are to be expected from established stability series and weathering reactions (Jackson, 1964). The occurrence of smectite in Ae horizons is common in Canadian Podzols (Brydon *et al.*, 1968), and vermiculite and interstratified minerals have been found in soils developed from similar materials in similar environments (Gjems, 1960).

A second approach to weathering reactions and mineral stability within soils is to analyse an equilibrium soil solution for the concentration of the cations and anions which are critical in controlling the stability of different minerals. This has been performed on 13 representative horizons, using the laboratory method of Weaver *et al.* (1971): water:soil ratio of 10:1, an equilibrium period of 3 weeks after which the suspension was filtered and the filtrate analysed for pH, and Ca, Mg, K, Na and SiO_2 concentration. The data can be plotted into stability phase diagrams constructed from the thermodynamic data of important minerals. Figure 2 shows the location of each sample within the K-feldspar-clay mineral system defined by Garrels and Christ (1965) in terms of silica concentration and the hydrogen and potassium ion concentrations. All the samples fall within the kaolinite section of the diagram, and similar locations occur in the other feldspar (Na-, Ca-) diagrams. This suggests that the weathering sequences are progressing towards kaolinite as the stable end-member, but it may also reflect the formation of crypto crystalline kaolinite on the weathering surface of plagioclase feldspars (Busenberg and Clemency, 1976).

It is interesting to compare the clay mineral assemblages of these soils with those observed by Andrews and Miller (1972) in weathered tills of different age in the Maktak/Narpaing fiord area of Baffin Island (Figure 1). Clay mineral assemblage was one of the criteria they used to differentiate between the three tills. The youngest till, exposed for 10,000 to 100,000 years contained very few clay minerals and a low free iron content. The second till, dated between 200,000 and 500,000 years contained illite and hydrobiotite and mixed chlorite/vermiculite clay minerals, as well as a higher free iron content. Similar clay minerals and higher free iron contents were found in the oldest till, dated as 400,000 to 1,000,000 years old.

Although the above clay mineral differences may be useful in differentiating tills on poorly vegetated surfaces, the data quoted above for these soils suggest that strong clay mineral changes can occur within 10,000 years on well vegetated surfaces. Gjems (1960) has noted the formation of smectite, vermiculite and hydrobiotite in Scandinavian podzols within a period of 300 years. Similarly, differences in free iron within these arctic and subarctic soils can be observed, as a function of differences in parent material, texture and pedogenic processes. Such clay mineral and free iron analyses may be applicable in establishing differences in poorly vegetated tills, but it must be appreciated that changes can occur quickly on vegetated surfaces and the full range of environmental variations must be recognised.

Podzolisation

Podzolisation is a process which involves the translocation of iron, aluminum and organic matter from the surface and Ae horizons, and their accumulation in the subsoil horizons (*eg.* Bf, Bh). The effects of podzolisation can be recognised by the occurrence of a bleached Ae horizon underlain by a reddish brown Bf, Bfh or Bh horizon, and by

Table 2. Clay mineral[a] assemblages in selected soil profiles.

Soil	Horizon	Chlorite	Illite	Kaolinite	Smectite	Vermiculite	Interstrat.	Amorphous
1. Ortstein Humo-Ferric Podzol	Ae	0	1	0	1	3 (Al)	1	0
	Bfc	0	0	0	0	3 (Al)	2	1
	C	0	0	0	0	3 (Al)	1	0
2. Ortstein Humo-Ferric Podzol	Ae	0	1	0	1	3 (Mg-Al)	2	0
	Bfc	1	1	1	0	3 (Al)	3	0
	C	1	1	0	1	3 (Al)	1	0
3. Gleyed Humo-Ferric Podzol	Aeg	0	2	1	0	3 (Mg)	2	0
	Bfhc	2	2	0	0	3 (Al)	1	2
	Cg	1	1	0	0	3 (Mg)	1	0
5. Orthic Humo-Ferric Podzol	Ae	0	1	1	3	0	1	0
	Bf	2	2	2	0	0	0	0
	C	2	2	2	0	0	0	0
6. Fera Gleysol	Aeg	1	3	2	2	0	0	0
	Bgf	2	3	2	1	0	0	0
	C	2	3	2	0	0	0	0
7. Degraded Dystric Brunisol	Bm	2	1	0	0	3 (Al)	1	0
	C	2	1	0	0	3 (Al)	1	0
8. Degraded Dystric Brunisol	Bm1	2	1	1	0	3 (Al)	1	0
	Bm2	2	1	0	0	3 (Al)	2	0
9. Orthic Humo-Ferric Podzol	Ae	0	1	0	0	3 (Mg)	1	0
	Bf	0	0	0	0	0	0	3
	C	3	0	1	0	1 (Al)	1	0
10. Cryic Humic Gleysol	Bg	3	1	0	0	1 (Al)	1	0
	Cz	2	2	0	1	3 (Mg)	1	0

13. Cryic Dystric Brunisol	Bm	2	0	0	1	2 (Al)	2	0
	C1	1	0	0	1	2 (Al)	2	0
15. Rego Gleysol	Afg	2	0	1	1	3 (Al)	3	0
	Cg	2	0	1	1	2 (Al)	1	0
16. Cryic Dystric	Bm	1	1	0	1	3 (Mg)	2	0
	C	1	2	0	1	3 (Mg)	2	0
17. Cryic Regosol	Bmj	1	3	0	0	1 (Mg)	2	0
	C	1	3	0	0	1 (Mg)	2	0

a 0 - absent 2 - minor
1 - trace 3 - major

analyses of the amount of iron, aluminum and organic matter present in the B horizons.

A northwards decrease in the morphological and chemical evidence of podzolisation is one of the most pronounced changes observed in the soils of this transect, and this change has been recognised by Tedrow and Cantlon (1958). It has often been assumed that the podzol profile is closely related, both genetically and spatially, to the boreal forest, and evidence of buried or fossil podzol soils has been used as indicators of previous forest cover (*eg.* Sorenson *et al.*, 1971).

One factor which affects the development of podzol features is the parent material. Other factors being equal, acidic parent materials are more readily able to develop podzol characteristics (*eg.* soils 7 and 9 on gneiss and quartzite at Cambrian Lake) and morphological evidence of podzolisation is suppressed on the reddish brown parent materials of the Schefferville area. The analytical data show that all three components (iron, aluminum and organic matter) have been involved in the translocation process; their relative importance varies. In soils developed from coarse igneous and metamorphic materials (Soils 1 and 2), iron and aluminum are translocated in approximately equal amounts. In the Schefferville soils (Soil 5), iron translocation dominates over aluminum translocation. This probably reflects the difference in availability of these two elements as weathering products of the parent material.

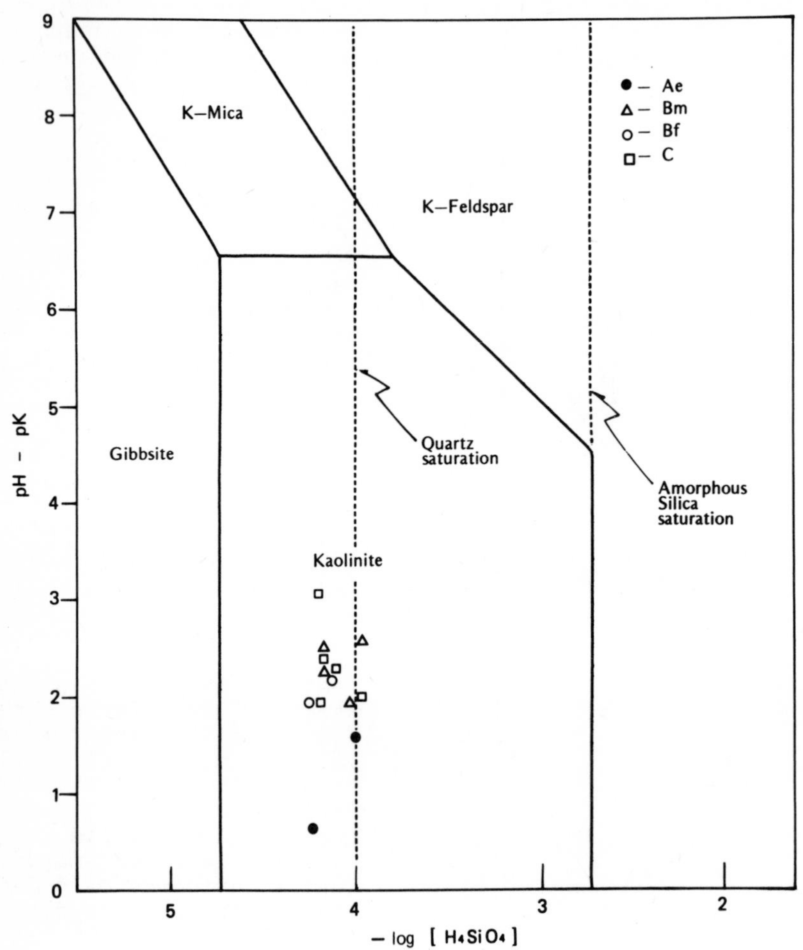

Figure 2 Equilibrium soil solutions plotted within the stability phase diagram for the K_2O-Al_2O_3-SiO_2-H_2O system at 25°C and 1 atmosphere (after Garrels and Christ, 1965).

In all of the podzolised soils, the subsoil organic matter content is low ($<$ 2% organic carbon: Table 3), except in the Schefferville soil (Soil 5). Oxalate:pyrophosphate extractable iron and aluminum ratios are generally high ($(Fe_o + Al_o) : (Fe_p + Al_p) > 2$). This suggests that either organic matter is not an important component of the translocation process in these soils, or that much of the translocated organic matter has been decomposed in the B horizons. The very low pH values suggests that inorganic movement of sesquioxides might be possible.

Vegetation cover is another important factor: podzolisation is best developed beneath the dense spruce-fir boreal

Table 3. Organic matter fractionation in selected profiles and horizons.

Soil	Horizon	Org. C %	Humic Acid %	Fulvic Acid %	FA/C_t %
1. Ortstein Humo-Ferric Podzol	LFH	52.78	2.46	4.10	7
	Ae	0.27	0.04	0.04	15
	Bf	1.33	0.08	0.57	43
	Bfc	1.01	0.06	0.64	64
	Bmc	0.35	0.04	0.14	40
2. Ortstein Humo-Ferric Podzol	H	25.71	0.91	3.48	14
	Ae	1.32	0.17	0.19	14
	Bfc	1.86	0.08	1.05	57
Boreal forest floor-Sept Iles	L	37.66	1.90	4.30	11
	F	35.14	1.50	2.00	6
	H	29.13	1.10	1.80	6
8. Degraded Dystric Brunisol	O	3.73	0.55	1.50	40
	Ae	2.17	0.28	0.53	24
	Bm1	0.36	0.09	0.05	14
	Bm2	0.29	0.00	0.14	48
9. Orthic Humo-Ferric Podzol	O	26.23	1.64	3.23	12
	Ah	5.59	1.63	0.00	0
	Ae	0.43	0.22	0.00	0
	Bf1	1.03	0.08	0.47	46
	Bf2	1.85	0.17	1.29	70
	C	0.50	0.04	0.26	52
13. Cryic Dystric Brunisol	Ah	15.48	1.64	2.32	15
	Bm	0.18	0.04	0.11	61
	C1	0.19	0.04	0.00	0
	C2	0.10	0.05	0.02	20
16. Cryic Dystric Brunisol	Ah	1.01	0.22	0.31	31
	Bm	0.60	0.12	0.18	30
	C	0.21	0.06	0.00	0
	Cz	0.27	0.05	0.06	22

stands, but it is also extensive beneath open spruce-lichen woodlands, spruce-moss forests, alder/willow shrub and lichen-heath tundra. Soil disturbance, such as cryoturbation, will also tend to inhibit the development of podzol morphology. Finally, time is important: soils developed on a chronosequence of beach deposits near Sept Iles show better development of profile morphology and chemical translocations in the older soils (Moore, in press). To better identify the importance of individual soil forming factors, it is necessary to examine the process of podzolisation in more detail.

Most current hypotheses of podzolisation involve the translocation of iron and aluminum as a soluble metal-organic complex. One school recognises the importance of specific organic compounds as mobilising agents, such as polyphenols, carboxylic acids and polysaccharides (Bloomfield, 1963; Davies, 1971). The identification of such compounds is difficult and cannot be tested readily under field conditions, except through careful experimentation and analyses.

A second school recognises the importance of specific organic compounds, but believes that larger, arbitrary groupings of soil organic compounds are more meaningful. In particular, fulvic acid (the organic component of soils which is soluble in both alkali and acid, and which shows relatively low molecular weight and condensation) has been identified as a sesquioxide-mobilising group, since fulvic acid forms a large proportion of the organic matter of Bf horizons and fulvic-metal complexes have been studied in the laboratory (*eg.* Schnitzer and Desjardins, 1969). The humin and humic acid components are relatively unimportant in the podzolisation process.

Results of humic and fulvic acid extractions for 6 representative profiles and a boreal forest floor are shown in Table 3. The data show that frequently 50 to 70% of the organic matter in the B horizons of Podzols can be extracted as fulvic acid. This supports the hypothesis of fulvic acid as a major translocation mechanism in these soils. Where podzolisation is weak or non-existent, subsoil organic matter contents are low and fulvic acid rarely forms more than 30% of the organic matter.

The mobilising agents can be produced either by the leaching of vegetation (Malcolm and McCracken, 1968), from decomposing litter on the soil surface, or by the decomposition of organic matter in the surface organic horizons. In soils which show strong evidence of podzolisation (*eg.* 1, 2, 9 and the boreal forest floor; Appendix 1), the surface organic horizons contain only 5 to 15% of their organic carbon as fulvic acid. The nothern soils with little or no evidence of podzolisation (Soils 8, 13 and 16) have surface horizons in which 20 to 30% of the organic carbon exists as fulvic acid, a finding which is similar to that reported by Kononova (1975). The organic matter content of the soil surface must also be taken into account. In the boreal forest, surface organic matter contents range from 5 to 15 kg/m^2, 5 to 10 kg/m^2 in the open spruce-lichen woodlands and 3 to 10 kg/m^2 in the lichen-heath tundra.

Thus, it appears that there are approximately equal amounts of fulvic acid in boreal, woodland and tundra surface horizons, and that a relatively high proportion of organic matter in tundra sites exists as fulvic acid. Primary productivity and biomass decreases from the boreal, through woodland to tundra sites. Organic matter decomposition pathways are probably similar in these soils, with fungi playing an important role (Holding *et al.*, 1974).

The above data on podzolisation and fulvic acid distribution suggest that the decrease in podzolisation along the transect cannot be related to changes in one specific soil forming factor. The decrease from the boreal forest through the spruce-lichen woodlands and into the lichen-heath tundra is probably a function of:

(1) decreasing primary productivity, biomass and soil organic matter decomposition;

(2) decreasing water availability for soil leaching;

(3) decreasing length of active season for organic matter decomposition and organic matter mobilisation.

The possibility of sesquioxide translocation in inorganic forms in these soils cannot be ignored. In a Podzol beneath Douglas Fir in Washington, Windsor (1970) found that much of the iron and aluminum presently being translocated in the soil was in particulate and inorganic forms.

SUMMARY AND CONCLUSIONS

The environmental gradients within the transect have produced different types of soils. Profile morphology and chemical analyses have revealed leaching, gleying and podzolisation to be the most important pedogenic processes. The expression of these processes within the soil profile decreases northwards, and is a function of the lower biotic component on the soil surface, the lower rate of water percolation and the colder temperatures and shorter active season. Podzolisation involves the translocation of iron, aluminum and organic matter and appears to be related to fulvic acid. Many of the arctic and subarctic soils contain relatively large amounts of fulvic acid, yet show little evidence of podzolisation. Clay mineral formation and transformation are weak in these soils. The strongest transformations are observed in soils developed from materials of the Labrador Trough which possess well defined clay minerals. In other soils, vermiculite, chlorite and interstratified minerals dominate, with smectite being formed in the Ae horizon of Podzols.

It is worthwhile to examine the role of this type of study in the overall structure of pedology (Dijkerman, 1974). A methodological framework for pedogenic studies is presented in Figure 3. This study has dealt with soils within an essentially static approach, and has covered the methods indicated in the first four stages of the framework. From morphological descriptions of soil profiles and laboratory analyses we can identify the major pedogenic processes operating in these soils and hypothesize on the relationship between these processes and the environmental factors of climate, topography, vegetation, parent material and time.

The next stage in research should be the simulation of these pedogenic processes under laboratory, or, better, the

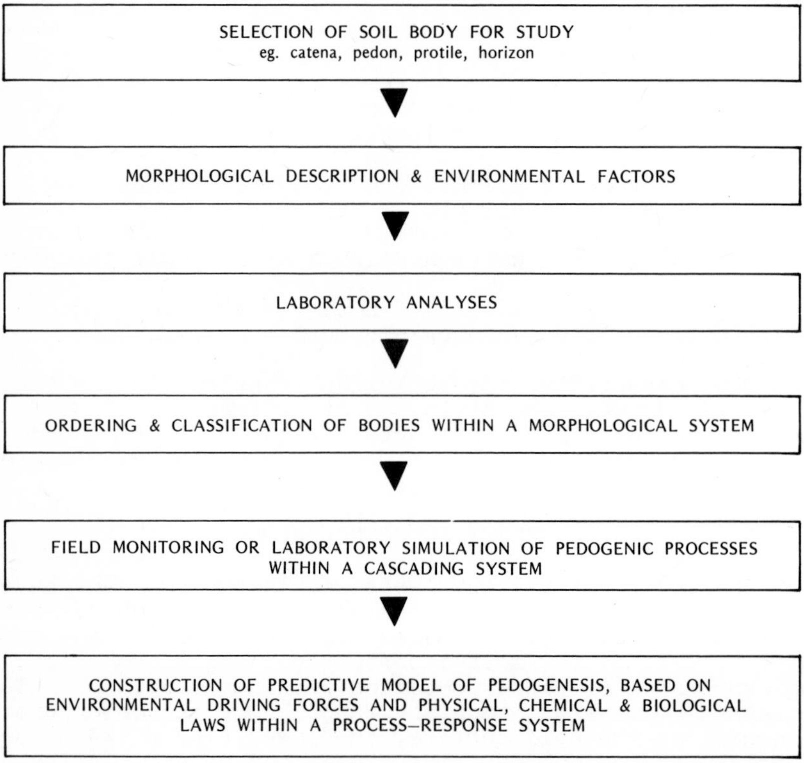

SELECTION OF SOIL BODY FOR STUDY
eg. catena, pedon, protile, horizon

▼

MORPHOLOGICAL DESCRIPTION & ENVIRONMENTAL FACTORS

▼

LABORATORY ANALYSES

▼

ORDERING & CLASSIFICATION OF BODIES WITHIN A MORPHOLOGICAL SYSTEM

▼

FIELD MONITORING OR LABORATORY SIMULATION OF PEDOGENIC PROCESSES WITHIN A CASCADING SYSTEM

▼

CONSTRUCTION OF PREDICTIVE MODEL OF PEDOGENESIS, BASED ON ENVIRONMENTAL DRIVING FORCES AND PHYSICAL, CHEMICAL & BIOLOGICAL LAWS WITHIN A PROCESS–RESPONSE SYSTEM

Figure 3 A methodological framework for pedogenic studies.

monitoring of them under field conditions. This allows us to test our hypotheses and to check whether the contemporary processes are identical to those recorded in and interpreted from the soil profile. From this monitoring, more sophisticated models of pedogenesis can be developed, using the environmental driving forces (water and air movement, radiation balance, organic matter accumulation and decomposition, *etc.*) and utilising physical, chemical and biological models. As a step in this direction, studies on organic matter accumulation and decomposition, nutrient cycling and pedogenic processes are being carried out on typical soil/vegetation units in the Schefferville area.

ACKNOWLEDGMENTS

This work has been supported by grants received from the National Research Council of Canada and Direction Generale de L'enseignement Superieur du Gouvernement du Quebec.

REFERENCES CITED

Andrews, J.T., and Miller, G.H., 1972, Chemical weathering of tills and surficial deposits in east Baffin Island, N.W.T., Canada: International Geography, v. 1, p. 5-7.

Bird, J.B., 1972, The Natural Landscapes of Canada: Toronto, Wiley, 191 p.

Bloomfield, C., 1963, Mobilisation phenomena in soils: Rothamsted, England, Report Rothamsted Exp. Sta. for 1963, p. 226-239.

Brydon, J.E., Kodama, H., and Ross, G.J., 1968, Mineralogy and weathering of the clays in Orthic Podzols and other podzolic soils in Canada. Trans. 9th Internatl. Congr. Soil Sci., v. 3, p. 41-51.

Bryson, R.A., Wendland, W.M., Ives, J.D., and Andrews, J.T., 1969, Radiocarbon isochrones on the disintegration of the Laurentide ice sheet: Arctic and Alpine Research, v. 1, p. 1-14.

Busenberg, E., and Clemency, C.V., 1976, The dissolution kinetics of feldspars at 25°C and 1 atm CO_2 partial pressure: Geochim, Cosmochim. Acta, v. 40, p. 41-49.

Canada Department of Transport, 1962, The Climate of Canada: Ottawa, Queen's Printer, 74 p.

Canada Department Energy, Mines and Resources, 1974, The National Atlas of Canada: Toronto, Macmillan Co. of Canada, 253 p.

Canada Soil Survey Committee, 1974, The System of Soil Classification for Canada: Canada Dept. Agriculture, 249 p.

Davies, R.I., 1971, Relation of polyphenols to decomposition of organic matter and to pedogenetic processes: Soil Sci., v. 111, p. 80-85.

Douglas, R.J.W., ed., 1970, Geology and Economic Minerals of Canada: Ottawa, Ont., Department Energy, Mines and Resources, 838 p.

Dijkerman, J.C., 1974, Pedology as a science: the role of data, models and theories in the study of natural soil systems: Geoderma, v. 11, p. 73-93.

Garrels, R.M., and Christ, C.L., 1965, Solutions, Minerals and Equilibria: N.Y., Harper and Row, 450 p.

Gersper, P.L., and Challinor, J.L., 1975, Vehicle perturbation on a tundra soil-plant system: I Effects on morphological and physical environmental properties of the soils: Soil Sci. Soc. America Proc., v. 39, p. 737-744.

Gjems, O., 1960, Some notes on clay minerals in podzol profiles in Fennoscandia: Clay Min. Bull., v. 4, p. 208-211.

Hare, F.K., 1959, A Photo-Reconnaissance Survey of Labrador - Ungava: Ottawa, Geographical Branch, Department of Mines and Technical Surveys, Mem. 6.

Hare, F.K., and Hay, J.E., 1971, Anomalies in the large-scale annual water balance over northern North America: Can. Geog., v. 15, p. 79-94.

Holding, A.J., Heal, O.W., Maclean, S.F., and Flanagan, 1974, Soil Organisms and Decomposition in Tundra: Stockholm, Sweden, Tundra Biome Steering Committee, 398 p.

Jackson, M.L., 1964, Chemical Composition of Soils, *in* Bear, F.E., ed., Chemistry of the Soil: New York, Reinhold, p. 71-141.

Kittrick, J.A., 1969, Soil minerals in the Al_2O_3 - SiO_2 - H_2O system and a theory of their formation: Clays Clay Miner., v. 17, p. 157-167.

Kononova, M.M., 1966, Soil Organic Matter: Oxford, England, Pergamon Press, 544 p.

______________, 1975, Humus of virgin and cultivated soils, *in* Gieseking, J.E., ed., Soil Components, v. 1: New York, Springer-Verlag, p. 475-526.

Malcolm, R.L., and McCracken, R.J., 1968, Canopy drip: a source of mobile soil organic matter for mobilization of iron and aluminum: Soil Sci. Soc. America Proc., v. 32, p. 834-838.

McKeague, J.A., 1965, A laboratory study of gleying: Can. Jour. Soil Sci., v. 45, p. 199-206.

______________, 1967, An evaluation of 0.1M pyrophosphate and pyrophosphate-dithionite in comparison with oxalate as extractants of the accumulation products in podzols and some other soils: Can. Jour. Soil Sci., v. 47, p. 95-99.

McKeague, J.A., and Cann, D.B., 1969, Chemical and physical properties of some soils derived from reddish brown materials in the Atlantic Provinces: Can. Jour. Soil Sci., v. 49, p. 65-78.

Mehra, O.P, and Jackson, M.L., 1960, Iron Oxide removal from soils and clays by a dithionite-citrate system buffered with sodium bicarbonate, *in* Swineford, A., ed., Nat. Conf. on Clays and Clay Minerals 1958: London, Pergamon Press, p. 317-327.

Moore, T.R., 1974, Pedogenesis in a subarctic environment: Cambrian Lake, Quebec: Arctic and Alpine Research, v. 6, p. 281-291.

__________, in press, Sesquioxide-cemented soil horizons in northern Quebec: their distribution, properties and genesis: Can. Jour. Soil Sci.

Nicholson, H.M., and Moore, T.R., in press, Pedogenesis in a subarctic iron-rich environment: Schefferville, Quebec: Can. Jour. Soil Sci.

Payette, S., 1968, Etude pedologique on 60e parallele nord, Ungava, Quebec: Centre d'Etudes Nordiques, Universite Laval, Quebec, Collection Nordicana No. 23, 90 p.

__________, 1973, Contribution a la pedologie de la zone hemi-arctique: region de Poste-de-la Baleine, Nouveau-Quebec: Naturaliste Can., v. 100, p. 123-163.

Payette, S., and Morisset, P., 1974, The Soils of the Sleeper Islands, Hudson Bay, N.W.T., Canada: Soil Sci., v. 117, p. 352-368.

Peech, M., 1965, Hydrogen-ion activity, *in* Black, C.A., ed., Methods of Soil Analysis: Madison, Wisc., Amer. Soc. Agron., Agronomy 9, p. 914-925.

Pettapiece, W.W., 1975, Soils of the subarctic in the lower Mackenzie Basin: Arctic, v. 28, p. 35-53.

Pettapiece, W.W., and Zoltai, S.C., 1974, Soil environments in the Western Canadian Subarctic, *in* Mahaney, W.C., ed., Quaternary Environments: Proceedings of a Symposium, Geographical Monographs: Toronto, York Univ. Ser. Geog., p. 279-292.

Retzer, J.L., 1965, Present soil-forming factors and processes in arctic and alpine regions: Soil Sci., v. 99, p. 38-44.

Richard, P., 1975, Histoire postglaciaire de la vegetation dans la partie centrale du Parc des Laurentides, Quebec: Naturaliste can., v. 102, p. 669-681.

Rieger, S., 1966, Dark well drained soils of tundra regions in Western Alaska: Jour. Soil Science, v. 17, p. 264-273.

Rodin, L.E., and Bazilevec, N.I., 1966, The biological productivity of the main vegetation types in the northern hemisphere: For. Abstr., v. 27, p. 369-372.

Rouse, W.R., and Kershaw, K.A., 1971, The effects of burning on the heat and water regimes of lichen-dominated subarctic surfaces: Arctic and Alpine Research, v. 3, p. 291-304.

Rousseau, J., 1968, The vegetation of the Quebec-Labrador peninsula between 55° and 60°N: Naturaliste Can., v. 95, p. 469-563.

Schnitzer, M., and Desjardins, J.G., 1969, Chemical characteristics of a natural soil leachate from a Humic Podzol: Can. Jour. Soil Sci., v. 49, p. 151-158.

Sorenson, C.J., Knox, J.C., Larson, J.A., and Bryson, R.A., 1971, Palesols and the forest border in Keewatin, N.W.T.: Quaternary Research, v. 1, 468-473.

Tedrow, J.C.F., 1968, Polar desert soils: Soil Sci. Soc. America Proc., v. 30, p. 381-387.

Tedrow, J.C.F., and Cantlon, J.C., 1958, Concepts of soil formation and classification in Arctic regions: Arctic, v. 11, p. 166-179.

Tinsley, J., 1950, The determination of organic carbon in soils by dichromate mixtures: Trans. 4th Internatl. Congr. Soil Sci., v. 1, p. 161-164.

Viereck, L.A., 1973, Wildfire in the taiga of Alaska: Quaternary Research, v. 3, p. 465-495.

Weaver, R.M., Jackson, M.L., and Syers, J.K., 1971, Magnesium and silicon activities in matrix solutions of montmorillonite-containing soils in relation to clay mineral stability: Soil Sci. Soc. America Proc., v. 35, p. 823-830.

Whittig, L.D., 1965, X-ray diffraction techniques for mineral identification and mineralogical composition, *in* Black, C.A., ed., Methods of Soil Analysis: Madison, Wisc., Amer. Soc. Agron., Agronomy 9, p. 671-698.

Windsor, G.J., 1970, Dynamics of P, Si, Fe and Al movement in gravitational water in a Douglas Fir ecosystem: Diss. Abstr., v. 31, p. 471-472.

Appendix 1. Major Morphological Properties of the 17 Selected Soil Profiles.

Soil, Location, Parent Material and Vegetation	Horizon	Depth (cm)	Color (moist)	Features
1. Ortstein Humo-Ferric Podzol at Sept Iles on fluvioglacial sand beneath black spruce-balsam fir forest	LFH	10-0	5YR2/1	forest floor, moist;
	Ae	0-5	10YR6/1	coarse sand, single grain, loose, moist;
	Bf	5-10	5YR4/6-7.5YR5/6	sand, single grain to weak crumb, loose to friable, moist;
	Bfc	10-30	2.5YR2/2-5YR 3/3-3/4	sand, moderate platy, firm to hard, moist;
	Bmc	30-45	5YR4/4-4/6	sand, moderate platy, firm, moist;
	C1	45-65	10YR5/8-6/8	fine sand, single grain, loose, moist;
	C2	65-100	10YR6/8	sand, single grain, loose, moist;
2. Ortstein Humo-Ferric Podzol at Ross Bay Junction on fluvio-glacial sand and gravel beneath open spruce-lichen woodland	H	5-0	7.5YR3/2-2/2	organic mat beneath lichen;
	Ae	0-4	10YR6/1-N6/0	fine sand, single grain, loose friable, moist;
	Bf	4-20	7.5YR3/6-5YR3/4	coarse sand, single grain to weak crumb, friable, moist;
	Bfc	20-55	10YR5/4-5YR3/4	coarse sand, moderate platy in pale brown but hard in cemented bands, moist;
	C	55-100	10YR5/4-6/4	coarse sand, single grain, loose, moist;
3. Gleyed Humo-Ferric Podzol at Ross Bay Junction on fluvio-glacial sand beneath mosses and lichens in seasonally waterlogged depression.	A	3-0	10YR2/1	mixed humus and fine sand moist;
	Aeg	0-10	10YR6/1-6/2	fine sand, single grain, loose, moist;
	Bfhc	10-15	5YR2/2-3/2	fine sand, moderate platy, firm, moist to wet;
	Bf	15-25	5YR3/2-10YR4/3	fine sand, mottled, weak platy, friable to firm, moist to wet;
	Bg	25-45	10YR4/3-5YR3/2	fine sand, mottled, single grain to weak crumb, loose to friable, wet;

Soil	Horizon	Depth	Colour	Description
	Cg	45-80	10Y4/2-5YR3/2	fine sand, weakly mottled, single grain to crumb, loose to friable water table at 60 cm.
4. Degraded Dystric Brunisol at Schefferville on iron-rich till beneath old spruce-lichen woodland burn	LH	4-0	5YR2/1	organic mat, moist;
	Ae	0-5	5YR5/2	gravelly sandy loam, fine granular, friable, moist;
	Bm	5-30	2.5YR4/6	gravelly silty loam, fine granular, friable, moist;
	C	30-80	2.5YR4/6	gravelly silty loam, fine blocky, firm, moist;
5. Orthic Humo-Ferric Podzol at Schefferville on iron-rich deposits beneath spruce-fir-moss closed canopy forest	L	15-7	7.5YR4/4	
	F	7-1	5YR3/3	forest floor;
	H	1-0	5YR3/1	
	Ae	0-2	5YR5/3	silty loam, single grain, friable, moist;
	Bfh	2-4	5YR2/2	silty loam, single grain, friable, moist;
	Bf	4-15	5YR4/6	silty loam, single grain, friable, moist;
	Bm	15-35	5YR4/6	silty loam, single grain, friable, moist;
	C	35-55	2.5YR4/4	silty loam, single grain, friable, moist;
6. Fera Gleysol at Schefferville on iron-rich deposits beneath mosses and sedges in seasonally waterlogged depression	Oh1	35-18	10YR3/2	decomposing organic matter, wet;
	Oh2	18-0	10YR2/2	
	Aeg	0-5	10YR4/2	gravelly loam, fine blocky, firm, wet;
	Bgf	5-15	5YR4/6	gravelly loam, fine blocky, firm, waterlogged;

7. Degraded Dystric Brunisol at Cambrian Lake on gneiss beneath lichen-heath tundra	A	5-0	10YR3/4	decomposing organic mat, moist;
	Aej	0-1	10YR6/2	sand, single grain, loose, moist;
	Bfj	1-2	10YR4/4-5YR3/4	sand, single grain, loose, moist;
	Bm	2-30	10YR5/4	gravelly sand, single grain, loose, moist;
	C	30-45	10YR5/4	gravelly sand, single grain, loose, moist;
8. Degraded Dystric Brunisol at Cambrian Lake on fluvioglacial sand beneath open spruce-lichen woodland	O	3-0	10YR2/1-2/2	decomposing lichen mat, moist;
	Ae	0-3	7.5YR6/2	sand, single grain, loose, moist;
	Bm1	3-10	7.5YR4/4	sand, single grain to weak granular, friable, moist;
	Bm2	10-40	7.5YR5/6	sand, single grain, friable, moist;
	C	40-50	7.5YR5/6-2.5YR 3/2	sand, single grain with occasional cemented band, friable, moist;
9. Orthic Humo-Ferric Podzol at Cambrian Lake on quartzite beneath lichen-heath tundra	O	2-0	10YR2/1	decomposing lichen mat, moist;
	Ah	0-10	10YR2/1-2/2	humose sand, weak granular, friable, moist;
	Ae	10-18	7.5YR6/2	sand, single grain, loose moist;
	Bf1	18-25	7.5YR3/2	sand, single grain to weak granular, loose to friable, moist;
	Bf2	25-35	5YR3/2-3/3	sand, weak granular, friable, moist;
	C	35-45	10YR4/3	sand with quartzite gravel, single grain, loose, moist;
10. Cryic Humic Gleysol at Cambrian Lake on shale beneath spruce-moss forest	O	10-0	5YR2/2	decomposing moss mat, wet;
	Ah	0-5	10YR2/1	humose sand, weak granular, loose to friable, saturated;
	Bmg	5-10	10YR3/4-4/4	sandy loam, weak platy, friable to firm; saturated;
	Cz	10-30	10YR5/4	sand frozen beneath 20 cm.

11. Degraded Dystric Brunisol at Payne Bay on fluvioglacial deposits beneath grass-lichen-heath tundra	O	1-0	5YR2/1	decaying plant material, moist;
	Ah	0-7	10YR2/2	humose sand and gravel, weak granular, friable, moist;
	Aej	7-8	10YR6/1	sand, single grain, loose, moist;
	Bm	8-30	10YR4/3	sand and gravel, single grain to weak granular, loose to friable, moist;
	C	30-40	10YR5/4	sand and gravel, single grain, loose moist;
12. Orthic Dystric Brunisol Brunisol at Payne Bay on fluvioglacial sand beneath grass-willow-heath tundra	Ah	0-2	10YR3/2	humose sand, weak granular, friable, moist;
	Bm	2-27	10YR4/3-3/3	sandy loam, weak granular, loose to friable, moist;
	C	27-40	10YR5/4	sand, single grain, loose, moist;
13. Cryic Dystric Brunisol at Frobisher Bay on glacial till beneath grass-lichen-heath tundra, adjacent to frost scar	Ah	0-3	10YR2/1	humose sand, granular, loose to friable, few bleached sand grains at base, moist;
	Bm	3-20	7.5YR3/2	coarse sand, weak granular, loose friable, moist;
	Ahb	20-25	10YR3/1	buried humose sand, weak platy, loose to friable, moist;
	C1	25-40	7.5YR4/4	coarse sand, weak granular, loose to friable, moist;
	C2	40-50	7.5YR4/4	coarse sand, weak granular, loose to friable, wet, becoming permafrost below 50 cm.
14. Orthic Dystric Brunisol at Frobisher Bay on fluvioglacial sand and gravel beneath grass-lichen-heath tundra	Ah	0-5	10YR2/1	humose sand, weak crumb, loose to friable, moist;
	Bm	5-20	7.5YR4/4	sand and gravel, single grain, loose, moist;
	C	20-120	7.5YR4/4-4/6	sand and gravel, single grain, loose, moist;

15. Rego Gleysol at Frobisher Bay on alluvial sand beneath grasses.and mosses	Ag	0-3	10YR4/1	thick mat of roots and decomposing organic matter, mottled, wet;
	Ahg	3-10	10YR2/2-2.5YR 4/6	decomposing organic matter with strong mottling, fibrous, wet;
	Afg	10-15	10YR3/2	decomposing root mat with strong brown mottling, fibrous, wet;
	Cg	15-40	2.5Y5/4-10YR5/4	coarse sand, mottled, single grain, loose, saturated.
16. Cryic Dystric Brunisol at Clyde River on fluvioglacial sand beneath lichen-heath tundra, adjacent to frost scars	O	3-0	10YR2/1	decomposing organic mat, moist;
	Ah	0-4	10YR3/2	humose sand, weak granular, loose to friable, moist;
	Bm	4-13	10YR5/4-7.5YR 3/2	sand and gravel, weak granular, friable, moist to wet;
	C	13-19	10YR4/1	loamy sand, single grain, loose, partially frozen, wet;
	Cz	19-30	10YR4/2	sand, frozen, wet;
17. Cryic Regosol at Clyde River on beach sand beneath sparse lichens and grasses	Bmj	0-7	10YR6/6	coarse sand and gravel, single grain, moist;
	C	7-20	10YR4/2	sand, single grain, moist; permafrost at 50 cm.

Appendix 2. Analytical Data of the Selected Soil Profiles.

	Profile	Horizon	pH H_2O	pH $CaCl_2$	org.C %	Cp %	Fe_d %	Fe_o %	Fe_p %	Al_o %	Al_p %	Mn_d ppm
1.	Ortstein Humo-Ferric Podzol at Sept Iles	LFH	3.7	2.6	52.78	8.04	0.08	0.04	0.38	0.14	0	3
		Ae	4.3	3.4	0.27	0.68	0.08	0.02	0.02	0.02	0.01	10
		Bf	4.6	3.8	1.33	0.94	0.55	0.23	0.43	0.27	0.26	10
		Bfe	5.0	3.9	1.01	0.80	0.46	1.00	0.31	0.37	0.26	9
		Bmc	5.5	4.4	0.35	0.32	0.33	0.17	0.18	0.14	0.12	30
		C1	5.8	4.4	0.22	0.14	0.22	0.05	0.13	0.14	0.08	18
		C2	6.2	4.6	0.18	0.12	0.17	0.03	0.07	0.08	0.07	12
2.	Ortstein Humo-Ferric Podzol at Ross Bay Junction	H	3.7	3.0	25.71	2.81	-	0.13	0.11	0.23	-	-
		Ae	4.2	3.4	1.32	0.37	0.41	0.01	0.05	0.11	0.09	11
		Bf	4.8	4.1	1.86	0.97	4.75	2.55	0.52	0.77	0.50	71
		Bfc	6.1	4.7	0.53	0.24	1.15	0.35	0.07	0.28	0.19	93
		C	5.5	4.5	0.39	0.29	1.13	0.09	0.12	0.06	0.05	105
3.	Gleyed Humo-Ferric Podzol at Ross Bay Junction	A	4.0	3.3	2.99	1.21	0.73	0.23	0.34	0.10	0.13	9
		Aeg	4.3	3.7	0.48	0.11	0.15	0.05	0.06	0.04	0.06	9
		Bfhc	4.3	3.0	2.78	2.00	5.40	4.49	0.80	0.46	0.48	403
		Bf	4.8	3.1	1.30	1.19	2.50	1.55	0.41	0.33	0.32	3
		Bg	5.0	4.2	0.84	0.73	1.44	0.61	0.28	0.23	0.23	115
		Cg	5.4	4.3	0.83	0.44	1.39	0.42	0.26	0.22	0.25	45
4.	Degraded Dystric Brunisol at Schefferville	LH	-	-	-	-	-	-	-	-	-	-
		Aej	4.6	3.3	1.03	-	2.18	0.32	-	0.06	-	160
		Bm	4.8	3.6	1.48	-	5.31	0.78	-	0.22	-	780
		C	5.1	4.0	1.11	-	4.34	0.33	-	0.21	-	1590
	Orthic Humo-Ferric Podzol at Schefferville	L	3.6	2.6	47.61	-	-	-	-	-	-	-
		F	3.6	2.6	-	-	-	-	-	-	-	-
		H	3.3	2.6	41.13	-	-	-	-	-	-	-
		Ae	3.6	3.2	2.95	-	3.99	0.58	-	0.23	-	160

5.	Orthic Humo-Ferric Podzol at Schefferville	L	3.6	2.6	47.61	-	-	-	-	-	-	-
		F	3.6	2.6	-	-	-	-	-	-	-	-
		H	3.3	2.6	41.13	-	-	-	-	-	-	-
		Ae	3.6	3.2	2.95	-	3.99	0.58	-	0.23	-	160
		Bfh	3.8	3.1	4.04	-	5.78	1.57	-	0.30	-	380
		Bf	4.1	3.7	1.67	-	5.91	1.05	-	0.37	-	870
		Bm	4.5	3.8	0.53	-	4.62	0.61	-	0.39	-	1340
		C	4.5	3.8	0.17	-	4.57	0.40	-	0.20	-	1350
6.	Fera Gleysol at Schefferville	Oh1	4.3	3.6	52.20	-	-	-	-	-	-	-
		Oh2	4.2	3.4	51.85	-	-	-	-	-	-	-
		Aeg	4.4	3.5	2.53	-	2.29	0.39	-	0.21	-	30
		Bgf	4.6	3.7	0.67	-	4.62	0.99	-	0.13	-	130
		C	4.7	3.9	1.03	-	3.33	0.50	-	0.13	-	130
7.	Degraded Dystric Brunisol at Cambrian Lake	O	5.0	3.9	26.40	0.58	0.61	0.32	0.09	0.13	0.29	-
		Aej	4.4	3.5	1.28	0.22	0.26	0.06	0.05	0.12	0.12	14
		Bfj	5.9	4.8	1.73	0.77	4.44	0.72	0.37	0.35	0.62	30
		Bm	5.1	4.2	0.26	0.16	1.01	0.46	0.14	0.18	0.30	35
		C	5.0	4.1	0.27	0.13	1.01	0.50	0.17	0.13	0.22	25
8.	Degraded Dystric Brunisol at Cambrian Lake	O	4.2	2.7	37.30	0.96	0.32	0.05	0.09	0.05	0.25	29
		Ae	4.5	4.0	2.17	0.36	0.42	0.02	0.06	0.06	0.19	10
		Bm1	4.8	4.2	0.14	0.28	1.44	0.14	0.12	0.14	0.27	30
		Bm2	5.2	4.3	0.30	0.26	0.91	0.53	0.12	0.20	0.26	169
		Bmg	5.6	4.4	0.29	0.25	1.25	0.69	0.13	0.38	0.54	347
		C	5.5	4.4	0.04	0.05	0.61	0.30	0.03	0.16	0.18	75
9.	Orthic Humo-Ferric Podzol at Cambrian Lake	O	4.0	2.9	26.23	2.57	0.54	0.08	0.15	0.11	0.27	88
		Ah	4.5	2.8	5.59	0.42	0.42	0.04	0.08	0.11	0.27	9
		Ae	4.5	3.2	0.43	0.07	0.37	0.03	0.02	0.05	0.10	15
		Bf1	4.6	3.8	1.03	0.48	1.36	0.95	0.19	0.49	0.54	29
		Bf2	4.8	3.9	1.85	1.48	1.31	1.22	0.39	0.76	1.32	39
		C	5.0	4.1	0.50	0.26	1.07	0.29	0.07	0.30	0.39	40

10. Cryic Humic Gleysol at Cambrian Lake	O	4.2	3.4	36.01	-	-	-	-	-	-	-
	Ah	5.2	4.6	25.43	-	1.65	1.08	-	0.29	-	313
	Bmg	4.9	4.1	0.72	0.41	1.35	1.45	0.37	0.28	0.23	14
	Cz	6.3	5.6	0.32	0.10	1.00	0.38	0:05	0.09	0.04	138
11. Degraded Dystric Brunisol at Payne Bay	O	4.9	4.6	26.41	2.21	0.14	0.02	0.10	0.27	0.43	158
	Ah	-	-	-	-	-	-	-	-	-	-
	Bm	5.9	4.9	0.61	0.07	0.10	0.05	0.06	0.07	0.08	37
	C	6.2	5.0	0.12	0.01	0.10	0.08	0.03	0.02	0.06	40
12. Orthic Dystric Brunisol at Payne Bay	Ah	5.0	4.7	11.31	1.78	0.44	0.08	0.27	0.50	1.09	1175
	Bm	5.5	3.9	1.12	0.26	0.40	0.02	0.13	0.04	0.29	50
	C	5.8	5.0	0.41	0.08	0.17	0.05	0.07	0.10	0.14	26
13. Cryic Dystric Brunisol at Frobisher Bay	Ah	5.7	4.9	15.56	1.13	0.16	-	0.07	-	0.07	370
	Bm	5.9	4.8	0.21	0.22	0.70	-	0.07	-	0.04	180
	C1	5.8	5.0	0.20	0.12	0.17	-	0.06	-	0.04	90
	C2	6.1	5.0	0.15	0.06	0.10	-	0.04	-	0.04	110
14. Orthic Dystric Brunisol at Frobisher Bay	Ah(0-5cm)	4.6	4.0	2.91	1.08	0.62	0.33	0.71	-	0.20	130
	C(5-15 cm)	5.1	4.3	0.30	0.14	0.48	0.28	0.12	-	0.10	200
	C(20-25 cm)	5.5	4.4	0.16	0.06	0.35	0.20	0.26	-	0.10	180
	C(30-35 cm)	5.4	4.4	0.11	0.05	0.15	0.19	0.04	-	0.10	180
	C(50-55 cm)	5.7	4.7	0.10	0.07	0.33	0.25	0.04	-	0.12	150
	C(70-75 cm)	5.6	4.8	0.10	0.07	0.22	0.20	0.03	-	0.09	130
	C(95-100 cm)	5.7	4.8	0.10	0.12	0.24	0.13	0.05	-	0.07	200
15. Rego Gleysol at Frobisher Bay	Ag	5.9	5.0	1.61	2.29	0.85	-	0.44	-	0.18	300
	Ahg	4.8	4.5	8.12	1.67	0.13	-	0.18	-	0.89	10
	Afg	5.0	4.5	10.45	0.07	0.76	-	0.06	-	0.15	350
	Cg	5.0	4.7	0.21	0.04	0.16	-	0.06	-	0.10	15
16. Cryic Dystric Brunisol at Clyde River	Ah	4.7	4.0	1.08	0.46	0.12	0.05	0.08	0.09	0.10	30
	Bm	5.7	4.3	0.61	0.29	0.12	0.04	0.07	0.08	0.07	32
	C	5.6	4.4	0.25	0.05	0.15	0.03	0.03	0.07	0.04	28
	Cz	5.6	4.4	0.31	0.08	0.17	0.06	0.04	0.07	0.02	26

17. Cryic Regosol at Clyde River	Bmj	6.8	6.2	0.40	0.21	0.10	0.07	0.07	0.12	0.08	39
	C	7.2	6.5	0.15	0.06	0.08	0.04	0.06	0.06	0.04	25

Development of Polar Desert Soils

J. C. F. Tedrow

Journal Series Contribution
N. J. Agric. Exp. Stn.

ABSTRACT

The polar desert soil zone includes the Queen Elizabeth Islands, northern Greenland, northern Svalbard, Severnaya Zemlya, Franz Josef Land, and some other nearby islands. The subpolar desert soil zone of the Western Hemisphere is comprised of the group of islands of the Canadian Arctic Archipelago which extend from Banks Island in the west to Baffin Island in the east, plus the mid latitudes of the ice-free sectors of Greenland. In the Eastern Hemisphere the subpolar desert zone includes parts of Svalbard, Novaya Zemlya, New Siberian Islands, and Wrangel Island, plus some northern fringes of the mainland, particularly northern Yamal and northern Taymyr. Barren features are dominant in the polar desert soil zone but only part of the landscape is mantled with Polar desert soil. The modal Polar desert soil consists of a deep, well-drained soil with an A-B-C sequence of horizons, a deep permafrost table, a sparse scattering of vascular plants, and a desert pavement. In sectors of frost-shattered rock and recent landforms, the deep well-drained sites show little, if any, genetic profile development. The Råmark soils of the Scandinavian mountains are primarily erosional surfaces, and likewise show little genetic development, and accordingly should not be considered as being equivalent to Polar desert soil. On well-drained limestone sites soil development is noticeably restricted and profiles do not develop genetic horizonation as in the case with silicate rocks.

INTRODUCTION

Since introduction of the term Polar desert soil to North American pedologists (Tedrow, *et al.*, 1958; Tedrow and Brown, 1963; Tedrow, 1966) there have been additional investigations resulting in a more complete understanding of the term as well as an evaluation of soil processes and soil properties on well-drained landscapes within the high arctic. When the term Polar desert soil was first used by Soviet investigators it was a general term without a high degree of specificity (Gorodkov, 1939). The term was somewhat analogous to prairie, steppe, or desert. Korotkevich (1967) delineated polar desert zones and subzones for the northern polar lands as well as Antarctica. The Queen Elizabeth Islands, northern Greenland, northern Svalbard, Severnaya Zemlya, Franz Josef Land, and a few other islands make up the polar desert zone of the Northern Hemisphere (Figure 1). Between the polar desert zone and the tundra zone is a transition zone with a few distinct pedologic properties of its own. This circumpolar zone, referred to as the subpolar desert zone, extends some 300 to 400 miles in a north-south direction and blends conditions of the predominantly mineral gley soils of the tundra zone at the south with the desert soils of the north. The subpolar desert zone approximates the term mid arctic as used by some North American geographers and naturalists. The subpolar desert zone would be included as a part of the arctic tundra zone by some Soviet investigators. This zonation is basically similar to those proposed by Korotkevich (1967), Alexandrova (1970), and Tedrow (1973).

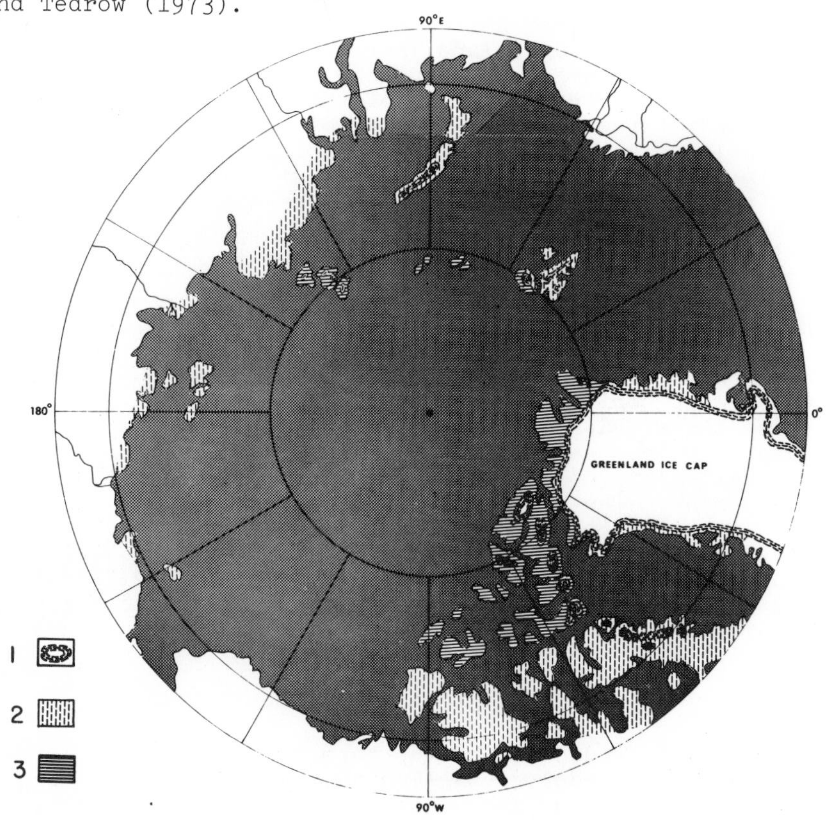

Figure 1 Zonation of polar deserts of the Northern Hemisphere. (1) Existing glaciers. (2) Subpolar desert zone. (3) Polar desert zone.

THE POLAR DESERT SOIL PROCESS

Polar desert soils are deep and well drained and usually have a desert pavement (Figure 2). Only a sparse colonization of vascular plants is usually present. The amount of organic matter in the soil is usually small-especially in coarse-textured material. Primary production of organic matter in the polar desert zone is at a very slow rate. Alexandrova (1970) reports an average of 40 kg/ha of organic matter is produced annually in the polar desert zone compared to 1240 kg/ha in the arctic tundra. Apparently lichens and algae are the main organic matter contributors. Permafrost is deep in Polar desert soils, and the entire solum as well as part of the C horizon exists within the active layer. The period of significant vital activity is generally limited to about two months each year--from mid June until mid or late August. During this two-month period the landscape receives only about 2.5 to 5 cm (1 to 2 in) of precipitation. Ambient temperatures for July approximate 7°C (45°F) in the southern fringes of the polar desert zone and about 1.5°C (35°F) to 2.8°C (37°F) in the extreme northern ice-free sectors.

Figure 2 Polar desert soil formed on glacial drift on Bathurst Island, Canada.

Whereas there is a range of acidity levels in Polar desert soils, usually they are slightly alkaline. With coarse-textured silicates, however, the pH values are as low as 4.4. Conductivity values show considerable range, and salt efflorescence, mainly in the form of thenardite (Na_2SO_4), takes place during the dry periods, especially during late summer. Pedogenic carbonate in the form of travertine may form throughout the solum, especially in the B horizon.

ROCKY AREAS WITHOUT SOIL PROFILE DEVELOPMENT

While most of the high arctic has a scant covering of vascular plants in the well-drained to xeric sites, not all of such sites can be considered mantled with Polar desert soil. Some of the latter conditions are present in areas of exposed bedrock, frost-shattered rock, blockfields, scree deposits, felsenmeer, and others. Figure 3 shows an example of a polar desert condition on Bathurst Island but the soil would not be considered polar desert because the materials consist largely of shattered rock. The Soil Map of the World (1972) refers to such conditions as Gleyic regosols and Lithosols. This depiction provides for conditions where there is no development of genetic horizons but is not realistic for sites where there is complete soil development as exemplified on the Beaufort Plain of the western Queen Elizabeth Islands.

Figure 3 Polar desert on Bathurst Island. In sites such as these there is no genetic soil horizonation.

POORLY DRAINED SOILS

In discussions of polar deserts the dry, frigid conditions are generally emphasized but in addition to the well-drained sites the high arctic is also punctuated with wet mineral and Bog soils (Figure 4). Such poorly drained conditions are present in sectors having negative relief such as basins, drainageways, silty clay flats, and sites where late snow persists. In the low-lying sectors Tundra and other wet mineral soils, as well as Bog soils are usually present (Tedrow, 1970).

Figure 4 Wet meadow of Tundra soil in the polar desert soil zone of northern Greenland.

SOIL DEVELOPMENT ON LIMESTONE

In some sectors of the globe the influence of carbonate rock results in somewhat different pathways of soil development as compared to those of acid, silicate rocks. This difference is generally ascribed to different acidity levels within the soils. In the podzolic zone, for example, on carbonate rock a Rendzina → Brown forest → Gray brown podzolic soil sequence is generally present, but with silicate rocks it approximates a Ranker → Brown podzolic → Podzol sequence. More detailed sequential development of soils on various other parent materials has been outlined by Kubiena (1970). The main reason why there are different stages of soil evolution on the different substrates apparently is related to the supply of bases and the resultant acidity levels which in turn affect the biogeochemical processes within the soil.

Soil formation on high-purity limestone of the high arctic presents a special situation. Limestone rocks tend to have a fresh, lichen-free surface that may be pitted, grooved, but otherwise unweathered, except for the small solution cavities where there are residues collecting from impurities in the limestone (Figure 5). On the other hand, silicate rocks are partially, and in some places completely, covered with lichens. Apparently solution of the limestone is at a sufficiently rapid rate to inhibit lichen colonization but silicate rock surfaces are far more stable insofar as chemical weathering is concerned. Beschel (1970) stated that rock deserts made up of limestone in the Queen Elizabeth Islands are more barren than the nonlimestone ones because surfaces of the limestone boulders weather faster than the epipetric lichens can grow there. Walton (1972) extended Beschel's views to the soil and recognized that limestone soils of the high arctic displayed a set of unique characteristics. Apparently the solution effect of the dissolving limestone carries over into the stages of terrestric soil formation. In situations in which the soil forms from the weathering of high-carbonate limestone there is silt and clay produced which results in a matrix conducive to a higher moisture content which in turn induces more intense frost action. Many of these high-purity limestone areas have developed relief features whose surface morphologies resemble those of gilgai and have variously sorted patterns present (Figure 6). The soil profile itself generally consists of a dull gray, silty clay and there is virtually no evidence of genetic horizonation except for the accumulation of stones at the surface, and possibly salt efflorescence. On the other hand, Polar desert soil formed on silicate matrices will display color and other pedogenic changes with depth (Figure 2).

A major point here is that most Polar desert soils are alkaline or near alkaline in reaction even though the parent materials may have originally been acid. We may encounter soils on carbonate-bearing sandstone and on feldspathic or even quartzose sandstone with many similar genetic features. On the other hand, at <u>high</u>-carbonate sites--ones in which there is a high degree of solution effect of the entire matrix--the developed soil features are considerably different from those on the low-carbonate or carbonate-free sites.

Figure 7 shows sectors within the high arctic and mid arctic in which native carbonate rock may be present. The glacial drift cover in the higher elevations may frequently be shallow or even absent. In the Western Hemisphere limestone rock is present throughout much of the Canadian Arctic Archipelago. In the Eastern Hemisphere the picture is more uncertain. The generalized areas in which some carbonate-bearing rock may be present include Novaya Zemlya, northern Taymyr, Severnaya Zemlya, New Siberian Islands, and Wrangel

Figure 5 Lichen-free boulder of limestone among partially lichen-covered silicate rocks in northern Greenland.

Island. The northern Siberian mainland, particularly from the Kanin Peninsula eastward to the Taymyr Peninsula and the mainland between the lower Lena and Kolyma Rivers are covered with rather thick Quaternary deposits of glacio-fluvial-marine origin. Few of these unconsolidated deposits are lime-bearing.

THE TIME FACTOR IN DEVELOPMENT OF POLAR DESERT SOIL

Polar desert soils do not display well-developed features on the more recent deposits as compared to the older ones. Recent alluvium and more recent glacial deposits tend to have dull, grayish colors throughout the "solum" (Figure 8). In Inglefield Land, Greenland, there appear to be 3 different-aged glacial deposits, the oldest of which shows more highly developed colors than do the younger deposits contiguous to the present-day margins of the Greenland Ice Cap (Tedrow, 1970). The oldest deposits have developed

Figure 6 A limestone area of Cornwallis Island showing an irregular surface pattern. There is no formation of genetic soil horizons.

yellowish brown to strong brown colors in the B horizons but the youngest deposits have retained their inherited gray colors.

A number of raised beaches and strand lines throughout the high arctic have been examined to study possible correlations of soil development with time. These studies have not produced much positive information because there appear to be few if any recognizable changes between the older, higher deposits as compared to the younger deposits near present-day sea level.

If we consider the polar desert region on a circumpolar basis we find in North America (Figure 9) that this region was affected by some 3 major glacial complexes. First is the Wisconsin-Laurentide glacial complex which was of great importance areally. This ice sheet disappeared only 8,000 to 15,000 years ago and therefore we are not dealing with an unusually long time frame with respect to

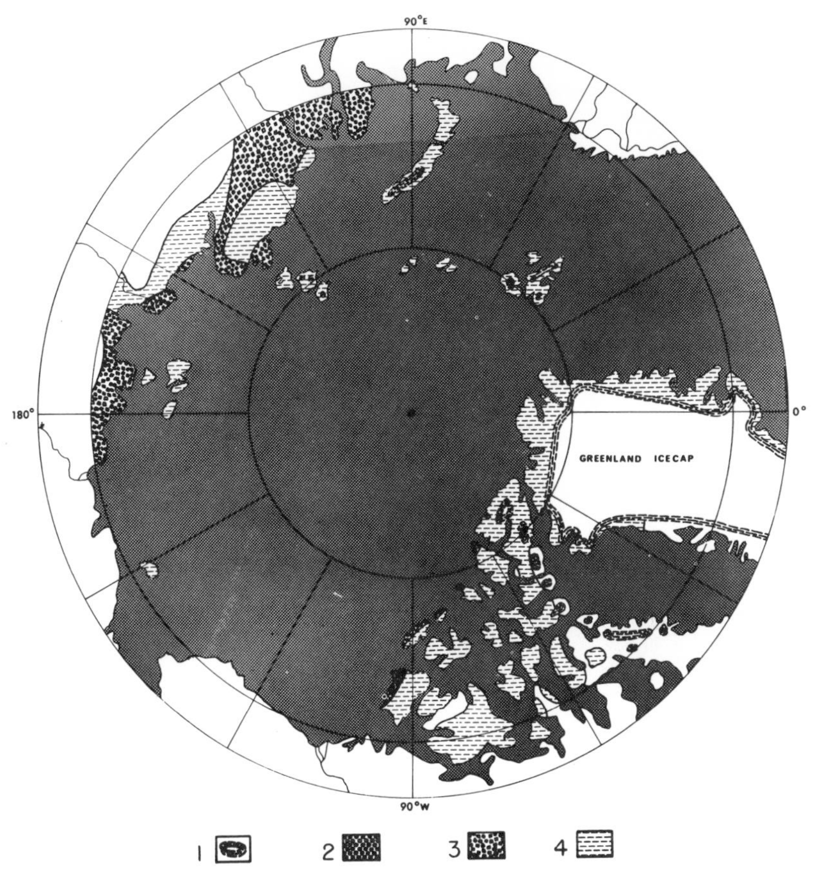

Figure 7 Map of the far northern areas showing: (1) Existing glaciers. (2) The Beaufort formation of Canada. Polar desert soils probably reach their greatest development on this formation. (3) Thick deposits of glacial drift with some marine facies. In general these sectors are non-lime-bearing. (4) Sectors of the polar deserts in which the soil material may contain considerable carbonate-bearing rock.

the ice-free landscapes. On the northwest margin of the Queen Elizabeth Islands where there was little if any important glacial activity, the Polar desert soils have stronger developed colors than is the case in the well-defined glaciated regions. The Ellesmere-Baffin and the Greenland Glacial Complexes have left deposits that have undergone little change with time, except the condition already described in Inglefield Land, Greenland. The

Figure 8 Well-drained soil on an alluvial deposit of Prince Patrick Island. There is little if any genetic horizonation in this soil.

Cordillera ice sheet of western North American did not extend into the polar desert zone.

In Eurasia the ice sheets probably extended over all of the present-day northern land masses now occupied by the polar deserts (Sachs and Strelkov, 1961; Hoppe, 1970). As shown in Figure 9, the major ice sheet covered virtually all of the arctic landscapes from the Norwegian sea eastward to approximately the Lena River Delta. East of the Lena River Delta the glacial picture is incomplete. There is no evidence to suggest, however, that any of the Eurasian polar desert zone escaped glaciation and/or marine transgression. In northern Europe the Würm (Wisconsin) and in northern Siberia the Zyryanka (also Wisconsin) aged deposits plus the complex glacial condition in northeastern Siberia indicate that there are probably no preglacial deposits present with modern soils on them.

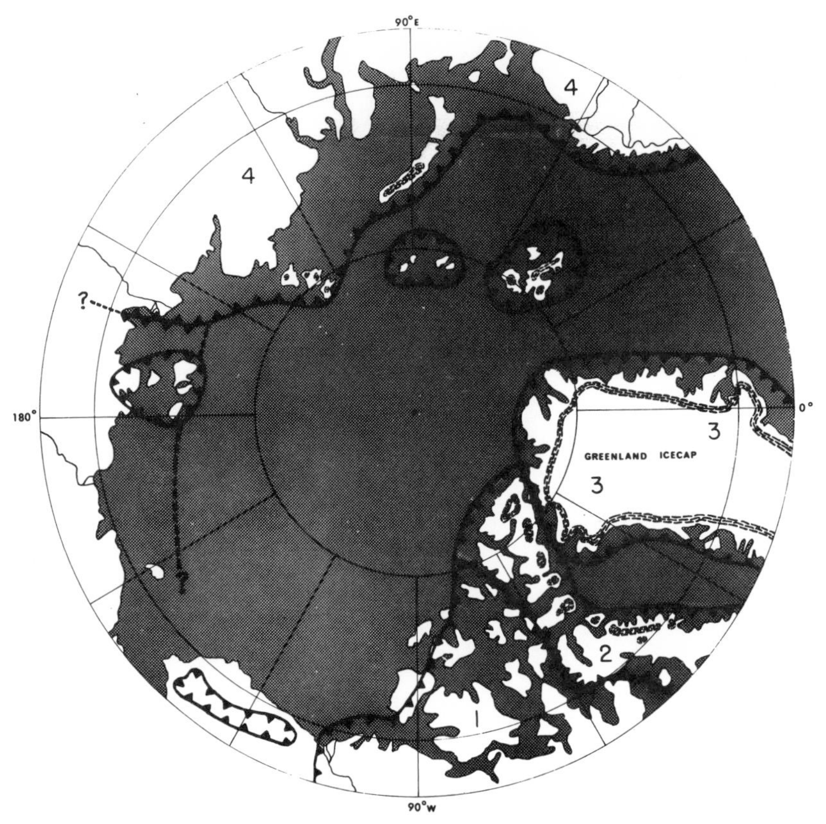

Figure 9 Major glacial complexes of the Northern Hemisphere. (1) Wisconsin-Laurentide. (2) Ellesmere-Baffin. (3) Greenland. (4) Baltic.

POLAR DESERT SOIL AND RÅMARK SOIL

The term Råmark has existed for some time in the literature but the soil's characteristics and position within various classification systems has never been completely clarified. Kubiena (1953) described this soil as a "Raw soil in the region of the melting soils of the arctic, appearing just beyond the tundra girdle as a climate form but partly reaching into the tundra zone with scanty chemical weathering....." There are many such raw appearing earthy conditions within the polar sectors. The higher elevations within the Scandinavian mountains typify the raw soil conditions. The high summer rainfall in the Scandinavian mountains produces unstable conditions resulting in downslope movement of the soil material. Therefore, soil in this sector is more of a dynamic erosional surface than one with genetic horizons. In the Brooks Range of Alaska the higher elevations are largely free of vascular plants. Above an altitude of some 5,000 feet there is very little fine (sand, silt, and clay) material present. In some respects this can be considered a paravariety of Polar desert soil but the lichen-covered matrix is composed largely of shattered bedrock. Immediately north of the Brooks Range the isolated mountains within the Foothills Province attain altitudes of up to 3,000 to 4,000 feet. These mountaintops in most instances have a barren appearance. The only semblance of a genetic soil process I have been able to detect in this sector is the formation of a discontinuous desert pavement and the accumulation of carbonates on the undersides of the surface gravels.

The Lake Hazen area of northern Ellesmere Island, which is well within the Polar desert zone, marks one of the most northerly sectors of the globe (*ca.* 81° 50' N) where terrestrial soils occur. Land in the vicinity of Lake Hazen was completely glaciated, with the ice receding probably within the last 10,000 years. Day (1964) showed that there is no significant development of genetic horizons, and accordingly the well-drained soils were classed as Regosol or more specifically as Subarctic saline regosol and Subarctic orthic regosol.

REFERENCES CITED

Alexandrova, V.D., 1970, Vegetation and primary productivity in the Soviet Subarctic, *in* Fuller, W.A., and Kevan, P.G., eds., Productivity and Conservation in Northern Circumpolar Lands: Morges, Switzerland, I.U.C.N., New Series, no. 16, p. 93-114.

Beschel, R.E., 1970, The diversity of tundra vegetation, *in* Fuller, W.A., and Kevan, P.G., eds., Productivity and Conservation in Northern Circumpolar Lands: Morges, Switzerland, I.U.C.N., New Series, no. 16, p. 85-92.

Day, J.H., 1964, Characteristics of the soils of the Hazen Camp area, Northern Ellesmere Island, N.W.T.: Ottawa, Defence Res. Bd., (Hazen 24), 15 p.

Gorodkov, B.N., 1939, Peculiarities of the arctic topsoil: Izvest. Gos. Geogr. Obshchestva, v. 71, p. 1516-1532 (In Russian).

Hoppe, G., 1970, The Würm ice sheets of northern and Arctic Europe: Acta Geographica Lodziensia, v. 24, p. 205-215.

Korotkevich, Y.S., 1967, Polar deserts: Sov. Antarct. Exped. Inf. Bull., v. 6, no. 65, p. 459-472 (Translated from the Russian by Scripta Technica).

Kubiena, W.L., 1953, Soils of Europe: London, Thomas Murey, 317 p.

____________, 1970, Micromorphological Features of Soil Geography: New Brunswick, N.J., Rutgers Univ. Press, 254 p.

Sachs, V.N., and Strelkov, S.A., 1961, Mesozoic and Cenozoic of the Soviet Arctic, *in* Raasch, G.O., ed., Geology of the Arctic: Toronto, Univ. Toronto Press, p. 48-68.

Soil Map of the World, 1972: Paris, UNESCO, 1:5,000,000.

Tedrow, J.C.F., 1966, Polar desert soils: Soil Sci. Soc. America Proc., v. 30, p. 381-387.

____________, 1970, Soil investigations in Inglefield Land, Greenland: Meddelelser om Grønland, v. 188, no. 3, 93 p.

____________, 1973, Soils of the polar region of North America: Biuletyn Peryglacjalny, v. 23, p. 157-165.

Tedrow, J.C.F., Brown, J., 1963, Soils of the northern Brooks Range, Alaska: Weakening of the soil-forming potential at high arctic altitudes: Soil Sci., v. 93, p. 254-261.

Tedrow, J.C.F., Drew, J.V., Hill, D.E., and Douglas, L.A., 1958, Major genetic soils of the Arctic Slope of Alaska: Jour. Soil Sci., v. 9, p. 33-45.

Walton, G.F., 1972, The high arctic environment and Polar desert soils, (Ph.D. Dissert): New Brunswick, Rutgers Univ., 479 p.

A Comparison of the Morphology and Genesis of Arctic Brown and Alpine Brown Soils in North America

J. G. Bockheim

ABSTRACT

The well-drained Arctic Brown and Alpine Brown (Alpine Turf) soils are the zonal soils of arctic and alpine tundras, respectively. These two Great Soil Groups were compared and contrasted by 1) providing new data from two arctic and two alpine locations, and 2) compiling data from the North American literature. Study areas include Pangnirtung, Baffin Island; Cornwallis Island, N.W.T.; Saddleback Mountain, Maine; and Mt. Baker, Washington.

The Alpine Brown soil is darker in color than its counterpart in the Low Arctic. A1 (Ah) horizons are crumb-like in Alpine Brown soils and granular in Arctic Brown soils. Roots are concentrated in the A1 horizon of both Great Soil Groups, giving the horizon a "turfy" nature. Roots seldom occur below 50 cm in Arctic Brown soils.

Chemical data from 30 Alpine Brown and 21 Arctic Brown soils were summarized from the literature. Additional data from five Alpine Brown and three Arctic Brown soils are provided from the present study. The two soil groups are medium acid; pH increases progressively with depth. Organic matter and total organic nitrogen are concentrated in the A1 horizon and are two-fold greater in Alpine Brown soils. Cation-exchange capacities are similar for both soils and are greatest at the surface, diminishing with depth. CEC correlates well with organic matter content.

Depth-distribution and magnitude of dithionite-extractable iron are similar in the two Great Soil Groups. Fe_2O_3 is maximized slightly in the B2 horizon. Pyrophosphate-extractable iron is most abundant in the B2 horizon of each soil group. Cheluviation may occur in arctic and alpine soils: alternatively, the iron and organic matter may be released *in situ* and complexed in the B2 horizon.

Sandy loam textures prevail in both soil groups. Clay contents are most commonly 9-12% and do not change substantially with depth. Silt is concentrated at the surface in Arctic Brown soils.

No single clay mineral species or group of related minerals is dominant in Arctic and Alpine Brown soils. Weathering sequences in the clay fraction include: 1) formation of goethite, 2) feldspar → kaolinite, 3) allophane → kaolinite, 4) 10Å mica → vermiculite, 5) chlorite ⇄ aluminum-interlayed vermiculite ⇄ vermiculite, 6) 10Å mica → montmorillonite, and 7) reactions involving mixed-layer phyllosilicate minerals.

Continuous permafrost occurs throughout the area featuring Arctic Brown soils, and discontinuous permafrost is present in areas containing Alpine Brown soils, except in the Cascade and Olympic Mountains. Permafrost does not appear directly to influence the properties of these well-drained soils.

Arctic and Alpine Brown soils appear in four orders and seven suborders of the U.S. classification system: Psamments, Orthents, Ochrepts, Umbrepts, Andepts, Borolls, and Orthods. In the Canadian system, two orders and four great groups are represented: Melanic, Eutric, and Dystric Brunisols and Regosols (Cryic). In addition, Alpine Brown soils occur in the Humo-ferric Podzol great group.

INTRODUCTION

Whereas numerous studies have compared and contrasted the climate, vegetation, and other characteristics of arctic and alpine regions (*cf.* Ives and Barry, 1974), there have been few attempts to compare the soils of these two life zones.

The zonal soil in the Low Arctic is the Arctic Brown (Tedrow and Hill, 1955). Its counterpart in alpine areas is well-drained Alpine Turf (Retzer, 1956) or Alpine Brown soil. The objective of this study is to compare and contrast well-drained arctic and alpine soils by: 1) providing new data from two arctic and two alpine locations in North America (Figures 1 and 2), and 2) compiling data from the North American literature to support or negate the comparisons

Nomenclature of arctic and alpine environments has been discussed by Polunin (1951) and Barry and Ives (1974) and will not be reviewed here. However, several definitions need to be given. Arctic tundra here is defined as the zone at high latitudes beyond the limit of erect arborescent growth and comprised of dwarf shrubs, sedges, mosses, and lichens. This is analogous to the Low Arctic vegetational zone of Polunin (1951). The arctic tundra is south of the arctic desert or stony sedge-moss-lichen tundra of Porsild (1957). Alpine tundra refers to the region above the altitudinal limit of erect arborescent growth and features sedges, grasses, and shrubs, and occasionally small patches of tree *krummholz*.

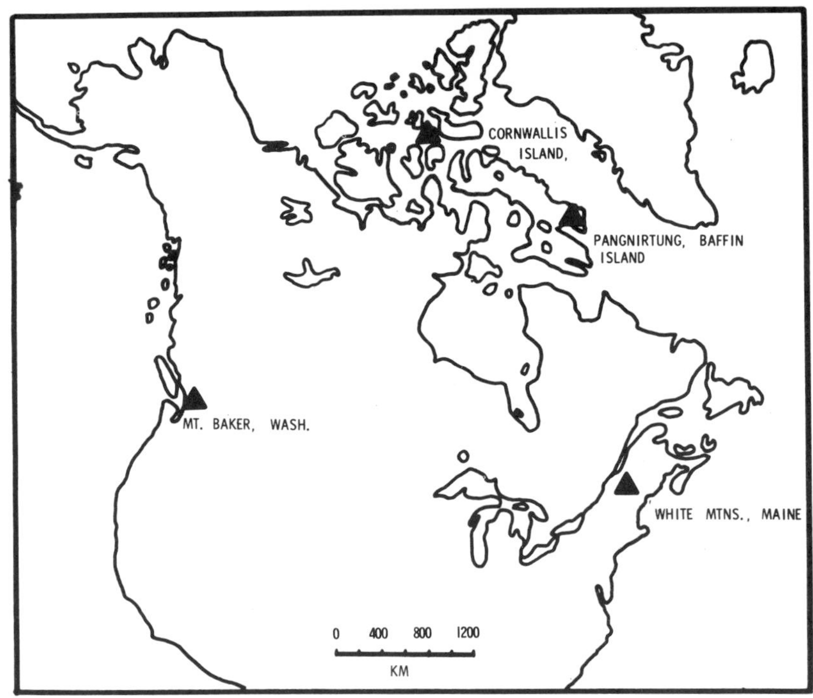

Figure 1 Location of sites sampled in this study.

The dark areas in Figure 2 show the probable occurrence of alpine vegetation and hatches estimate areas of arctic vegetation. In Alaska and central Canada, it becomes difficult to distinguish between alpine and arctic vegetation where altitudinal and latitudinal treelines merge. Alpine vegetation in Alaska here includes *Dryas* meadows and barrens (*Dryas-Carex-Betula*) and Aleutian heath and barrens (*Empetrum-Vaccinium)* (U.S. Dept. Interior, 1970, p. 89). In central Canada, alpine vegetation includes arctic bell *(Cassiope tetragona)*, heath *(Ericaceae)*, sedges *(Carex)*, bluegrass *(Poa)*, mountain avens *(Dryas)*, and dwarf willows *(Salix)* (Dept. Energy Mines and Resources, 1974, p. 45-46).

Arctic vegetation in Alaska includes cottongrass-sedge tundra *(Eriophorum)* and watersedge tundra *(Carex)*. In central Canada, arctic vegetation is comprised of shrubby birch *(Betula)* and willows *(Salix)*, sedges *(Carex)*, blueberry *(Vaccinium)*, crowberry *(Empetrum)*, and Labrador tea *(Ledum)*.

In preparing Figure 2, alpine and arctic vegetation were distinguished from one another primarily on the basis of topography in conjunction with latitude. For example, high-

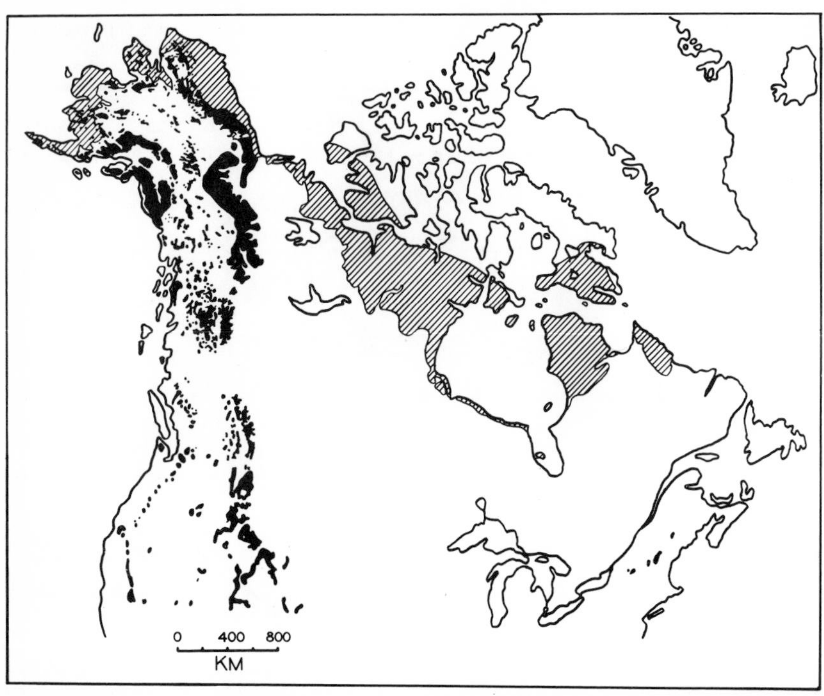

Figure 2 Alpine (dark areas) and arctic (hatched areas) regions of northern North America (see text for discussion). Alpine areas feature Alpine Brown and associated soils; arctic areas feature Arctic Brown and related soils (Compiled from Dept. Energy Mines and Resources, 1974; U.S. Dept. Interior, 1970; and Zwinger and Willard, 1972.)

mountain areas in the Alaska Range feature *Dryas* meadows and barrens. Therefore, areas with similar vegetation above timerline in the southern Brooks Range were designated as alpine. Low, flat, poorly drained areas on the Arctic Slope feature mainly cottongrass-sedge tundra. Areas of similar vegetation in the northern Brooks Range, therefore, were designated as arctic. The separation of arctic and alpine vegetation in these areas is admittedly arbitrary but hopefully points to the need of better defining arctic and alpine ecosystems.

The dark areas on Figure 2 likewise feature the well-drained Alpine Brown and associated soils (see also Table 1). The hatched areas contain Arctic Brown and related soils.

LITERATURE REVIEW

Table 1. Great Soil Groups of arctic and alpine tundra, North America

Arctic Tundra, *i.e.* Low Arctic (Tedrow *et al.*, 1958)	Drainage	Alpine Tundra (Retzer, 1956)
Arctic Brown	Well	Alpine Turf (Brown)
Upland Tundra	Somewhat poor	Alpine Meadow
Meadow Tundra	Poor	--
Bog	Very poor	Alpine Bog

Climate

Early geographers and climatologists considered climates of arctic and alpine environments to be similar, *e.g.* ET or tundra type of Køppen. Our knowledge of alpine and arctic climates has improved (*cf.* Vowinkle and Orvig, 1970; Barry and Hare, 1974; Flohn, 1974). Using data from Billings and Mooney (1970), Webber (1974) compared an arctic tundra climate (Point Barrow, Alaska) with that of an alpine tundra (Niwot Ridge, Colorado). Table 2 provides climatic data for the four sites considered in this study. Bliss (1956) compared plant development in microenvironments of arctic and alpine tundras.

Mean annual precipitation and precipitation during the growing season are generally much greater in alpine than in arctic environments of North America. Snowfall likewise is more abundant in alpine environments. Mean annual and mean July temperatures are dependent upon latitude and altitude but are lower at arctic stations considered in this study. Average length of growing season is similar for the two life zones. Annual mean wind velocity is greater in alpine environments, perhaps compensating for the somewhat higher temperatures. Summer cloudiness is common to many alpine areas and is universal in the Low Arctic. For example, mean relative humidity during the growing season is similar for Resolute and Mt. Washington (Table 2).

Vegetation

As in the case of climate, many investigators refer to alpine and arctic tundra synonymously. Indeed, many features of the flora of the two areas are similar. Species such as *Oxyria digyna, Silene acaulis, Trisetum spicatum,* and *Saxifraga oppositifolia* are common to both regions (Billings, 1974). Holm (1927) has shown that 37% of Colorado alpine

Table 2. Comparison of arctic and alpine climates for the four sampling sites in this study

	Arctic		Alpine	
	Pangnirtung, Baffin Island	Resolute, N.W.T.	Mt. Washington, N.H.	Mt. Baker, Wash.
Latitude	66°08'	74°43'	44°16'	48°51'
Altitude (m)	13	64	1917	1265
Mean annual precipitation (cm)	33.5	13.0	187	279
Precip. for June, July, Aug. (cm)	12.2	6.6	49	27.4
Annual snowfall (cm)	178	68	495	1310
Mean annual temp (C)	-8.9	-16.2	-2.8	4.5
Mean July temp (C)	7.8	4.6	8.7	12.2
Frost-free days	--	60	69	--
Permafrost	universal	universal	sporadic	nonexistent
Mean wind velocity (m sec^{-1})	--	5.2	16	--
Mean relative humidity for June-Aug. (%)	--	85	90	--

Data from:

Canadian Hydrographic Service (1970).
U.S. Weather Bureau (1964).
U.S. Weather Bureau (no date).

species also occur in the arctic. Bliss (1956) and Billings and Mooney (1968) compared the vegetation of arctic and alpine environments. The number of vascular plants was 40 and 100 in an arctic and an alpine tundra region, respectively (Billings and Mooney, 1968). Average areal vascular production was 100 and 200 g m^{-2} yr^{-1} in arctic and alpine areas. Billings (1974) compared and contrasted life forms, general morphology, and physiological adaptation in the two areas. Average areal net photosynthetic efficiency was similar. A major difference was length of photoperiod, which is 84 days at Point Barrow, Alaska versus 15 hours (maximum) on Niwot Ridge, Colorado.

Relief

In the Low Arctic, Arctic Brown soils occupy dry sites, including coarse-textured raised beach ridges, escarpments, and other narrow ridges in a region otherwise featuring low, poorly drained relief (Tedrow and Hill, 1955). Alpine Brown soils occupy ridge crests and steep mountain slopes and other well drained areas in the alpine tundra (Retzer, 1965).

Parent Material

Both Great Soil Groups are found on a variety of transported parent materials, including colluvium, loess, volcanic ash, till and outwash. In the Canadian Arctic, Arctic Brown soils have formed from glaciomarine sediments on raised beaches. Residual Alpine Brown soils probably are more common than residual Arctic Brown soils. Both soil groups have been described as occurring on materials ranging from calcareous to siliceous.

Time

Arctic Brown soils range from early Neoglacial (a minimum of about 4,500 years in age) to middle Wisconsin (?), *ca.* 60,000 years BP on Baffin Island (Bockheim, 1974). Older Arctic Brown soils probably exist in unglaciated portions of Alaska and possibly the N.W.T. Alpine Brown soils require 4,500 years to form in the central Front Range, Colorado (Mahaney, 1974). Alpine Brown soils in the Rocky Mountains may be as old as Bull Lake, *circa,* 145,000 years BP (Birkeland, 1973).

Soils of Arctic and Alpine Areas

Alpine soils (Bockheim, 1972; Retzer, 1974) and arctic soils (Tedrow, 1970; Rieger, 1974) recently have been reviewed and will only be highlighted here. Arctic Brown soils were first described in North America by Tedrow and Hill (1955) in northern Alaska. Drew and Tedrow (1957) and Hill and Tedrow (1961) provided additional chemical, physical, and mineralogical data on several profiles. Later comprehensive studies on Arctic Brown soils were published by Brown and Tedrow (1964), Holowaychuk *et al.* (1966), and

Rieger (1966). Data summarized in this study originated from these sources and Ugolini (1966), Tedrow and Douglas (1964), and Bockheim (1974).

In the U.S., alpine soils were initially placed in one group, Alpine Meadow soils (Thorp and Smith, 1949). Retzer (1956) objected to the usage of "meadow" for alpine soils other than those which are poorly drained, because the term implies moist, damp conditions in the U.S. Accordingly, he identified three Great Soil Groups: The Alpine Turf soil represents well drained conditions, the Alpine Meadow is somewhat poorly to poorly drained, and the Alpine Bog occurs in peaty, very poorly drained areas. The Alpine Turf soil is synonymous with the Alpine Humus soil in Europe (Jenny, 1930) and in Australia (Costin, 1955).

Van Ryswyk (1969) used the term Alpine Brown for well-drained soils with dark colored A1 horizons in south-central British Columbia. He reported two phases, one with continuous volcanic ash and the other with discontinuous accumulations of pyroclastic materials. Other investigators have referred to alpine soils collectively as alpine tundra soils.

In this study, Alpine Brown, Alpine Turf, and Alpine Humus are viewed as the same Great Soil Group. Data compiled in this study originated from Retzer (1956, 1965), Nimlos and McConnell (1962), van Ryswyk (1969), Bockheim (1972), Sneddon *et al.* (1972), Birkeland (1973), Knapik *et al.* (1973), Mahaney (1974), Miller and Birkeland (1974), and Pawluk and Brewer (1975).

The similarity of Arctic Brown and Alpine Brown soils was mentioned by Drew and Tedrow (1957), Nimlos and McConnell (1962), Brown and Tedrow (1964), and Retzer (1965). However, no quantitative comparisons were made.

METHODS

Four study areas were selected for the study. In addition, relevant data from the literature was summarized to evaluate arguments presented. Two arctic study areas include Pangnirtung, Baffin Island (66°08') and Resolute, Cornwallis Island, N.W.T. (74°43') (Figure 1). The Pangnirtung site occurs in the Polar Desert-Arctic Brown transition zone (Tedrow, 1970) or Sub-polar Desert zone (Tedrow, 1973). The Cornwallis Island area is within the Polar Desert zone.

Two alpine tundra study areas include Saddleback Mountain, western Maine (44°56'), and Mount Baker, northwestern Washington State (48°51').

Soil descriptions were prepared at each site. Samples were collected from each horizon. Chemical analyses were performed on the $<$ 2mm fraction and included pH on a 1:1.5 or 1:2 soil:water suspension; total organic nitrogen by the semimicrokjeldahl or macrokjeldahl methods (Bremner, 1960); and cation exchange capacity by saturation with 1 N NH_4OAc (pH 7.0) (Chapman, 1965). The Mt. Baker samples were saturated with pH 4.0 and 8.0 NaOAc, in order to avoid ammonium fixation by vermiculite (Bower *et al.*, 1952; modified by deVilliers and Jackson, 1967). CEC values determined by pH 8.0 are reported, because they more closely approximate CEC determined with 1 N NH_4OAc. Organic matter was determined by the modified Walkley-Black method (Allison, 1965). Organic matter in the Baffin samples was digested with chromic acid at 90°C for 1-1/2 hours and detected colorimetrically against standards of known organic matter content (determined by Walkley-Black) (Genson *et al.*, 1976).

Carbon was extracted with 0.1 M sodium pyrophosphate at pH 10 (McKeague, 1968). Free iron was determined in the extract (McKeague, 1967). In addition, humic-acid and fulvic-acid fractions were separated from the extract by acidification with 6 N H_2SO_4 and centrifugation. The modified Walkley-Black (Allison, 1965) method was used to determine the amount of carbon in aliquots of the humic- and fulvic-acid fractions previously evaporated to dryness on a hot plate.

Particle-size distribution was determined by the modified Bouyoucos method (Bouyoucos, 1951; Day, 1965). Cornwallis and Baffin samples were pretreated and analyzed for clay minerals by X-ray diffraction according to the methods of Jackson (1956). Mt. Baker samples were pretreated using the Kittrick and Hope (1963) modification of Jackson's method. Slides for X-ray analysis of the Mt. Baker samples were prepared using the paste method (Thiesen and Harward, 1962). Minerals in the clay fraction (-2 μm) of all the samples were identified according to criteria summarized by Warshaw and Roy (1961).

RESULTS AND DISCUSSION

Soil Morphology

Morphologically, the three Arctic Brown and six Alpine Brown soils from the four study areas are similar (*cf.* Figures 3 and 4). In addition, analysis of morphologic data from 30 Alpine Brown and 21 Arctic Brown profiles compiled from the literature verifies the similarity of these two Great Soil Groups. Average thickness of sola for Alpine Brown soils is 40 and for Arctic Brown soils is 41 cm (Table 3). Sola of Arctic Brown soils from Baffin and Cornwallis Islands average 36 cm, and weathering profiles of Alpine Brown soils from Mt. Baker and Saddleback Mountain average 41 cm in thickness (Table 4).

Profiles examined from the literature indicate that Alpine Brown soils contain somewhat thicker A1 horizons than Arctic Brown soils (13 vs. 10 cm). However, Arctic Brown soils feature slightly thicker B2 horizons (29 vs. 27 cm) (Table 3). Similar trends are apparent in soils from the four study areas (Table 4).

Figure 3 Alpine Brown soil, Mt. Baker, Washington (profile 70-9)

Both Great Soil Groups feature mollic (chernozemic) or umbric (sombric) surface horizons over cambic (Bm) or spodic (Bfh, Bhf, Bf) subsurface horizons.

Colors of horizons in Alpine Brown soils usually are darker than in Arctic Brown soils. In Alpine Brown soils, the A1 is most commonly black (5YR, 10YR 2/1, -- moist) and B2 horizons are yellowish brown (10YR 5/4, 5/8) to brown (7.5YR 4/4, 5/4; 10YR 4/3, 5/4). In arctic areas, the A1 generally is dark brown (7.5YR 3/2, 4/2; 10YR 3/3, 4/3) to very dark greyish brown (10YR 2/2, 3/2), and the B2 horizon is dark brown (10YR 3/3,4/3). C horizons are yellowish brown (10YR 5/4) to dark yellowish brown (10YR 4/4) in Alpine areas and olive grey (2.5Y 4/2, 5Y 5/2) to brown (10YR 5/3) in arctic areas.

Whereas A1 horizons of Arctic Brown soils generally have granular structure, the corresponding horizons in

Figure 4 Arctic Brown soil, Cumberland Peninsula, Baffin Island (profile 74-4)

Alpine Brown soils are crumb-like.[1] Blocky structure commonly is reported in illuvial horizons of Alpine Brown soils; granular structure or single grains are most often evident in Arctic Brown soils.

Roots are concentrated in the A1 horizon of both soil groups, giving that horizon a "turfy" nature. Roots usually are micro (-0.075) to fine (1-2mm) in size. Whereas roots seldom occur below 50 cm in Arctic Brown soils, roots often extend beyond this depth in Alpine Brown soils, except where bedrock is close to the surface.

Chemical Properties

Arctic and Alpine Brown soils from the literature are medium acid (Table 3). Soil reaction increases from the A1 to the C horizon. Soils analyzed in the current study showed the same trends but range from 0.5 to 1.5 pH units more acid (Table 4). Acid values are easier to understand in alpine soils, in view of the copious amount of precipitation, usually from 85 to 375 cm, and leaching.

---

[1] Crumb structure is not recognized in the Canadian system of soil horizon nomenclature. In the U.S. system, crumbs are porous aggregates and granules are soil grains bound by rootlets, microbial gums and other substances. The latter units are less porous than crumbs.

Table 3. Summary of physical and chemical properties of Arctic and Alpine Brown soils compiled from the literature.

Property, Alpine Brown soils	$\bar{X}^a$	R	N	$\bar{X}^a$	R	N	$\bar{X}^a$	R	N
	A1			B2			C		
Thickness (cm)	13	4-30	21	27	11-65	22	--	--	--
pH	5.5	4.5-7.7	30	5.7	4.8-8.3	31	5.7	4.5-8.4	28
Organic matter (%)	13	2.8-25	30	5.1	0.17-17	30	1.4	0.38-4.1	22
CEC (me/100 g)	20	4.8-52	26	11	2.0-35	26	7.0	1.6-22	20
Base sat. (%)	65	2-100	24	48	6-99	23	58	2.99	18
N(%)	0.53	0.08-0.93	24	0.28	0.02-0.46	22	0.07	0.0.23	14
Fe_2O_3 (%)	1.1	0.13-2.3	17	1.7	0.47-4.0	18	1.6	0.39-5.4	13
Sand (%)	55	26-85	22	56	22-74	22	58	23-85	19
Silt (%)	33	11-57	22	33	18-70	22	31	10-55	19
Clay (%)	12	1-21	22	11	1-25	22	11	1-31	19
S/Si+C	1.2	0.4-2.8	19	1.3	0.3-2.8	20	1.4	0.3-5.7	17
Arctic Brown soils	A1			B2			C		
Thickness (cm)	10	2-20	19	29	0-56	21	--	--	--
pH	5.7	3.9-8.3	21	5.8	4.6-8.6	20	6.4	5.0-8.9	18
Organic matter (%)	7.8	3.6-19	11	3.0	0.89-6.6	11	0.72	0.37-2.0	8
CEC (me/100 g)	25	9.2-50	7	11	3.4-21	4	3.6	1.8-7.8	4
Base sat. (%)	25	7-58	7	16	6-40	4	25	16-41	3
N (%)	0.34	0.28-0.38	3	--	--	--	0.02	0.02	1
Fe_2O_3 (%)	1.1	0.7-1.8	11	1.3	0.5-2.5	13	0.8	0.4-2.0	13
Sand (%)	61	28-81	12	66	15.89	8	68	16-98	12
Silt (%)	28	6-53	12	22	2-61	8	23	1-62	12
Clay (%)	11	0-24	12	12	1-24	8	9	0-22	12
S/Si+C	1.6	0.4-2.8	12	1.9	0.2-8.1	8	2.1	0.8-16	12

[a] $\bar{X}$ is the mean, R the range, and N the number of observations.

Table 4. Selected chemical and physical properties of Arctic and Alpine Brown soils from the four study areas.

Horizon	Depth (cm)	pH	Organic matter (%)	CEC (me/100 g)	N (%)	Fe_p (%)	Sand (%)	Silt (%)	Clay (%)
			Profile 70-3, Mt. Baker						
A1	0-10	4.0	24.9	37.1	0.76	0.80	45	45	10
B2H	10-22	4.4	16.8	33.0	0.58	0.92	43	41	16
			Profile 70-8, Mt. Baker						
A1	0-8	4.5	19.4	35.6	0.59	0.48	61	22	17
B21h	8-17	4.8	12.6	30.6	0.36	0.82	61	19	20
B22h	17-31	5.0	8.6	21.7	0.31	0.62	57	20	23
B3	31-67	5.0	3.8	15.6	0.18	0.22	69	12	19
C	67-126	5.0	0.8	8.9	0.06	0.10	63	26	11
			Profile 70-9, Mt. Baker						
A1	0-8	4.5	12.7	24.7	0.38	0.82	61	21	18
B21h	8-16	4.8	6.5	18.0	0.24	0.72	62	20	18
B22	16-37	4.8	4.0	15.4	0.16	0.70	59	25	16
IIC	37-66	4.5	1.0	11.1	0.06	0.25	48	36	16
			Profile 70-11, Mt. Baker						
A1	0-9	4.6	10.9	25.2	0.36	0.90	49	34	17
B21hir	9-17	4.9	8.4	25.2	0.27	1.25	50	29	21
IIB22h	17-32	4.7	6.2	21.0	0.24	0.75	76	10	14
IIC	32-54	4.8	3.0	12.1	0.11	0.28	68	21	11
			Profile 67-1, Saddleback Mtn.						
A1	0-13	3.8	20.2	--	--	--	68	18	14
B2hir	13-23	4.4	12.4	--	--	--	44	41	15
			Profile 67-2, Saddleback Mtn.						
A1	0-10	3.5	19.0	--	--	--	70	20	10
B21hir	10-33	4.4	22.0	--	--	--	59	33	8
B22hir	33-61	4.3	12.5	--	--	--	40	52	8
			Profile 74-25, Baffin Island						
A1	0-3 1/2	4.0	5.0	11.3	--	0.10	64	36	0
IIB21	3 1/2-25	4.7	1.8	3.8	--	0.14	84	7	9
IIB22	25-42	4.8	1.6	3.6	--	0.13	82	17	1
IIC	42-57	5.0	0.5	2.1	--	0.08	74	24	2
			Profile 74-29, Baffin Island						
A1	0-8	3.9	4.5	9.2	--	0.14	66	34	0
IIB21	8-18	4.8	1.6	3.4	--	0.12	85	6	9
IIB22	18-41	5.0	1.5	3.6	--	0.15	84	15	1
IIB3	41-64	4.9	1.2	3.3	--	0.17	72	27	1
IIC1	64-90	5.3	0.7	2.5	--	0.07	78	21.5	0.5
IIC2	90-100	5.7	0.2	1.7	--	0.07	72.5	27	0.5
			Profile 73-5, Cornwallis Island						
A1	0-4 1/2	7.8	2.5	8.8	0.15	--	72.5	19.5	8
AC	4 1/2-9	7.3	1.5	5.1	0.06	--	81.5	10	8.5
C1	9-17	8.2	0.57	3.4	0.03	--	82	9	9
C2	17-48	7.9	0.75	3.7	0.03	--	77	11.5	11.5

In the Low Arctic, precipitation is low, 10 to 50 cm. Apparently, the degree of leaching is not so important as is the chemistry of decomposition products from arctic and alpine vegetation.

Organic matter contents are greatest in the A1, decreasing progressively with depth (Tables 3 and 4). Organic matter values are nearly twice as great in Alpine Brown soils. This probably reflects a greater below-ground biomass productivity for alpine tundra ecosystems. Whereas alpine tundra produced about 2,400 g m^{-2} yr^{-1}, arctic tundra ecosystems generate about 800 g m^{-2} yr^{-1} of below-ground biomass (Webber, 1974).

Total nitrogen is greatest in the A1 horizon in both soil groups and diminishes with depth (Tables 3 and 4). N values in Alpine Brown soils are nearly twice those in Arctic Brown, probably due to the fact that organic matter contents are two-fold greater in the alpine soils.

Cation-exchange capacities are similar for both soil groups (Table 3). CEC is greatest at the surface and diminishes regularly with depth. Exchange properties correlate well with organic matter content (Table 5). CEC values are lower for Arctic Brown soils analyzed in the present study than those reported in the literature, because the former contain lower amounts of organic matter (Table 4).

Depth-distribution and magnitude of dithionite-extractable iron are similar in the two Great Soil Groups (Table 3). In each case, Fe_2O_3 increases slightly in the B2, then decreases in the C1 horizon. Depth-distribution of iron may suggest a slight podzolization process (Drew and Tedrow, 1957). However, the Fe_2O_3 may have been released by weathering *in situ*. Many of the soils feature lithologic discontinuities which makes pedogenic interpretations difficult.

Pyrophosphate-extractable iron (Fe_p) is used in defining a spodic (podzolic) horizon in the U.S. soil classification system (Soil Survey Staff, 1970);

$$\text{spodic horizon} = \% \; Fe_p + \% \; Al_p / \% \; \text{clay} \quad 0.2 \qquad (1)$$

The two Alpine Brown soils from Mt. Baker and the two Arctic Brown soils from Baffin Island feature spodic horizons and are therefore Spodosols. Albic horizons are occasionally reported in Arctic Brown soils (Tedrow and Hill, 1955; Drew and Tedrow, 1957). Bhir (BFh) horizons have not been reported in the Low Arctic except where latitudinal treeline has retreated leaving relict palaeosols (Spodosols) (Sorenson *et al.*, 1971). Tedrow (1973) reports "Dwarf" Podzol or podzol-like soils on sandy, quartzose deposits as far north as 69°44' in Alaska.

Table 5. Correlation between cation-exchange capacity and organic matter content for Arctic and Alpine Brown soils.

	Literature		Current Study	
	R	df	R	df
Alpine Brown	0.75	83	0.95	9
Arctic Brown	0.91	30	0.99	10

R = correlation coefficient; df = degrees of freedom

Sandy loam textures prevail in both Great Soil Groups. Clay contents are most commonly 9-12% and do not change substantially with depth (Tables 3 and 4). Silt contents are generally 23 to 33%. Silt is concentrated at the surface in Arctic Brown soils. Percent sand increases slightly with depth in both soil groups. Sand/silt + clay ratios increase with depth in each soil, indicating a progressive increase in coarseness with depth.

Clay mineral analysis of four Alpine Brown and three Arctic Brown soils and examination of the literature indicates that no single mineral or suite of minerals consistently prevails in either Great Soil Group. In Arctic Brown soils, the clay fraction is dominated by chlorite and hydrous mica in a variety of soils derived mainly from siliceous sedimentary rocks (Holowaychuk *et al.*, 1966); hydrous mica from sandstone, and kaolinite from gabbro (Drew and Tedrow, 1957); and a mixed-layer hydrous mica-vermiculite-montmorillonite in a variety of soils derived from sandstone, siltstone, and shales (Tedrow *et al.*, 1968). Quartz and feldspar commonly are present in small quantities in the clay fraction.

Depth-distribution of clay minerals in soils of arctic regions may be controlled by some vertical frost sorting as well as lithology of parent materials and pedogenesis. Loess and aeolian sand commonly overlie residual soils and soils of transported materials such as till and outwash. Hill and Tedrow (1961) reported goethite in surface horizons of Arctic Brown soils derived from a variety of parent materials in northern Alaska. These investigators also reported a "claylike" substance associated with feldspar grains. The material was identified by differential thermal analysis as a poorly crystallized form of kaolinite. Tedrow *et al.* (1968) reported weathering of illite (hydrous mica) to vermiculite and aluminum-interlayered vermiculite to chlorite in Arctic Brown soils of Prince Patrick Island, N.W.T. A1 horizons featured slightly more montmorillonite than C horizons in Arctic Brown soils of northeast Greenland (Tedrow, 1970).

Alpine Brown soils commonly feature an abundance of vermiculite (Nimlos and McConnell, 1965; van Ryswyk, 1969; Bockheim, 1972). Chlorite, amorphous material, hydrous mica, and montmorillonite may prevail in soils developed from specific parent materials. As in the case of Arctic Brown soils, quartz and feldspars are often present in small quantities.

Synthesis of goethite in Alpine Brown soils was reported by Sneddon *et al.* (1972) and Pawluk and Brewer (1975). "Chloritization" of vermiculite was reported by Sneddon *et al.* (1972). These investigators proposed that montmorillonite may have formed at the expense of a 10Å mica in Alpine Brown soils of British Columbia. Chemical weathering of vermiculite to montmorillonite via an intermediate mixed-layer vermiculite-montmorillonite also was proposed by Sneddon *et al.* (1972). Alteration of volcanic ash to kaolinite was suggested as occurring in alpine soils of British Columbia (Sneddon *et al.*, 1972). Biotite appears to have been converted to vermiculite in the Ptarmigan soils of the Rocky Mountains (Retzer, 1956; Nimlos and McConnell, 1965).

Bockheim (1972) reported weathering of chlorite to vermiculite, especially in the Al horizon of Alpine Brown soils in the northern Cascade Range. He was able to duplicate the weathering reaction in the laboratory by complexing interlayer aluminum of dioctahedral chlorite from the subsoil with 1 N NH_4F. Similarly, vermiculite from the surface soil was "chloritized" by aging with 0.1 N aluminum chloride solution.

Bockheim (1972) and Reynolds (1972) considered chemical weathering in alpine environments of western North America to be advanced, considering the age ($<$14,000 yrs in the North Cascades) and susceptibility of the soils to erosion. Weathering indices (Jackson, 1968) of clay-size minerals range between 7 and 12, which likewise suggest rapid weathering in alpine environments.

It is difficult to compare weathering rates in alpine and arctic environments. Much of the clay mineralogy from arctic regions is qualitative. Clay mineral weathering sequences are similar for Arctic and Alpine Brown soils (Table 6). The magnitude of the reactions may be slightly greater in Alpine Brown soils of equivalent age, parent material, and relief. The differences in chemical weathering in the two soil groups may be due to differences in climate and/or vegetation. Bockheim (1972) has shown that transformation of chlorite to vermiculite is more pronounced under *Lupinus-Valeriana-Polygonum/Luetkea* vegetation than under *Carex nigricans* sedge on Mt. Baker, Washington. Likewise, Knapik *et al.* (1973) found a strong relationship between soil development and alpine plant communities in Banff National Park.

Permafrost

Table 6. A summary of weathering reactions occurring in clay fractions (<2 µm) of Arctic and Alpine Brown soils.

a. Olivine-pyroxene-amphibole (3)[a] ⟶ Goethite (12)
b. Feldspar (5) ⟶ Kaolinite (10)
c. Volcanic glass (5) ⟶ Kaolinite (10)
d. 10 Å (4) ⟶ Vermiculite (8)
e. Dioctahedral chlorite (4) ⇌ Aluminum-interlayed vermiculite ⇌ Vermiculite (8)
f. 10 Å (4) ⟶ Montmorillonite (9)

[a]Weathering index (Jackson, 1968)

Continuous permafrost occurs throughout the area featuring Arctic Brown soils. Alpine Brown soils north of 60° in western North America also feature permafrost (Figure 5). Permafrost occurs at high elevations throughout British Columbia (Brown, 1970). Retzer (1965) showed discontinuous permafrost as far south as 36° in the Rocky Mountains. Permafrost also exists in isolated areas of the White Mountains in New England (Antevs, 1932) and may occur in the Shickshock Mountains of northern New Brunswick (Ives, 1974, Figure 4A.7).

The active layer in well-drained Arctic Brown soils is generally deep and below the solum. Unlike the somewhat poorly drained Upland Tundra and poorly drained Meadow Tundra soils, Arctic Brown soils generally are not subject to congelipedoturbation, or frost churning (Tedrow *et al.*, 1958). Similarly, Alpine Meadow soils feature permafrost in the Rocky Mountains but Alpine Brown soils do not (Retzer, 1956; 1965).

A comparison of Figures 2 and 5 illustrates that the distribution of alpine areas (*i.e.* Alpine Brown and associated soils) corresponds to the areas of discontinuous permafrost south of 60°N. Notable exceptions are the Olympic and Cascade Mountains. At least two explanations may be offered for the lack of permafrost in the Pacific Northwest: 1) The mild winters of exceptionally high snowfall may insulate the ground and preclude permafrost formation. 2) Geothermal heat associated with recent and past volcanic activity may have melted away or prevented formation of permafrost.

Regarding the first possibility: mild, snowy winters are not characteristic of the eastern Cascades, which have a more continental climate. Permafrost has not been reported in areas east of the Cascade Crest. Secondly, permafrost is reported for the equally maritime, snowy Coast

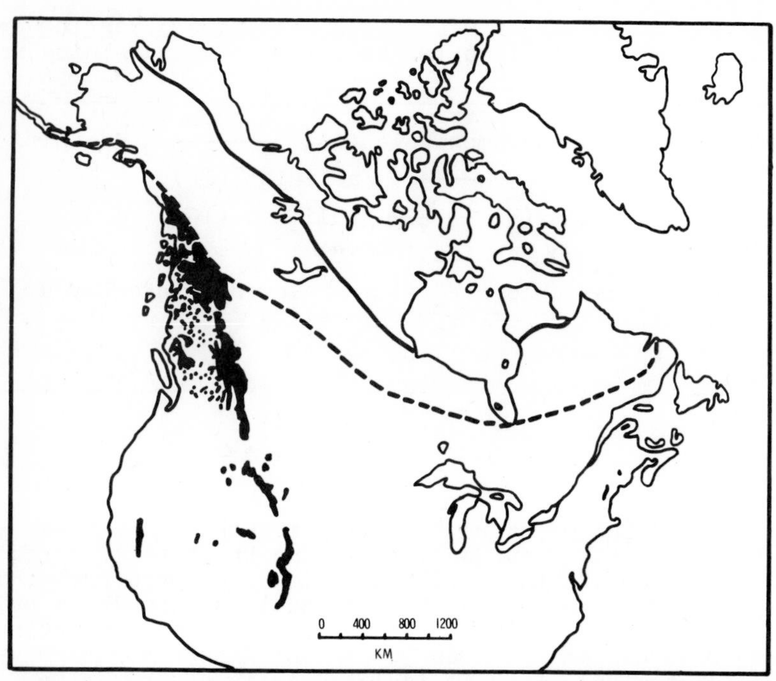

Figure 5 Distribution of permafrost in northern North America. Solid line denotes southern boundary of continuous permafrost zone. Dashed line represents southern limit of permafrost. Shaded areas portray permafrost areas at high altitudes south of the permafrost limit (Compiled from Brown, 1970; Péwé, 1973; Retzer, 1965; Ives, 1974; and author's experience in New England).

Range of western British Columbia as far south as 50°N (Brown, 1970). The second hypothesis is attractive from the standpoint that the Cascades have been volcanically active throughout the Pleistocene. The present cone of Mt. St. Helens (elev. 2,948 m) in southwestern Washington State, is believed to have formed as recently as 2,000 years ago (Mullineaux and Crandell, 1962). Heat generated during volcanism may have melted existing permafrost or may have prevented its formation throughout the late Quaternary. Alternatively, ameliorative climates associated with widespread volcanism in the Pacific Rim belt may have melted any permafrost existing at the time. A major problem with the geothermal hypothesis is that it doesn't account for the lack of permafrost in the Olympic Mountains, which are not of volcanic origin. Permafrost was encountered at 0.75 m depth on a 1825-m cinder cone near Mt. Garibaldi, southwestern British Columbia (Mathews, 1951). The permafrost extended to a depth of 12 m.

A third, less likely, hypothesis is that insufficient permafrost investigations have been carried out in the Cascade and Olympic Mountains. Evidence of present or relict permafrost features were never encountered by the author during three summers of soils investigations in the Cascade Range.

The effects of deep-lying permafrost on properties of Arctic and Alpine Brown soils are not known. In that the permafrost table is commonly below 1 m in areas featuring these well-drained soils, the present-day effect is probably minimal. During earlier, cooler climatic intervals, the permafrost table may have risen and had an effect on the moisture and temperature regimes of Arctic and Alpine Brown soils in North America.

Classification

Arctic Brown and Alpine Brown soils appear in four of the 10 orders in the United States soil classification system (Soil Survey Staff, 1970) (Table 7). Seven suborders are represented: Psamments, Orthents, Ochrepts, Umbrepts, Andepts, Borolls, and Orthods. Because of the temperature regime (MAST 0 to 8°C), Arctic and Alpine Brown soils are cryic at the subgroup level. Where these soils are underlain by permafrost, the soils become pergelic (MAST <0°C) at the great group level. Otherwise, the soils are typic, or lithic, if a lithologic contact is within 50 cm of the soil surface.

In the Canadian system (Canada Dept. of Agriculture, 1970), Arctic Brown soils are represented in two of the seven orders and four great groups: Melanic, Eutric, and Dystric Brunisols and Regosols (Cryic) (Table 8). Alpine Brown soils appear in these great groups and also in the Humo-ferric Podzol great group. In the proposed system for classifying cryosolic soils, Arctic and Alpine Brown soils would become Brunisolic and Regosolic Static Cryosols (Anonymous, 1973).

Future Research

A major need is to expand upon the program initiated (and since discontinued) by the International Biological Program -- Tundra Biome. Additional research sites should be established to supplement the Barrow, Prudhoe Bay, Eagle Summit, and Niwot Ridge sites. Climatic data at alpine stations in North America is especially lacking.

Alpine Brown soils should be examined along a transect up the Western Cordilleran chain from about 55°N to the southern Brooks Range. Critical areas for comparison of Alpine Brown and Arctic Brown soils are in western Alaska, including the mountains south of Norton Sound and north of the Yukon River, on the Seward Peninsula, and in the DeLong Mountains. An additional area of importance is the Richardson Mountains, Yukon Territory. Rieger's (1966) study in

Table 7. Classification of Arctic and Alpine Brown soils according to the U.S. system (U.S. Soil Survey Staff, 1970).

Order	Suborder
Entisol	Orthent
	Psamment
Inceptisol	Andept
	Ochrept
	Umbrept
Mollisol	Boroll
Spodosol	Orthod

Table 8. Classification of arctic and alpine soils according to the Canadian system (Can. Dept. of Agri., 1970)

Arctic Brown		Alpine Brown	
Order	Great Group	Order	Great Group
Brunisolic	Melanic	Podzolic	Humo-ferric
	Eutric	Brunisolic	Melanic
	Dystric		Eutric
Regosolic	Regosol		Dystric
		Regosolic	Regosol

the Yukon-Kuskokwim delta, western Alaska, provided useful data on dark well-drained soils of tundra regions, but did not separate the soils according to specific landscapes or geographic regions, *i.e.* alpine vs. arctic.

The examination of Alpine and Arctic Brown soils should be conducted in the company of a plant ecologist familiar with polar and alpine vegetation. Synecological work is needed in the critical areas mentioned. The effect of vegetation on development of Arctic Brown soils is not well understood.

Compared to western conterminous United States, our knowledge of Quaternary history and stratigraphy is deficient in the Low Arctic and in alpine areas in the Pacific Northwest and in the mountains of western Canada.

The effect of past and present permafrost on properties of Arctic and Alpine Brown soils needs elucidation. Finally, quantitative clay mineralogical analysis is needed for Arctic Brown and associated soils.

ACKNOWLEDGMENTS

Jerry Genson of the Wisconsin Soil Testing and Plant Analysis Laboratory and John Pastor, graduate student in soils, provided chemical analyses of the Baffin Island samples. Dr. T.M. Ballard, Department of Soil Science, University of British Columbia, provided chemical analysis of the Cornwallis Island samples.

REFERENCES CITED

Allison, L.E., 1965, Organic carbon, *in* Black, C.A. *et al.*, ed., Methods of Soil Analysis, Pt. 2, Agron. No. 9: Madison, WI., Amer. Soc. Agron., p. 1367-1378.

Anonymous, 1973, Tentative classification system for Cryosolic soils: Proc. of the Canada Soil Survey Committee, Nov., 1973, p. 350-358.

Antevs, E., 1932, Alpine zone of Mt. Washington Range: Auburn, ME., Merrill and Webber, 118 p.

Barry, R.G. and Hare, F.K., 1974, Arctic climate, *in* Ives, J.D. and Barry, R.G., ed., Arctic and Alpine Environments: London, Methuen, p. 17-54.

Barry, R.G. and Ives, J.D., 1974, Introduction, *in* Ives, J. D. and Barry, R.G., ed., Arctic and Alpine Environments: London, Methuen, p. 1-13.

Billings, W.D., 1974, Arctic and alpine vegetation: plant adaptations to cold summer climates, *in* Ives, J.D. and Barry, R.G., ed., Arctic and Alpine Environments: London, Methuen, p. 403-443.

Billings, W.C. and Mooney, H.A., 1968, The ecology of arctic and alpine plants: Biol. Rev., v. 43, p. 481-529.

Birkeland, P.W., 1973, Use of relative age-dating methods in a stratigraphic study of rock glacier deposits, Mt. Sopris, Colorado: Arctic Alpine Research, v. 5, p. 410-416.

Bliss, L.C., 1956, A comparison of plant development in microenvironments of arctic and alpine tundras: Ecol. Monog., v. 26, p. 303-337.

Bockheim, J.G., 1972, Effects of alpine and subalpine vegetation on soil development, Mount Baker, Washington, (Ph.D. Dissert): Seattle, Univ. of Washington, 171 p.

____________, 1974, Soils of Cumberland Peninsula, Baffin Island, N.W.T., Canada: Final rept. submitted to Parks Canada, Dept. of Indian & Northern Affairs, Ottawa, (Contract 74-112 to A.S. Dyke), 107 p.

Bouyoucos, G.J., 1951, A recalibration of the hydrometer method for making mechanical analysis of soils: Agron. Jour., v. 43, p. 434-438.

Bower, C.A., Reitemeier, R.F., and Fireman, M., 1952, Exchangeable cation analysis of saline and alkali soils: Soil Sci., v. 73, p. 251-261.

Bremner, J.M., 1960, Determination of nitrogen in soil by the kjeldahl method: Jour. Agr. Sci., v. 55, 1-23.

Brown, J. and Tedrow, J.C.F., 1964, Soils of the northern Brooks Range, Alaska: 4. Well-drained soils of the glaciated valleys: Soil Sci., v. 97, p. 187-195.

Brown, R.J.E., 1970, Permafrost in Canada: Toronto, Univ. of Toronto Press, 234 p.

Canadian Hydrographic Service, 1970, Climate of the Canadian Arctic: Can. Hydro. Ser., Marine Service Branch, Dept. Energy, Mines and Resources.

Canada Department of Agriculture, 1970, The system of soil classification for Canada, Publ. 1455: Ottawa, Infor. Canada, 255 p.

Chapman, H.D., 1965, Cation-exchange capacity, *in* Black, C.A., ed., Methods of Soil Analysis, Pt. 2, Agron. No. 9: Madison, WI., Amer. Soc. Agron., p. 891-901.

Costin, A.B., 1955, Alpine soils in Australia with reference to conditions in Europe and New Zealand: Jour. Soil Sci., v. 6, p. 35-50.

Day, P.R., 1965, Particle fractionation and particle-size analysis, *in* Black, C.A., ed., Methods of Soil Analysis, Pt. 1, Agron. No. 9: Madison, WI., Amer. Soc. Agron., p. 545-567.

deVilliers, J.M. and Jackson, M.L., 1967, Cation exchange capacity variations with pH in soil clays: Soil Sci. Soc. America Proc., v. 31, p. 473-476.

Drew, J.V. and Tedrow, J.C.F., 1957, Pedology of an Arctic Brown profile near Point Barrow, Alaska: Soil Sci. Soc. America Proc., v. 21, p. 336-339.

Energy, Mines and Resources, 1974, National Atlas of Canada; Toronto, MacMillan, 254 p.

Flohn, H., 1974, Contribution to a comparative meteorology of mountain areas, *in* Ives, J.D. and Barry, R.G., ed.,

Arctic and Alpine Environments: London, Methuen, p. 55-71.

Genson, J.J., Liegel, E. and Schulte, E.E., ed., 1976, Wisconsin soil testing and plant analysis procedures: Madison, Univ. of Wisconsin, 38 p.

Hill, D.E. and Tedrow, J.C.F., 1961, Weathering and soil formation in the arctic environment: Amer. Jour. Sci., v. 259, p. 84-101.

Holm, T.H., 1927, The vegetation of the alpine region of the Rocky Mountains in Colorado: Nat. Acad. Sci. Memoirs, v. 19, p. 1-45.

Holowaychuk, N., Petro, J.H., Finney, H.R., Farnham, R.S. and Gersper, P.L., 1966, Soils of Ogotoruk Creek Watershed, *in* Wilmovsky, N.J. and Wolfe, J.N., ed., Environment of the Cape Thompson region, Alaska: Oak Ridge, TN, U.S. Atomic Energy Comm., Div. Tech. Infor., p. 221-273.

Ives, J.D., 1974, Permafrost, *in* Ives, J.D. and Barry, R.G., ed., Arctic and Alpine Environments: London, Methuen, p. 159-194.

Ives, J.D. and Barry, R.G., ed., 1974, Arctic and Alpine Environments: London, Methuen, 999+ p.

Jackson, M.L., 1956, Soil chemical analysis -- advanced course: Madison, Soils Dept., Univ. of Wisconsin, 894 pp.

____________, 1968, Weathering of primary and secondary minerals in soils: 9th Internatl. Congr. Soil Sci. Trans., v. IV (Paper 30), p. 281-292.

Jenny, H., 1930, Hochgebirgsboden, *in* Handbuch der Bodenlehne: Berlin, Springer Verlag, p. 96-118.

Kittrick, J.A. and Hope, E.W., 1963, A procedure for the particle-size separation of soils for X-ray diffraction analysis: Soil Sci. v. 96, p. 319-325.

Knapik, L.J., Scotter, G.W. and Pettapiece, W.W., 1973, Alpine soil and plant community relationships of the Sunshine Area, Banff National Park: Arctic Alpine Research, v. 5, no. 3, pt. 2, p. 161-170.

Mahaney, W.C., 1974, Soil stratigraphy and genesis of Neoglacial deposits in the Arapaho and Henderson Cirques, central Colorado Front Range, *in* Mahaney, W.C., ed., Quaternary Environments: Proceedings of a Symposium: Geographical Monographs, No. 5: Toronto, York Univ., p. 197-240.

Mathews, W.H., 1951, Historic and prehistoric fluctuations of alpine glaciers in the Mount Garibaldi map-area, southwestern British Columbia, Jour. Geol., v. 59, p. 357-380.

McKeague, J.A., 1967, An evaluation of 0.1 M pyrophosphate and pyrophosphate-dithionite in comparison with oxalate extractants of the accumulation products in Podzols and some other soils: Can. Jour. Soil Sci., v. 47, p. 95-99.

_______________, 1968, Humic-fulvic acid ratio, Al, Fe and C in pyrophosphate extracts as criteria of A and B horizons: Can. Jour. Soil Sci., v. 48, p. 27-35.

Miller, C.D. and Birkeland, P.W., 1974, Probable pre-Neoglacial age of the type Temple Lake Moraine, Wyoming: discussion and additional relative-age data: Arctic Alpine Research, v. 6, p. 301-306.

Mullineaux, D.R. and Crandell, D.R., 1962, Recent lahars from Mt. St. Helens, Washington: Geol. Survey Am. Bull., v. 73, p. 855-870.

Nimlos, T.J. and McConnell, R.C., 1962, The morphology of alpine soils in Montana: Northwest Science, v. 36, p. 99-112.

____________, 1965, Alpine soils in Montana: Soil Sci., v. 99, p. 310-321.

Pawluk, S. and Brewer, R., 1975, Micromorphological, mineralogical and chemical characteristics of some alpine soils and their genetic implications: Can. Jour. Soil Sci., v. 55, p. 415-437.

Péwé, T.L., 1973, Ice wedge casts and past permafrost distribution in North America: Geoforum, v. 15, p. 15-26.

Polunin, N., 1951, The real arctic: suggestions for its delimitation, subdivision and characterization: Jour. Ecol., v. 39, p. 308-315.

Porsild, A.E., 1957, Illustrated flora of the Canadian Arctic Archipelago: Nat. Mus. Canada Bull. 146, 209 p.

Retzer, J.L., 1956, Alpine soils of the Rocky Mountains: Jour. Soil Sci., v. 7, p. 22-32.

_____________, 1965, Present soil-forming factors and processes in arctic and alpine regions: Soil Sci., v. 99, p. 38-44.

_____________, 1974, Alpine soils, *in* Ives, J.D. and Barry, R.G., ed., Arctic and Alpine Environments: London, Methuen, p. 771-802.

Reynolds, R.C., Jr., 1972, Clay mineral formation in an alpine environment: Clays Clay Minerals, v. 19, p. 361-374.

Rieger, S., 1966, Dark well-drained soils of tundra regions in western Alaska: Jour. Soil Sci., v. 17, p. 264-273.

__________, 1974, Arctic soils, *in* Ives, J.D. and Barry, R.G., ed., Arctic and Alpine Environments: London, Methuen, p. 749-769.

Sneddon, J.I., Lavkulich, L.M. and Farstad, L., 1972, The morphology and genesis of some alpine soils in British Columbia, Canada: Soil Sci. Soc. America Proc., v. 36, p. 100-110.

Soil Survey Staff, 1970, Selected chapters from the unedited text of the soil taxonomy of the National Cooperative Soil Survey: Soil Conserv. Serv., U.S. Dept. Agric., Wash., D.C. (variable pages).

Sorenson, C.J., Knox, J.C., Larsen, J.A. and Bryson, R.A., 1971, Paleosols and the forest border in Keewatin, N.W.T.: Quaternary Res. v. 1, p. 468-473.

Tedrow, J.C.F., 1970, Soil investigations in Inglefield Land, Greenland: Meddelelser om Grønland, v. 188, 93 p.

__________, 1973, Soils of the polar region of North America: Biuletyn Peryglacjalny, v. 23, p. 157-165.

Tedrow, J.C.F. and Douglas, L.A., 1964, Soil investigations on Banks Island: Soil Sci., v. 98, p. 53-65.

Tedrow, J.C.F. and Hill, D.E., 1955, Arctic Brown soil: Soil Sci. v. 80, p. 265-275.

Tedrow, J.C.F., Drew, J.V., Hill, D.E. and Douglas, L.A., 1958, Major genetic soils of the Arctic Slope of Alaska: Jour. Soil Sci., v. 9, p. 33-45.

Tedrow, J.C.F., Bruggemann, P.F. and Walton, G.F., 1968, Soils of Prince Patrick Island: Arctic Inst. of North Amer., Res. Pap. No. 44, Washington, D.C., 82 p.

Thiesen, A.A. and Harward, M.E., 1962, A paste method for preparation of slides for clay mineral identification by X-ray diffraction: Soil Sci. Soc. America Proc., v. 26, p. 90-91.

Thorp, J. and Smith, G.D., 1949, Higher categories of soil classification: Order, suborder, and great soil groups: Soil Sci., v. 67, p. 117-126.

Ugolini, F.C., 1966, Soils of the Mesters Vig district, northeast Greenland. 1. The Arctic Brown and related soils: Medd. om Gronland, v. 176, 22 p.

U.S. Dept. Interior, 1970, National Atlas of the U.S.: Washington, D.C., U.S. Geol. Survey, 335+p.

U.S. Weather Bureau, 1964, Local climatological data with cooperative data 1963: Mount Washington, New Hampshire: U.S. Dept. Comm., Weather Bureau, Ashville.

U.S. Weather Bureau, (no date), Mount Baker, Washington Station, Climatological summary: means and extremes for period 1927-1951, 1 sheet.

van Ryswyk, A.L., 1969, Forest and alpine soils of south-central British Columbia (Ph.D. Dissert.): Pullman, Wash., Wash. State Univ., 178 p.

Vowinkle, E. and Orvig, S., 1970, The climate of the North Polar Basin, *in* Orvig, S., ed., Climates of the Polar Regions: World Survey of Climatology, v. 14: N.Y., Elsevier, p. 129-252.

Warshaw, C.M. and Roy, R., 1961, Classification and a scheme for the identification of layer silicates: Geol. Soc. Amer. Bull., v. 72, p. 1455-1492.

Webber, P.J., 1974, Tundra primary productivity, *in* Ives, J.D. and Barry, R.G., ed., Arctic and Alpine Environments: London, Methuen, p. 445-473.

Zwinger, A.H. and Willard, B.E., 1972, Land above the trees: a guide to American alpine tundra: N.Y., Harper and Row, 489 p.

Genesis of Organic Soils in Manitoba and the Northwest Territories

C. Tarnocai

Contribution No. SRI 601

ABSTRACT

Organic soil formation began several thousand years (4-6 thousand BP) after the glacial ice retreated from the continental portion of the study area. On the arctic islands, however, it began shortly after the glacial ice melted (8.5-9 thousand BP). These time differences are due to the fact that the climate was too dry and warm for organic soil formation on the continent while the cooler and moister climate of the arctic islands was much more favorable for peat deposition. The conditions for optimum organic soil development now occur in the boreal and subarctic regions where the climate is cool and moist.

Two basic processes have been identified as relating to the deposition of organic soil materials in the study area: the filling-in process is associated with shallow lakes where peat deposition slowly fills in the lake basin while the gradual build-up process commonly occurs on level and gently sloping areas.

Time is the dominant soil-forming factor and all other factors (climate, water properties, relief, vegetation and organisms) are time-dependent or subdominant factors. The interaction of the various subdominant factors and time produces the individual organic soil layers resulting in the organic soil profile. The effects of time and the subdominant factors on the genesis of both the Organic and Cryosolic soils is discussed in detail in the paper.

The type of organic soil material (sphagnum, fen, forest, aquatic and mixed peats) resulting from the above processes will also be given along with the dominant organic soils belonging to the Organic and Cryosolic Orders.

INTRODUCTION

Organic soils, developed from peat materials, compose

a large percentage of the area in Manitoba and the Northwest Territories. In Manitoba, approximately 30 percent of the land area is covered by organic soils and what is probably the world's largest concentration of organic soils is found in the Hudson Bay Lowland. In the Northwest Territories, organic soils are one of the major soils in the boreal and subarctic regions and they are also very common in the low arctic region (see Figure 1). The total area of organic soils in Canada is 339 x 10^6 acres (358 x 10^3 square miles) as given by Clayton *et al.* (in press).

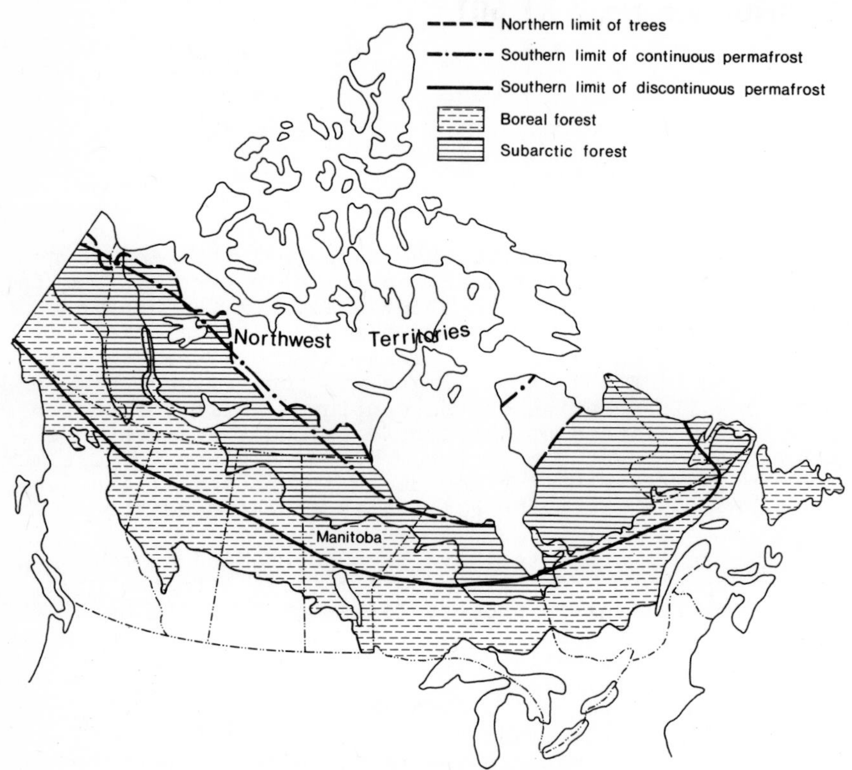

Figure 1 The arctic (north of the tree-line), subarctic, boreal and prairie (in Manitoba, south of the boreal) regions are shown on the map of the study area.

The term "organic soil," as used in this paper, refers to all of the soils which are included in the Organic Order (The System of Soil Classification for Canada, 1974) and to the Organic Cryosols (C.S.S.C., 1973). The soils in the Organic Order contain 30 percent or more organic matter to a depth of 60 cm or greater if the surface layer is fibric peat or 40 cm or greater if the surface layer is of another kind of peat. Organic Cryosols contain 30 percent or more organic matter to a depth of 40 cm or greater.

The objectives of this paper are to describe the basic processes of peat deposition, the type of materials associated with these processes and the genesis of the organic soils developed from these peat materials in Manitoba and the Northwest Territories.

PEAT DEPOSITION PROCESSES

Peatlands are dynamic systems formed by the interaction of biological and physical environmental factors. The stratigraphy of peat materials deposited through long periods of time is evidence of the environmental conditions operating throughout that period of time. The depositional sequence indicates the dynamic conditions in the peatland as a result of environmental changes, *eg*. climatic and hydrological variation or the development of permafrost.

Radiocarbon dates of basal peat materials indicate that peat deposition did not begin until several thousand years after the continental ice melted on the mainland. The oldest date in the Mackenzie River area is 8190 ± 60 BP (Table 1) and in Manitoba 7220 ± 110 BP (Table 1). When all of the radiocarbon dates from Manitoba and the Districts of Mackenzie and Keewatin in the Northwest Territories are compared, it appears that the majority of peat deposits began to develop between 4 and 6 thousand BP (2,000 to 4,000 BC). On the arctic islands, however, the basal peat dates are much older - 8.5 to 9 thousand BP (6,500 to 7,000 BC) - indicating that peat development began much earlier than on the continent.

There is evidence that there was a relatively warm and dry period, lasting several thousand years, after the retreat of glacial ice from the North American continent (Nichols, 1969; Ritchie, 1971; Terasmae, 1972). It is possible that during this period it was generally too warm and dry for optimum peat development. In the high arctic areas the climate was assumed to have been cooler and moister (although warmer than at present) after the retreat of glacial ice and, consequently, conditions were favorable for peat development. As the climate became colder, peat development ceased in the high arctic and the boreal and subarctic regions became established as the areas of optimum peat development.

Examination of peat deposits suggests that two basic types of peatland development take place: the filling-in process (Zoltai and Tarnocai, 1975; Dansereau *et al.*, 1952) and the gradual build-up process (Zoltai and Tarnocai, 1975).

The filling-in process (Figure 2) begins in a shallow lake or pond. The basal peat may contain marl or gastropods in the aquatic peat and this is followed by fen peat. Fen peat begins to form along the lakeshore and eventually forms

Table 1. Radiocarbon dates of basal peat.

Location	Age of Basal Peat (years)	Source
Porcupine Mountain, Manitoba	6770 ± 70 BP	Nichols, 1969
Clearwater Bog, Manitoba	1280 ± 75 BP	Nichols, 1969
Grand Rapids, Manitoba	7220 ± 110 BP	Richie and Hadden, 1975
52°53'N & 99°08'W, Manitoba	4670 ± 130 BP	Klassen, 1967
Lynn Lake, Manitoba	6530 ± 130 BP	Nichols, 1967
67°16'N & 135°14'W, NWT	8190 ± 60 BP	Zoltai and Tarnocai, 1975
67°41'N & 132°05'W, NWT	7200 ± 60 BP	Zoltai and Tarnocai, 1975
66°13'N & 130°52'W, NWT	5600 ± 70 BP	Zoltai and Tarnocai, 1975
69°07'N & 132°56'W, NWT	6020 ± 100 BP	Zoltai and Tarnocai, 1975
65°50'N & 129°05'W, NWT	6120 ± 120 BP	Mackay *et al.*, 1973
65°15'N & 126°42'W, NWT	3960 ± 50 BP	Korpijaakko *et al.*, 1972
69°30'N & 135°47'W, NWT	4140 ± 140 BP	Lowdon *et al.*, 1971
Ennadai Lake, NWT	5780 ± 110 BP	Nichols, 1967
Bathurst Island, NWT	9210 ± 170 BP	Blake, 1964
75°50.5'N & 98°02.5'W, NWT	8420 ± 80 BP	Lowdon and Blake, 1975.
Sherard Valley, Mellville Is., NWT	9040 ± 160 BP	Barnet, 1973

a floating mat which slowly extends into the center. The water space between this floating fen mat and the aquatic peat layer on the lake bottom is then slowly replaced by peat material. The peatland at this stage is still influenced by mineral-rich waters (minerotrophic). These peat deposits are sometimes capped by sphagnum peat when the peatland is no longer influenced by mineral-rich waters. At this stage it has become elevated above the regional water table and hence is ombrotrophic.

The gradual build-up process (Figure 3) begins in moist areas where the water table is close to the surface or an increase in moisture regime results in the invasion of mosses, particularly *Sphagnum* species. This process does not produce marl of aquatic peat. There may be a thin organic-rich mineral layer at the base of the peat deposit

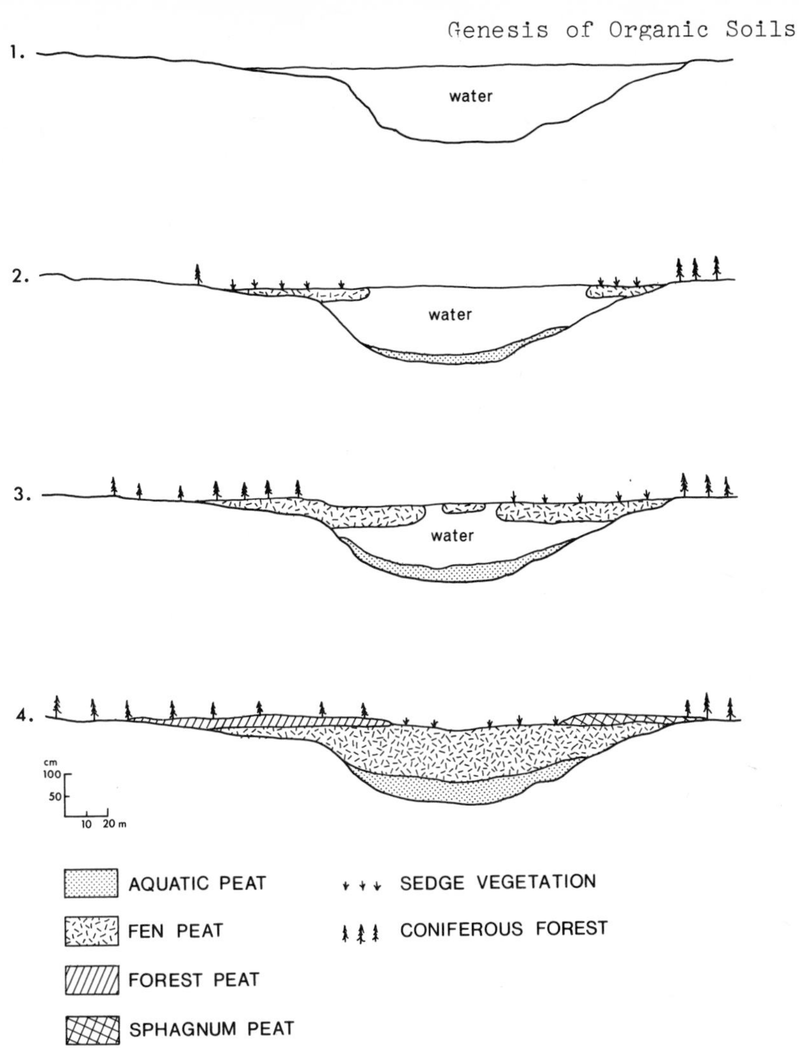

Figure 2   A schematic diagram showing the probable stages of the filling-in process.

but it is followed by layers of either forest or sphagnum peat or both. Very often this process produces a blanket-like peat deposit on the terrain, covering not only the level areas but also the gentle slopes and very often the uplands as well. Such peat deposits, in most cases, are not very deep and their depth varies greatly according to the topography. This process operates under ombrotrophic conditions even though it is initiated on extremely calcareous till materials, *e.g.* in central Manitoba.

In most climatic regions of the study area both of these processes are present although, in certain regions,

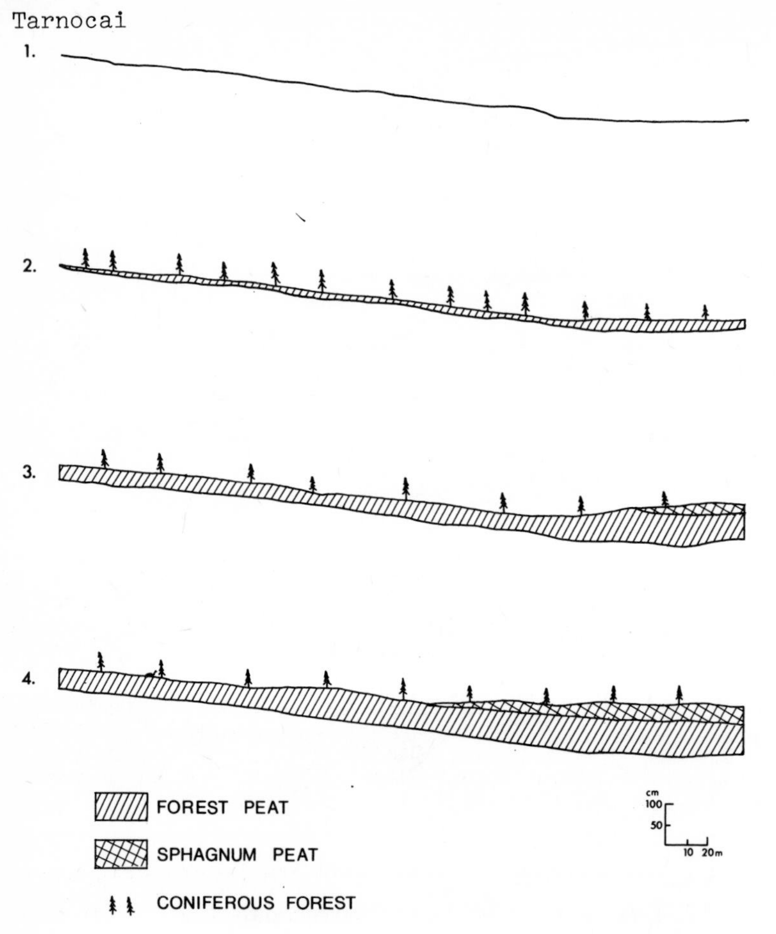

Figure 3 A schematic diagram showing the probable stages of the gradual build-up process.

one or the other may be absent. Very few peatlands are found in the prairie region. In most cases the organic deposit is not deep enough for it to be classified as peatland or organic soil. If they are present in this region, they have developed as a result of the filling-in process.

The boreal and subarctic regions are the optimum areas, under present climatic conditions, for peat development to take place. In this regions both the filling-in and the gradual build-up processes are very active. Most of the peat deposits in these areas are now associated with permafrost but, regardless of which process formed the deposit, most of the deposition took place in a permafrost-free environment. Thus, the development of permafrost in these regions represents a secondary process resulting in permafrost-related peat landforms (*e.g.* palsas, peat plateaus and polygonal peat plateaus) on deposits developed by the filling-in or gradual build-up processes.

In the arctic region, however, peat deposition takes place in a permafrost environment (Zoltai and Tarnocai, 1975). In the initial stage, on low centered lowland polygons where ice wedges have already developed in the mineral materials, the filling-in process is probably the most active. The peat deposit, however, reaches a stage at which, because of the filling-in by organic materials, the centers are no longer depressed. From then on a gradual build-up process takes place along with a lateral growing of ice wedges, which results in excessive mixing of organic and mineral materials. On the arctic islands very little or no active peat deposition is taking place at the present time and the existing peat deposits are mainly relict features in a strongly eroded state.

GENESIS OF ORGANIC SOILS

In the previous section two basic processes were described for peat deposition. In some cases, the peat deposition processes produce an organic soil with very little or no further chemical or physical changes having to take place. There is, however, generally further decomposition. In either of these instances subsequent major, chemical, physical and morphological changes occur when permafrost develops (Tarnocai, 1972). A majority of the organic deposits in the study area had reached an advanced state of development in a permafrost-free environment with permafrost forming at a much later date. Those soils that are in a permafrost-free environment are classified as Organic Soils while those that are associated with permafrost are classified as Organic Cryosols.

The genesis of organic soils began at the time when the basal peat was deposited and continues to the present. During this time physical and chemical changes have taken place in the soil but the changes are not as great, even in the lower or older layers, as in some mineral soils. Organic soils in a sense represent a high energy balance system where a great deal of energy is stored and very little is released (by degradation). The energy which is released is mainly from the surface layers with an increasingly smaller amount released from the lower layers.

The peat parent material is continuously being added to the surface by vegetation litter. Thus, the parent material of organic soil reflects the succession of vegetation, characterized by layers differing not only as to their degree of decomposition but also as to the nature of the parent materials.

Organic soil, in most cases, is composed of more than one peat layer. These peat layers are the reflection of the type of vegetation contributing to the organic layer rather than of the later soil-forming processes, as in the case of the development of soil horizons in mineral soils. Thus, to show the genesis of organic soils with a single equation, as is done for mineral soils, is not possible. The approach taken here is to show that the genesis of organic soil is greatly time-dependent and began at time zero when the basal peat was deposited. Time is the dominant soil-forming factor and all other factors are time-dependent.

A single, homogeneous peat layer is the result of the interaction of biological and physical soil-forming factors. Thus, the equation for a single peat layer (Sp) can be written:

$$Sp = f\ (cl, w, r, v, o, t \ldots) \qquad (1)$$

where the variables are defined as: cl-climate; w-water properties; r-relief or landform; v-vegetation; o-organisms; and t-time. The dots stand for unspecified components such as permafrost and dust or water pollution.

The organic soil, as indicated above, is composed of several peat layers, and all soil-forming factors are time-dependent. Thus, the equation for an organic soil (So) is:

$$So = Sp1 + Sp2 \_ Sp3 \_ \ldots + Spn = f(t)_{cl,\ w,\ r,\ v,\ o\ \ldots} \qquad (2)$$

where Spl, Sp2, Sp3 .... Spn represent the individual layers of peat materials. Time (t) is the dominant factor and stands for the total time. The subdominant factors could change with time and are listed as subscripts.

This relationship is probably better shown in a schematic model (Figure 4) illustrating the organic profile development and the relationship between the various subdominant factors and between the subdominant factors and time. Figure 4 indicates that soil genesis is active at the Sp3 level. Processes such as removal of soluble organic materials by ground water, further decomposition, or chemical and physical changes resulting from permafrost development could also be active at the Spl and Sp2 levels.

FACTORS AFFECTING THE GENESIS OF ORGANIC SOILS

Time

As has already been indicated, time is the dominant factor in the genesis of organic soils. Time zero, the point at which peat development began (basal peat), is 4-6 thousand years BP in the continental regions and 8.5-9 thousand years BP in the arctic islands. From time zero onwards the interaction of the soil-forming factors produces

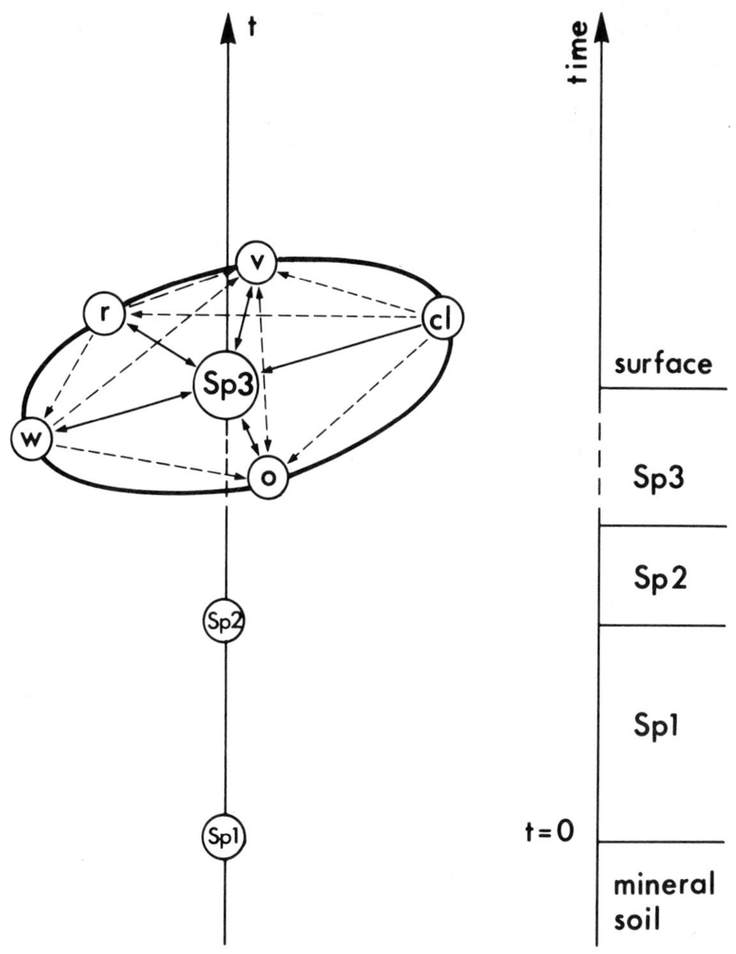

Figure 4 Organic soil profile development and the relationship between the subdominant factors as a function of time, where t is total time. cl is climate; w is water properties; r is relief or landform; v is vegetation; o is organisms; and Sp a single peat layer.

the different layers of peat materials, the most recent one being on the surface. Any of the soil-forming factors, however, could change with time producing different organic layers or changing the rate of peat deposition. In the extreme case, the change (*e.g.* in climate) could be so drastic that organic soil development would cease, as was the case with the arctic islands.

Climate

There is very little organic soil development in either the prairie region because the climate is dry and warm or in the high arctic islands because the climate is dry and cold. The most favorable climate for the development of organic soils coincides with that of the boreal and subarctic regions. Here, the climate is cool and moist, the optimum climatic conditions for organic soil development (Terasmae, 1972).

The majority of organic soils in the study area are associated with permafrost. Permafrost development probably took place in the cooler climatic period. For example, in central Manitoba between 600 BP and 200 BP (Thie, 1974) a drastic cooling in soil climate resulted in permafrost-related changes in the physical and chemical properties of the organic soils (Zoltai and Tarnocai, 1971; Tarnocai, 1972). In addition to this, an elevated (plateau or domed) surface morphology developed due to ice lense formation in the perennially frozen soil materials.

Water Properties

The chemical composition of the ground waters, especially the cation content, is a very important factor influencing the floristic composition of peatland vegetation, hence, the type of peat or the organic soil development. This hypothesis is well demonstrated by Sjors, 1963; Heinselman, 1970; Tarnocai, 1973; and Mills *et al.*, 1974.

In Table 2 the chemistry of surface waters from various peatlands and organic soils of northern Minnesota, southeastern Manitoba, Hudson Bay Lowland and the upper Mackenzie River area is given. This data indicates that Fibrisols (bogs) have developed in areas where the water is low in pH (3.3-4.7), low in calcium (0.5-4.0 ppm) and magnesium (0.1-0.7 ppm) and very low in sodium and potassium. The anion content of these waters is also very low. The waters associated with Mesisols (fen type of peatland), on the other hand, are much higher in pH (5.3-7.8) and calcium (5.0-42 ppm), higher in magnesium (0.1-22.2 ppm) and sodium (6.0-7.6 ppm), low in potassium and medium in anion content.

Any changes in the hydrology of the peatland (excessive drainage or inundation) will also affect the vegetation and hence the formation of the organic soil.

Relief or Landform

Landform types associated with peatlands in the study area have been studied and described by Tarnocai (1970) and Zoltai *et al.*, (1975). They recognized three basic types of peat landforms: bogs, fens and swamps. Subdivisions of these are based on the surface morphology (*e.g.* domed, plateau, flat, sloping and patterned).

Table 2. Chemical composition of surface waters from various types of peatlands and organic soils.

Peatland Type and Organic soil	Sample No.	pH	Ca^{+2}	Mg^{+2}	Na^{+}	K^{+}	Cl^{-}	SO_4^{-2}	HCO_3^{-}	Source
					ppm					
Bog-pool	1	4.6	0.5	0.2	0.3	0.1	1.2	1.1	0	Sjors, 1963
Ombrotrophic bogs (bog-pool)	67-68,156-157,188-189, 232-233	3.3-3.8	1.0-1.6	0.1-0.4	-	-	-	-	-	Heinselman, 1970
Domed bog (Typic Fibrisol)	S106	4.5	1.9	0.6	-	-	-	-	-	Mills *et al.*, 1974
Domed bog (Typic Fibrisol)	S55	4.1	4.0	0.7	-	-	-	-	-	Mills *et al.*, 1974
Bog plateau (Fibrisol)	108	4.7	2.0	0.7	Tr[a]	-	Tr	Tr	Tr	Tarnocai, 1973
Polygonal peat plateau (Fibric Organic Cryosol)	102	3.9	1.4	0.1	Tr	-	Tr	Tr	Tr	Tarnocai, 1973
	111	4.0	2.8	0.1	Tr	-	Tr	Tr	Tr	Tarnocai, 1973
Fen-flark	3	6.8	8.9	1.8	0.9	0.3	2.4	Tr	24.1	Sjors, 1963
Patterned fen	117-18,-119, -120,-176,-177, -182,-183,-260	5.3-6.4	5.0-10.6	0.1-2.8	-	-	-	-	-	Heinselman, 1970
Lowland fen (Typic Mesisol)[b]	S18	6.9	36.0	4.6	-	-	-	-	-	Mills *et al.*, 1974
	S19	7.2	42.0	6.0	-	-	-	-	-	Mills *et al.*, 1974
Patterned fen (Typic Mesisol)	100	7.8	18.6	22.2	7.6	-	17.7	29.7	97.0	Tarnocai, 1973
Flat fen (Typic Mesisol)	119	6.9	37.1	3.3	6.0	-	3.5	1.9	164.1	Tarnocai, 1973

[a]Traces, less than 0.1 ppm

[b]periodically burned

The landform type determines the moisture regime and the watersource for the peatland and thus, the type of vegetation growing on the peat deposit. In general, fen and swamp types of peat landforms are associated with a minerotrophic environment and the resulting organic soils are Mesisols or Humisols. They are characterized by saturated conditions and the water table is above or just at the surface for most of the growing season. The water supply of fens and swamp type of landforms is mainly from mineral-rich ground waters. On the other hand, bog type peat landforms are ombrotrophic (water supply is mainly from rain) and the resulting organic soils are Fibrisols, Mesisols and Organic Cryosols. The water of the bogs is below the surface and, in the extreme case of the domed bogs, especially those associated with permafrost, there is a very dry surface peat cover.

Vegetation

Vegetation plays a very important part in the development of organic soils since the organic soil material originates from vegetation and reflects the succession of vegetation by its peat layers. The properties of the soil, *e.g.* degree of decomposition and chemistry, are largely related to the type of vegetation from which the organic material was derived.

All peat materials described in a later section are grouped together according to their botanical origin. They are derived, basically, from four types of vegetation: the forest, sedge, *Sphagnum* and aquatic vegetation communities. In general, *Sphagnum* vegetation produces fibric peat materials and Fibrisols; forest and sedge vegetation produces dominantly mesic and humic peat materials and Mesisols and Humisols; and aquatic vegetation produces humic peat materials and Humisols.

The peatland vegetation communities in the boreal region have been studied and described according to species composition by Heinselman (1970), Dansereau *et al.*, (1952) and Moss (1953). They delineated the floristic composition of the peatland environment and the successional stages resulting from environmental changes.

Organisms

The decomposition rates of organic soils are frequently much slower than those of mineral soils. This is basically due to the following factors:

1. Low oxygen content associated with waterlogging
2. Low nutrient content and pH
3. Low soil temperatures.

Latter *et al.* (1967) studied the microbiological activity in organic (peat) soils and compared the results with those obtained from mineral soils from the grassland region.

They found that the total numbers of bacteria are half as much in the peat soil ($14-35 \times 10^8/cm^3$) as in the grassland soil ($16-79 \times 10^8/cm^3$). They also estimated the total length of living fungi hyphae and they found that the peat soil contained 15-180 m/cm^3 and the grassland soil contained 160-580 m/cm^3. The ratio of bacteria to length of fungal mycelium is 1:300 in grassland soil and 1:1300 in peat soil. They also indicated that nitrogen-fixing bacteria, both aerobic and nitrifying, are virtually absent in peat soil.

Decomposition occurs most rapidly in the surface layer of the organic soil profile. The studies of Clymo (1965) show that the greatest loss in dry weight, which indicates the rate of decomposition, occurs in the surface 20 cm and becomes very low or disappears completely below this depth. This is due to the anaerobic conditions under which very few decomposer organisms can operate.

There are also indications that the rate of decomposition differs depending on the botanical origin of the organic soil material. *Sphagnum papillosum* decomposes at only about half the rate of *S. cuspidatum* (Clymo, 1965). The more easily metabolized compounds will be used up by the organisms most rapidly, leaving less palatable compounds to be degraded more slowly. This means that the decomposition rate is rapid in the initial stages and becomes slower as time proceeds (Waksman and Stevens, 1929; Theander, 1954).

The soil animals also play an important role in the decomposition process (Cragg, 1961; Macfadyen, 1963). There is, however, too little data available for a complete assessment of their role in the decomposition of organic soils.

PEAT MATERIALS

The peat materials that occur in the study area can be grouped into five broad types according to their botanical origin and physical and chemical properties. These peat materials are described in detail in the works of Tarnocai (1973), Mills (1974) and Zoltai and Tarnocai (1975). A brief description of the peat types and subtypes is given below:

Sphagnum Peat

This type of peat material develops on very wet to wet bogs with the dominant peat former being the *Sphagnum* spp. *Sphagnum* peat is usually undecomposed (fibric), light yellowish-brown to very pale brown in color and loose and spongy in consistency with the entire *Sphagnum* plant being readily identifiable. In the north somewhat decomposed (mesic) *Sphagnum* peat also occurs, characterized by darker colors. Two subtypes of *Sphagnum* peat (*recurvum* and *fuscum*) were separated based on the dominant species and the peat environment.

Sphagnum (recurvum) Peat. *Sphagnum* growing submerged

in acid water in pure colonies. The main peat formers are *Sphagnum recurvum* P.-Beauv., *S. squarrosum* Crome and *S. cuspidatum* Ehrh.

Sphagnum (*fuscum*) Peat. *Sphagnum* growing in moist to wet acid peat, associated with scattered ericaceous shrubs consisting of cushion-forming *Sphagnum* species, such as *S. fuscum* (Schimp.) Klinggr. and *S. rubellum* Wils.

Fen Peat

This material develops on very wet fens with the dominant peat formers being *Carex* sp., *Drepanocladus* sp. and tamarack. Fen peat is usually moderately well-decomposed (mesic to humic), dark brown to very dark brown, and the fibers are fine to medium with a horizontally matted or layered structure. The subtypes of fen peat were separated based on the dominance of the plant material.

Drepanocladus (Brown Moss)-Fen Peat. The dominant peat formers are dark-colored mosses of the Genera *Drepanocladus, Calliergon* and *Aulacomnium.*

Carex (Sedge)-Fen Peat. The dominant peat formers are *Carex* sp. and some *Eriophorum* sp.

Woody-Fen Peat. The dominant peat formers are *Larix laricina* (Du Roi) K. Koch, *Betula glandulosa* Michx. and *Kalmia polifolia* Wang.

Forest Peat

This material develops on slightly better-drained bogs under black spruce with the dominant peat formers being ericaceous shrubs, feather mosses and lichens. The forest peat is usually moderately decomposed (mesic), has a very dark brown to dark reddish-brown matrix, an amorphous to very fine-fibered structure and may have a somewhat layered macrostructure. The subtypes of forest peat were separated based on the dominance of the plant material.

Woody-Forest Peat. The dominant peat formers are *Picea mariana* and ericaceous shrubs (*Ledum* spp., *Vaccinium vitis-idaea* L., *Empetrum nigrum* L. and *Larix laricina).*

Feather Moss-Forest Peat. The dominant peat formers are the feather mosses (*Hypnum* spp., *Dicranum* spp., *Pleurozium* spp. and *Hylocomium* spp.) associated with locally dense black spruce forest cover.

Lichen-Forest Peat. The dominant peat formers are lichens (*Cladonia* spp. and *Cetraria nivalis* (L.) Ach.), feather mosses and ericaceous shrubs.

Aquatic Peat

This material usually develops in shallow lakes and

ponds. The peat is primarily derived from various aquatic mosses, plants and algae. The material is slightly sticky, dark brown to black in color and is usually well-decomposed (humic). Aquatic peat is usually found at the bottom of the peat deposits.

Mixed Peat

This material is found in collapsed areas where, due to the thawing of permafrost, the peat banks are eroding and slumping into a water saturated area. During this process the different peat layers are mixed extensively resulting in a peat deposit where two or more types are usually intermixed.

SUMMARY OF ORGANIC SOILS DEVELOPED IN THE STUDY AREA

Organic soils have developed from the five basic peat materials which dominate the boreal, subarctic and continental arctic (low arctic) portions of the study area. They are rare in the prairie region and cover only small areas on the arctic islands.

In the prairie region, a very small amount of Humisols have developed from well decomposed fen and aquatic peat materials. The optimum organic soil development occurs in the boreal and subarctic regions. The dominant organic soils in the permafrost-free portion of the boreal region are Fibrisols developed from *Sphagnum* and some forest peat, and Mesisols developed from fen and forest peat materials. North of the discontinuous permafrost boundary the above-indicated Fibrisols and Mesisols are still the major organic soils. Organic Cryosols (C.S.S.C., 1973) have developed, however, on all peat materials and cover increasingly larger areas as one moves northward, becoming the dominant organic soils in the northern part of the subarctic region. In the arctic region only Organic cryosols are found and they have developed mainly from fen-type peat materials.

ACKNOWLEDGMENTS

I wish to thank S.C. Zoltai, M.A. Zwarich and J.H. Day for their technical advice and for reviewing the manuscript.

REFERENCES CITED

Barnet, D.B., 1973, Radiocarbon Dates from eastern Melville Island, Report of Activities Part B, Geol. Survey Can., Paper 73-1B: Ottawa, p. 137-140.

Blake, W., 1964, Preliminary account of the glacial history of Bathurst Island, Arctic Archipelago, Geol. Survey Can., Paper 64-30: Ottawa, 8 p.

Clayton, J.S., Ehrlich, W.A., Cann, D.B., Day, J.H., and Marshal, I.B. (in press), Soils of Canada V.2 Soil inventory and two 1:5000000 maps: Ottawa, Can. Soil Surv. Comm. and Soil Res. Inst.

Clymo, R.S., 1965, Experiments on the breakdown of *Sphagnum* in two bogs: Jour. Ecol. v. 53, p. 747-758.

Cragg, J.B., 1961, Some aspects of the ecology of moorland animals: Jour. Anim. Ecol., v. 30, p. 205-234.

C.S.S.C., 1973, Tentative Classification for Cryosolic Soils, *in* Proc. of the Ninth Meeting of the Canada Soil Survey Committee: Saskatoon, Univ. of Saskatchewan, p. 350-358.

Dansereau, P., and Segadas-Vianna, F., 1952, Ecological Study of the Peat Bogs of Eastern North America: Can. Jour. Botany, v. 30, p. 490-520.

Heinselman, M.L., 1970, Landscape evolution, peatland types, and the environment in the Lake Agassiz Peatlands natural area, Minnesota: Ecol. Monographs, v. 40, p. 235-261.

Klassen, R.W., 1967, Surficial Geology of the Waterhen-Grand Rapids Area, Manitoba: Ottawa, Geol. Survey Can. Paper 66-36, 6 p.

Korpijaakko, M., Korpijaakko, M.L., and Radforth, N.W., 1972, Ice-biotic relationship in frozen peat, *in* Proc. 14th Muskeg Res. Conf., 1971: Nat. Res. Counc., Ass. Comm. Geotech. Res., Tech. Mem. No. 102, p. 111-120.

Latter, P.M., Cragg, J.B. and Heal, O.W., 1967, Comparative studies on the microbiology of four moorland soils in the North Pennines: Jour. Ecol., v. 55, p. 445-482.

Lowdon, J.A., Robertson, I.M., and Blake, W., Jr., 1971, Geological Survey of Canada radiocarbon dates, XI: Radiocarbon, v. 13, p. 255-324.

Lowdon, J.A., and Blake, W., Jr., 1975, Geological Survey of Canada radiocarbon dates, XV: Ottawa, Geol. Survey Can. Paper 75-7, 32 p.

Macfadyen, A., 1963, The contribution of the microfauna to soil metabolism, *in* Doeksen, J., and Van Der Drift, J., eds., Soil Organisms: Amsterdam, 3.

Mackay, J.R. and Mathews, W.H., 1973, Geomorphology and Quaternary history of the Mackenzie River valley near Fort Good Hope, N.W.T., Canada: Can. Jour. Earth Sci., v. 10, p. 26-41.

Mills, G.F., 1974, Organic Soil Parent Materials, *in* Proc. of the Canada Soil Survey Committee Organic Soil Mapping Workshop, Winnipeg, Man., June 3-7, 1974, p. 21-43.

Mills, G.F., Hopkins, L.A., and Smith, R.E., 1974, Inventory and assessment for agriculture of the organic soils of the Roseau River watershed in Manitoba, Canada-Manitoba Soil Survey, 174 p.

Moss, E.H., 1953, Marsh and bog vegetation in northwestern Alberta: Can. Jour. Botany, v. 31, p. 448-470.

Nichols, H., 1967, Pollen diagrams from subarctic central Canada: Science, v. 155, p. 1665-1668.

__________, 1969, The Late Quaternary History of Vegetation and Climate at Porcupine Mountain and Clearwater Bog, Manitoba: Arctic and Alpine Res., v. 1, p. 155-167.

Ritchie, J.C., and Hare, F.K., 1971, Late Quaternary vegetation and climate near the arctic tree line of northwestern North America: Quaternary Research, v. 1, p. 331-342.

Ritchie, J.C., and Hadden, K.A., 1975, Pollen stratigraphy of Holocene sediments from the Grand Rapids area, Manitoba, Canada: Review of Palaeobotany and Palynology, v. 19, p. 193-202.

Sjörs, H., 1963, Bogs and fens on Attawapiskat River, northern Ontario, Cont. to Botany, 1960-61: Nat. Museum of Canada, Bulletin No. 186, p. 45-133.

Tarnocai, C., 1970, Classification of peat landforms in Manitoba: Canada Dept. of Agriculture, Res. Station, Pedology Unit, Winnipeg, 45 p.

__________, 1972, Some characteristics of Cryic Organic Soils in Northern Manitoba: Can. Jour. Soil Sci., v. 52., p. 485-496.

__________, 1973, Soils of the Mackenzie River area: Environmental-Social Program, Northern Pipelines, Report No. 73-26, 136 p.

Terasmae, J., 1972, Muskeg as a climate-controlled ecosystem, Proc. 14th Muskeg Res. Conf., Kingston, 1971: N.R.C. Tech. Memo. No. 102, p. 147-158.

Theander, O., 1954, Studies on *Sphagnum* peat. 3. A quantitative study of the Carbohydrate constituents of *Sphagnum* mosses and *Sphagnum* peat: Acta Chem. Scand., v. 8, p. 989-1000.

Thie, J., 1974, Distribution of thawing of permafrost in the Southern Part of the discontinuous permafrost zone in Manitoba: Arctic, v. 27, p. 189-200.

Waksman, S.A., and Stevens, K.R., 1929, Contributions to the chemical composition of peat, V., The role of microorganisms in peat formation and decomposition: Soil Sci., v. 27, p. 271-281

Zoltai, S.C., Pollett, F.C., Jeglum, J.K., and Adams, G.D., 1975, Developing a Wetland Classification for Canada, *in* Proc. of the Fourth North American Forest Soils Conference, p. 497-511.

Zoltai, S.C., and Tarnocai, 1971, Properties of a Wooded Palsa in Northern Manitoba: Arctic and Alpine Research, v. 3, p. 115-129.

__________, 1975, Perennially frozen peatlands in the Western Arctic and Subarctic of Canada: Can. Jour. Earth Sciences, v. 12, p. 28-43.

Pedogenesis of Canadian Cordilleran Alpine Soils

L. M. Lavkulich, J. I. Sneddon

ABSTRACT

The Canadian Cordillera occupies almost all of British Columbia and southwestern Alberta, and extends north to include the Yukon and western Northwest Territories. This broad physiographic region is characterized by a variety of climates, geology, topography, hydrology and biotic communities. The landscapes have been influenced by repeated glaciation, nivation, solifluction, congeliturbation and depositions of volcanic ash. In the alpine areas the above processes have influenced the development of a variety of soils, many of which are unique to the Canadian Cordillera. Only limited studies have been conducted on the alpine soils of the region. Some of these studies have been oriented towards resources development, fewer have specifically addressed soil morphology and genesis. In the area, pedogenic processes have given rise to soils ranging from those with very weakly expressed characteristics to those exhibiting strong pedogenic features, the result of weathering, organo-clay complexing, organic matter dynamics, accumulation and translocation of clays and other soil components. The distribution of the different soils in the alpine are related to the environmental factors, especially climate.

INTRODUCTION

The Cordillera is one of the best defined physiographic regions of Canada extending 800 km wide in the south from western Alberta to the islands of British Columbia. Further north the region extends from the Mackenzie River in the Northwest Territories through the Yukon into Alaska. The length of the Canadian Cordillera is about 2560 km from near the Arctic Ocean to the Canadian-United States border (Figure 1). The Cordillera stretches another 14,400 km to the tip of South America. It is an area of complex geological landforms and the resulting abundance of environments for living things. In the Cordillera, climates vary

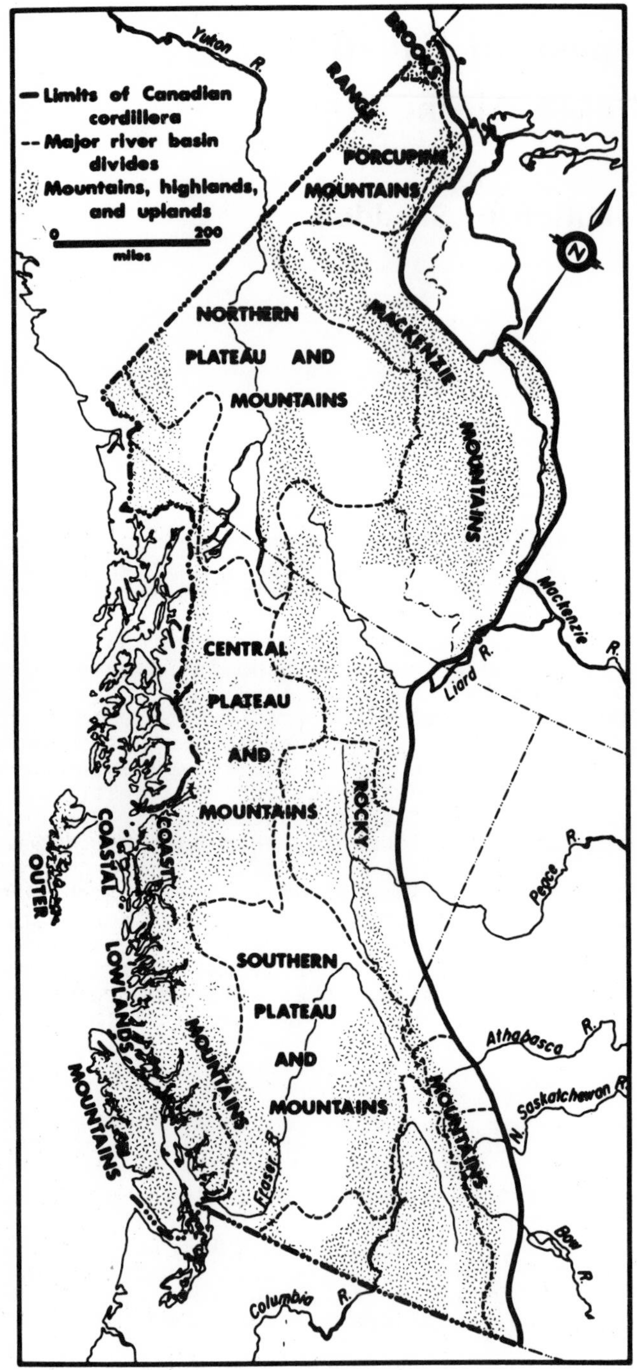

Figure 1. The Canadian Cordillera

from hot valley bottoms with sage and cactus, to the permanent snows of the glaciers and high mountain masses, and in between a belt of forest. Thus climates too dry to grow trees to climates too cold to grow trees are encountered. It is an area also of complex cultural features, highly prized valley bottoms for agricultural development, transportation corridors and for centers for urban growth, long forested slopes and serene alpine landscapes. These areas are characterized by high relief, steep slopes and altitudinally varying local climates. These areas of rugged topography give rise to countless topoclimates which differ widely in response to slope and aspect presenting such a mosaic of heterogeneous facets that conventional climatological analysis is likely to be misleading (Barry and van Wie, 1974). The result of these combinations is that a highly complex mosaic of soils is formed which in many cases is unique.

Mountain areas are commonly termed high energy environments. The sources of these high energy conditions are identified as being relief energy that is potential energy due to gravitational stress and extremes of climate (Hewitt, 1972). This high energy situation results in a variety of landscapes formed by weathering, nivation, solifluction, congeliturbation, mass movement and pedoturbation. In the Canadian Cordillera this has been further complicated by the successive deposition of volcanic ash.

This paper will restrict discussion to selected studies on the pedogenesis of alpine soils of the Canadian Cordillera. Only limited studies have been conducted on the pedogenesis of alpine soils in the Canadian Cordillera, although many studies have been conducted in the realm of resource development.

Alpine areas once considered remote are becoming increasingly accessible to exploitation and are important areas for wildlife, and as catchment basins. In addition, because of their high energy potential, they are important as snow avalanche, soil erosion and mass wasting areas. Short growing seasons, cold soil temperatures, wind, frost action and an unstable rooting medium contribute to low vegetative productivity. Generally alpine may be defined as the area above timberline below the line of permanent snow, which on a global basis has been approximated to occur near the 10°C mean isotherm for the warmest month. As climatic belts shift so does the timberline, thus it is highly probable that areas now above timberline were once forested. Thus the resultant soils often have a complex mode of origin.

In order to place alpine soils of the Canadian Cordillera into perspective, a brief description of the region is presented.

HISTORY OF THE CORDILLERA

The term Cordillera implies a group of mountain ranges,

together with their associated plateaus and intermontane basins. Eardley (1962) and Gilluly (1963) summarized much of the history of the Cordillera.

In Canada, the Cordillera is made up of two geologically different mountain systems, the eastern Cordillera and the western Cordillera (Nelson, 1970). The dividing line between the two systems is the Rocky Mountain Trench. In the north, beyond the Rocky Mountain Trench the eastern Cordillera is made up of the Mackenzie, Richardson and British Mountains and associated ranges, while the western Cordillera in the north is composed of the Coast Mountains and the Yukon and Porcupine Plateaus.

The older western Cordillera was formed about 140 million years ago. Continual erosion since that time has exposed the resistant granitic dioritic rocks of the Coast Range and the metamorphic, intrusive, volcanic and sedimentary rocks of the remainder of the western system. The eastern Cordillera formed about 57 million years ago, comprise the Rocky Mountains and the associated Foothills. The Rockies are composed largely of limestone and dolomite while the Foothills are overlain by younger sandstones and shales which are more heavily vegetated than the Rockies.

The Cordillera was affected by glaciation. In southwestern British Columbia Mathews (1950) reports that ice surfaces over-rode all but the highest summits in the area. The form of nunatak horns and subdued ice rounded summits indicate that the ice surface, during the Vashon stade of the Fraser Glaciation, sloped gradually westward and southward from above 2400 m in the central and eastern parts of the Coast Mountains to around 1800 m at the Tantalus Mountains at the head of Howe Sound in southwestern British Columbia. Holland (1964) states that in general ice did not cover mountain peaks in the Cordillera above about 2130 m. Peaks and ridges above this elevation show the effects of intense alpine glaciation by the presence of cirque basins and currently filled cirques.

Westgate *et al.* (1970), Wilcox (1965), Okazaki *et al.* (1972), Westgate and Fulton (1975) and Sneddon and Lavkulich (1976) have reviewed the occurrence of known volcanic ash deposits in the northern Cordillera. The major volcanic ash deposits are Glacier Peak dated at 12,000 years BP, Mazama, 6600 BP, Mount St. Helens 500 and 3000 to 3500 BP, Mt. Rainer 2000 to 2300 BP and Bridge River 2120 to 2670 BP. Mazama ash appears to be the most extensive.

Further north volcanic ash believed to be from Mount Edgecumbe in southeastern Alaska dated at 9000 BP may be present in Canada. The White River ash in the southern Yukon, extending into the Northwest Territories and northern British Columbia is dated at 1220 and 1900 years BP. In addition, other volcanic ash deposits and sources will no doubt be discovered as research progresses, especially in the northern Cordillera.

The importance of volcanic ash can not be underestimated in the genesis of alpine soils as many of the unique soils found in these environments have been influenced by volcanic ash and its weathering products.

ALPINE SOILS

Alpine soils are extremely variable in morphology and genesis as a result of the complex environment in which they are found. Alpine soils belong to the Brunisolic, Podzolic, Gleysolic, Organic and Regosolic Order of soils (Canada Soil Survey Committee, 1974). In addition, processes characteristic of Chernozemic and Luvisolic soils have been described in the literature, and soluble salt accumulations have been observed as evaporite deposits on the underside of rocks. Thus all processes that occur in Canadian soils occur in the alpine.

In general, parent materials of alpine soils are relatively thin and coarse-textured as a result of the rapid denudation of these high energy environments. Often the mode of origin of the parent materials is unknown or of multiple origin resulting from glaciation, wind action and *in situ* weathering. Most authors feel that physical weathering is more important than chemical weathering in alpine areas although the latter is probably less well understood in these environments. (Costin *et al.*, 1972; Nimlos and McConnell, 1962; Retzer, 1956; Drew *et al.*, 1958; Johnson and Cline, 1965; White, 1976).

In general the climate of alpine soils is harsh with a long duration of the snow pack, cold soil temperatures, high solar radiation input for short periods and wind swept. The biota reflect these conditions as only the more adapted and pioneer vegetative species are found in these environments. In fact, the biota of alpine areas are similar to those reported in Arctic areas or the Tundra.

There are repeated fluctuations of temperature around the freezing point of water in alpine areas. The freezing and thawing appear to do the plants little harm. The upper horizons of the soil warm quickly during the summer but only to shallow depths, thus plants tend to have rather shallow root systems (plant roots do not grow towards low temperatures) leading to the formation of turfy surface horizons in alpine areas.

Soil fauna have been little studied and in the Canadian Cordillera only one study has been found that reports on ant communities in alpine areas (Wiken *et al.*, 1976). Other animals living in the soil, at least during part of their lives, include marmots, pikas (rock rabbits), ground squirrels and ptarmigans. Their effects on pedogenesis are little understood, although they must contribute to pedoturbation and organic matter additions.

Topography of alpine areas is also highly variable

ranging from depressions to steep slopes. Thus soils are very poorly drained to excessively drained. The cold winter temperatures and rapid melt in these areas result in limited water being available for leaching.

Pedogenically alpine soils are young. Alpine soils, however, often exhibit greater apparent pedogenic development as a result of additions of wind blown silt and volcanic ash. Volcanic ash, in particular, weathers readily near the surface with the liberation of large amounts of sesquioxides giving the resultant soils characteristics of soils pedogenically older.

CORDILLERAN ALPINE SOILS OF CANADA

Until recently, little work of a pedogenic nature has been carried out on Canadian alpine soils. In the 1950's and early 1960's reference to alpine soils was made by Farstad and Leahey (1953), Rowles *et al.* (1956), Farstad and Rowles (1960) and Farstad *et al.* (1964). Similarly research with reference to alpine soils, but mainly of an ecological nature, was reported by Archer (1963), Brooke (1966), Krajina (1965) and Peterson (1964). In 1969, two workers reported directly on genesis of alpine soils of the Canadian Cordillera (van Ryswyk, 1969 and Sneddon, 1969). Both of these studies were located in the southern portion of British Columbia.

Van Ryswyk studied alpine soils and adjacent forested soils in the Cascade Mountains of British Columbia, particularly Lakeview Mountain which rises to 2620 m in elevation. He found a nearly continuous but thin (1 to 30 cm) cover of volcanic ash (probably Mazama) up to about 2300 m whereafter the ash became discontinuous on the soil surface and within the pedon. In this area the upper tree line occurs at 2170 m. He classified the soils as Orthic Humo Ferric Podzol under forest vegetation with continuous volcanic ash; Sombric Humo-Ferric Podzol under alpine vegetation on continuous volcanic ash; and Alpine Dystric Brunisol under alpine vegetation on discontinuous ash.

The Orthic Humo-Ferric Podzol soils contained numerous black fecal pellets up to 1.0 mm in diameter in the upper 8 cm of the soil. Occasionally these pellets were found to depths of 40 cm, probably the result of pedoturbation. The pellets also contained plant remains and volcanic ash particles. Charcoal fragments occurred to depths of 60 cm.

The Sombric Humo-Ferric Podzols also contained black fecal pellets and charcoal fragments but in lesser amounts and to depths of about 25 cm. The Ah horizons were turfy and stones present in the soil were strongly stained by organic matter. The author postulates that the Sombric Humo-Ferric Podzols Ah formed after the vegetation changed from coniferous to ericaceous.

The Alpine Dystric Brunisol also showed a turfy Ah

horizon but fewer organic stains on the coarse fragments. The occasional fecal pellets were found to depths of about 18 cm as were the occasional charcoal fragments. The Alpine Dystric Brunisol soil exhibited a buried Ah and Bm horizon sequence beginning at a depth of about 15 cm and at a depth of about 120 cm high silt and very fine sand contents were encountered that appear to have accumulated by illuviation. In general, the Brunisol had lower ash content than the Podzols and the former showed less segregation of ash materials but more intense redistribution probably the result of burial and intense movements by frost heaving or solifluction.

Van Ryswyk (1969) found that there is more turbation taking place under the forest vegetation than under alpine vegetation. The presence of high percentages of silt and fine sand that do not contain volcanic ash grains in the C horizons of all the soils studied indicate that these soil separates have been accumulating for up to 6600 BP by illuviation or translocation by geological processes. The importance of meso-fauna in mineral organic bridging as evidenced by fecal pellets is alluded to but the significance in terms of soil genesis is not known. The authors postulate that the present Sombric Podzol and Alpine Brunisol were once occupied by coniferous vegetation and following about 3000 BP the climate became harsher to permit growth of only alpine vegetation and to allow the development of features of solifluction and frost heaving leading to horizon distortion and burial.

Sneddon (1969) and later Sneddon *et al*. (1972 a, b) reported on five soils occurring in three alpine areas of British Columbia characteristic of three distinct types of alpine environments. The study sites included Fitzsimmons Ridge in the Coast Mountains and the Camelsfoot Range and Yanks Peak of the Quesnel Highlands (Holland, 1964). In the Camelsfoot Range and Fitzsimmons Ridge areas, localized accumulations of volcanic ash (probably Mazama in the former case and Bridge River in the latter) were observed. No such localized accumulations were evident at Yanks Peak.

The vegetation at the Camelsfoot Range site was of the arctic type. Yanks Peak and Fitzsimmons Ridge supported a subalpine type of vegetative community with a wider range of species including grasses, sedges, forbs and the occasional alpine fir, with krummholz life form in more protected areas. The Camelsfoot Range was the more arid of the three sites examined also being the more windswept.

The soils were classified as Mini Humo-Ferric Podzols at the Camelsfoot and Fitzsimmons study areas. Yanks Peak had soils representative of both Orthic Humo-Ferric Podzols and Alpine Dystric Brunisols (C.S.S.C., 1974).

Although all of the soils were found above timberline, the authors suggested that at Yanks Peak and Fitzsimmons Ridge conifers were once present as evidenced by the presence of charcoal and the krummholz life forms of the vegetation. It was postulated that either the effects of a cooler climate or wildfire had removed the forest cover. Although most of the soils did not meet the criteria for a sombric surface horizon (C.S.S.C., 1974) they had dark colored surface horizons, especially at Fitzsimmons and Yanks Peak, where volcanic ash was prevalent. The authors suggested that the dark surface horizons had been influenced by the ash.

The soils studied were derived from gravelly parent materials with an enrichment of silt and clay at the surface. This was attributed to physical weathering and to the additions of aeolian materials. Chemical weathering was evidenced by abundant amorphous iron and aluminum in all horizons and by the pedogenic formation of smectite, chlorite and traces of kaolinite at the expense of vermiculite and mica.

Petrographic analysis of the fine sand fraction demonstrated the presence of volcanic ash, especially in surface horizons, and mixing of soil layers by congeliturbation.

The authors concluded that in true alpine areas much of the morphology associated with podzol soil development is masked by organic matter accumulation caused by the low temperatures which inhibit organic matter breakdown with a net accumulation occurring in the surface horizon.

Lord and Luckhurst (1974) reported on some alpine soils in northeastern British Columbia. The study area is a ridge in the Rocky Mountain Foothills called Nevis Mountain reaching a height of 2160 m. The mineral soils were developed from acid sandstone and shale and limestone and calcareous siltstone derived colluvium. Six alpine grassland communities and two forest communities were studied. The climate is continental with long, cold winters and brief cool summers. The area is utilized by a band of 50 to 60 stone sheep (*Ovis dalli stonei*).

On the basic parent materials on south and west exposures, under the influence of alpine climate, some soils have "Black Chernozemic" morphologies. These soils developed under grassland vegetation and according to the Canada Soil Survey Committee (1974) are classified as Alpine Eutric Brunisols. In the U.S. System (Soil Survey Staff, 1960) the soils would be classified as Typic Cryoborolls or Pergelic Cryoborolls. These alpine soils cause problems with equivocable classification as currently defined in the System of Soil Classification for Canada.

The remaining alpine soils of the study area were classified as Eutric Brunisol and Humisol on calcareous parent materials; as Dystric Brunisol on acid parent materials and Gleysol and Mesisol on weakly calcareous parent materials. The two associated forest soils were developed on calcareous deposits and classified as Chernozem-like under the poplar community and Humo Ferric Podzol under the white spruce community. From the study it is apparent that soils in the alpine may approach classical chernozems in characteritstics.

Although not mentioned by Lord and Luckhurst the Chernozemic soils may be relicts of a previous era when the climate was warmer and drier, especially the soils developed under the poplar community. The importance of parent material affecting the resultant soil is well illustrated by the study.

Beke and Pawluk (1971) reported on the pedogenic influence of volcanic ash in the Marmot Creek area of Alberta. They concluded that in the study area receiving 25 to 33 cm of rainfall during the summer, podzol processes are reflected to shallow depths. Podzolic soils in the area were found to be restricted to these soils with volcanic ash deposits in the C horizon. In the true alpine sites of their study area they classified the soils as Regosols (above 2280 m).

Knapik *et al.* (1973) in an investigation of alpine soils of the Sunshine area, Banff National Park, noted the relationships between microclimate, geomorphic processes, plant communities and soil development. Soils with deep (10 to 12 cm) Ah horizons below well rooted turf layers were described. These soils were classified as Alpine Dystric Brunisols where base saturation was low and Alpine Eutric Brunisols where base saturation was high. Large amounts of amorphous aluminum and iron compounds were extracted from the better drained, more acid soils on stable slope positions. The authors suggested this was largely the result of the presence of volcanic ash and was not entirely the result of the podzolic process. Subsequent work by Sneddon (1973) showed that the acid ammonium oxalate extraction procedure extracts comparatively large amounts of aluminum compounds from soils when volcanic ash is present.

Walmsley and Lavkulich (1975 a and b) reported on a toposequence of soils occurring on Cap Mountain near Wrigley, Northwest Territories. The area is within the intermittent permafrost zone. The climate of the area is considered to be subarctic. Geographically the transect extended from a semiflat alpine meadow area at the top of Cap Mountain (1170 m) to an area of stone stripes and stone rings at approximately 1000 m, which further extended into a rocky area consisting mainly of broken rock fragments and colluvium at 800 m. From this area, the transect extended to the foot of the slope which consisted of an extensive area of coalescing fans, dissected by drainage channels at an elevation of 700 to 550 m, which gradually extended into organic terrain supporting stunted black spruce beyond which was an area of lichen covered polygonal bogs at 500 m.

The soils were classified into the U.S. System (Soil Survey Staff, 1960). In the alpine meadow area, a Pergelic

Cryohemist or Cryic Typic Mesisol (C.S.S.C., 1974) was found. This area is poorly drained with organic matter frozen at 30 cm. The stone stripe area was characterized by a Pergelic Cryochrept or Cryic Eutric Brunisol. The soil is well drained on 10 to 15% slopes, and parent material is calcareous and loamy. Cryoturbation had caused a large amount of mixing and convoluting of the soil horizons. On the basis of morphology of the site and the position of the buried organic pockets and the location of the B horizon, it was concluded that maximum cryoturbation occurs to about 15 cm.

On the colluvial slope area the dominant soil was a Lithic Cryochrept or Lithic Alpine Eutric Brunisol. The soil was moderately well drained developed from calcareous colluvium which overlies a noncalcareous shale bedrock. The soil profile exhibited the presence of buried Ah horizons, indicative of successive downslope movement. The profile developed only to a depth of about 25 cm in comparison to the stone stripe area where profile development extended to 48 cm.

On the coalescing fan area a Typic Cryaquent or Orthic Gleysol dominated on the prevailing 7% slope. The parent material consisted of a mixture of colluvial and alluvial material composed of shattered noncalcareous shale bedrock. Soil development was restricted to about 20 cm. It is believed that solifluction processes are active in this area.

The polygonal bog area is very poorly drained and supports a Pergelic Cryofibrist or Cryic Fibrisol. The soil was found to be frozen at 35 cm.

Thus this area exhibits considerable cryoturbation. Particle size analysis showed that the stone stripe area had a coarser surface soil texture than at depth, illustrating the effects of wind and frost action in these alpine environments. Soil formation on mineral sites is shallow and retarded by mixing of soil horizons as a result of cryoturbation and solifluction processes. Organic matter distribution with depth was variable. The organic matter accumulated in the alpine meadow study area was at a higher stage of decomposition than found at the lower elevation in the polygonal bog area. The elevation of the sites and distribution of snow has affected permafrost distribution, the type of soil and associated vegetation. The study also demonstrates the fact that as one progresses further north in the Canadian Cordillera depth of soil development decreases, cryoturbation increases and gleysols and organic soils become more frequent.

Pawluk and Brewer (1975) studied alpine soils at two locations in Alberta, namely Luscar Mountain at an elevation of 2550 m, and two pedons from Wa Wa Ridge, in Sunshine Basin, at Banff National Park at an elevation of 2400 m. The soils were classified as Orthic Melanic Brunisols even though the two soils at Wa Wa Ridge met the requirements of a podzol B on the basis of sesquioxide content. The soils were developed from colluvium overlying conglomerate at Luscar and loess/colluvium overlying weathered sandstone at Wa Wa.

Both sites are believed to have been influenced by the deposition of volcanic ash (Mazama). On the basis of micromorphological investigations intensive soil fauna activity was found in the upper horizons. Stratigraphic discontinuities coincided and clearly marked the division between the upper and middle and lower sola. A and upper B horizons were developed from materials of variable composition of volcanic ash and vermiculite. The weathering of volcanic ash resulted in strong chloritization of vermiculite, in the formation of organo-alumina and ferra complexes and very fine nodules of goethite.

The fabric of the soils was described as being composed of granic units, frequently densely packed into granoidic or vugy porphyroskelic fabric. In the upper horizons of the soils some of the granic units were fecal pellets and some remains of plant fragments in agreement with the findings of van Ryswyk (1969). Faunal activity was found to decrease with depth.

Silty argillans observed in the lower solum of one of the pedons at Wa Wa suggested illuviation, which has been postulated by other workers but demonstrated by Pawluk and Brewer (1975). In these studies pyrophosphate extractable iron and aluminum often met the criterion of an f horizon, even in surface A horizons, but often the B horizons do not have the red hues commonly associated with Bf horizons. On the basis of micromorphology indicating biologically incorporated organic matter and other criteria, none of the soils were classified as Podzolic.

SUMMARY

On the basis of the limited studies on alpine soils of the Canadian Cordillera, it is obvious the soils are highly complex and probably the product of several cycles of pedogenesis as affected by geomorphic processes, climatic and vegetative shifts, fire, cryoturbation, faunal activity, solifluction, wind and volcanic ash. The soils truly represent a high potential energy environment product. Almost all orders of soils recognized in Canada are approached in characteristics in alpine soils. The limited study that these soils have received has nonetheless raised several questions of classical models of soil morphology, genesis, and classification. These include the questioning of the concept of horizons being approximately parallel to the land surface, the criteria for separation of Chernozemic Ah horizons from other Ah or indeed B horizons, the influence of parent material on soil genesis (especially volcanic ash), the separation of Brunisols from Podzols, the importance of soil fauna in soil genesis and the decomposition of organic matter.

Considering the vast area involved and the importance of alpine soils for watershed protection, wildlife, recreation, grazing, mining and for aesthetic values, little is known about the genesis of alpine soils of the Canadian Cordillera.

REFERENCES CITED

Archer, A.C., 1963, Some synecological problems in the alpine zone of Garibaldi Park (M.Sc. Thesis): Vancouver, University of British Columbia.

Barry, R.D., and Van Wie, C.C., 1974, Topo and microclimatology in alpine areas, *in* Ives, J.D., and Barry, R.G., eds., Arctic and Alpine Environments: London, Methuen, p. 73-84.

Beke, G.J., and Pawluk, S., 1971, The pedogenic significance of volcanic ash layers in the soils of the East Slopes (Alberta) watershed basin: Can. Jour. Earth Sci., v. 6, p. 664-675.

Brooke, R.C., 1966, Vegetation-environment relationships of subalpine mountain hemlock zone ecosystems (Ph.D. Dissert.): Vancouver, University of British Columbia.

Canada Soil Survey Committee, 1974, The system of soil classification for Canada: Ottawa, Agriculture Canada.

Costin, A.B., Hallsworth, E.G., and Woof, M., 1952, Studies in pedogenesis in New South Wales. III. The alpine humus soil: Jour. Soil Sci., v. 3, p. 190-218.

Drew, J.V., Tedrow, J.C.F., Shanks, R.E., and Koranda, J.J., 1958, Rate and depth of thaw in Arctic soils: Trans. Amer. Geophys. Union, v. 39, p.697-701.

Eardley, A.J., 1962, Structural geology of North America: N.Y., Harper and Row, 743 p.

Farstad, L., and Leahey, A., 1953, Soils of the Canadian Cordillera in British Columbia, *in* Proc. 7th Pacific Science Congress 6, p. 10-16.

Farstad, L., and Rowles, C.A., 1960, Soils of the Cordilleran Region: Agr. Inst. Rev., v. 15, p. 33-36.

Farstad, L., Laird, D.G., and Keser, N., 1964, Soil classification and land use studies in British Columbia, *in* Inventory of the Natural Resources of British Columbia: B.C. Natural Resources Conference, Victoria, p. 78-82.

Gilluly, J., 1963, The tectonic evolution of the western United States: Quart. Jour. Geol. Soc. London, v. 119, p. 133-174.

Hewitt, K., 1972, The mountain environment and geomorphic

processes, *in* Slaymaker, H.O., and McPherson, H.J., eds., Mountain Geomorphology: Vancouver, Tantalus, p. 17-34.

Holland, S.S., 1964, Landforms of British Columbia--a physiographic outline: British Columbia Dept. Mines and Petr. Res. Bull. 48, Victoria.

Johnson, D.D., and Cline, A.J., 1965, Colorado mountain soils: Adv. Agron., v. 17, p. 233-281.

Knapik, L.J., Scotter, G.W., and Pettapiece, W.W., 1973, Alpine soils and plant community relationships of the Sunshine area, Banff National Park: Arctic and Alpine Research, v. 5, p. 161-170.

Krajina, V.J., 1965, Biogeoclimatic zones and classification of British Columbia: Ecol. Western North America, v. 1, p. 1-17.

Lord, T.M., and Luckhurst, A.J., 1974, Alpine soils and plant communities of a stone sheep habitat in northeastern British Columbia: Northwest Sci., v. 48, p. 38-51.

Mathews, W.H., 1950, Historic and prehistoric fluctuations of alpine glaciers in the Mount Garibaldi map area; southwestern British Columbia: Jour. Geol., v. 59, p. 357-380.

Nelson, S.J., 1970, The geological history of western Canada, in The Face of Time: Calgary, Alberta Soc. of Petr. Geol.

Nimlos, T.J., and McConnell, R.C., 1962, The morphology of alpine soils in Montana: Northwest Sci., v. 36, p. 99-112.

Okazaki, R., Smith, H.W., Gilkeson, R.A., and Franklin, J., 1972, Correlation of West Blacktail ash with pyroclastic layer T, from the 1800 AD eruption of Mount St. Helens: Northwest Sci., v. 46, p. 77-89.

Pawluk, S., and Brewer, R., 1975, Micromorphological, mineralogical and chemical characteristics of some alpine soils and their genetic implications, Can. Jour. Soil Sci., v. 55, p. 415-437.

Peterson, E.B., 1964, Plant associations in the subalpine mountain hemlock zone in southern British Columbia (Ph.D. Dissert.): Vancouver, University of British Columbia.

Retzer, J.L., 1956, Alpine soils of the Rocky Mountains, Jour. Soil Sci., v. 7, p. 22-32.

Rowles, C.A., Farstad, L., and Laird, D.G., 1956, Soil

resources of British Columbia, in Trans. 9th B.C. Natural Resources Conference, Victoria, p. 85-112.

Sneddon, J.I., 1969, The genesis of some alpine soils in British Columbia (M.Sc. Thesis): Vancouver, University of British Columbia.

_____________, 1973, A study of two soils derived from volcanic ash in southwestern British Columbia and a review and determination of ash distribution in western Canada (Ph.D. Dissert.): Vancouver, University of British Columbia.

Sneddon, J.I., and Lavkulich, L.M., (in press), Volcanic ash: its significance and distribution in western Canada, Can. Jour. Earth Sci.

Sneddon, J.I., Lavkulich, L.M., and Farstad, L., 1972a, The morphology and genesis of some alpine soils in British Columbia, Canada. I. Morphology, classification and genesis: Soil Sci. Soc. America Proc., v. 36, p. 100-104.

_____________, 1972b, The morphology and genesis of some alpine soils in British Columbia, Canada. II. Physical, chemical, and mineralogical determinations and genesis: Soil Sci. Soc. America Proc., v. 36, p. 104-110.

Soil Survey Staff, 1960, Soil Classification, A Comprehensive System, 7th Approximation: Washington, U.S.D.A., 265 p.

Van Ryswyk, A.L., 1969, Forest and alpine soils of south-central British Columbia (Ph.D. Dissert.): Pullman, Washington State University.

Walmsley, M.E., and Lavkulich, L.M., 1975a, Landform-soil-vegetation-water chemistry relationships, Wrigley area, N.W.T. I. Morphology, classification and site description: Soil Sci. Soc. America Proc., v. 39, p. 84-88.

______________, 1975b, Landform-soil-vegetation-water chemistry relationships, Wrigley area, N.W.T.: II. Chemical, physical and mineralogical determinations and relationships: Soil Sci. Soc. America Proc., v. 39, p. 89-93.

Westgate, J.A., and Fulton, R.J., 1975, Tephrostratigraphy of Olympia interglacial sediments in south-central British Columbia, Canada: Can. Jour. Earth Sci., v. 12, p. 489-502.

Westgate, J.A., Smith, D.G.W., and Tomlinson, M., 1970, Late Quaternary tephra layers in southwestern Canada, in Early man and environments in northwestern North

America: Alberta, University of Calgary, p. 13-34.

White, S.E., 1976, Is frost action really only hydration shattering? - a review: Arctic and Alpine Research, v. 8, p. 1-6.

Wiken, E.B., Broersma, K., Lavkulich, L.M., and Farstad, L., (in press), Biosynthetic alteration in a British Columbia soil by ants: Soil Sci. Soc. America Proc.

Wilcox, R.E., 1965, Volcanic ash chronology, *in* Wright, H.E. Jr., and Frey, D.G., eds., Quaternary of the U.S.: Princeton, N.J., Princeton Univ. Press., p. 807-816.

Field Guide

Surface Geology and Soils, Toronto Centred Region

D. W. Hoffman

FIELD TRIP

The effects of glaciation on two major bedrock formations have resulted in a number of different soils that occupy the area around Toronto (Figures 1 and 2). In general, the soils south of a line joining the village of Schomberg to Aurora to Musselman's Lake are developed on materials that show the influence of the shaley Lorraine Formation. To the north of this line the soil materials contain much more calcium carbonate and little shale is evident. These and other characteristics of the Trenton Formation have influenced soil morphology (Figure 1).

There are 5 glacial tills in the area which differ in texture, color, and mineral composition. These combined with lacustrine and outwash deposits have given rise to a number of different soils (Figures 2 and 3). The following field tour has been organized to provide participants the opportunity of investigating a few of the major soils in the region so that some knowledge of their genesis, morphology and classification might be gained.

The tour starts from York University and we proceed to #7 highway and then east through Thornhill to stop 1 about 2 miles east of town (Figure 4). We have been travelling through the Peel Plain which is a level to undulating tract of clay soils covering 300 square miles across the central portions of York and Peel Counties. The general elevation is from 500 to 850 feet above sea level and there is a gradual and fairly uniform slope toward Lake Ontario.

The soils in the Peel Plain are developed on shallow deposits of lacustrine clay underlain by clay till. Lake Peel was contemporaneous with early Lake Algonquin (Figure 2). The evidence for submergence is not as strongly marked as in some lake plains nor is it always present.

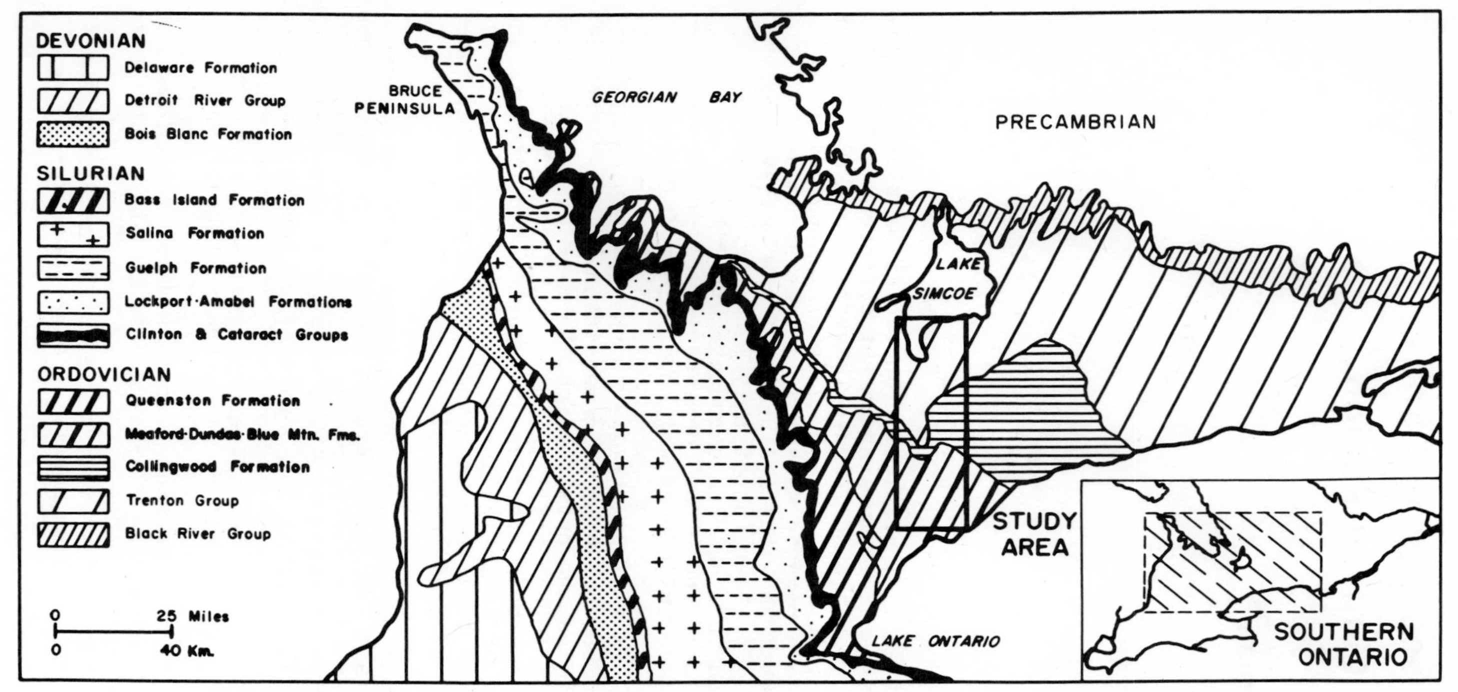

Figure 1 Bedrock geology of South-Central Ontario (modified from Mahaney, W.C., and Ermuth, H.F., 1975, The Effects of Agriculture and Urbanization on the Natural Environment: Toronto, Geographical Monographs no. 7, York University Series in Geography), courtesy Ontario Dept. Mines, Ministry of Natural Resources.

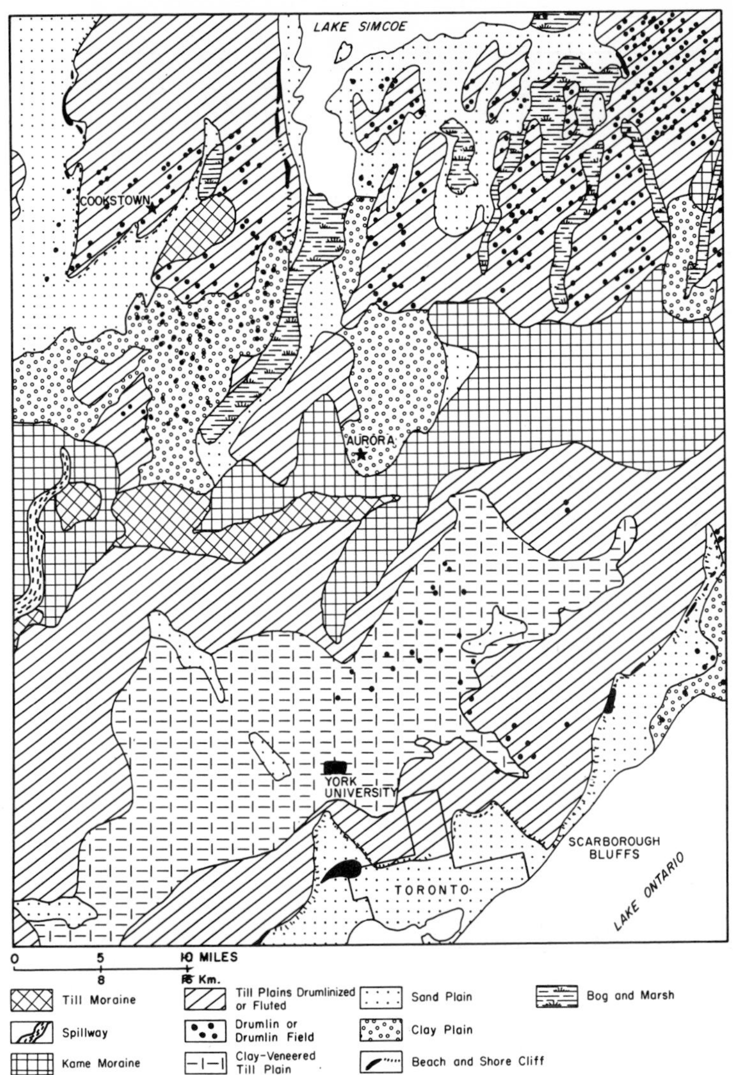

Figure 2 Surficial geology of South-Central Ontario (modified from Watt, A.K., *et al.*, 1953, Geology of the Toronto-Orangeville Area Ont.: Boulder, Colo., Geol. Soc. America Guidebook to trip no. 3.)

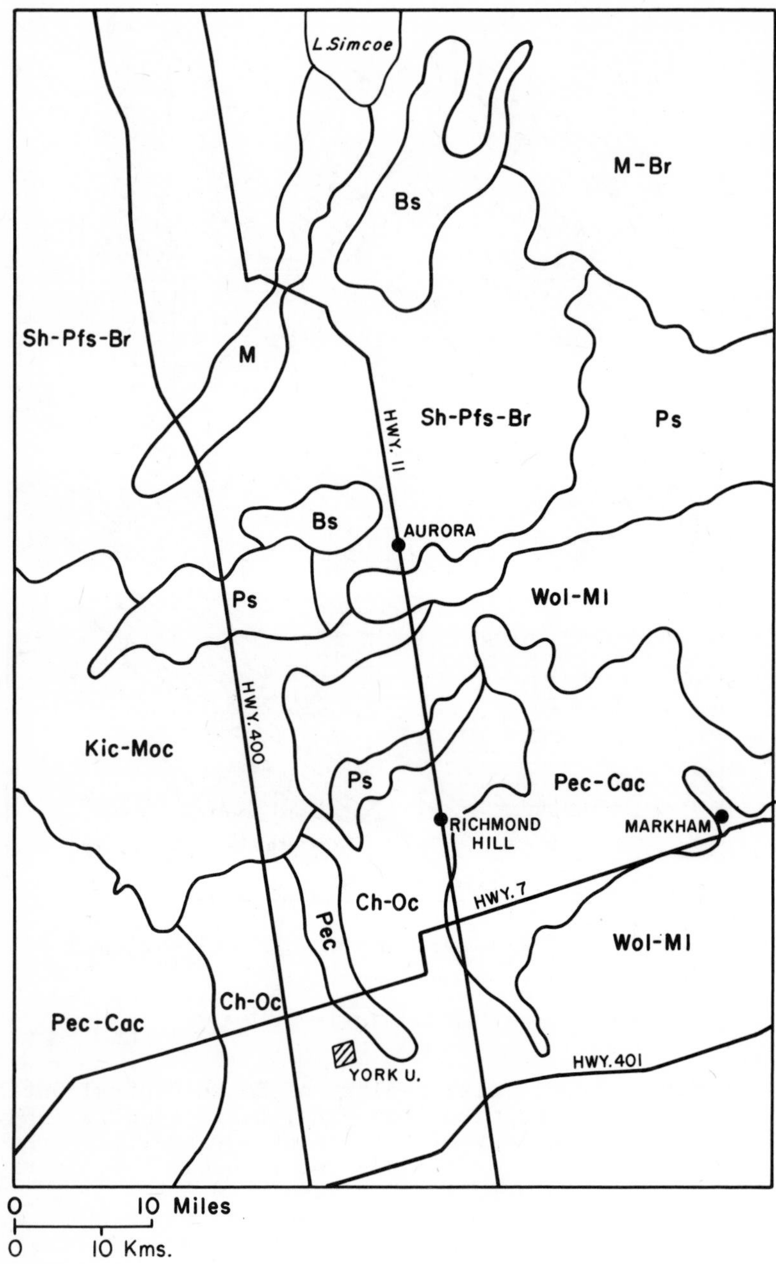

Figure 3 Soil associations, South-Central Ontario.

Surface Geology and Soils of the Toronto Area

KEY

CARTOGRAPHIC UNIT	SOIL SERIES	DESCRIPTION
Bs	Bondhead	Dominantly well drained soils on sandy loam till.
Ch-Oc	Chinguacousy Oneida	Dominantly imperfectly and well drained soils developed on brown shaly clay or clay loam till.
Kic-Moc	King Monaghan	Dominantly well and imperfectly drained soils developed on gray-brown, shaly clay or clay loam till.
M-Br	Muck Brighton	Organic soils and well drained sands with a significant amount of poorly drained sands (Granby).
Ps	Pontypool	Dominantly rapidly drained soils on fine and medium sands.
Wol-Ml	Woburn Milliken	Dominantly well and imperfectly drained soils on brown, shaly loam till.
Sh-Pfs-Br	Schomberg Percy Brighton	Dominantly well drained soils developed on lacustrine silty clay loams, clay loams, and fine sands, and outwash medium sands.
Pec-Cac	Peel clay Cashel clay	Dominantly imperfectly drained soils on shallow lacustrine deposits underlain by clay till.

Stop #1 - Peel clay

Classification

Order:	Luvisolic
Great Group:	Gray Brown Luvisol
Subgroup:	Gleyed Orthic Gray Brown Luvisol
Family:	fine, mixed, neutral, strongly calcareous; climate 3H

Figure 4 Index map showing generalized topography and stops.

Soil Description

Horizon	Depth (cms.)	
Ah	0-15	Dark grayish brown (10YR 4/2m) clay; strong, fine and medium subangular blocky; firm; plentiful fine roots; clear, smooth boundary; 8 to 20 cms. thick; neutral.

Horizon	Depth (cms.)	
Aeg	15-20	Light yellowish brown (10YR 6/4m) clay; many medium dark yellowish brown (10YR 4/4m) mottles; strong medium blocky; hard; clear, smooth boundary; 7.5 to 17.5 cms. thick; neutral.
Btg	20-45	Dark brown (7.5YR 3/2m) clay; medium, faint, dark yellowish brown (10YR 4/4m) mottles; strong, coarse blocky; hard; clear, smooth boundary; 10 to 17.5 cms. thick, neutral.
Ckg	45-50	Grayish brown (10YR 5/2m) clay; many, distinct, yellowish brown (10YR 5/6m) mottles; hard when dry; plastic when wet; abrupt, smooth boundary; 0 to 15 cms. thick; moderately effervescent; moderately calcareous.
IICg	50+	Light grayish brown (10YR 6/2m) clay; many, medium, distinct yellowish brown (10YR 5/8m) mottles; strong fine and medium prismatic; hard; moderately effervescent; strongly calcareous.

Horizon	Depth cms.	Sand %	Silt %	Clay %	pH	Organic Matter %
Ah	0-15	24.2	40.0	35.8	6.7	8.3
Aeg	15-20	25.8	42.4	31.8	6.4	1.2
Btg	20-45	16.4	31.6	52.0	7.0	1.8
Ckg	45-50	17.3	35.5	47.2	7.5	0.2
II CkG	50+	18.5	36.2	45.3	7.8	0.1

From stop #1 proceed east about 7 miles and then turn north. Proceed through Markham in a northerly direction for approximately 10 miles to stop #2.

Stop #2 - Woburn loam

Classification

Order: Luvisolic
Great Group: Gray Brown Luvisolic
Subgroup: Brunisolic Gray Brown Luvisolic
Family: medium, mixed, acid to neutral, moderately to strongly calcareous; climate 3H-G

Soil Description

Horizon Depth (cms.)

D.W. Hoffman

Horizon	Depth (cms.)	
Ap	0-15	Very dark grayish brown (10YR 3/2m) loam; moderate, medium and coarse granular; friable; plentiful fine roots; abrupt, smooth boundary; 12.5 to 20.0 cms. thick; slightly acid.
Ae1	15-35	Yellowish brown (10YR 5/4m) loam; weak fine and medium subangular blocky; friable; few fine roots; gradual, wavy, boundary; slightly acid.
Ae2	35-42	Reddish yellow (7.5YR 6/6m) loam; weak fine and medium platy; friable; few fine roots; clear, wavy boundary; 5 to 15 cms. thick; slightly acid.
Bt	42-62	Dark brown (7.5YR 4/4m) clay loam; moderate medium and coarse subangular blocky; firm; abrupt wavy, boundary; few fine roots; gradual wavy boundary; 12 to 25 cms. thick; neutral.
B/C	62-85	Brown (10YR 5/3m) clay loam; moderate, coarse subangular blocky; firm; clear, wavy boundary; neutral.
Ck	85+	Brown (7.5YR 4/4m) loam till; moderate, coarse subangular blocky and weak medium prismatic; firm to hard; moderately calcareous, moderately alkaline.

Horizon	Depth cms.	Sand %	Silt %	Clay %	pH	Organic Matter %	Bulk Density
Ap	0-15	32.4	44.8	22.8	6.5	7.8	1.29
Ae1	15-35	33.2	46.4	20.4	6.3	2.1	1.65
Ae2	35-42	35.3	43.3	21.4	6.5	1.2	1.68
Bt	42-62	28.7	37.7	33.6	6.7	1.8	1.70
B/C	62-85	26.1	42.4	31.5	7.0	0.7	1.81
Ck	85+	29.8	47.1	23.1	7.5	0.2	1.92

Stop #2 is located in what Chapman and Putnam (1966) refer to as the South Slope physiographic region. This lies between the Oak Ridges moraine and the Peel Plain. Indeed, the area between Lake Ontario and the interlobate moraine can be divided into 3 regions. The Iroquois lake plain occupies the lowest land, generally under 600 feet above sea level in the York County area and the Peel Plain occupies the central position. The South Slope rises to the line of contact with the moraine at 1,000 to 1,200 feet above sea level. The South Slope contains a variety of soils most of which are developed on tills characterized by a sprinkling of shale. The shale content increases, as does the clay content, from east to west.

The next stop is reached by continuing north about 2.5 miles to Ballantrae, turning west to the next cross-road and from there just a short distance north. Stop #3 is in a small outwash plain which is part of the interlobate moraine.

The Oak Ridges interlobate moraine is one of the most distinctive physiographic units in Ontario extending from the Niagara Escarpment to the Trent River. The surface of the moraine is hilly with a few more level areas interspersed at intervals. The hills are composed of sands with pockets of gravel, sandy loam till and lacustrine silt loam occurring in some places. The soil profile at Stop #3 is characteristic of the sandy soils found in much of the moraine.

Stop #3 - Brighton sandy loam

Classification

Order:	Brunisolic
Great Group:	Melanic Brunisol
Subgroup:	Orthic Melanic Brunisol
Family:	coarse, mixed, moderately to very strongly calcareous; climate 3H-K

Soil Description

Horizon	Depth (cms.)	
Ah	0-15	Dark yellowish brown (10YR 3/4m) sandy loam; single-grain; loose; few fine roots; some fine gravel; clear, smooth boundary; 6 to 8 inches thick; neutral.
Bm1	15-37	Brown (7.5YR 4/3m) loamy sand; single grain; loose; some fine and medium gravel; abrupt, wavy boundary, with shallow tongues extending into the underlying horizon; neutral.
Bm2	37-55	Light yellowish brown (10YR 6/3m) loamy sand; single-grain; loose; weakly calcareous; mildly alkaline.
Bm3	55-60	Brown (7.5YR 4/4m) loamy sand; single-grain; loose; clear, wavy boundary, 1 to 3 inches thick; moderately calcareous; mildly alkaline.
Ck	60+	Light gray (10YR 7/2) loamy sand; single-grain; loose; extremely calcareous; moderately alkaline.

Horizon	Depth cms.	Sand %	Silt %	Clay %	pH	Organic Matter %
Ah	0-15	83.9	6.9	9.2	6.3	6.42
Bm1	15-37	88.8	5.1	6.1	6.5	1.27
Bm2	37-55	93.6	6.1	0.3	6.5	0.44
Bm3	55-60	94.8	3.0	2.2	7.1	0.37
Ck	60+	95.5	1.0	3.5	8.0	0.10

From Stop #3 proceed west to Aurora for lunch. Aurora is located in what was the basin of glacial Lake Schomberg. Because it is superimposed upon a strongly drumlinized till plain the Schomberg plain is not nearly so level as the Peel plain but it is characterized by more extensive and deeper beds of varved clay. There are places in the Schomberg plain, particularly at the top of the larger drumlins, which were not submerged. These have no varved clay deposits but contain loam and sandy loam till. The soil series developed on the well drained sites of the varved clays is called Schomberg while that developed on the till is called Bondhead. Stops 4 and 5 provide an opportunity to study these two soils. These stops are located 1 mile northeast and 2 miles northwest of Aurora respectively (Figures 2 and 4).

Stop #4 - Schomberg silt loam

Classification

Order:	Luvisolic
Great Group:	Gray Brown Luvisol
Subgroup:	Orthic Gray Brown Luvisol
Family:	moderately fine and fine, mixed, moderately to very strongly calcareous; climate 3H-K

Soil Description

Horizon	Depth (cms.)	
Ap	0-15	Very dark grayish brown (10YR 3/2m) silt loam; fine granular; friable; plentiful, fine and very fine roots; clear, smooth boundary; 10 to 15 cms. thick, weakly calcareous; mildly alkaline.
Ae	15-20	Grayish brown (10YR 5/2m) silt loam; compound, weak, medium subangular blocky and weak, medium platy; friable; few, fine roots and pores; some fine gravel; abrupt, irregular boundary, with narrow tongues extending into underlying horizon; 5 to 25 cms thick; neutral.
Bt	20-45	Dark brown (10YR 3/3m) silty clay loam; compound, weak, medium subangular blocky

Horizon	Depth (cms.)	
		and moderate, medium angular blocky and medium platy; firm; abrupt, wavy boundary with shallow tongues extending into underlying horizon; 15 to 30 cms. thick; neutral.
Ck	45+	Brown (10YR 5/3m) silty clay loam; moderate, medium, angular blocky and medium platy; firm; varved silt loam and clay; strongly calcareous; moderately alkaline.

Stop #5 - Bondhead sandy loam

Classification

Order:	Podzolic (Luvisolic)
Great Group:	Gray Brown Podzolic (Gray Brown Luvisol)
Subgroup:	Orthic Gray Brown Podzolic (Orthic Gray Brown Luvisol)
Family:	medium, mixed, alkaline, moderately to very strongly calcareous; climate 4H-K

Soil Description

Horizon	Depth (cms.)	
Ap	0-15	Dark brown (10YR 3/3m) sandy loam; weak, fine and medium granular; friable; plentiful, fine roots; abrupt, smooth boundary; 10 to 15 cms. thick; weakly calcareous; mildly alkaline.
Ae1	15-25	Brown (10YR 4.5/3m) sandy loam; weak, fine and medium platy; friable; few, fine roots; small amount of fine gravel, usually near base of horizon; abrupt, wavy, broken boundary, with some deep tongues extending into the underlying horizon; 0 to 30 cms. thick; neutral.
Ae2	25-35	Very pale brown (10YR 7/3m) sandy loam; weak, fine and medium platy; friable; few fine roots; small amount of fine gravel; abrupt, wavy, broken boundary, with some tongues extending into underlying horizon; 0 to 10 cms. thick; slightly acid.
Bt	35-45	Dark reddish gray (5YR 4/2m) clay loam; moderate, medium subangular blocky and platy; firm; some strongly weathered

Horizon	Depth (cms.)	
		dolomite gravel; clear, wavy boundary; 10 to 15 cms. thick; mildly alkaline.
Ck	45+	Light brownish gray (10YR 6/2m) sandy loam; weak, coarse platy; friable; abundant, angular gravel; some cobbles, mostly concentrated in stoneline along top of horizon; strongly calcareous; moderately alkaline.

Horizon	Depth cms.	Sand %	Silt %	Clay %	pH	Organic Matter %
Ap	0-15	51.5	41.7	16.8	6.5	12.90
Ae1	15-25	51.0	35.3	13.7	6.2	1.73
Ae2	25-35	53.0	37.2	9.8	6.0	1.25
Bt	35-45	40.3	30.5	29.2	7.4	0.85
Ck	45+	51.6	45.3	3.1	8.2	0.20

From Stop #5 return to highway #11 and proceed north to Holland Marsh. Turn left on metro road 8 immediately after crossing the Schomberg River and proceed in a westerly direction to highway #400. Take the sideroad parallel to the highway to the south to the Ontario Ministry of Agriculture and Food Muck Research Station. Stop #6 is located here (Figure 4).

Stop #6 provides a good example of organic terrain that would have been designated as "Peat" in early soil surveys. The material is brown in color and distinctly fibrous throughout the profile. Microscopic examination of the material shows that it has a moss-grass-sedge-reed plant derivation. Occasionally some wood fragments are present but is generally not a prominent component.

These plant species are not equally distributed and the layers possess various combinations. In general it appears that "moss" is confined to the surface layers and "reeds" appear most prominently in the lowest layers or those adjacent to the mineral soil. The absence of *Ledum, Chamaedaphne* and *Vaccinium*, in the surface horizons is curious since they are well established in other sites where mosses are also a major component.

The site possesses little mineral admixture, although the Schomberg river flows through the center of the bog, and although floods such as those produced by "Hurricane Hazel" have swept over the entire bog.

Stop #6 - Fennel Series

Classification

Order Organic

Great Group: Mesisol
Subgroup: Terric Humic Mesisol
Family: Mesic, climate 3H-K, loamy

ORGANIC SOIL PROFILE

Bradford Bog

Depth cms.

0-15	Dark brown 7.5YR 3/2 (NaP color - light gray 10YR 7/3). Recognizable fibers consist of grass stems, moss leaves together with masses of fine semi-transparent rootlets. (Moss - Fen peat).
15-30	Dark brown 7.5YR 3/2 (NaP color- light gray 10YR 7/2). *Polytrichum* and grass stems together with masses of fine rootlets. Prominent aggregations of amorphous gelatinous material. (Moss - Fen peat).
30-40	Very dark brown 10YR 2/2 (NaP color - light gray 10YR 7/2). Many flattened stems of grasses or sedges together with brown gelatinous aggregates and fine roots. Some moss structures. (Fen peat).
40-60	Very dark grayish brown 10YR 3/2 (NaP color - light gray 10YR 7/1). Moss stems and flattened grass stems with brown gelatinous aggregates. (Moss - Fen peat).
60-75	Very dark grayish brown 10YR 3/2 (NaP color - light gray 10YR 7/2). Material mainly gelatinous aggregates and rootlets. Some moss, and grass stems and branches of shrubby plants. (Fen Peat).
75-90	Very dark brown 10YR 2/2 (NaP color - light gray 10YR 7/2). Complex mass of fine rootlets, moss stems, decomposing branches of shrubby plants. (Woody-Moss-Fen peat).
90-105	Very dark brown 10YR 2/2 (NaP color - light gray 10YR 7/2). Material consists of fine rootlets, gelatinous aggregates, woody stems, grass and some moss stems. (Woody-Fen peat).
105-120	Black 10YR 2/1 (NaP color - light gray 10YR 7/2). Material consists of woody stems, tree bark, grassy stems and gelatinous aggregates. (Woody-Fen peat).
120-135	Black 10YR 2/1 (NaP color - light gray 10YR 7/2). Woody stems, grassy stems and dark

gelatinous aggregates. (Woody-Fen peat).

135-150 Very dark brown 10YR 2/2 (NaP color - light gray 10YR 7/2). Wood fragments are numerous. (Woody-Fen peat).

150+ Mineral with marl.

Depth cms.	Water holding Capacity %	Volume Water at zero suction %	Bulk Density gms/cc.	Ash (Natural sample) %	Organic Matter %	Unrubbed Fibre .15 mm %	Fibre Ash %	pH
0-15	621	74.9	.24	8.7	81.3	71.7	6.3	6.6
15-30	1287	74.3	.15	8.0	82.0	81.6	6.7	6.1
30-45	1152	84.4	.14	5.9	94.1	62.5	5.4	5.8
45-60	1260	83.4	.11	5.5	94.5	61.6	5.2	5.7
60-75	1244	84.9	.11	6.2	93.8	68.4	6.6	5.5
75-90	1141	87.6	.12	8.1	91.9	71.4	8.4	5.6
90-105	1254	86.9	.11	6.8	93.2	68.8	7.0	5.6
105-120	1315	84.4	.11	8.9	91.1	72.6	7.8	5.7
120-135	1126	83.4	.12	6.8	93.2	71.0	6.9	5.6
135-150	981	85.1	.16	8.7	91.3	72.6	8.0	5.7

REFERENCES CITED

Chapman, L.J., and Putnam, D.F., 1966, The Physiography of Southern Ontario: Toronto, Univ. of Toronto Press.

Mahaney, W.C., and Ermuth, H.F., 1975, The Effects of Agriculture and Urbanization on the Natural Environment, Geographical Monographs, no. 7: Toronto, York Univ. Ser. Geog., 152 p.

Index